Handbook of Aquatic Microbiology

This comprehensive handbook covers the different aspects of the aquatic environment, microbiology, and microbial applications. The world's aquatic environment is facing a serious threat due to inappropriate planning, implementation, and management. This book compiles effective strategies for managing the aquatic environment. It highlights the role of microorganisms as pollution indicators, in bioremediation, and as bio-control agents. The book also covers the impact of pollution on microorganisms, biofilms, cyanobacterial blooms, and the metagenomics approach to isolate microbes. This book is essential for students and researchers of microbiology, environmental sciences, and biotechnology.

Pramod Kumar Pandey is the director of ICAR-Directorate of Coldwater Fisheries Research, Anusandhan Bhavan, Bhimtal, Uttarakhand, India.

Sumanta Kumar Mallik is a scientist (SS) at the ICAR-Directorate of Coldwater Fisheries Research, Anusandhan Bhavan, Industrial Area, Bhimtal, Uttarakhand, India.

Rameshori Yumnam is an assistant professor in the Department of Zoology at the Manipur University, Imphal, India.

Handbook of Aquatic Microbiology

Edited by

Pramod Kumar Pandey, Sumanta Kumar Mallik, and
Rameshori Yumnam

CRC Press is an imprint of the
Taylor & Francis Group, an **informa** business

Designed cover image: Shutterstock

First edition published 2025
by CRC Press
2385 NW Executive Center Drive, Suite 320, Boca Raton FL 33431

and by CRC Press
4 Park Square, Milton Park, Abingdon, Oxon, OX14 4RN

CRC Press is an imprint of Taylor & Francis Group, LLC

ISBN: 9781032526805 (hbk)
ISBN: 9781032528205 (pbk)
ISBN: 9781003408543 (ebk)

DOI: 10.1201/9781003408543

Typeset in Times
by Deanta Global Publishing Services, Chennai, India

Dedicated
to
Dr. Panjab Singh
in recognition of his unwavering service to the pursuit of knowledge and his remarkable impact on Indian agriculture

Contents

Foreword

The world's aquatic ecosystems are vital to the health of our planet, sustaining a myriad of life forms and serving as a critical component of the Earth's biogeochemical cycles. As our understanding of these intricate environments deepens, so does our appreciation for the pivotal role played by microorganisms in shaping the dynamics of aquatic ecosystems. It is with great pleasure that I would like to introduce the *Handbook of Aquatic Microbiology* edited by Dr. Pramod Kumar Pandey and others. This comprehensive volume will serve as an indispensable guide to the multifaceted world of aquatic microbiology, offering a profound exploration of the microbial life that thrives beneath the surface of our oceans, lakes, rivers, and other water bodies. The importance of this field cannot be overstated, as the microbial inhabitants of aquatic environments influence fundamental ecological processes, drive nutrient cycling, and impact water quality.

In this handbook, the editors and authors have come together to provide a treasure trove of knowledge, encompassing a wide range of topics related to aquatic microbiology. From pristine freshwater systems to the most extreme aquatic habitats, the diversity and adaptability of aquatic microorganisms are showcased in intricate detail. Readers will have knowledge of the latest advancements in research, gaining insights into the vital roles played by aquatic microorganisms in nutrient cycling, bioremediation, and the maintenance of ecosystem health. It serves as an essential reference not only for researchers and professionals in the field but also for academicians, students, and anyone with a passion for understanding and protecting our planet's aquatic treasures. As we confront the challenges posed by climate change, pollution, and the ever-increasing demands on our water resources, a deep understanding of aquatic microbiology has never been more critical. The *Handbook of Aquatic Microbiology* is a beacon of knowledge in this endeavour, providing a compass to navigate the intricate and awe-inspiring world of microorganisms that inhabit our waters.

I commend and congratulate the authors and editors for their dedication to advancing our understanding of aquatic microbiology, and I am confident that this volume will be a source of inspiration and guidance for all who seek to explore and safeguard our planet's aquatic ecosystems.

Anupam Mishra
28 October 2023

Preface

The aquatic realm, comprising the vast expanses of oceans, rivers, lakes, and wetlands, is an intricate and dynamic ecosystem teeming with life, much of which is hidden from the naked eye. Within these watery realms, an invisible world of microorganisms plays a pivotal role in shaping the very fabric of our planet. From the blue depths of the open ocean to the serene tranquillity of freshwater ponds, microbial life forms the foundation of aquatic ecosystems, impacting not only the health of these environments but also the global biogeochemical cycles that sustain life on Earth.

The *Handbook of Aquatic Microbiology* is a culmination of the tireless efforts of numerous scientists, researchers, and experts in the field, each driven by an insatiable curiosity to unravel the mysteries of aquatic microbiology. In this comprehensive volume, we delve into the microscopic wonders that inhabit our waters, examining their ecological significance, their contributions to biogeochemical processes, their role in shaping aquatic food webs, and their interactions with human activities.

This book is intended as a compendium of knowledge, a guide for both seasoned researchers and newcomers to the field. It presents a synthesis of the current state of aquatic microbiology, offering a multidisciplinary perspective that bridges the gap between the fundamental principles of microbiology and the complex, dynamic aquatic environments in which these microorganisms thrive. Throughout these pages, readers will find a wealth of information ranging from microbial physiology to the ecological consequences of microbial activities in aquatic ecosystems.

Our aim is to provide a valuable resource for scientists, educators, policymakers, and anyone with an interest in the intricate web of life that flourishes beneath the water's surface. By exploring the world of aquatic microbiology, we hope to inspire curiosity and foster a deeper appreciation for the crucial role that these tiny organisms play in sustaining life on Earth. As editors, we are indebted to the dedicated contributors who have shared their expertise and insights, making this handbook a reality. Their commitment to advancing our understanding of aquatic microbiology is evident in the wealth of knowledge presented within these pages. We also extend our gratitude to the reviewers whose meticulous scrutiny ensured the quality and accuracy of the content.

In closing, we invite you to embark on a journey into the unseen world of aquatic microbiology. We hope that this handbook will serve as a valuable reference and a source of inspiration for all those who seek to explore the fascinating realm of aquatic microorganisms.

Pramod Kumar Pandey
Sumanta Kumar Mallik
Rameshori Yumnam

Contributors

Arur Anand
Regional Remote Sensing Centre
National Remote Sensing Centre
Indian Space Research Organisation
Department of Space
Nagpur, Maharashtra, India

Raja Aadil Hussain Bhat
Indian Council of Agricultural Research
Directorate of Coldwater Fisheries Research
Bhimtal
Nainital, Uttarakhand, India

Rajive Kumar Bhramchari
College of Fisheries
Dholi
Dr. Rajendra Prasad Central Agricultural University
Bihar, India

Suresh Chandra
Indian Council of Agricultural Research
Directorate of Coldwater Fisheries Research
Bhimtal
Nainital, Uttarakhand, India

Basanta Kumar Das
Aquatic Environmental Biotechnology and Nanotechnology Division
Indian Council of Agricultural Research
Central Inland Fisheries Research Institute
Barrackpore
Kolkata, West Bengal, India

Pragyan Dash
Indian Council of Agricultural Research
Directorate of Coldwater Fisheries Research
Bhimtal
Nainital, Uttarakhand, India

Parvaiz Ahmad Ganie
Indian Council of Agricultural Research
Directorate of Coldwater Fisheries Research
Bhimtal
Nainital, Uttarakhand, India

Garima
Indian Council of Agricultural Research
Directorate of Coldwater Fisheries Research
Bhimtal
Nainital, Uttarakhand, India

Abhay Kumar Giri
Indian Council of Agricultural Research
Indian Agricultural Research Institute
Gauria Karma
Hazaribagh, Jharkhand, India

Nupur Joshi
Himalayan School of Biosciences
Swami Rama Himalayan University
Jollygrant
Dehradun, Uttarakhand, India

Ganesan Kantharajan
Indian Council of Agricultural Research
National Bureau of Fish Genetic Resources
Lucknow, Uttar Pradesh, India

Ayyathurai Kathirvelpandian
Centre for Peninsular Aquatic Genetic Resources
Indian Council of Agricultural Research
National Bureau of Fish Genetic Resources,
Kochi, Kerala, India

Deepak Kumar
Indian Council of Agricultural Research
Directorate of Coldwater Fisheries Research
Bhimtal
Nainital, Uttarakhand, India

H. Sanath Kumar
Indian Council of Agricultural Research
Central Institute of Fisheries Education
Mumbai, India

Kundan Kumar
Aquatic Environment & Health Management Division
Indian Council of Agricultural Research
Central Institute of Fisheries Education
Mumbai, India

Saurav Kumar
Aquatic Environment & Health Management Division
Indian Council of Agricultural Research
Central Institute of Fisheries Education
Mumbai, India
and
National Bureau of Fish Genetic Resources
Kochi
Kerala, India

V. Santhana Kumar
Aquatic Environmental Biotechnology and Nanotechnology Division
Indian Council of Agricultural Research
Central Inland Fisheries Research Institute
Barrackpore
Kolkata, West Bengal, India

Kishor Kunal
Indian Council of Agricultural Research
Directorate of Coldwater Fisheries Research
Bhimtal
Nainital, Uttarakhand, India

Sumanta Kumar Mallik
Indian Council of Agricultural Research
Directorate of Coldwater Fisheries Research
Bhimtal
Uttarakhand, India

Aslah Mohamad
Laboratory of Aquatic Animal Health and Therapeutics
Institute of Bioscience
University of Putra Malaysia
Serdang
Selangor, Malaysia

Snatashree Mohanty
Indian Council of Agricultural Research
Central Institute of Freshwater Aquaculture
Bhubaneswar
Odisha, India

Pramod Kumar Pandey
Indian Council of Agricultural Research
Directorate of Coldwater Fisheries Research
Bhimtal
Uttarakhand, India

Ajey Kumar Pathak
Indian Council of Agricultural Research
National Bureau of Fish Genetic Resources
Lucknow
Uttar Pradesh, India

Richa Pathak
Indian Council of Agricultural Research
Directorate of Coldwater Fisheries Research
Bhimtal
Uttarakhand, India

Anirban Paul
Indian Council of Agricultural Research
Central Institute of Freshwater Aquaculture
Bhubaneswar
Odisha, India

Tapas Paul
Department of Aquatic Environment Management
College of Fisheries
Bihar Animal Sciences University
Patna
Bihar, India

M. Rajkumar
Indian Council of Agricultural Research
Central Marine Fisheries Research Institute
Kochi
Kerala, India

V. Ramasubramanian
Indian Council of Agricultural Research
National Academy of Agricultural Research Management
Hyderabad, India

R. Bharathi Rathinam
Indian Council of Agricultural Research
Indian Agricultural Research Institute
Hazaribagh
Jharkhand, India

Sanjay Rathod
Indian Council of Agricultural Research
Central Institute of Fisheries Education
Mumbai, India

Pramoda Kumar Sahoo
Indian Council of Agricultural Research
Central Institute of Freshwater Aquaculture
Bhubaneswar
Odisha, India

Satya Narayan Sahoo
Fish Health Management Division
Indian Council of Agricultural Research
Central Institute of Freshwater Aquaculture
Kausalyaganga
Bhubaneswar
Odisha, India

Dhruba Jyoti Sarkar
Aquatic Environmental Biotechnology and Nanotechnology Division
Indian Council of Agricultural Research
Central Inland Fisheries Research Institute
Barrackpore
Kolkata
West Bengal, India

Pritam Sarkar
Aquatic Environment & Health Management Division
Indian Council of Agricultural Research
Central Institute of Fisheries Education
Mumbai, India

Soma Das Sarkar
Fisheries Resource Assessment and Informatics Division
Indian Council of Agricultural Research
Central Inland Fisheries Research Institute
Barrackpore
Kolkata
West Bengal, India

Uttam Kumar Sarkar
Indian Council of Agricultural Research
National Bureau of Fish Genetic Resources
Lucknow
Uttar Pradesh, India

Debajit Sarma
Indian Council of Agricultural Research
Central Institute of Fisheries Education
Mumbai
Maharashtra, India

Amit Seth
Department of Life Sciences (Botany)
Manipur University
Canchipur
Imphal, India

Neetu Shahi
Indian Council of Agricultural Research
Directorate of Coldwater Fisheries Research
Bhimtal
Uttarakhand, India

Satya Prakash Shukla
Aquatic Environment & Health Management Division
Indian Council of Agricultural Research
Central Institute of Fisheries Education
Mumbai, India

M. Junaid Sidiq
Department of Aquaculture
Indian Council of Agricultural Research
Central Institute of Fisheries Education
Mumbai, India

Bhupendra Singh
Indian Council of Agricultural Research
Directorate of Coldwater Fisheries Research
Bhimtal
Uttarakhand, India

Maibam Birla Singh
Blue-Green Agro-tech
KhabamLamkhai
Imphal, Manipur, India

Mamta Singh
College of Fisheries
Bihar Animal Sciences University
DKAC Campus
Arrabari
Kishanganj
Bihar, India

Mohan Singh
Indian Council of Agricultural Research
Directorate of Coldwater Fisheries Research
Bhimtal
Uttarakhand, India

Kapil S. Sukhdhane
Aquaculture Division
Indian Council of Agricultural Research
Central Institute of Fisheries Education
Mumbai, India

Ritesh Shantilal Tandel
Indian Council of Agricultural Research
Directorate of Coldwater Fisheries Research
Bhimtal
Nainital, Uttarakhand, India

Rameshori Yumnam
Department of Zoology
Canchipur
Manipur University
Manipur, India

1 Microbiomes of the Aquatic Environment

Pragyan Dash, Ritesh Shantilal Tandel, and Raja Aadil Hussain Bhat

1.1 INTRODUCTION

Freshwater and marine ecosystems are home to a complex and diverse array of microorganisms, including viruses, bacteria, fungi, and algae. These microorganisms play an important role in aquatic ecosystem by acting as decomposers, as producers, and in nutrient cycles. Microorganisms interact with one another and their surroundings to influence the health and function of aquatic ecosystems. The community of microorganisms and combined genetic material, inhabiting a particular environment is summarized in the term 'microbiome.' For a million years, the aquatic microbiome has shaped the microbiome of each aquatic plant and animal; however, the host-associated microbiome is not a direct reflection of the water's microbial community. Aquatic animals are subjected to multiple environmental, nutritional, and genetical selection forces, which impact the coevolved alliances between microbiome and organism (Song et al., 2017).

The functions of aquatic ecosystems depend on interaction between microorganisms and aquatic animals. Bacterial strains competing for nutrients can form more diverse and stable communities (Ghoul and Mitri, 2016). Symbiotic associations between bacteria and algae can improve photosynthesis and primary production efficiency in aquatic environments (Zhou et al., 2018).

The distribution and abundance at higher trophic levels, such as fish and other aquatic creatures, can be affected by changes in the composition of the microbial community. This can also have significant effects on nutrient cycling and ecosystem function. Therefore, for efficient ecosystem management and conservation, a deeper comprehension of the function of microbial communities in aquatic ecosystems is necessary. Aquatic habitats are projected to see considerable changes in microbial community structure and function as global temperatures rise. These changes will have a significant impact on ecosystem services and human well-being. Monitoring these changes and developing targeted management strategies can mitigate the impacts of climate change on aquatic ecosystems. This chapter emphasizes the diversity of microbiomes in aquatic algae, vertebrates, and macrophytes, and their interaction in the aquatic environment (Figure 1.1).

1.2 ALGAE-ASSOCIATED MICROBIOME

Algae, the phytoplankton, are a diverse group of aquatic organisms that play significant roles in maintaining the ecological balance of aquatic ecosystems. Interactions between phytoplankton, microalgae, and their associated microorganisms can be influenced by the production and release of various chemical mediators (Cirri and Pohnert, 2019).

The chemical signals exchanged between algae and microorganisms are associated with their survival, functioning, and health. There are several compounds, such as orfamide A, oxylipins, amino acids, vibrioferrin, methylamine, s-containing osmolytes, auxin, and vitamins, which play a role as defense metabolites, nutrients, growth promoters, antibiotics, signaling molecules, in the binding and uptake of iron, demobilization and killing, as osmolytes, etc. (Cirri and Pohnert, 2019).

DOI: 10.1201/9781003408543-1

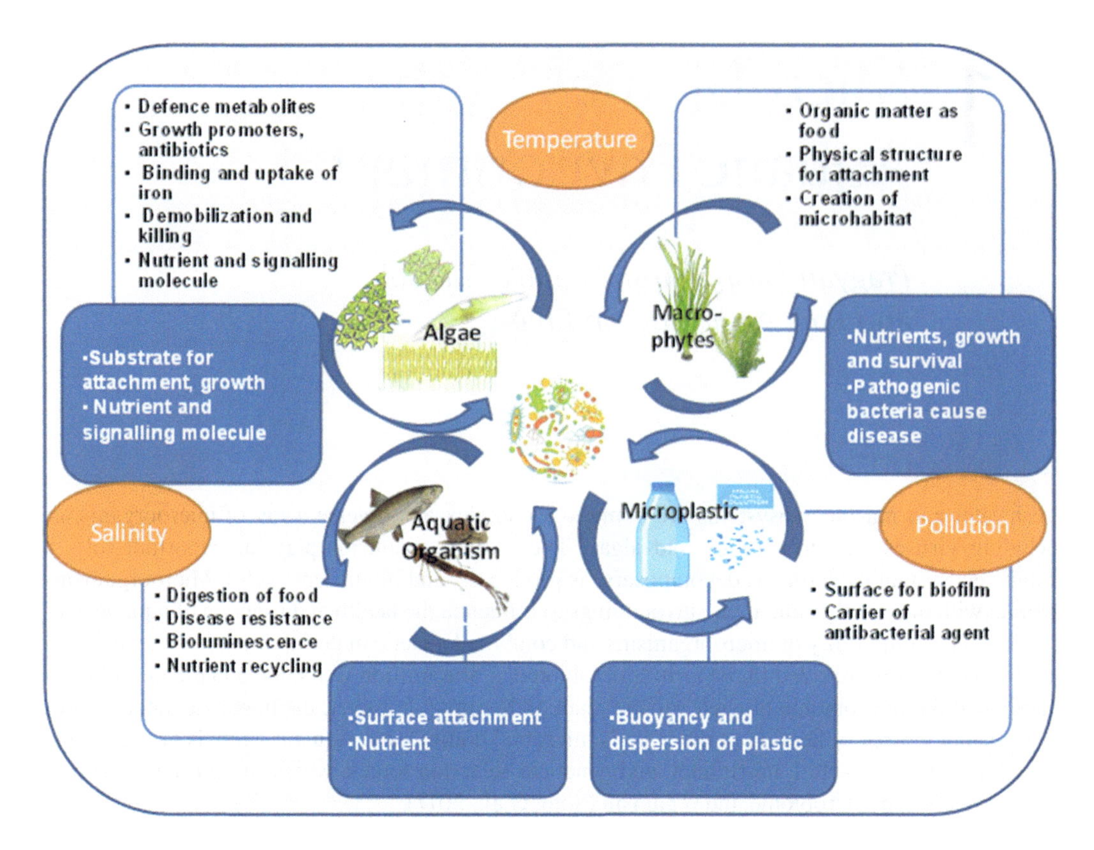

FIGURE 1.1 Interactions of microbiome in an aquatic ecosystem.

For example, nitrogen-fixing bacteria and methylamine-degrading bacteria release ammonia (NH_3), which increases nitrogen availability for microalgae. Tryptophan produced by microalgae serves as a precursor for the biosynthesis of auxins in bacteria and promotes mutualism (Amin et al., 2015). Bacteria from the *Roseobacter* clade (such as *Phaeobacter inhibens* and *Dinoroseobacter shibae*) can trade vitamins or vitamin precursors with algae that cannot synthesize them de novo (Croft et al., 2005; Wienhausen et al., 2017). In exchange for the amino acid tryptophan, *P. inhibens* and *Sulfitobacter* supply auxins and ammonium to diatoms and coccolithophores (Amin et al., 2015; Segev et al., 2016). *Ruegeria pomeroyi* can sense various sulfuric compounds released by microalgae and respond by producing auxins that support algal growth as well as quorum-sensing molecules that promote bacterial proliferation (Johnson et al., 2016; Durham et al., 2015).

Algicidal bacteria can even control the algal bloom by inducing the lysis of algal cells. A typical example is the bacterium *Pseudomonas protegens*, which surrounds the microalga *Chlamydomonas reinhardtii*. *P. protegens* deflagellates the alga and disrupts its Ca^{2+} homeostasis (Aiyar et al., 2017). Further, *P. inhibens* initially increases the coccolithophore's growth by releasing the growth-promoting indole-3-acetic acid (IAA), but later on, it triggers algal cell death by activation of pathways of oxidative stress.

Diatoms and most microalgae require inorganic nitrogen such as nitrate or ammonium, provided by cyanobacteria and nitrogen-fixing bacteria (Foster et al., 2011). Accordingly, the association of algae with nitrogen-fixing bacteria is seen in an ecosystem. Further, other bacteria, like the α-proteobacterium *Donghicola*, can convert the organic nitrogen source methylamine to ammonia. The ammonia is readily taken up by the diatom *Phaeodactylum tricornutum*, and in return, bacteria proliferate with the exudates of the diatom (Suleiman et al., 2016).

Microbiome composition and diversity can be influenced by factors such as water quality, nutrient availability, and habitat. Algae living in polluted water have a less diverse microbiome, dominated by potentially pathogenic bacteria, than algae living in clean water. Some algae, like *Navicula accomoda* and *Stigeoclonium tenue*, can comfortably live in the most heavily polluted zones, while some, like *Cocconeis* and *Chamaesiphon*, are reported to occur only in unpolluted parts of streams (Sen et al., 2013).

Algal-associated microbial populations can control an ecosystem's primary production, carbon sequestration, and nutrient cycling. Additionally, algae are a crucial part of aquaculture, and the microbiomes linked to algae can have a direct impact on the species used for aquaculture in terms of nutrient availability, water quality, and disease resistance.

1.3 AQUATIC ORGANISMS–ASSOCIATED MICROBIOME

Aquatic organisms, including fish, crustaceans, and other fauna, rely on their associated microbiomes for their physiology and health (Colston and Jackson, 2016; Sehnal et al., 2021). Microbiomes related to aquatic organisms are found in various organs, including the skin, gills, gut, and mucus. Skin and gill of fish are the critical site for microbial colonization and play a significant role in protecting fish from pathogenic microbes (Ross et al., 2019), while the gut microbiome contributes significantly to the digestion and nutrient utilization, immune function, health, and overall behavior and ecology of aquatic organisms.

The gut microbial community is broadly divided into allochthonous and autochthonous microbiota based on their interaction with the host. The allochthonous microbiota comprises microorganisms that passes through the lumen along with the food, while the autochthonous microbiota resides permanently in the fish gut without causing any harm to it (Rajeev et al., 2021).

Microorganisms found in the gut of fish and other aquatic organisms can break down complex carbohydrates, proteins, and lipids that are otherwise indigestible by the host (Ringø et al., 2016; Givens et al., 2015). For example, a study on Atlantic salmon (*Salmo salar*) indicated that the gut microbiome produced enzymes that facilitated the digestion of complex carbohydrates, such as cellulose and chitin (Ringø et al., 2016). Similarly, in zebrafish (*Danio rerio*), the microbiota stimulated fatty acid uptake and lipid droplet formation in the intestinal epithelium and liver (Carmody and Turnbaugh, 2012; Semova et al., 2012). The gut microbiome is also known to modulate the host's immune response by producing antimicrobial compounds and promoting the development of immune cells (Gómez and Balcázar, 2008). In Nile tilapia (*Oreochromis niloticus*), certain bacteria in the gut microbiome were reported to be associated with increased resistance to bacterial infections (Tan et al., 2019). Li et al. (2018) and Yao et al. (2021) also reported that the gut microbiome of shrimp (*Litopenaeus vannamei*) played a crucial role in regulating the expression of immune-related genes and protecting the host against viral infections. In addition to digestion and immune function, the microbiome can also influence the behavior and ecology of aquatic organisms (Sullam et al., 2012). In killifish (*Fundulus heteroclitus*), the gut microbiome influenced their reproductive success, with certain microbial communities associated with higher reproductive output (Lombardo et al., 2011; Scott et al., 2020).

Interaction between host physiology, gut architecture, and ecological and environmental factors determines the diversity of the gut microbiome (Ghanbari et al., 2015). Also, the gut microbiome shows a direct relationship when feeding habit changes from herbivore to omnivore to carnivore in finfish species (Givens, 2012; Larsen et al., 2014). As reported in many studies, the early colonization of gut microbiota starts from the egg stage, based on the binding glycoproteins on the egg surface (Larsen et al., 2014). Upon hatching, a bacterium linked with egg chorions first emerges as a colony in the gastrointestinal system. Following that, different kinds of gut microbiota begin to colonize when fish drink water and consume food from the environment (Hansen and Olafsen, 1999). A healthy fish has a more varied alpha gut microbiota than a sick fish, and this variation can be used as a diagnosis to manage fish health (Clarke et al., 2014). Moreover, varied and abundant populations of gut microbes improve fish defense and protection mechanisms (Johnson et al., 2008).

The microbiome composition and diversity of these organs can be influenced by various factors such as water quality, diet, and habitat. Larsen et al. (2014) found that fish living in polluted water had a less diverse microbiome than fish living in clean water. The study also found that potentially pathogenic bacteria dominated the microbiome of fish living in polluted water. Similarly, Ingerslev et al. (2014) found that the gut microbiome of rainbow trout (*Oncorhynchus mykiss*) fed on a plant-based diet was composed of *Leuconostoc, Weissella, Streptococcus*, etc. The gut microbiome composition of fish fed on an animal-based diet consisted of Bacteroidetes, Proteobacteria, and Actinobacteria, which was different from the fish fed with a plant-based diet. Also, the microbiome of fish from freshwater habitats differs from that of fish from marine habitat (Fan et al., 2019). Among freshwater fish microbiome species, *Acinetobacter, Aeromonas, Flavobacterium, Lactococcus*, and *Pseudomonas*, obligate anaerobes *Bacteroides, Clostridium*, and *Fusobacterium*, and members of the family *Enterobacteriaceae* persist, whereas in marine fish, *Aeromonas, Alcaligenes, Alteromonas, Carnobacterium, Flavobacterium, Micrococcus, Moraxella, Pseudomonas*, and *Vibrio* dominate (Gomez and Balcázar, 2008).

1.4 AQUATIC MACROPHYTES–ASSOCIATED MICROBIOME

Aquatic macrophytes are dominant primary producers in many freshwater and marine environments, providing essential ecosystem services such as nutrient cycling, sediment stabilization, and habitat provision. Aquatic macrophytes interact with the environmental microbiome through various pathways, including the provision of organic matter, changes in water quality, and creation of microhabitats. Aquatic macrophytes release organic matter, which acts as a food source for the microbial community, increasing their abundance and diversity. The composition of the microbiome can be altered due to the water quality, which includes increased nutrient availability and light penetration, with some taxa flourishing in nutrient-rich habitats and others in nutrient-poor environments (Bárta et al., 2021). Moreover, the creation of microhabitats by aquatic macrophytes, such as roots and leaves, can provide a physical structure for the attachment and growth of microbes, leading to the formation of specialized microbial communities.

Furthermore, the microbiome is shown to significantly impact the growth and health of aquatic macrophytes. Some bacteria can form a symbiotic relationship with aquatic macrophytes, providing essential nutrients and improving their growth and survival (Zhu et al., 2021). On the other hand, pathogenic microorganisms can cause diseases in aquatic macrophytes, leading to declines in their populations and altering the structure and function of the ecosystem (Saha and Weinberger, 2019).

1.5 MICROPLASTIC-ASSOCIATED MICROBIOME

The makeup of microbial communities on the surface of microplastics or nanoplastics in an aquatic environment, referred to as the 'plastisphere,' has been an interest of research for several decades (Amaral-Zettler et al., 2020). The ability of bacteria and fungi to digest plastic particles, as well as the interaction between the microbial community and microplastics, has piqued scientists' interest in environmental biotechnology (Barros et al., 2021).

Microplastics provide a surface for microbial attachment, leading to the formation of biofilms. The microbial communities associated with microplastics in aquatic ecosystems are diverse and depend upon the attachment on different types of plastics and water bodies (Table 1.1). Biofilms of microplastic in ponds had higher relative abundances of the phyla Proteobacteria, Firmicutes, and Chloroflexi, while river biofilms of microplastic had higher relative abundances of the phylum Acidobacteria and family Nitrospiraceae, Nitrosomonadaceae, and lower relative abundances of Proteobacteria (Hoellein et al., 2014).

Microbiomes belonging to bacterial genera such as *Klebsiella, Pseudomonas*, and *Sphingomonas* can degrade plastic (Kelly et al., 2021). *Rhodococcus ruber* C208 can degrade polyethylene film by up to 8% within 30 days of incubation, which increases with mineral oil up to 50% (Hadar and Sivan

TABLE 1.1
Microorganisms Associated with Microplastics in an Aquatic Environment

Microorganism	Type of Microplastic	Source of Isolation	Reference
Vibrio alginolyticus	Polyethylene	Mangrove	Tan et al. (2022)
Vibrio harveyi, *Enterococcus faecalis*	Polyethylene	Seawater	Hchaichi et al. (2020)
Vibrio parahaemolyticus	Polyethylene, polystyrene, polypropylene	North/Baltic Sea	Kirstein et al. (2016)
Pseudomonas monteilii, *P. mendocina*, *P. syringae*	Polyvinyl chloride (PVC)	River water in controlled conditions	Wu et al. (2019)
Bacillus cereus and *Bacillus gottheilii*	Polystyrene	Mangrove ecosystems	Auta et al. (2017)
Acinetobacter	Polypropylene	River	Mughini-Gras et al. (2021)
Flavobacterium	Polyethylene terephthalate (PET)	River	Mughini-Gras et al. (2021)
Aspergillus, *Cladosporium*, *Wallemia*	Polyethylene (PE), Polyamide (PA), polyurethane (PU), polypropylene (PP), polystyrene (PS), and cellulose acetate (CA)	Western South Atlantic and Antarctic Peninsula	Lacerda et al. (2020)
Aspergillus niger, *Penicillium* spp., *Rhizopus* spp., *Mucor* spp., *Aspergillus nidulans*, *Fusarium* spp., *Microsporum canis*, and *Aspergillus fumigatus*	Polypropylene, polyester, polyvinyl chloride	Estuary	Williams et al. (2021)

2004). Changes in microbial communities in microplastics can influence the biomass of plastics, as reported in polyethylene terephthalate (PET) and polyvinyl chloride (PVC), and hence, can affect the buoyancy and dispersion of plastics in water (Miao et al., 2021).

Due to their small size, microplastics are ingested by a wide range of aquatic organisms, including bacteria, zooplankton, and fish. Plastic debris has been found in fishes, shellfish, and mussels, which after consumption, can enter the human body through trophic transfer, biomagnification, and bioaccumulation (Giani et al., 2019; Hermabessiere et al., 2019). Microplastic intake can adversely affect the health of aquatic organisms, as microplastics are potential carriers for pathogenic bacteria, antimicrobials, and toxic compounds (Campanale et al., 2020).

1.6 IMPACT OF ENVIRONMENTAL FACTORS ON THE STRUCTURE AND FUNCTION OF THE AQUATIC MICROBIOME

The aquatic microbiome is particularly vulnerable to environmental stressors, i.e., pollutants and changes in water quality. In recent years, researchers have explored the impact of various environmental stressors on the microbiome in aquatic environments, including pollution, temperature, and nutrient availability.

1.6.1 Temperature

Temperature is a key factor that shapes the diversity and composition of aquatic microbial communities, and it has been shown to impact the metabolic rates of these organisms significantly.

As temperature increases, the microbial community structure shifts, with thermophilic bacteria becoming dominant (Chiriac et al., 2017). In a study on a eutrophic lake, temperature increase was associated with a decrease in the abundance of cyanobacteria and an increase in the abundance of other phytoplankton taxa (Paerl and Paul, 2012; O'Neil et al., 2012).

Temperature can also influence the diversity of the microbial community, affecting the competitive dynamics between various species. In research on a freshwater reservoir, warming was linked to increased dominance of a single bacterial taxon, which displaced other taxa and reduced total diversity (Wang et al., 2021). On the contrary, another report on the freshwater lake shows an increase in microbial diversity and a more significant number of rare and low-abundance species with increasing temperatures (Islam et al., 2019).

Temperature profoundly impacts how the microbial community functions by affecting microbial metabolic rates. Changes in community composition, such as a decrease in the abundance of cyanobacteria and a rise in the quantity of other phytoplankton species, were linked to changes in nutrient cycling rates, such as enhanced rates of nitrogen fixation and denitrification at higher temperatures (Paerl and Paul, 2012).

Temperature can also influence microbial community function by affecting interactions between various microbial populations. In one freshwater lake, warming was linked to a decrease in the population of bacteria that consume methane, resulting in higher methane emissions from the lake (Schulz and Conrad, 1996). Similarly, higher temperature in a coastal marine ecosystem was linked to a decrease in the abundance of microbial populations engaged in the breakdown of petroleum hydrocarbons, implying that rising temperature may limit the ability of these communities to respond to oil spills (Chong et al., 2018).

1.6.2 Salinity

Many aquatic habitats, including estuaries, salt marshes, and hypersaline lakes, experience changes in salinity due to fluctuations in water flow, evaporation, and precipitation, which can significantly change the composition and diversity of the aquatic microbiome. In the Great Salt Lake in Utah, it was found that extreme halophilic bacteria prevail in the north arm, which has a salinity range from 20 to 30 ppt, whereas moderate halophiles prevail in the south arm at around 15–17 ppt (Baxter et al., 2005). A similar finding was also observed in the lakes of China; *Bacteroidetes*, *Cyanobacteria*, and *Alpha proteobacteria* were the dominant bacterial phyla in low saline lakes, whereas *Gamma proteobacteria* and *Euryarchaeota* were abundant in moderately saline lakes (50–250 ppt). The hypersaline lakes (>250 ppt) were dominated by extreme halophiles such as *Halorubrum*, *Halohasta*, and *Natronomonas* from *Euryarchaeota* (Banda et al., 2021). Changes in salinity are shown to be connected with changes in the abundance of pathogenic Vibrio bacteria in the Baltic Sea and shrimp farms in Vietnam (Baker-Austin et al., 2013).

1.6.3 Pollution

Pollution, due to anthropogenic activities such as agriculture, industry, and urbanization, can introduce a wide range of pollutants into aquatic environments. These include nutrients, heavy metals, and organic chemicals. Several reports have shown that pollution can seriously affect the composition and performance of the aquatic microbiome. For example, a study conducted in a polluted river in Northeastern China revealed that the pollution decreased the overall bacterial diversity. However, the community was dominated by opportunistic and pathogenic bacteria such as *Acinetobacter johnsonii*, *Clostridium cellulovorans*, and *Trichococcus pasteurii* (Zhao et al., 2014). This shift in microbial composition can significantly impact ecosystem functioning, including nutrient cycling and water quality. In addition to changes in microbial composition, pollution can also lead to changes in microbial metabolism. In an estuary of China polluted with polycyclic aromatic hydrocarbons (PAHs), the microbial community was dominated by proteobacteria containing genes encoding for dehydrogenases, which are the key enzymes for pyrene degradation (Zhang et al., 2019). However, these adaptations can also negatively impact ecosystem functioning, such as releasing greenhouse gases.

Pollution in the form of eutrophication is also a threat to aquatic microbiomes, leading to algal blooms and oxygen depletion. Eutrophication can lead to changes in the composition of microbial communities, most notably a shift from an autotrophic to a heterotrophic population (Meyer-Reil and Köster, 2000; Andersson et al., 2006). Furthermore, increased nutrient availability can lead to increased abundance of harmful bacteria, such as *Vibrio* (Baker-Austin et al., 2013).

1.7 CONSERVATION AND MANAGEMENT IMPLICATIONS FOR THE ECOSYSTEM

Understanding the interactions between the aquatic microbiome and its environment has important implications for conserving and managing aquatic ecosystems. Management of water quality, such as temperature, pH, and salinity, including controlling the levels of pollutants, can help to protect the diversity and stability of the aquatic microbiome.

One of the critical challenges in managing aquatic ecosystems in a changing climate scenario is predicting how microbial communities will respond to temperature changes. Many factors can influence the sensitivity of microbial communities to temperature, including the specific microbial taxa present, the availability of nutrients and other resources, and the presence of other stressors, such as pollution or invasive species. As a result, it can be challenging to predict the exact nature and magnitude of the impacts of temperature on microbial community structure and function.

However, several strategies can be employed to manage and conserve aquatic ecosystems in a changing climate scenario. One approach is to monitor microbial community structure and function over time, using high-throughput sequencing and other molecular techniques to track changes in community composition and metabolic activity. These facts can be used to identify critical microbial taxa and processes that may be particularly sensitive to temperature changes and to develop targeted management strategies to mitigate the impacts of climate change on ecosystem function.

Developing bioremediation strategies, which utilize microorganisms to remove pollutants from aquatic environments, is another important consideration for conserving and managing aquatic ecosystems (Gao et al., 2018). Another approach is to manage other stressors that can exacerbate temperature impacts on aquatic microbial communities. For example, reducing nutrient pollution can help maintain the resilience of microbial communities to temperature changes by reducing the risk of harmful algal blooms and other negative impacts on ecosystem function. Similarly, reducing the impact of invasive species can help maintain the diversity and function of native microbial communities, which may be more resilient to temperature changes than introduced species.

1.8 CONCLUSION

Microbiomes can have beneficial and detrimental effects on aquatic organisms, and host genetics, environmental conditions, and the surrounding microbiome community can influence their composition and function. However, there are still many knowledge gaps and future research directions in aquatic microbiome research, including understanding the mechanisms underlying microbiome–host interactions, identifying key microbial taxa and functional genes involved in these interactions, and investigating the effects of anthropogenic activities on aquatic microbiomes.

The microbiome plays a crucial role in maintaining the health of aquatic ecosystems and is shaped by a complex interplay of physical, chemical, and biological factors. Despite this, the aquatic microbiome is vulnerable to disturbance from eutrophication and global warming. Further research is needed for proper understanding of the impact of these disturbances on the overall ecosystem health. Ultimately, this type of study will help chart out the management and conservation strategies for protecting aquatic ecosystems and the microorganisms that inhabit them.

REFERENCES

Aiyar, Prasad, Daniel Schaeme, María García-Altares, David Carrasco Flores, Hannes Dathe, Christian Hertweck, Severin Sasso, and Maria Mittag. "Antagonistic bacteria disrupt calcium homeostasis and immobilize algal cells." *Nature Communications* 8, no. 1 (2017): 1756.

Amaral-Zettler, Linda A., Erik R. Zettler, and Tracy J. Mincer. "Ecology of the plastisphere." *Nature Reviews Microbiology* 18, no. 3 (2020): 139–151.

Amin, S. A., L. R. Hmelo, H. M. Van Tol, B. P. Durham, L. T. Carlson, K. R. Heal, R. L. Morales et al. "Interaction and signalling between a cosmopolitan phytoplankton and associated bacteria." *Nature* 522, no. 7554 (2015): 98–101.

Andersson, Agneta, Kristina Samuelsson, Pia Haecky, and Jan Albertsson. "Changes in the pelagic microbial food web due to artificial eutrophication." *Aquatic Ecology* 40 (2006): 299–313.

Auta, H. S., C. U. Emenike, and S. H. Fauziah. "Screening of Bacillus strains isolated from mangrove ecosystems in Peninsular Malaysia for microplastic degradation." *Environmental Pollution* 231 (2017): 1552–1559.

Baker-Austin, Craig, Joaquin A. Trinanes, Nick G. H. Taylor, Rachel Hartnell, Anja Siitonen, and Jaime Martinez-Urtaza. "Emerging Vibrio risk at high latitudes in response to ocean warming." *Nature Climate Change* 3, no. 1 (2013): 73–77.

Banda, Joseph Frazer, Qin Zhang, Linqiang Ma, Lixin Pei, Zerui Du, Chunbo Hao, and Hailiang Dong. "Both pH and salinity shape the microbial communities of the lakes in Badain Jaran Desert, NW China." *Science of the Total Environment* 791 (2021): 148108.

Barros, Juliana, and Sahadevan Seena. "Plastisphere in freshwaters: An emerging concern." *Environmental Pollution* 290 (2021): 118123.

Bárta, Jiří, Caio Cesar Pires de Paula, Eliška Rejmánková, Qiang Lin, Iva Kohoutová, and Dagmara Sirová. "Complex phyllosphere microbiome aids in the establishment of the invasive macrophyte Hydrilla verticillata (L.) under conditions of nitrogen scarcity." *BioRxiv* (2021): 2021–01.

Baxter, Bonnie K., Carol D. Litchfield, Kevin Sowers, Jack D. Griffith, Priya Arora Dassarma, and Shiladitya Dassarma. "Microbial diversity of great salt lake." In *Adaptation to Life at High Salt Concentrations in Archaea, Bacteria, and Eukarya*, pp. 9–25. Springer Netherlands, 2005.

Campanale, Claudia, Carmine Massarelli, Ilaria Savino, Vito Locaputo, and Vito Felice Uricchio. "A detailed review study on potential effects of microplastics and additives of concern on human health." *International Journal of Environmental Research and Public Health* 17, no. 4 (2020): 1212.

Carmody, Rachel N., and Peter J. Turnbaugh. "Gut microbes make for fattier fish." *Cell Host & Microbe* 12, no. 3 (2012): 259–261.

Chiriac, Cecilia M., Edina Szekeres, Knut Rudi, Andreea Baricz, Adriana Hegedus, Nicolae Dragoş, and Cristian Coman. "Differences in temperature and water chemistry shape distinct diversity patterns in thermophilic microbial communities." *Applied and Environmental Microbiology* 83, no. 21 (2017): e01363–17.

Chong, Chun Wie, Santha Silvaraj, Yasoga Supramaniam, Ian Snape, and Irene Kit Ping Tan. "Effect of temperature on bacterial community in petroleum hydrocarbon-contaminated and uncontaminated Antarctic soil." *Polar Biology* 41 (2018): 1763–1775.

Cirri, Emilio, and Georg Pohnert. "Algae– bacteria interactions that balance the planktonic microbiome." *New Phytologist* 223, no. 1 (2019): 100–106.

Clarke, Gerard, Roman M. Stilling, Paul J. Kennedy, Catherine Stanton, John F. Cryan, and Timothy G. Dinan. "Minireview: Gut microbiota: The neglected endocrine organ." *Molecular Endocrinology* 28, no. 8 (2014): 1221–1238.

Colston, Timothy J., and Colin R. Jackson. "Microbiome evolution along divergent branches of the vertebrate tree of life: What is known and unknown." *Molecular Ecology* 25, no. 16 (2016): 3776–3800.

Croft, Martin T., Andrew D. Lawrence, Evelyne Raux-Deery, Martin J. Warren, and Alison G. Smith. "Algae acquire vitamin B12 through a symbiotic relationship with bacteria." *Nature* 438, no. 7064 (2005): 90–93.

Durham, Bryndan P., Shalabh Sharma, Haiwei Luo, Christa B. Smith, Shady A. Amin, Sara J. Bender, Stephen P. Dearth et al. "Cryptic carbon and sulfur cycling between surface ocean plankton." *Proceedings of the National Academy of Sciences* 112, no. 2 (2015): 453–457.

Fan, Lanfen, Zhenlu Wang, Miaoshan Chen, Yuexin Qu, Junyi Li, Aiguo Zhou, Shaolin Xie, Fang Zeng, and Jixing Zou. "Microbiota comparison of Pacific white shrimp intestine and sediment at freshwater and marine cultured environment." *Science of the Total Environment* 657 (2019): 1194–1204.

Foster, Rachel A., Marcel M. M. Kuypers, Tomas Vagner, Ryan W. Paerl, Niculina Musat, and Jonathan P. Zehr. "Nitrogen fixation and transfer in open ocean diatom–cyanobacterial symbioses." *The ISME Journal* 5, no. 9 (2011): 1484–1493.

Gao, Hong, Yuebo Xie, Sarfraz Hashim, Alamgir Akhtar Khan, Xiaolin Wang, and Huiyong Xu. "Application of microbial technology used in bioremediation of urban polluted river: A case study of Chengnan River, China." *Water* 10, no. 5 (2018): 643.

Ghanbari, Mahdi, Wolfgang Kneifel, and Konrad J. Domig. "A new view of the fish gut microbiome: Advances from next-generation sequencing." *Aquaculture* 448 (2015): 464–475.

Ghoul, Melanie, and Sara Mitri. "The ecology and evolution of microbial competition." *Trends in Microbiology* 24, no. 10 (2016): 833–845.

Giani, Dario, Matteo Baini, Matteo Galli, Silvia Casini, and Maria Cristina Fossi. "Microplastics occurrence in edible fish species (Mullus barbatus and Merluccius merluccius) collected in three different geographical sub-areas of the Mediterranean Sea." *Marine Pollution Bulletin* 140 (2019): 129–137.

Givens, Carrie Elizabeth. "A fish tale: Comparison of the gut microbiome of 15 fish species and the influence of diet and temperature on its composition." PhD diss., University of Georgia, 2012.

Givens, Carrie E., Briana Ransom, Nasreen Bano, and James T. Hollibaugh. "Comparison of the gut microbiomes of 12 bony fish and 3 shark species." *Marine Ecology Progress Series* 518 (2015): 209–223.

Gómez, Geovanny D., and José Luis Balcázar. "A review on the interactions between gut microbiota and innate immunity of fish." *FEMS Immunology & Medical Microbiology* 52, no. 2 (2008): 145–154.

Hadar, Y., and A. Sivan. "Colonization, biofilm formation and biodegradation of polyethylene by a strain of Rhodococcus ruber." *Applied Microbiology and Biotechnology* 65 (2004): 97–104.

Hansen, G. H., and J. A. Olafsen. "Bacterial interactions in early life stages of marine cold water fish." *Microbial Ecology* 38 (1999): 1–26.

Hchaichi, Ilef, Francesca Bandini, Giulia Spini, Mohamed Banni, Pier Sandro Cocconcelli, and Edoardo Puglisi. "Enterococcus faecalis and Vibrio harveyi colonize low-density polyethylene and biodegradable plastics under marine conditions." *FEMS Microbiology Letters* 367, no. 15 (2020): fnaa125.

Hermabessiere, Ludovic, Ika Paul-Pont, Anne-Laure Cassone, Charlotte Himber, Justine Receveur, Ronan Jezequel, Maria El Rakwe et al. "Microplastic contamination and pollutant levels in mussels and cockles collected along the channel coasts." *Environmental Pollution* 250 (2019): 807–819.

Hoellein, Timothy, Miguel Rojas, Adam Pink, Joseph Gasior, and John Kelly. "Anthropogenic litter in urban freshwater ecosystems: Distribution and microbial interactions." *PloS One* 9, no. 6 (2014): e98485.

Ingerslev, Hans-Christian, Mikael Lenz Strube, Louise von Gersdorff Jørgensen, Inger Dalsgaard, Mette Boye, and Lone Madsen. "Diet type dictates the gut microbiota and the immune response against Yersinia ruckeri in rainbow trout (Oncorhynchus mykiss)." *Fish & Shellfish Immunology* 40, no. 2 (2014): 624–633.

Islam, M. M. M., Sana Shafi, Suhaib A. Bandh, and Nowsheen Shameem. "Impact of environmental changes and human activities on bacterial diversity of lakes." In *Freshwater Microbiology*, pp. 105–136. Academic Press, 2019.

Johnson, Pieter T. J., Richard B. Hartson, Donald J. Larson, and Daniel R. Sutherland. "Diversity and disease: Community structure drives parasite transmission and host fitness." *Ecology Letters* 11, no. 10 (2008): 1017–1026.

Johnson, Winifred M., Melissa C. Kido Soule, and Elizabeth B. Kujawinski. "Evidence for quorum sensing and differential metabolite production by a marine bacterium in response to DMSP." *The ISME Journal* 10, no. 9 (2016): 2304–2316.

Kelly, John J., Maxwell G. London, Amanda R. McCormick, Miguel Rojas, John W. Scott, and Timothy J. Hoellein. "Wastewater treatment alters microbial colonization of microplastics." *PloS One* 16, no. 1 (2021): e0244443.

Kirstein, Inga V., Sidika Kirmizi, Antje Wichels, Ale Garin-Fernandez, Rene Erler, Martin Löder, and Gunnar Gerdts. "Dangerous hitchhikers? Evidence for potentially pathogenic Vibrio spp. on microplastic particles." *Marine Environmental Research* 120 (2016): 1–8.

Lacerda, Ana L. D. F., Maíra C. Proietti, Eduardo R. Secchi, and Joe D. Taylor. "Diverse groups of fungi are associated with plastics in the surface waters of the Western South Atlantic and the Antarctic Peninsula." *Molecular Ecology* 29, no. 10 (2020): 1903–1918.

Larsen, A. M., H. H. Mohammed, and C. R. Arias. "Characterization of the gut microbiota of three commercially valuable warmwater fish species." *Journal of Applied Microbiology* 116, no. 6 (2014): 1396–1404.

Li, Erchao, Chang Xu, Xiaodan Wang, Shifeng Wang, Qun Zhao, Meiling Zhang, Jian G. Qin, and Liqiao Chen. "Gut microbiota and its modulation for healthy farming of Pacific white shrimp Litopenaeus vannamei." *Reviews in Fisheries Science & Aquaculture* 26, no. 3 (2018): 381–399.

Lombardo, F., G. Gioacchini, and O. Carnevali. "Probiotic-based nutritional effects on killifish reproduction." *Fisheries and Aquaculture Journal* 27 (2011): 33.

Meyer-Reil, Lutz-Arend, and Marion Köster. "Eutrophication of marine waters: Effects on benthic microbial communities." *Marine Pollution Bulletin* 41, no. 1–6 (2000): 255–263.

Miao, Lingzhan, Yuxuan Gao, Tanveer M. Adyel, Zongli Huo, Zhilin Liu, Jun Wu, and Jun Hou. "Effects of biofilm colonization on the sinking of microplastics in three freshwater environments." *Journal of Hazardous Materials* 413 (2021): 125370.

Mughini-Gras, Lapo, Rozemarijn Q. J. van der Plaats, Paul W. J. J. van der Wielen, Patrick S. Bauerlein, and Ana Maria de Roda Husman. "Riverine microplastic and microbial community compositions: A field study in the Netherlands." *Water Research* 192 (2021): 116852.

O'Neil, Judith M., Timothy W. Davis, Michele Astrid Burford, and Christopher J. Gobler. "The rise of harmful cyanobacteria blooms: The potential roles of eutrophication and climate change." *Harmful Algae* 14 (2012): 313–334.

Paerl, Hans W., and Valerie J. Paul. "Climate change: Links to global expansion of harmful cyanobacteria." *Water Research* 46, no. 5 (2012): 1349–1363.

Rajeev, Riya, K. K. Adithya, G. Seghal Kiran, and Joseph Selvin. "Healthy microbiome: A key to successful and sustainable shrimp aquaculture." *Reviews in Aquaculture* 13, no. 1 (2021): 238–258.

Ringø, E. Z. Z. V., Zhigang Zhou, J. L. Gonzalez Vecino, S. Wadsworth, Jaime Romero, Åshild Krogdahl, Rolf Erik Olsen et al. "Effect of dietary components on the gut microbiota of aquatic animals. A never-ending story?." *Aquaculture Nutrition* 22, no. 2 (2016): 219–282.

Ross, Ashley A., Aline Rodrigues Hoffmann, and Josh D. Neufeld. "The skin microbiome of vertebrates." *Microbiome* 7 (2019): 1–14.

Saha, Mahasweta, and Florian Weinberger. "Microbial 'gardening' by a seaweed holobiont: Surface metabolites attract protective and deter pathogenic epibacterial settlement." *Journal of Ecology* 107, no. 5 (2019): 2255–2265.

Schulz, Silke, and Ralf Conrad. "Influence of temperature on pathways to methane production in the permanently cold profundal sediment of Lake Constance." *FEMS Microbiology Ecology* 20, no. 1 (1996): 1–14.

Scott, Jarrod J., Thomas C. Adam, Alain Duran, Deron E. Burkepile, and Douglas B. Rasher. "Intestinal microbes: An axis of functional diversity among large marine consumers." *Proceedings of the Royal Society B* 287, no. 1924 (2020): 20192367.

Segev, Einat, Thomas P. Wyche, Ki Hyun Kim, Jörn Petersen, Claire Ellebrandt, Hera Vlamakis, Natasha Barteneva et al. "Dynamic metabolic exchange governs a marine algal-bacterial interaction." *elife* 5 (2016): e17473.

Sehnal, Ludek, Elizabeth Brammer-Robbins, Alexis M. Wormington, Ludek Blaha, Joe Bisesi, Iske Larkin, Christopher J. Martyniuk, Marie Simonin, and Ondrej Adamovsky. "Microbiome composition and function in aquatic vertebrates: Small organisms making big impacts on aquatic animal health." *Frontiers in Microbiology* 12 (2021): 567408.

Semova, Ivana, Juliana D. Carten, Jesse Stombaugh, Lantz C. Mackey, Rob Knight, Steven A. Farber, and John F. Rawls. "Microbiota regulate intestinal absorption and metabolism of fatty acids in the zebrafish." *Cell Host & Microbe* 12, no. 3 (2012): 277–288.

Sen, Bulent, Mehmet Tahir Alp, Feray Sonmez, Mehmet Ali Turan Kocer, and Ozgur Canpolat. "Relationship of algae to water pollution and waste water treatment." *Water Treatment* 14 (2013): 335–354.

Song, Hao, Jiahui Xu, Michel Lavoie, Xiaoji Fan, Guangfu Liu, Liwei Sun, Zhengwei Fu, and Haifeng Qian. "Biological and chemical factors driving the temporal distribution of cyanobacteria and heterotrophic bacteria in a eutrophic lake (West Lake, China)." *Applied Microbiology and Biotechnology* 101 (2017): 1685–1696.

Suleiman, Marcel, Karsten Zecher, Onur Yücel, Nina Jagmann, and Bodo Philipp. "Interkingdom cross-feeding of ammonium from marine methylamine-degrading bacteria to the diatom Phaeodactylum tricornutum." *Applied and Environmental Microbiology* 82, no. 24 (2016): 7113–7122.

Sullam, Karen E., Steven D. Essinger, Catherine A. Lozupone, Michael P. O'Connor, Gail L. Rosen, R. O. B. Knight, Susan S. Kilham, and Jacob A. Russell. "Environmental and ecological factors that shape the gut bacterial communities of fish: A meta-analysis." *Molecular Ecology* 21, no. 13 (2012): 3363–3378.

Tan, Baoyi, Yibin Li, Huifeng Xie, Zhenqing Dai, Chunxia Zhou, Zhong-Ji Qian, Pengzhi Hong et al. "Microplastics accumulation in mangroves increasing the resistance of its colonization Vibrio and Shewanella." *Chemosphere* 295 (2022): 133861.

Tan, Herng Yih, Sai-Wei Chen, and Shao-Yang Hu. "Improvements in the growth performance, immunity, disease resistance, and gut microbiota by the probiotic Rummeliibacillus stabekisii in Nile tilapia (Oreochromis niloticus)." *Fish & Shellfish Immunology* 92 (2019): 265–275.

Wang, Shang, Weiguo Hou, Hongchen Jiang, Liuqin Huang, Hailiang Dong, Shu Chen, Bin Wang, Yongcan Chen, Binliang Lin, and Ye Deng. "Microbial diversity accumulates in a downstream direction in the Three Gorges Reservoir." *Journal of Environmental Sciences* 101 (2021): 156–167.

Wienhausen, Gerrit, Beatriz E. Noriega-Ortega, Jutta Niggemann, Thorsten Dittmar, and Meinhard Simon. "The exometabolome of two model strains of the Roseobacter group: A marketplace of microbial metabolites." *Frontiers in Microbiology* 8 (2017): 1985.

Williams, Janet Olufunmilayo, and Nosayame Thomas Osahon. "Assessment of microplastic degrading potential of fungal isolates from an estuary in rivers state, Nigeria." *South Asian Journal of Research in Microbiology* 9, no. 2 (2021): 11–19.

Wu, Xiaojian, Jie Pan, Meng Li, Yao Li, Mark Bartlam, and Yingying Wang. "Selective enrichment of bacterial pathogens by microplastic biofilm." *Water Research* 165 (2019): 114979.

Yao, Wenxiang, Xiaoqin Li, Chunyan Zhang, Jing Wang, Youwang Cai, and Xiangjun Leng. "Effects of dietary synbiotics supplementation methods on growth, intestinal health, non-specific immunity and disease resistance of Pacific white shrimp, Litopenaeus vannamei." *Fish & Shellfish Immunology* 112 (2021): 46–55.

Zhang, Shuangfei, Zhong Hu, and Hui Wang. "Metagenomic analysis exhibited the co-metabolism of polycyclic aromatic hydrocarbons by bacterial community from estuarine sediment." *Environment International* 129 (2019): 308–319.

Zhao, Jun, Xin Zhao, Lei Chao, Wei Zhang, Tao You, and Jie Zhang. "Diversity change of microbial communities responding to zinc and arsenic pollution in a river of northeastern China." *Journal of Zhejiang University-Science B* 15 (2014): 670–680.

Zhou, Jin, Mindy L. Richlen, Taylor R. Sehein, David M. Kulis, Donald M. Anderson, and Zhonghua Cai. "Microbial community structure and associations during a marine dinoflagellate bloom." *Frontiers in Microbiology* 9 (2018): 1201.

Zhu, Hai-Zhen, Min-Zhi Jiang, Nan Zhou, Cheng-Ying Jiang, and Shuang-Jiang Liu. "Submerged macrophytes recruit unique microbial communities and drive functional zonation in an aquatic system." *Applied Microbiology and Biotechnology* 105 (2021): 7517–7528.

2 Microbial Planktonic Communities

Garima, Kishor Kunal, and Parvaiz Ahmad Ganie

2.1 INTRODUCTION

Plankton are tiny organisms that serve as the foundation for food chains in the aquatic ecosystem. The term "plankton" refers to all freshwater and marine creatures that are immobile and unable to swim against the flow of the water. They float in the water column during their entire life. According to the common interpretation of the phrase, it describes a group of organisms that have evolved to spend the most of their life in apparent suspension in open bodies of water like the ocean, lakes, ponds and rivers. Organisms including bacteria, algae, crustaceans, protozoans, coelenterates, molluscs and members of other phyla are all collectively referred to as plankton. They vary widely in size, from 0.2 μm to more than 20 cm (a few may be smaller than 0.2 μm), from microscopic bacteria or viruses to large organisms such as jellyfish. Their distribution varies with the availability of light and nutrients. Plankton play a significant role in the ecosystem as a pioneer component of the food web, in biogeochemical cycles and as a major carbon sequestrant. Phytoplankton account for 50% of the total oxygen (O_2) produced by photosynthesis on the Earth according to one estimate. In principle, the phytoplankton contain chlorophyll "a" (including bacteriochlorophyll "a" and divinyl chlorophyll "a") as photosynthetic pigment. The accessory photosynthetic pigments in phytoplankton include chlorophyll b, c and d, carotenoids (carotene, xanthophyll) and phycobilins (phycoerythrin, phycocyanin, allophycocyanin, phycoerythrocyanin).

2.2 ADAPTATIONS OF PLANKTON

In order to maintain a planktonic existence in the continuously changing water surface, all plankton species have a diversity of floating systems. Flat bodies, lateral spines, oil droplets, gas-filled floats, sheaths constructed of gel-like substances, and ion replacement are some of the most significant modifications.

2.2.1 Surface Area to Volume Ratio

Some plankton species can prevent sinking by expanding the surface area of their bodies while reducing the volume due to their flat bodies and spines. The quick exchange of gases by diffusion is favoured by a large surface area to volume ratio, which also produces a high frictional resistance and causes the sink to move slowly. Additionally, it speeds up the removal of waste from the body's surface and aids in photosynthesis and other nutrient absorption.

2.2.2 Increased Buoyancy

Plankton also possess buoyancy apparatus like gas-filled floats and vesicles, while a few plankton excrete heavy ions (Mg^{2+}, SO_4^{2-}) and contain lighter ions (NH^{4+}, Cl^-).

 DOI: 10.1201/9781003408543-2

2.2.3 Body Shape

Plankton body forms are crucial for their suspension in the water column. The body's flattened shape helps to slow down sinking by rocking back and forth like a falling leaf. The surface area is increased by the long spine and the projection. Additionally, they act as a protection mechanism against predators. Individual chains can be used to navigate spiral paths.

2.3 CLASSIFICATION OF PLANKTON

Planktons are categorised on the basis of lifecycle, size, trophic level and various functional characteristics.

a) On the basis of the lifecycle, planktons are classified as:
   1. Holoplankton – Organisms that remain in planktonic form for their entire lifetime, e.g., algae, jellyfish (*Cyanea, Cassiopea*), etc.
   2. Meroplankton – Organisms that live as plankton at only some stages of their lifecycle, e.g., larvae of starfish, worms, sea urchins, fish, etc.

b) On the basis of size, plankton are classified as:
   1. Megaplankton – They are large in size, >20 cm, e.g., jellyfish, tunicates, pyrosomes, etc.
   2. Macroplankton – Their size ranges from 2 to 20 cm.
   3. Mesoplankton – Includes organisms with size 0.2 to 20 mm, e.g., copepods (*Mesocyclops and Macrocyclops*).
   4. Microplankton – Their size varies from 20 to 200 μm, e.g., most of the phytoplankton, protozoans and large protists.
   5. Nanoplankton – Size ranges from 2 to 20 μm, e.g., protists, diatoms (*Skeletonema* and *Nitzschia*) and algae (*Chlorella* and *Nanochloropsis*).
   6. Picoplankton – Size ranges from 0.2 to 2 μm, e.g., bacteria (*Methanospirillum and Methanobacterium*), Cyanobacteria (*Synechococcus, Merismopedia*), chrysophytes.
   7. Femtoplankton – Includes archaea and viruses with size <0.2 μm, e.g., *Thermococcus* and *Thermoplasma*.

c) On the basis of various trophic levels and characteristics, plankton are classified as:
   1. Phytoplankton – They are autotrophs or producers, e.g., cyanobacteria, algae, diatoms, dinoflagellates, etc.
   2. Zooplankton – Includes primary consumers that feed on other plankton, e.g., small protozoans, larvae of fish and shellfish.
   3. Mycoplankton – Includes fungi.
   4. Bacterioplankton – Includes bacteria and is important for nutrient recycling.
   5. Virioplankton – Includes viruses.
   6. Mixotrophs – They act as both producers and consumers according to environmental conditions. When abundant light and nutrients are available, they perform photosynthesis, and in the absence of limited resources, they feed on other plankton to fulfil their nutrient requirements.

d) On the basis of various functional characteristics, plankton are classified as:
   1. Nitrogen fixers – They fix or convert nitrogen from the air into ammonia or other nitrogenous compounds, e.g., *Trichodesmium*.
   2. Calcifiers – These organisms form calcium carbonates. e.g., Coccolithophores.
   3. Silicifiers – They use dissolved silicic acid ($Si(OH)_4$) to build their shells of biogenic silica (bSiO2), e.g., diatoms.
   4. DMS producers – They produce dimethyl sulphide, e.g., *Phaeocystis*.

2.4 TAXONOMIC TREE OF PLANKTON

Various authors have classified plankton into different classifications. Table 2.1 summarises the major recognised classifications of plankton.

2.5 MICROBIAL PHYTOPLANKTON

Phytoplankton, which is derived from the Greek words *phyto* (plant) and *plankton* (made to wander or drift), are microscopic organisms that exist in freshwater and saltwater habitats.

Phytoplankton, the ecological group of microalgae adapted to live in apparent suspension in water masses, are much more than ecosystem engineers. Protists, bacteria and single-celled plants make up some phytoplankton. Cyanobacteria, silica-encased diatoms, dinoflagellates, green algae and chalk-coated coccolithophores are a few examples of frequent types.

Phytoplankton obtain their energy through photosynthesis, as do trees and other plants on land. As a result, phytoplankton lives on the well-lit surface layers (euphotic zone) of seas and lakes because they need sunlight to survive. Phytoplankton have far faster turnover rates than trees (days versus decades), are dispersed across a broader surface area than terrestrial plants, and are subject to less seasonal change. As a result, phytoplankton react to changes in the climate quickly on a global scale. The primary focus of research has been on changes in phytoplankton community structure, which may have far-reaching consequences for the entire ecosystem via the food web and biogeochemical cycles. A major factor in the dynamics of the phytoplankton community is the environmental state of the research area. In aquatic habitats, phytoplankton are the earliest bioindicators of pollution.

Since environmental factors always affect phytoplankton assemblage and aquatic ecosystems, it is important to comprehend how environmental changes and risks affect each ecosystem. The condition of the environment's natural ecology is tracked using bioindicators. Plankton, which respond swiftly to environmental changes and act as an important biomarker for measuring water quality and as an indicator of water contamination, are one example of an organism within a biological system that provides information about the health of its surroundings. The utilisation of phytoplankton and zooplankton as bioindicators for determining aquatic body health and trophic status, as well as the considerable association between ecosystem abiotic and biotic components, may be deduced.

Some planktonic species can withstand challenging abiotic conditions and flourish in polluted habitats, indicating a high level of tolerance, whereas sensitive species are lacking, indicating a low level of tolerance. Therefore, adopting these species can enhance the use of these creatures in water quality monitoring research. Aquatic ecosystems contain a significant amount of plankton, and changes in its population and community structure have a direct impact on how well the ecosystem works (Yuan et al., 2014). According to Arvanitidis and Eleftheriou (2003), modern research has mostly focused on changes in phytoplankton community structure that may have significant effects on the ecosystem as a whole through the food web and biogeochemical cycles.

The foundation of both marine and freshwater food webs is phytoplankton, which also play a significant role in the global carbon cycle. Despite making up only approximately 1% of the world's plant biomass, they produce at least half of the oxygen produced globally and around half of the photosynthetic activity. Phytoplankton are incredibly diverse, ranging from photosynthesising bacteria to algae that resemble plants to coccolithophores with armour-like plates. Although many additional types are present, significant phytoplankton groups include the diatoms, cyanobacteria and dinoflagellates (Karlusich et al., 2020).

The majority of phytoplankton are too tiny to be distinguished separately by the naked eye. However, due to the existence of auxiliary pigments (such as phycobiliproteins or xanthophylls in some species) and chlorophyll within their cells, some types may be detectable as coloured patches on the water surface when present in sufficient quantities.

TABLE 2.1
Classification of Plankton

Linnaeus (1735)	Haeckel (1866)	Chatton (1925)	Copeland (1938)	Whittaker (1969)	Woese et al. (1990)	Cavallier-Smith (1998)	Cavellier-Smith (2015)
Two Kingdoms	Three Kingdoms	Two Empires	Four Kingdoms	Five Kingdoms	Three Domains	Two Empires, Six Kingdoms	Two Empires, Seven Kingdoms
	Protista	Prokaryota	Monera	Monera	Bacteria	Bacteria	Bacteria
					Archaea		Archaea
			Protoctista	Protista	Eucarya	Protozoa	Protozoa
		Eukaryota				Chromista	Chromista
			Plantae	Plantae		Plantae	Plantae
Vegetabilia	Plantae			Fungi		Fungi	Fungi
Animalia	Animalia		Animalia	Animalia		Animalia	Animalia

2.5.1 Diversification of Phytoplankton

The phytoplankton groups can be mainly divided into three domains, i.e., Bacteria, Archaea and Eucarya (Woese et al., 1990). The domain "Bacteria" encompasses two main divisions, i.e., Cyanobacteria and Anoxyphotobacteria, and one exempted division, Prochlorobacteria. The division "Cyanobacteria" (also called blue-green algae) contains three orders, namely, Chroococcales, Oscillatoriales and Nostaocales. The order "Prochlorales" is included under division "Cyanobacteria" by most authors, but may be sometimes referred to under division "Prochlorobacteria". The most fundamental photosynthetic organisms of phytoplankton belong to the bacterial kingdom (previously known as the Eubacteria), even though plankton also contain acellular microorganisms (viruses) and a variety of well-characterised Archaea (the halobacteria, methanogens and sulfur-reducing bacteria, which formerly comprised the Archaebacteria). In aquatic ecosystems, planktonic bacterial populations are essential for nutrient cycling and energy transfer. Environmental heterogeneity (e.g., different depths of a stratified water column, littoral versus pelagic zone, inner pond in a lake covered with macrophytes) can cause differences in the composition of the bacterioplankton within a lake (Wetzel, 2001). Microbial communities hold the majority of the Earth's biological variety. Small organic compounds produced by microbial metabolism show the biochemical and physiological diversity of these biological processes as well as their taxonomic specificity. These nanomolecules act as a pathway for exchange and signalling that is taxon-specific (Reynolds, 2006).

Another early occurrence that happened about 3000 million years ago was the emergence of phototrophic forms, which were defined by their essential capacity to utilise light energy in order to synthesise food or energy. Some of these species are photoautotrophs, which utilise CO_2 to produce energy or food. Some of these species are photoheterotrophs, needing to manufacture their own cells from organic substrates. Modern variants of photoheterotrophs include purple non-sulfur bacteria (Rhodospirillaceae) and green flexibacteria (Chloroflexaceae), both of which contain pigments that resemble chlorophyll (bacteriochlorophyll a, b or c). An electron donor substance is stripped of its electrons using light radiation during photosynthesis. Water is the main source of reductant electrons in the majority of modern plants, and oxygen is released as a byproduct (oxygenic photosynthesis). Modern-day representatives of the Anoxyphotobacteria, which performed anoxygenic photosynthesis, are the purple and green sulfur bacteria, found in anoxic sediments. Some of these are planktic in the sense that they live in anoxic, densely stratified layers deep in little lakes with the appropriate levels of stability (Reynolds, 2006). The characteristic could be viewed as the byproduct of having evolved in a completely anoxic environment. However, oxic marine habitats have produced aerobic, anoxygenic phototrophic bacteria that possess bacteriochlorophyll a (Shiba et al., 1979). It has also become apparent that these bacteria's contribution to the oceanic carbon cycle is not necessarily negligible (Kolber et al., 2001; Goericke, 2002).

Nevertheless, the Cyanobacteria's development of oxygenic photosynthesis some 2800 million years ago has proven to be a significant step in the evolution of life in water and later, on land. In addition, the biological oxidation of water and the simultaneous removal and burial of carbon in marine sediments modified the composition of the atmosphere through time (Falkowski, 2002). Simple coccoids and rods have evolved into loose mucilaginous colonies, called coenobia, filamentous life forms, and pseudotissues among the Cyanobacteria (which have chlorophyll "a"–mediated photosynthesis). Three of the four main evolutionary lines, the chroococcalean, oscillatorialean and nostocalean (the stigonematalean line being the exception), have significant planktic representatives that have significantly diverged between marine and freshwater habitats. The oldest and most numerous category of still-existing photosynthetic organisms is also the one with the greatest number of individuals. The chroococcales are unicellular or coenobial cynobacteria without filaments. Most genera of these plankters form mucilaginous colonies inhabiting freshwater. The picophytoplanktic forms are the ones abundant in the ocean. Examples of this order include *Aphanocapsa, Aphanothece, Chroococcus, Cyanodictyon, Gomphosphaeria, Merismopedia, Microcystis, Snowella, Synechococcus, Synechocystis, Woronichinia*, etc. (Reynolds, 2006).

The order "Oscillatoriales" includes filamentous uniseriate cyanobacteria, found in both freshwater and marine habitats. Examples include *Arthrospira, Limnothrix, Lyngbya, Planktothrix, Pseudanabaena, Spirulina, Trichodesmium, Tychonema,* etc. The nostocales are filamentous cyanobacteria having unbranched cells. Examples include *Anabaena, Anabaenopsis, Aphanizomenon, Cylindrospermopsis, Gloeotrichia, Nodularia,* etc., which are mostly found in freshwater and brackish water habitats.

Links to eukaryotic protists, plants and animals from the Cyanobacteria had been impliedly sought after and openly assumed. The identification of a prokaryote that resembled the colouring of green plants by carrying chlorophyll a and b but missing phycobilins seems to match the context (Lewin, 1981). Prochloron, a salps symbiont, is not itself a planktonic organism but can be found in collections of marine plankton. The first description of Prochlorothrix from freshwater phytoplankton in the Netherlands (Burger-Wiersma et al., 1989) served to solidify the idea that bacteria possessing chlorophyll a and b are an evolutionary "missing link" (Reynolds, 2006).

Then, another astounding discovery was made: Prochlorococcus, an oxyphototrophic prokaryote with divinyl chlorophyll a and b pigments but no bilins, was found to be the most prevalent picoplankton in the low-latitude ocean, contrary to what had previously been thought (Chisholm et al., 1988, 1992). The discovery of a previously unknown organism's biospheric function is an accomplishment in and of itself (Pinevich et al., 2000); the fact that the organisms appear to occupy this transitional stage in the history of plant life is a great scientific achievement.

Prochlorococcus, however, was not grouped with the other Prochlorales or even clearly distinguished from Synechococcus in subsequent analyses of the phylogenetic relationships of the newly defined Prochlorobacteria using immunological and molecular techniques (Moore et al., 1998; Urbach et al., 1998). According to the current consensus, it is best to treat Prochlorales as abnormal cyanobacteria (Lewin, 2002).

The second domain, "Archaea", contains single-celled organisms. These microbes are prokaryotes because they lack cell nuclei. Although originally given the name archaebacteria (in the Archaebacteria kingdom), archaea were initially categorised as bacteria. Although some archaea, like the flat, square cells of *Haloquadratum walsbyi*, have quite diverse shapes, in general, bacteria and archaea are comparable in size and shape. Despite sharing a similar morphology with bacteria, archaea have genes and some metabolic pathways that are more similar to those of eukaryotes, particularly for the transcription and translational enzymes. Prokaryote classification in general, as well as the classification of archaea, is a hotly debated topic. Archaea are currently categorised into groupings of creatures that have similar structural characteristics and ancestry. These classifications heavily rely on molecular phylogenetics, which uses the ribosomal RNA gene sequence to infer relationships between organisms. The "Euryarchaeota" and the "Thermoproteota" (formerly Crenarchaeota), which are the two major phyla of archaea, contain the majority of culturable and thoroughly researched species. Archaea can be found in a wide variety of settings, are increasingly acknowledged as an important component of global ecosystems, and may make up 20% of the microbial cells in the ocean. However, extremophiles were the first archaeans to be found. As evidenced by their presence in geysers, black smokers and oil wells, some archaea do tolerate high temperatures, frequently above 100 C. However, archaea include mesophiles that thrive in temperate environments, in swamps and marshland, sewage, the oceans, the intestinal tract of animals and soils. Other typical habitats include extremely cold habitats and highly saline, acidic or alkaline water. At salinities higher than 20–25%, halophiles, including the genus *Halobacterium*, outnumber their bacterial counterparts. They thrive in extremely salty environments like salt lakes. Thermophiles thrive in environments like hot springs, where the temperature is over 45 C; hyperthermophilic archaea thrive in environments above 80 C. Early in the history of bacterial evolution, the archaeans and the ancestors of bacteria split apart (Woese, 1987; Woese et al., 1990).

It is now known that the origin of all eukaryotic algae and higher plants may be traced to early primary endosymbioses between early eukaryotic protistans and cyanobacteria (Margulis, 1970, 1981). The domain "Eucarya" may be divided into 12 phyla, namely, Glaucophyta, Prasinophyta,

Chlorophyta, Euglenophyta, Cryptophyta, Raphidophyta, Xanthophyta, Eustigmatophyta, Chrysophyta, Bacillariophyta, Haptophyta and Dinophyta. The phytoplankton belonging to phylum Glaucophyta bear cyanelles (instead of plastids) and are represented in freshwater ecosystems. Examples include *Cyanophora* (found in shallow, productive calcareous lakes), *Glaucocystis*, etc. The cyanelles are supposed to be an evolutionary intermediate between cyanobacterial cells and chloroplasts (which are evident in other more evolved phyla, like Chlorophyta or euglenophyta, etc.) (Reynolds, 2006).

The green eukaryotes with endosymbiotic cyanobacteria reflect the evolution of the bryophyte and vascular plant phyla, as well as the chlorophyte and euglenophyte phyla. The red eukaryotes encompass the evolution of the rhodophytes, chrysophytes and haptophytes with its secondary and even tertiary endosymbioses. The presence of chlorophyll b among the photosynthetic pigments and the normal accumulation of glucose polymers (such as starch and paramylon) as the primary byproduct of carbon absorption are significant characteristics that set the algae of the green line apart. Although both are thought to have a lengthy history on the planet (1500 million years ago), the divide of the green algae between the prasinophyte and the chlorophyte phyla reflects the evolutionary progress and anatomic diversification within the line.

The phytoplankton of phylum Parsinophyta are characterised by mostly green algae that are unicellular, primarily motile, have 1–16 flagella that are arranged laterally or apically, have finely scaled cell walls, have plastids that contain chlorophylls a and b, and have mannitol and starch assimilating substances. The phylum contains two classes and four orders. The first class, Pedinophyceae, with only one order, Pedinomonadales, is characterised by a single lateral flagellum on small cells (e.g., Pedinomonas). The second class, Parsinophyceae, encompasses three orders, namely, Chlorodendrales, Pyramimonadales and Scourfieldiales. The order Chlorodendrales is characterised by flattened, four-flagellated cells and includes both freshwater (*Nephroselmis*, *Scherffelia*) and marine (*Mantoniella*, *Micromonas*) examples. The order Pyramimonadales contains both freshwater and marine planktonic examples characterised by cells with 4 or 8 (rarely 16) flagella arising from an anterior depression, e.g., *Pyramimonas*. The order Scourfieldiales comprises mainly freshwater phytoplankton characterised by cells with two flagella, which may be sometimes unequal in length, e.g., *Scourfieldia* (Reynolds, 2006).

The phytoplankton of phylum Chlorophyta are commonly called green algae. The phylum comprises five orders (Tetrasporales, Volvocales, Chlorococcales, Ulotrichales, Zygnematales), all belonging to class Chlorophyceae. Oedogoniales, Chaetophorales, Cladophorales, Coleochaetales, Prasiolales, Charales, Ulvales and other contemporary chlorophyte groups lack contemporary planktic representation. The phylum Chlorophyta is characterised by colonial, filamentous, siphonaceous, thalloid and green-pigmented algae, having one or more chloroplasts with chlorophylls a and b, and starch or occasionally fat as an assimilation byproduct. The order Tetrasporales contains phytoplankton species with non-flagellate cells embedded in mucilaginous or palmelloid colonies, but with motile propagules, e.g., *Paulschulzia*, *Pseudosphaerocystis*. Phytoplankton of the order Volvocales are unicellular or colonial biflagellates having cup-shaped chloroplasts, and include both marine (*Dunaliella*, *Nannochloris*) and freshwater species (*Chlamydomonas*, *Eudorina*, *Pandorina*, *Phacotus*, *Volvox*). The order Chlorococcales is characterised by non-flagellate, unicellular or coenobial algae, including numerous phytoplankton genera such as *Ankistrodesmus*, *Ankyra*, *Botryococcus*, *Chlorella*, *Coelastrum*, *Coenochloris*, *Crucigena*, *Choricystis*, *Dictyosphaerium*, *Elakatothrix*, *Kirchneriella*, *Monorophidium*, *Oocystis*, *Pediastrum*, *Scenedesmus* and *Tetrastrum*. The other order, Ulotrichales, has single-celled or mostly filamentous unbranched algae with band-shaped chloroplasts, e.g., *Geminella*, *Koliella*, *Stichococcus*. The last order of phylum Chlorophyta is Zygnematales. The species of this order are exclusively freshwater habitant. The order is characterised by species that are unicellular or filamentous algal forms that show isogamous reproduction facilitated by conjugation. Most of the genera of this order belong to family Desmidaceae, characterised by primarily unicellular or (rarely) filamentous coenobia, where cells are essentially condensed into two semi-cells and connected by an isthmus, e.g., *Arthrodesmus*, *Closterium*,

Cosmarium, *Euastrum*, *Spondylosium*, *Staurastrum*, *Staurodesmus* and *Xanthidium* (Reynolds, 2006).

One of the most well-known subgroups of flagellates, which are excavate (group) eukaryotes of the phylum, are the euglenoids or euglenids. The cell structure of the euglenophyta is typical of that class. The phytoplankton of phylum euglenophyta are green-pigmented biflagellates with a single cell. They have numerous, atypical plastids that contain chlorophylls a and b. With a few marine and endosymbiotic members, they are mostly found in freshwater, particularly when it is rich in organic compounds. These species reproduce by fission that occurs along a longitudinal axis. These phytoplankton produce paramylon and oil as byproducts of assimilation. Many euglenids only eat by diffusion or phagocytosis. The phylum has two orders, Eutreptiales and Euglenales, belonging to the common class Euglenophyceae. These two orders differ in the sense that plankton of order Eutreptiales (e.g., *Eutreptia* sp.) have two emergent flagella of equal length, as compared with those of Euglenales (e.g., *Euglena*, *Lepocinclis*, *Phacus*, *Trachelmonas*, etc.), which have one very short and another long and emergent flagella on their cells. A monophyletic group with chloroplasts and the ability to manufacture its own food through photosynthesis includes the mixotrophic *Rapaza viridis* (1 species), Eutreptiales (24 species) and Euglenales (983 species).

The phytoplankton of phylum Cryptophyta are a group of algae, most of which have one or two large plastids. They are widespread in freshwater and can also be found in brackish and marine environments. Each cell has an anterior groove or pocket and is between 10 and 50 µm in size and flattened in shape. Two somewhat different flagella are often found at the edge of the compartment. Chlorophyll a and c, but not b, are present in the cells. The colour of the cells might be brown, blue, blue-green or red, depending on the presence of supplementary phycobiliproteins or other pigments. The assimilatory substance in these phytoplankton is starch. Common examples of the phylum include *Chilomonas*, *Chroomonas*, *Cryptomonas*, *Plagioselmis*, *Pyrenomonas*, *Rhodomonas* and others (Reynolds, 2006).

Next is the small group of single-celled flagellates, which are currently recognised in the phylum Raphidophyta despite sharing characteristics with cryptophytes, dinoflagellates and euglenophytes. The tiny subclass has both marine and freshwater species. All raphidophytes are unicellular and have huge, cell-wall-free cells that range in size from 50 to 100 m. Raphidophytes have two flagella that are arranged so that they both come from the same invagination (or gullet). While one flagellum is coated in hair-like mastigonemes and points forward across the cell surface, the other is located within a ventral groove and points backward across the cell surface. Numerous ellipsoid chloroplasts, which house chlorophylls a, c1 and c2, are present in raphidophytes (Reynolds, 2006). Chlorophyll a, which gives these algae their green colour, is partially concealed by a xanthophyll (in this case, diatoxanthin) to produce the somewhat yellowish colouring. One genus, *Gonyostomum*, belonging to order Raphidomonadales (syn. Chloromonadales) is widespread and can occasionally be found in large quantities in acidic, humic lakes. In acidic waters, like the pools in bogs, freshwater species are more prevalent. Large blooms of marine species are frequently produced in the summer, especially in coastal waters.

Similarly to Raphidophytes, the Xanthophytes and Eustigmatophytes also show yellow-green colouring. They are algae of phylum Xanthophyta that are coenocytic, filamentous, colonial and unicellular. Typically subapically and unequally biflagellated, motile species have two or more discoid plastids per cell that contain chlorophyll a. Due to the main supplementary pigment, diatoxanthin, and the assimilation product, lipid, cells are primarily yellow-green. There are multiple orders, two of which (Mischococcales and Tribonematales) have freshwater planktic representatives. Species of order Mischococcales are unicellular, sometimes colonial, rigid-walled phytoplankton, e.g., *Goniochloris*, *Nephrodiella*, *Ophiocytium*, while species of order Tribonematales are simple or branched uniseriate filamentous xanthophytes, e.g., *Tribonema* sp.

The phylum Eustigmatophyta is a small group of eukaryotic forms of algae (17 genera; 107 species), which contains organisms that live in freshwater, marine environments and soil. All eustigmatophytes have coccoid cells and polysaccharide cell walls, and they are all unicellular.

Eustigmatophytes have one or more yellow-green chloroplasts, which are made up of accessory pigments violaxanthin and beta-carotene in addition to chlorophyll a. Eustigmatophyte zoids (gametes) have one or two flagella that extend from the cell's apex. The end effect of their absorption is probably lipid. Typical genera of plankton include *Chlorobotrys*, *Monodus*, etc.

The phylum Chrysophyta (golden algae) is represented by unicellular, colonial and filamentous, many times uniflagellate, or unequally biflagellate algae. Masking of chlorophyll by fucoxanthin imparts a distinct golden colour to these algae. Chrysophytes are thought to be related to Phaeophyta in that they both exhibit a peculiar combination of chlorophylls a, c1, and c2, as well as a significant amount of another pigment such as xanthophyll fucoxanthin. Phaeophyta has all of the macrophytic brown seaweeds but no planktic vegetative forms. In contrast, the majority of chrysophytes are microphytes with a large number of planktic taxa. The bulk of these are found in freshwater, where they are typically thought to denote low productivity and nutrient condition. The major planktic forms include three classes with four orders. The class Chrysophyeae, with order Chromulinales, are mostly planktic, unicellular or colony-forming flagellates with one or two unequal flagella, which are occasionally naked and often in a hyaline lorica or gelatinous envelope. A few common examples of the order include *Chromulina*, *Chrysococcus*, *Chrysolykos*, *Chrysosphaerella*, *Dinobryon*, *Kephyrion*, *Ochromonas* and *Uroglena* species. The order Hibberdiales of this class (Chlorophyceae) is represented by mostly unicellular or colony-forming epiphytic algae, but has some planktic representatives, like *Bitrichia* sp. The second class, Dictyochophyceae, has one planktic order, i.e., Pedinellales, which is characterised by plankton with radial symmetry, unicellular or coenobial with two unequal flagella. The species are found in both freshwater (*Pedinella*) and marine (*Apedinella*, *Pelagococcus*, *Pelagomonas*, *Pseudopedinella*) habitats. The third class, Synurophyceae, with order Synurales, is represented by unicellular or colony-forming flagellated plankton, which bear distinct siliceous scales, e.g., *Mallomonas*, *Synura*. In recent years, Chrysophytes' phylogenies have undergone extensive taxonomic revision and reinterpretation. The choanoflagellates (formerly Craspedophyceae, Order Monosigales) are no longer considered to be allied to the Chrysophytes (Reynolds, 2006).

The last three phyla, Bacillariophyta, Haptophyte, Dinophyta, of the planktic forms are each notably represented in both freshwater and marine plankton. They constitute the majority of pelagic eukaryotes in the oceans. These phyla are believed to have relatively recent origins, in the Mesozoic period (Reynolds, 2006).

The Bacillariophyta (the diatoms) phylum is a large group comprising several genera of algae, specifically microalgae, conspicuously found in both marine and freshwater habitats. Diatoms generate about 20 to 50% of the oxygen produced on the planet each year and assimilate over 6.7 billion tonnes of silicon each year from their aquatic habitat. As approximately half of the organic matter in the oceans is made up of diatoms, they account for a sizeable amount of the biomass of the planet. The bacillariophytes or diatoms are unicellular or coenobial organisms, which can be found in a variety of shapes, such as ribbons, fans, zigzags, or stars. Their individual cells might be anywhere between 2 and 200 microns in size. An assembly of living diatoms doubles through asexual multiple fission roughly every 24 hours in the presence of sufficient nutrients and sunlight, with the maximum lifespan of individual cells being around 6 days. The bacillariophytes are non-motile algae with numerous discoid plastids containing chlorophyll a, c1 and c2, masked by the accessory pigment fucoxanthin. Bacillariophytes or diatoms have two distinct shapes: a few (centric diatoms) are radially symmetric, while most (pennate diatoms) are broadly bilaterally symmetric. The frustule, a silica (hydrated silicon dioxide) cell wall, that surrounds diatoms is a distinctive aspect of their anatomy. Species of the phylum are ascribed to one or other of the two main diatom orders (Biddulphiales and Bacillariales) in the class Bacillariophyceae. The plankters of order Biddulphiales (centric diatoms) are diatoms with cylindrical halves, sometimes well separated by girdle bands. A few species of the group have been known to form filaments (pseudo-filaments) by adhesion of cells at the end of their valves. A few common examples of centric diatoms include *Aulacoseira*, *Cyclotella*, *Stephanodiscus*, *Urosolenia* from freshwater habitat, and *Cerataulina*, *Chaetoceros*,

Detonula, *Rhizosolenia*, *Skeletonema*, *Thalassiosira* from marine habitat. The order Bacillariales (pinnate diatoms) have species with boat-like halves and an absence of girdle bands. A few species of the group are known to form coenobia by adhesion of cells on the edge of their girdles. Common examples of the order include *Asterionella*, *Diatoma*, *Fragilaria*, *Synedra*, *Tabellaria* (freshwater plankters) and *Achnanthes*, *Fragilariopsis*, *Nitzschia* (marine plankters) (Reynolds, 2006).

The planktic forms of phylum Haptophyta are divided into class Haptophyceae and further subdivided into three orders, namely, Pavlovales, Prymnesiales and Coccolithophoridales. The haptophytes are typically gold or yellow-brown algae, which are usually unicellular with two subequal flagella and one coiled haptonema, though having amoeboid, coccoid or palmelloid stages in some cases. Chlorophylls a, c1 and c2 with auxiliary fucoxanthin make up a pigment mixture in these plankters, which is similar to other gold-brown phyla. The haptophytes were first included with the Chrysophyceae (golden algae); however, ultrastructural findings have classified them separately. The order Pavlovales is characterised by cells with small haptonema and haired flagella, and includes both freshwater and marine species, e.g., *Diacronema*, *Pavlova*. The plankton of order Prymnesiales differ from those of Pavlovales in having smooth flagella. They are mostly marine or brackish water, but a few freshwater representatives also exist. Common examples of the order include *Chrysochromulina*, *Isochrysis*, *Phaeocystis*, *Prymnesium*. The third order, Coccolithophoridales, includes exclusively marine plankter species. The organisms are characterised by small and often complex, flat calcified scales (coccoliths) covering the cell surface. Common examples of the order include *Coccolithus*, *Emiliana*, *Florisphaera*, *Gephyrocapsa*, *Umbellosphaera*. Coccolithophores are the most famous and best-known haptophytes. They account for 673 of the 762 species of haptophytes that have been reported. They fossilise extremely effectively, and their accumulation is primarily responsible for the vast chalk deposits that gave the Cretaceous epoch, which lasted from 120,000 to 65,000 million years ago, its name (from the Greek *kreta*, chalk). There are still enough contemporary coccolithophorids in a few areas to cause "white water" episodes. Emiliana is one of the most extensively researched modern coccolithophorids (Reynolds, 2006).

The final group of phytoplankton in this classification is that of dinoflagellates: a monophyletic group of single-celled eukaryotes that make up the phylum Dinoflagellata or Dinophyta, are typically thought of as algae and are called dinoflagellates or dinophytes. Although dinoflagellates are mainly found in marine plankton, they are also widespread in freshwater environments. Many dinoflagellates are phototrophic, but a significant portion of these also engage in phagotrophy and myzocytosis, a combination of photosynthesis and prey consumption. The dinoflagellates are primarily unicellular, occasionally colonial algae with two unequally sized and oriented flagella. They have intricate plastids with chlorophylls a, c1 and c2, which are typically covered by auxiliary pigments. The cell walls are thick or include polygonal plates for reinforcement. Starch and oil are the assimilation byproducts of the dinoflagellates. The phylum Dinoflagellata has two main planktic classes, namely, Dinophyceae (four planktic orders) and Adinophyceae (one planktic order) (Reynolds, 2006).

The plankton of class Dinophyceae are biflagellates, with one transverse flagellum surrounding the cell and the other directed posteriorly, while those of class Adinophyceae and order Prorocentrales (e.g., *Exuviella*, *Prorocentrum*) are naked or cellulose-covered cells having two halves in the shape of a watchglass. The four orders of plankter belonging to class Dinophyceae are Gymnodiniales (mostly marine), comprised of free-living, free-swimming plankters with flagella located in well-developed transverse and sulcal grooves, without thecal plates (e.g., *Amphidinium*, *Gymnodinium*, *Woloszynskia*); Gonyaulacales (both marine and freshwater), comprising of plankters with armoured, plated, free-living unicells and asymmetrical apical plates (e.g., *Ceratium*, *Lingulodinium*); Peridiniales, differing from Gonyaulacales in having symmetrical apical plates (e.g., *Glenodinium*, *Gyrodinium*, *Peridinium*); and Phytodiniales, comprising coccoid dinoflagellates with thick cell walls and without thecal plates. Some genera of phytodiniales are non-planktic and even pass part of the life cycle as epiphytes, but few are plankton of humic freshwaters (e.g., *Hemidinium*) (Reynolds, 2006).

Diatoms, coccolithophorids and dinoflagellates are relatively recent additions to the fossil record, which clearly demonstrates how evolutionary diversification occurs. There is no question regarding these three groups' tremendous rise throughout the Mesozoic, even though it cannot be said with certainty that any of them did not previously exist. The great extinctions that occurred near the end of the Permian period around 250 million years ago may have been the catalyst. At that time, a gigantic eruption of volcanic lava, ash and shrouding dust from what is now northern Siberia caused global cooling. A buildup of atmospheric carbon dioxide and a period of extreme global warming (which, with the help of a positive feedback loop from methane mobilisation from marine sediments, increased ambient temperatures by as much as 10–11 C) soon reversed the trend. A terrible setback occurred, possibly bringing life on Earth as close to extinction as it has ever been. Many species became extinct in a time span of less than 100 million years, and the survivors underwent drastic changes. The rump biota, both on land and in water, were able to spread out and radiate into habitats and niches that were otherwise empty during the course of the following 20 or so million years as the planet cooled (Falkowski, 2002).

The early Triassic has fossilised dinoflagellates, whereas the late Triassic (around 180,000 million years ago) contains coccolithophorids. In the Jurassic and Cretaceous periods, numerous additional species also emerged in addition to the diatoms. With the exception of picocyanobacteria, these three groups arose to dominate most other species in the sea, a dominance that still exists today (Reynolds, 2006).

2.6 MICROBIAL ZOOPLANKTON

The word "zooplankton" refers to the group of creatures that float aimlessly in the water and can only partially swim against the currents of the water (Alcaraz and Calbet, 2003). By acting as an intermediary species for the energy transfer from planktonic algae to the larger invertebrate predators, these creatures play an important role in the food chain. For fish, prawns, molluscs, corals and other organisms from their larval stages through their terminal stages, zooplankton are frequently used as live food. They are heterotrophic organisms whose primary source of food is phytoplankton. Through a number of methods, primarily mucous feeding webs, faecal pellets, moulting, corpses after death and vertical migrations, they also contribute to the carbon export processes. A few zooplankton are macroscopic, which are readily visible to the naked eye and include jellyfish, fish and prawn larvae, but the majority of zooplankton are microscopic. Zooplankton respond quickly to changes in nutrient intake, making them excellent biological monitors of how various nutrients vary over time in aquatic ecosystems.

Since zooplankton exhibit nocturnal behaviour and high mobility, sampling them to determine their quantity and variety is a laborious task. Based on hydrological factors, predator pressure and primary production, zooplankton occurrence differs from one water body to another. Fish and zooplankton have a strong relationship; in fact, the prey–predator relationship between the two is crucial for fish recruitment (Lomartire, 2021). Common taxonomic groups of zooplankton in freshwater ecosystems are characterised by Rotifera (*Brachionus, Asplanchna, Filinia, Keratella, Philodina, Lepedella Lecane*), Cladocera (*Daphnia, Moina, Simocephalus, Ceriodaphina*) and Crustacea, while copepods (*Cyclops, Mesocyclops, Diaptomus*) are found abundantly in estuary and marine water. Zooplankton are herbivores, carnivores, omnivores, detritivores, as well as parasites of other planktonic animals. Their methods of feeding range from actively capturing motile prey to pumping water through special structures to filter it and retain the food particles (a process known as filter-feeding; Alcaraz and Calbet, 2003).

According to Steinberg et al. (2017), they can be divided into size classes, which vary in terms of morphology, diet, feeding methods and other characteristics both within and across classes. In general, heterotrophic and mixotrophic plankton make up microzooplankton. They primarily consist of phagotrophic protists (Krill, Salp and Appendicularians), including ciliates (Tintinnids) and mesozooplankton nauplii (Sieburth et al, 1978).

Microzooplankton graze more intensely than mesozooplankton, and as a result, they are crucial for maintaining an ecosystem's balance. In addition, microzooplankton play a crucial role in regenerating nutrients that support primary production and serve as food for metazoans (Liu et al., 2021). Despite being crucial to ecosystems, microzooplankton are still poorly understood.

2.6.1 Diversification of Microzooplankton

2.6.1.1 Protozoans

Protozoans are a class of protists that consume organic material, such as other microbes, organic tissues and other organic waste. Because they frequently exhibit animal-like activities, such as motility and predation, and because they lack a cell wall, unlike plants and many types of algae, protozoa were historically thought of as "one-celled animals". Although it is no longer accepted as appropriate to classify protozoa as belonging to the animal kingdom, the term is nonetheless used informally to refer to single-celled organisms that have the ability to move autonomously and feed through heterotrophy. Foraminiferans, radiolarians, zooflagellates and some dinoflagellates are examples of marine protozoans.

2.6.1.2 Radiolarians

Unicellular predatory protists known as radiolarians are covered in ornate globular shells with pores that are typically formed of silica. The Latin word meaning "radius" is the source of their name. They extend various body parts through the holes to capture their prey. When radiolarians die and get preserved as part of the ocean sediment, their shells, like the silica frustules of diatoms, can fall to the ocean floor. As microfossils, these remains shed light on historical maritime conditions (Wassilieff, 2006).

2.6.1.3 Foraminiferans

Foraminiferans are single-celled predatory protists with holes in their shells, just like radiolarians. Their name means "hole bearers" in Latin. Foraminiferans grow by adding new chambers to their tests, which are commonly known as their shells. The shells are often formed of calcite, although they can also contain silica (rarely) or agglutinated sediment particles called chiton. About 40 species of foraminiferans are planktic, although the majority are benthic (Hemleben et al., 1989). Because of their extensive research and reliable fossil records, scientists may deduce a lot about previous habitats and climates (Wassilieff, 2006).

2.6.1.4 Ciliates

The ciliates are a group of alveolates distinguished by the presence of cilia, which are hair-like organelles that are structurally identical to eukaryotic flagella but that are typically shorter, more numerous, and have a different undulating pattern. All members of the group have cilia, which are utilised for a variety of activities including swimming, crawling, attachment, feeding and feeling (although the unique Suctoria only have them for a portion of their life cycle).

A significant class of protists called ciliates is widespread practically everywhere there is water, including lakes, ponds, oceans, rivers and soils. A potential total of 27,000–40,000 species are known to exist, and approximately 4500 distinct free-living species have been described (Foissner et al., 2009).

2.6.1.5 Dinoflagellates

Dinoflagellates are a phylum of unicellular flagellates with about 2000 marine species (Gomez, 2012). Some dinoflagellates are predatory and thus, belong to the zooplankton community, e.g., *Gyrodinium*, *Protoperidinium*, *Nassellaria*.

2.6.1.6 Mixotrophs

A mixotroph is an organism that has more than one trophic mode, ranging from complete autotrophy at one extreme to heterotrophy at the other, and may utilise a variety of different sources of energy and carbon. Mixotrophs are thought to make up more than half of all tiny plankton. Eukaryotic mixotrophs can either have their own chloroplasts or endosymbionts, or they can take over the entire phototrophic cell or acquire them through kleptoplasty, e.g., Tintinnids, *Euglena mutabilis*, Zoochlorella, *Dinophysis acuta*.

2.7 ECOLOGICAL SIGNIFICANCE OF PLANKTON

Phytoplankton, the ecological group of microalgae adapted to live in apparent suspension in water masses, are much more than an ecosystem engineer. Ecosystem services or eco-services are defined as the goods and services provided by ecosystems to humans (Naselli-Flores and Padisák, 2022). The Millennium Ecosystem Assessment (MEA, 2003, 2005) identified about 30 ecosystem services and categorised them into 4 main groups, in which the ecosystem services provided by phytoplankton may also be categorised.

2.7.1 Provisioning Services

All the products acquired from ecosystems are grouped in this category: e.g., food, fuels, active ingredients and drugs, genetic resources.

2.7.1.1 Food Production

– Phytoplankton can be harvested and used as food. Phytoplankton are used as a source of food in space travel and also for carbon dioxide fixation. Chlorella is used as a protein supplement. As producers, they account for half of the total amount of oxygen evolved during photosynthesis. Fish are dependent on plankton for their food to a great extent.

2.7.1.2 Green Chemistry

By the process called bioremediation, phytoplankton help in decreasing the detrimental effects of compounds.

2.7.1.3 Bioactive Compounds and Fuels

Phytoplankton produce bioactive compounds, which can be used for benefits to human health. Biofuel production through phytoplankton or microalgae appears to be the only source of biodiesel, which has the potential to completely remove fossil fuel as the main energy source.

2.7.1.4 Genetic Resources

The great diversity of phytoplankton serves as a source of genetic material.

2.7.2 Supporting Services

This group includes all the services that are instrumental for the functioning of ecosystems and that thus allow the release of all the other services provided by ecosystems (e.g., oxygen production through photosynthesis, primary production, nutrient cycling). Unlike other categories of services, they generally occur over a long period of time.

2.7.2.1 Biomass and Oxygen Production

Phytoplankton are primary producers and constitute the base of the food chain in aquatic ecosystems. They serve as food for a few organisms directly, while in many cases, they get transferred to a higher trophic level by intermediaries. Supporting services provided by phytoplankton include almost half of the global primary and oxygen production.

2.7.2.2 Biogeochemical and Nutrient Cycling

Phytoplankton greatly promote biogeochemical cycles and nutrient (re)cycling, not only in aquatic ecosystems but also in terrestrial ones.

2.7.2.3 Sediment Formation

After death and decay, phytoplankton contribute to sediment formation.

2.7.3 Regulating Services

These include the benefits deriving from the regulation of ecosystem processes (e.g., climate regulation, water depuration).

2.7.3.1 Air Quality Maintenance

Phytoplankton serve as system for purifying polluted air by reducing carbon dioxide (CO_2), nitrogen oxide (NO_x) and/or sulfur oxide (SO_x) in the polluted air and generating oxygen.

2.7.3.2 Climate Regulation

By taking up carbon dioxide (CO_2) for the process of photosynthesis, phytoplankton play a large role in the natural carbon cycle, helping to regulate the amount of CO_2 in the atmosphere and keep the Earth's climate in balance.

2.7.4 Cultural Services

This group includes all the nonmaterial benefits that people receive from ecosystems, such as spiritual and aesthetic experiences, cognitive development and recreational activities.

2.7.4.1 Tourism

Phytoplankton support the food chain of aquatic ecosystems, which in turn, supports the survival of different fauna, which increases the aesthetic value of a water body and attracts tourists.

2.7.4.2 Education

Various aspects of phytoplankton need to be understood for conserving ecological processes and making better utilisation of available resources.

2.8 MODERN TECHNIQUES USED IN PLANKTON STUDIES

2.8.1 High Pressure Liquid Chromatography (HPLC)

Pigment analysis with HPLC is a rapid, very sensitive and objective method for determining the phytoplankton composition and for estimating the biomass of the different algal groups. HPLC analysis quantifies the composition and concentration of phytoplankton specific pigments, allowing chemotaxonomic characterisation of the phytoplankton community, based on established relationships between pigments and various taxonomic groups.

2.8.2 Flow Cytometry

Flow cytometry has routinely been used to study phytoplankton by taking advantage of innate fluorescence molecules that distinguish different species and their physiological conditions (Haynes et al., 2016). Advances in cytometer design and functionality have modernised certain aspects of marine biology applications, creating a more accurate, data-rich and timely assessment of microscopic marine organisms.

2.8.3 Metabarcoding

DNA metabarcoding is a developing approach that identifies multiple species from a mixed sample (bulk DNA or eDNA) based on high-throughput sequencing (HTS) of a specific DNA marker (Liu et al., 2020). Metabarcode data sets are more comprehensive, quicker to produce, and less reliant on taxonomic expertise than traditional methods. They also provide many opportunities for detailed research because the technique focuses on characterising a community from an environmental sample by using "universal" primer sets, which can amplify across species to examine a specific region of DNA that leads to the reading of its sequence or "barcode". Metabarcoding can be used to simultaneously determine species presence and diversity while enabling the detection of novel species.

2.8.4 Metagenomics

Metagenomics is a molecular tool used to analyse DNA acquired from environmental samples in order to study the community of microorganisms present without the necessity of obtaining pure cultures (Nazir, 2016). Functional metagenomics allows high-resolution genomic analysis of unculturable microbes and correlation of the genomes with particular functions in the environment.

REFERENCES

Alcaraz, Miguel, and Albert Calbet. "Zooplankton ecology." *Marine Ecology* (2003). Encyclopedia of Life Support Systems (EOLSS), eds C. Duarte and A. Lott Helgueras (Oxford: Developed under the Auspices of the UNESCO, Eolss Publishers): 295–318.

Arvanitidis, C., and A. Eleftheriou. "Executive summary: Marine biodiversity in the Mediterranean and the Black Sea." *MARBENA Proceedings* (2003).

Falkowski, Paul G. "The ocean's invisible forest." *Scientific American* 287.2 (2002): 54—61.

Foissner, Wilhelm, and David Leslie Hawksworth, eds. *Protist diversity and geographical distribution*. Vol. 8. Springer Science & Business Media, 2009.

Goericke, Ralf. "Top-down control of phytoplankton biomass and community structure in the monsoonal Arabian Sea." *Limnology and Oceanography* 47.5 (2002): 1307-1323.

Gómez, F. "A checklist and classification of living dinoflagellates (Dinoflagellata, Alveolata)." *Cicimar Oceánides* 27, no. 1 (2012): 65–140.

Haynes, Matthew, Brian Seegers, and Alan Saluk. "Advanced analysis of marine plankton using flow cytometry." *Biotechniques* 60, no. 5 (2016): 260.

Hemleben, Christoph, Michael Spindler, and O. Roger Anderson. *Modern planktonic foraminifera*. Springer Science & Business Media, (1989): 363

Karlusich, Juan Jose Pierella, Federico M. Ibarbalz, and Chris Bowler. "Exploration of marine phytoplankton: From their historical appreciation to the omics era." *Journal of Plankton Research* 42, no. 6 (2020): 595–612.

Lewin, Ralph A. "Prochlorophyta–a matter of class distinctions." *Photosynthesis research* 73 (2002): 59-61.

Liu, Kailin, Bingzhang Chen, Liping Zheng, Suhong Su, Bangqin Huang, Mianrun Chen, and Hongbin Liu. "What controls microzooplankton biomass and herbivory rate across marginal seas of China?." *Limnology and Oceanography* 66, no. 1 (2021): 61–75.

Liu, Mingxin, Laurence J. Clarke, Susan C. Baker, Gregory J. Jordan, and Christopher P. Burridge. "A practical guide to DNA metabarcoding for entomological ecologists." *Ecological Entomology* 45, no. 3 (2020): 373–385.

Lomartire, Silvia, João C. Marques, and Ana M. M. Gonçalves. "The key role of zooplankton in ecosystem services: A perspective of interaction between zooplankton and fish recruitment." *Ecological Indicators* 129 (2021): 107867.

Millennium Ecosystem Assessment. "Ecosystem and human well-being: A framework for assessment." World Resources Institute, Washington, DC, 2003.

Millennium Ecosystem Assessment, M. E. A. *Ecosystems and human well-being.* Vol. 5. Washington, DC: Island Press, 2005.

Naselli-Flores, Luigi, and Padisak Judit. "Ecosystem services provided by marine and freshwater phytoplankton." *Hydrobiologia* (2022): 850:2691-2706. DOI: 10.1007/s10750-022-04795-y(0123456789().,-volV)(01234567

Nazir, Asiya. "Review on metagenomics and its applications." *Imperial Journal of Interdisciplinary Research,* (2016). Vol: 2, Issue -3: 277-286

Reynolds, Colin S. *The ecology of phytoplankton.* Cambridge University Press, 2006.

Sieburth, John McN, Victor Smetacek, and Jürgen Lenz. "Pelagic ecosystem structure: Heterotrophic compartments of the plankton and their relationship to plankton size fractions." *Limnology and Oceanography* 23, no. 6 (1978): 1256–1263.

Steinberg, Deborah K., and Michael R. Landry. "Zooplankton and the ocean carbon cycle." *Annual Review of Marine Science* 9 (2017): 413–444.

Wassilieff, M. "Estuaries." Te Ara The Encyclopaedia of New Zealand, 2006.

Wetzel, Robert G. *Limnology: Lake and river ecosystems.* Gulf Professional Publishing, 2001.

Woese, Carl R., Otto Kandler, and Mark L. Wheelis. "Towards a natural system of organisms: Proposal for the domains Archaea, Bacteria, and Eucarya." *Proceedings of the National Academy of Sciences* 87, no. 12 (1990): 4576–4579.

Yuan, Mingli, Cuixia Zhang, Zengjie Jiang, Shujin Guo, and Jun Sun. "Seasonal variations in phytoplankton community structure in the Sanggou, Ailian, and Lidao Bays." *Journal of Ocean University of China* 13 (2014): 1012–1024.

3 Climate Change

A Strong Driving Force for Cyanobacterial Bloom

Snatashree Mohanty, Anirban Paul, and Pramoda Kumar Sahoo

3.1 INTRODUCTION

The surge in population has elevated and accelerated the momentum of intensification in aquaculture to meet the escalated demand for affordable animal protein. Aquaculture provides affordable and accessible animal protein; hence, it takes centre stage for ensuring absolute food security in the developing world (FAO, 2017). The lack of scope for horizontal expansion of the agricultural sector has put significant pressure on the aquatic ecosystem to harness more unit production from a limited water area. This challenge has been smartly addressed by intensive aquaculture with best management practices. Through the process, the water body receives a huge amount of inputs associated with nutrient investment, management and recycling. The delicate balance between these three aspects should co-exist for a healthy aquatic ambience in terms of microbial dynamics and maintaining optimum physico-chemical parameters. Moreover, nutrient enrichment in waterbodies is more popularly coined as eutrophication, which facilitates algal proliferation. Algae are integral component of the aquatic food chain. However, uncontrolled algal proliferation, termed as bloom, has various negative consequences on water quality and aquatic inhabitants as well. Besides the marine environment, freshwater aquaculture has now been threatened and is experiencing episodes of flourishing cyanobacterial blooms that have broken the sustainability and integrity of the ecosystem. This is due to the impressive transformation of culture practices towards supra-intensification.

3.2 HARMFUL ALGAL BLOOMS (HABS)

Harmful algal blooms (HABs) in freshwater are dominated by cyanobacteria, termed 9 including *Aphanizomenon*, *Cylindrospermopsis*, *Dolichospermum*, *Microcystis*, *Nodularia*, *Planktothrix* and *Trichodesmium*. These species are common bloom formers and cause nuisance in freshwater bodies. Cyanobacteria are capable of converting carbon dioxide (CO_2) into biomass by using sunlight as a source of energy. Adverse effects of blooms include deterioration of water quality, oxygen depletion by forming surface mats, and undermined aesthetic value. In addition, some species of cyanobacteria are capable of producing toxins known as cyanotoxins, which has restricted the use of drinking water, irrigation, aquaculture, and fish breeding and is a concern for public health (El Herry-Allani and Bouaïcha, 2013). In addition, they provide various ecological advantages as a part of their survival strategies, making the web interaction complex, which necessitates in-depth study to formulate promising control methods. Acute intoxication of cyanotoxins in human beings has been reported to cause hepatopancreatic, digestive, endocrine, dermal and nervous system disorders (Carmichael, 2001). Not only the rise of blooms but also their senescence and bacterial decomposition deplete oxygen and cause fish kill in aquatic bodies. Moreover, cyanobacterial composition and abundance have been modulated with the changed scenario of the climatic

DOI: 10.1201/9781003408543-3

conditions, making their control an even more challenging task. This chapter draws the attention of readers towards the significance of cyanobacterial blooms, driving forces, challenges and the impact of climatic change on algal proliferation.

3.3 CYANOBACTERIAL BLOOMS: SIGNIFICANCE, DRIVING FORCES AND CHALLENGES

Cyanobacteria have a global presence and naturally possess resting cells, sheaths, capsules and photo-protective cellular pigments, which enable them to resist heat and desiccation. They can successfully adjust their buoyancy in the water column for meeting optimum nutrient and light requirements for growth and expansion (Reynolds, 2006). They are also equipped with different physiological adaptations to survive, including fixing atmospheric nitrogen (N_2) into biologically available ammonia, sequestering iron, storing nutrients like phosphorus, nitrogen and other essential nutrients, and importantly, producing toxins (Gallon, 1992; Wilhelm and Trick, 1994; Reynolds, 2006). Cyanobacteria are the only prokaryotes that can undergo oxygenic photosynthesis to produce oxygen via photosystem I and II and also have the capacity to convert atmospheric nitrogen to the biologically available form of ammonium (NH_4^+) (Canfield, 2005). Cyanobacteria can massively grow and form dense toxic blooms in freshwater and marine environments. Eutrophication has been linked to cyanobacterial bloom amplification in freshwater by many researchers (Reynolds, 2006; Paerl and Fulton, 2006). Among the bloom-forming algae, the cyanobacterial bloom holds centre stage in aquatic ecosystems due to its potential adverse implications, not only on biomass loss but also resulting in additional ecological and economic costs. The level of the devastating impact of blooms varies in different places depending upon the nature of the affected water body, type and frequency of the blooms, sources of nutrient influx, and the available and practical remedial strategies at that site (Steffensen, 2008). There are various driving forces that co-exist to accelerate bloom formation. Some important factors are described in the following.

3.3.1 Co-Existence of Grazers and Cyanobacteria in the Food Web

Grazing is a common top-down ecological control phenomenon in the planktonic food web, which creates a kind of selection pressure and mortality of certain species. Phytoplankton like cyanobacteria and grazers co-exist for the maintenance of the food web (Smayda, 2008). However, for survival, both grazers and cyanobacteria employ various strategies in the ecosystem, and if one has strong defence, it can overcome the other. Cyanobacteria are very good bloom-forming phytoplankton, which are naturally equipped with various adaptations, including elongation in multiple dimensions (Vanderploeg et al., 1988), colony formation (Hessen and Van Donk, 1993), production of mucus, toxins and allelochemicals, and avoidance of grazers through jumping or migration (Porter, 1976; Jakobsen, 2002; Selander et al., 2012; and Strom et al., 2013), while grazers apply various means to avoid toxins, like moving to a depth where non-toxic cyanobacteria are more prevalent (Haney et al., 1994; DeMott and Moxter, 1991). However, non-toxic *Microcystis* has been reported to produce microcystin, the hepatotoxin, in the presence of zooplankton (Jang et al., 2003). Other species are given lower priority due to inadequate nutritional content (Martin-Creuzburg and von Elert, 2009). Thus, cyanobacteria successfully establish themselves to dominate over other phytoplankton due to less grazing activity. Nevertheless, the photoprotective pigments provide strength to some bloom-forming cyanobacteria to withstand extreme irradiation in surface water, leaving other phytoplankton in the subsurface (Paerl et al., 1983). Considering all these defensive activities, the relevance of grazing and protection from grazing determines the fittest for survival in the aquatic food web.

3.3.2 Extracellular Polymeric Substances (EPS) and Colonial Adaptation

Cyanobacteria secrete large amounts of EPS, which enables them to get protection from grazing and provides floating velocity (Passow, 2002) to super-exceed other phytoplankton as a strategy for dominance in the ecosystem. Floating velocity enables them to achieve greater migration and dominance. EPS are complex polymers, characterised by hydrophobicity, carbon and nutrient enrichment, metal adsorption and stickiness, which lead to various characteristics during bloom episodes (Liu et al., 2018). Among cyanobacteria, *Microcystis* sp. are potential bloom formers in freshwater globally. They are most harmful and frequently documented and of great concern due to their toxicity and also having the smartest adaptations to outcompete residents. Xiao et al. (2018) reported a very interesting feature of *Microcystis* related to its extraordinary survival strategy among the phytoplankton. *Microcystis* display significant phenotypic plasticity and colony-forming nature, which provides the required floating velocities (maximum recorded/single colony ~10.08 m h^{-1}). They can successfully modulate vertical migration and buoyancy by manipulating internal cell pressure through gas vesicles to effectively utilise light and nutrients (Šejnohová et al., 2012). This migration helps *Microcystis* to survive even at low P concentrations (Baldia et al., 2007; Saxton et al., 2012). Cyanobacteria use a light source to produce oxygen and are hence termed oxygenic photoautotrophs. According to recent literature, they typically produce sulfated extracellular polysaccharides to support phototrophic biofilms. Genetic engineering has revealed the presence of two sets of genes that can regulate the production level of EPS and control cell aggregation and floating capacity without the presence of internal gas vesicles (Maeda et al., 2021). The EPS of cyanobacteria also provides a defence to reduce grazing by protists (Liu et al., 2018).

3.3.3 Presence of Nutrients

Due to receiving nutrients from expanding agricultural activities, anthropogenic activity and intensification of fish culture systems, water bodies have become treasuries of nutrients, leading to eutrophication. Nutrient influx or eutrophication has resulted in proliferation of HABs. Among nutrients, inorganic phosphorus and nitrogen play important roles in algal growth. Hence, nutrient levels and their manipulation can certainly be a long-term approach to controlling the algal population in the ecosystem. There exists a substantial relationship between cyanobacterial abundance and availability of phosphorus, which is classically considered the limiting nutrient in freshwater systems for phytoplankton growth (Kotak et al., 2000; Gobler et al., 2016). As per recent literature, N can also play a significant role in controlling the density and level of toxicity of some non-diazotrophic cyanobacterial blooms (Gobler et al., 2016). Hence, redirected research has educated us that it is not a single nutrient that is important; rather, reduction of dual nutrients (N and P) must be the focus to mitigate cyanobacterial bloom.

For example, as a part of its survival strategy, *Microcystis* has several specific adaptations to withstand P-limited conditions. *Microcystis* can store P intracellularly for its survival during P deprivation (Carr and Whitten, 1982), and it has the capacity of storing P on its external surface (Saxton et al., 2012) as a part of its survival strategies. Hence, depending on nutrient availability, it can manage to meet the requirement for its survival and dominance over other phytoplankton. Moreover, *Microcystis* has the capacity to up-regulate phosphate-binding proteins and alkaline phosphatase under a P-deficit environment (Harke and Gobler, 2013; Gobler et al., 2016). Cyanobacteria can be either diazotrophic (e.g., *Anabaena and Dolichospermum*) or non-diazotrophic (e.g., *Microcystis*) based on nitrogen-fixing capabilities. They have special strategies to acquire N from the system for their survival. As per the literature, *Microcystis* can potentially use N from diazotrophs and effectively compete with N_2-fixing taxa under highly favourable N_2 fixation conditions (Paerl et al., 2014). These reports have attracted the attention of researchers to dual nutrient (N and P) management to control cyanobacterial growth. Nevertheless, we have zero control over the ingress of exogenous nutrients or the internal nitrogen availability in the water body due to nitrogen fixation, which has made the cyanobacterial growth study more complex.

3.3.4 Temperature and Carbon Dioxide

The abiotic factors, temperature and light, are interlinked and have substantial influence on cyanobacterial growth. Temperature cannot be controlled in large water bodies and is energy-consuming in smaller water bodies (Hudnell et al., 2010). HABs generally prefer warm water, which may result from climate change and hydrological disruptions. Climatic changes in the form of rising global temperature and CO_2 levels, and irregular precipitation patterns, have synergistically influenced freshwater ecosystems and eventually, *Microcystis* bloom dynamics (Paerl and Paul, 2012). Warmer temperatures and heat waves accelerate bloom formation, dominated by cyanobacteria (Paerl and Huisman, 2008; Joehnk et al., 2008). Atmospheric CO_2 diffuses into the water body through the natural air–water exchange process and has a certain role in cyanobacterial growth. Algae have the capability to fix atmospheric CO_2 and can produce biomass from CO_2. Hence, they are very commonly used for carbon sequestration. As per a recent review, they have the capacity to survive at both high and low CO_2 (Sandrini et al., 2015).

3.3.5 Sunlight

Sunlight has a great role in the production and maintenance of aquatic food webs, including phytoplankton, macrophytes and subsequent higher trophic-level organisms. This signifies the importance of sunlight for healthy aquatic life and a balanced ecosystem. However, adaptation to less sunlight by Cyanobacteria allows them to establish successfully over many other phytoplankton species in aquatic environments (Floder et al., 2002). Survival in the low-light area is further explained by the presence of the light-harvesting pigments (phycobilins) and light-harvesting antennae (phycobilisomes) (Canfield et al., 2005). Phycobilins are of two types, i.e., phycocyanin and phycoerythrin. The phycocyanin (blue pigment) is specific to cyanobacteria for capturing light for photosynthesis; hence, cyanobacteria are also called blue-green algae. Cyanobacteria can create shade from their colonial growth in order to create a low-light environment by reducing the availability of sunlight for other phytoplankton and can modulate their own phycobilisomes to meet their light requirements (Oliver and Ganf, 2000). In this way, cyanophytes can survive and successfully thrive under varied light conditions.

3.3.6 Condition of the Ecosystem (Lentic/Lotic)

The velocity and flow rate in the freshwater ecosystem have been decreasing in recent times as a consequence of global climate change (Paerl and Huisman, 2008). Hence, this scenario has created frequent stagnant conditions in waterbodies and inclined them towards more proliferation of HABs. The lentic/stagnant phase of the water body encourages more nutrient accumulation, causing eutrophication in the ecosystem. In order to address this issue, artificial circulation of water bodies may be an approach to reduce nutrient enrichment. However, increasing the flow rate may be the other way to address the same problem. However, the process is neither possible at all times nor cheap. Stratification and long water replacement time give stability and accelerate cyanobacterial growth (Joehnk et al., 2008). Cyanobacteria love stagnant water. Hence, disturbance of this condition by vertical mixing using destratification devices like bubblers or reducing water storage time can be effective in small lakes and ponds to control cyanobacteria, unlike bigger waterbodies (Visser, 1995). All the above-described factors directly or indirectly influence the occurrence of cyanoblooms. However, their extended adaptation to the varied climatic conditions, along with buoyancy according to the changing nutrient regimes, and light sources make them more prevalent in surface waters (Walsby et al., 1997). Hence, unlike the lotic condition, a lentic water body with a high retention time of water and low flowing velocity determines the nutrient storage time and subsequent algal bloom occurrences.

3.3.7 Current Mitigation Measures

The widespread presence and expansion of cyanobacteria have been a chronic problem in eutrophic water bodies. At present, there exist various temporary control measures, developed after a spate of research, and a few are described in the following.

3.3.8 Physical, Chemical and Biological Control

There are various methods, including physical, chemical, biological and ecological technologies, for controlling algae (Zhang et al., 2019). Physical methods are environment-friendly approaches that are equipment and labour intensive. Some of the effective physical control methods include the use of ultraviolet radiation, light shielding, membrane filtration and ultrasonic technology (Nawi et al., 2020). On the other hand, chemical control of algae generally involves the application of copper sulfate, herbicide, potassium permanganate, ferrate, ozone, hydrogen peroxide and bromine dioxide to kill algae (Tsai et al., 2019; Chen et al., 2021). Chemical algaecides immediately give control over algae. However, this approach has attracted controversies and debates for its toxicity to off-target organisms, residue problems and the environment as well. Among algaecides, copper sulfate has been widely used and documented, and a low dose can reduce off-flavour problems in fish ponds. As per the literature, copper sulfate concentrations of 0.5 and 1.0 mg L^{-1} can cause a significant reduction in the growth of *M. aeruginosa* (Tsai et al., 2019). Its toxicity is furthermore increased due to the strong interaction of environmental parameters, including pH, temperature, and the concentrations of calcium and dissolved organic matter (Schrader et al., 2005). Moreover, longer and more frequent use of copper sulfate leads to the establishment of copper-tolerant cyanobacteria (Tucker, 2000), which is another cause for its limited application in aquaculture ponds. The toxicity of this algaecide leads to the release of lethal microcystin (a potent hepatotoxin) into the water body and increases the risk to the aquatic habitat as well as causing drinking water pollution. In addition, accumulation of this algaecide in the sediment also causes serious problems over the years. After copper sulfate, hydrogen peroxide has been documented as an effective algaecide for multiple reasons. It successfully kills cyanobacteria, causes no pollution, and the oxidation process by peroxide is accelerated by light, which helps in the breakdown of microcystins into peptide residues. This process helps detoxify waterbodies from microcystins produced by blooms (Matthijs et al., 2012). However, caution is advised when using these chemicals, especially in small impoundments, as their application must be carefully regulated. The other method is a hybrid of physical and chemical methods, popular as the physico-chemical method, for example, coagulation and flocculation. This process involves the use of surfactants like aluminium sulfate, iron cetyltrimethyl ammonium bromide, chitosan and ferric chloride (Zhang et al., 2016). However, the selection of doses and appropriate flocculant for implementation in larger waterbodies is more costly. Biological methods for algal bloom control include the use of microorganisms such as bacteria, fungi, viruses and aquatic organisms to control algae (Bridgeman et al., 2013). Another effective method is the introduction of fish and benthic filter feeders that can feed on cyanobacteria, called biomanipulation. A recent report concludes that silver and bighead carp are two promising and appropriate biomanipulators to control *Microcystis* blooms in eutrophic waters where we intend to manipulate the food web and also nutrient suspension (Ke et al., 2009).

3.3.9 Nutrient Manipulation and Quenching

Ecological control encourages the use of aquatic vegetation restoration technology, constructed wetland technology and plant floating bed technology (Rajasekhar et al., 2012). All these measures are based on the common principle of competition for nutrients, which will reduce their availability for cyanobacterial growth. The ultimate target for cyano-bloom control is to reduce nutrients in the ecosystem. New research tells us that phosphorus and nitrogen input reductions in the water body are a promising cyano-bloom controlling strategy. Hence, establishing an appropriate N/P ratio as a threshold below which bloom can grow is important (Paerl, 2013). However, the persistence of nutrients depends on storage time, flow rate and mixing of water strata (Scott et al., 2013). Smith and Schindler, 2009) reported that total molar N:P ratios >15 do not favour Cyano-HAB proliferation. The point sources are expected to share a greater magnitude of N and P. Therefore, controlling and regulating point sources becomes more accessible, providing a successful avenue for management Smith and Schindler, 2009).

In recent decades, the use of P has substantially increased due to intensive farming, conversion of forests and grasslands into crop fields, and anthropogenic activities, which can account for >50% of annual P loads (Sharpley et al., 2003). Classically, P quenching is an effective means of controlling CyanoHABs in freshwater ecosystem (Smith and Schindler, 2009). Recent knowledge from literature reveals the potential application of Phoslock in small reservoirs. This is basically a bentonite clay blended with lanthanum, which is electrostatically bound to the bentonite and phosphate anions. However, shallow lakes where sediment resuspension may be a possible consequence of wind are not suitable for Phoslock application (Robb et al., 2003). P exists as dissolved inorganic P (DIP) in the form of orthophosphate (PO_4^{3-}), which is readily used by cyanobacteria, whereas the other form, i.e., dissolved organic P (DOP), is less rapidly used by cyanobacteria (Lean, 1973). However, some fractions of DOP in the presence of microbes are converted to DIP, and the P level increases. Cyanobacteria can store P intracellularly as polyphosphate as a strategy for future survival during a deficit in P (Healy, 1982). The management of phosphorus in every form has an important role in the ultimate curtailing of cyano-blooms.

Nitrogen exists in dissolved, particulate and gaseous forms (Galloway et al.,, 2004). The dissolved form includes dissolved inorganic N (DIN), including ammonium (NH_4^+), nitrate (NO_3^-) and nitrite (NO_2^-); dissolved organic N (DON) includes amino acids and peptides, urea, organonitrates; and particulate organic N (PON) includes polypeptides, proteins, organic detritus. All these forms come from non-point and point sources. Non-point sources include surface runoff, atmospheric deposition and groundwater, while point sources are dominated by municipal, agricultural and industrial wastewater. Recycled N, as NH_4^+ and urea, is a major source of N for the proliferation of CyanoHABs (Blomqvist, 1994). As with P, N cycling, distribution also directs bloom occurrence, which emphasises dualistic nutrient management strategies. Sediment removal, which includes dredging and disturbance of lake bottoms, can restrict the growth of CyanoHABs. Suction dredging of the upper 0.5 m of sediments during a 2-year period led to a significant decrease in nutrient concentrations and CyanoHABs (Peterson, 1982). However, the concern associated with dredging is the risk of nutrient leaching back into the system if this is not managed.

3.4 CYANO-BLOOM IN THE CLIMATE CHANGE FRONT

3.4.1 Elevated Temperature

Climatic change in the form of global warming generates a notable challenge in evaluating the dynamics of toxic cyanobacterial bloom's intensity, expansion, integrity, frequency and proliferation (Nwankwegu et al., 2019). Climate change is generally linked to an increase of the average temperature regime. The current warming trend anticipates that global warming will reach 1.5 C between 2030 and 2052 (Masson, 2018). Literature has successfully linked global warming with cyanobacterial abundance. Temperature is an important parameter to determine the seasonal dynamics of phytoplankton (Falkowski and Raven,2013) and directly influences the photosynthetic capacity, respiration and growth rate (Robarts and Zohary, 1987). Though there is a strong relation between nutrient availability and cyanobacterial growth, climate change, especially elevated temperature due to global warming, can further modulate the proliferation and composition of bloom by increasing the temperatures, altering precipitation and hydrologic properties (Paerl et al., 2011). Global warming has a positive effect on the cyanobacterial communities and their ecology. Rising water temperatures greatly favour the formation of CyanoHABs in numerous ways. It is well documented that relatively high temperatures, usually more than 25° C, facilitate accelerated cyanobacterial growth rates (Robarts and Zohary, 1987; Coles and Jones, 2000). Increased warming in recent days has resulted in more growth and a longer bloom season for toxin-producing HABs like *M. aeruginosa* (Davis et al., 2009; Gobler et al., 2017). Zhang et al. (2011) studied the effect of temperature on microcystin-LR (MC-LR) lethality in zebrafish (*Danio rerio*). Fish exposed to MC-LR under varied temperatures of 12°C, 22°C and 32° C recorded the highest mortality at the highest temperature, establishing the direct relationship between temperature and toxin production. Cyanobacteria are prokaryotes; they love warmer temperatures for rapid growth, often at more than 25° C, and can outcompete eukaryotic algae (Weyhenmeyer, 2001). Further, it has been observed that at elevated temperatures, cyanobacteria effectively compete with most of the eukaryotic primary producers, i.e., diatoms, cryptophytes, chlorophytes and dinoflagellates (Elliott et al., 2006; Joehnk et al., 2008). This trend of optimum growth rate of cyanobacteria and declining growth rates of selected eukaryotic taxa proves the dependency of cyanobacterial blooms on the temperature worldwide (Robarts and Zohary, 1987; Paerl, 1990; Kanoshina et al., 2003; Fernald et al., 2007).

In addition to the direct effects of increasing temperatures on higher cyanobacterial growth, the warming up of surface waters also exaggerates the vertical stratification of the water bodies. The density difference between the surface (epilimnion) and the deep water (hypolimnion) determines the intensity of vertical stratification (Wagner and Adrian, 2009). Increased water temperatures decrease the density of the upper layer, thereby encouraging the vertical stratification of water bodies. In addition, global warming elongates the period of stratification. Warmer waters encourage more intense vertical stratification. Interestingly, many species of cyanobacteria are competent to exploit and overcome the created stratified conditions. In contrast to eukaryotic phytoplankton species, some bloom-forming cyanobacteria can produce gas vesicles, which allow buoyancy regulation and increased concentration of cyanobacteria at the surface of the water in comparison to other phytoplanktonic species (Walsby, 1975; Walsby et al., 1997). As a consequence, the accumulation of large numbers of algal cells may lead to the production of large amounts of toxin. The concentrations of cyanotoxin in dense surface blooms are often several magnitudes higher than the toxin concentrations in the surrounding waters. In addition, the surface water cyanobacteria bloom forms shade over the non-buoyant eukaryotic phytoplankton deeper down in the water column, thus suppressing their competitors (Huisman et al., 2004; Jöhnk et al., 2008).

3.4.2 Elevated CO_2

The result of climate change is also evident from consistently increasing atmospheric CO_2 released from fossil fuel emissions, an important component of the greenhouse effect. In this scenario, cyanobacterial growth becomes multi-fold, benefiting from elevated CO_2, as it requires more CO_2 especially during active growth periods (Verspagen et al., 2014). Furthermore, growing concentrations of atmospheric CO_2 are also suitable for the growth of a few species of cyanobacteria. The amalgamation of warmer water and CO_2 absorption further creates favourable conditions for cyanobacterial growth. The ability of the harmful cyanobacterial bloom-forming genus *Microcystis* to adapt to the increased CO_2 levels has been illustrated in much literature from laboratory and field research. *Microcystis* spp. use CO_2 and HCO_3^- and accumulate inorganic carbon in specialised organ carboxysomes, and strain competitiveness was found to be determined by the concentration of inorganic carbon (Sandrini et al., 2016).

3.4.3 Irregular Precipitation and Rainfall

Another indication of climate change is irregular precipitation patterns, thunderstorms coupled with intense rainfall (Reichwaldt and Ghadouani, 2012). On the other hand, prolonged droughts are also experienced nowadays. These episodic events lead to changed hydrological properties and more nutrient storage and inflow to receiving water through runoff. When such a condition is followed by a drought without rainfall and a reduced flow rate, cyanobacterial abundance is mostly improved. A longer evaporation and stratification period also encourages eutrophication due to the minimum mixing of the water column. These irregular events in rainfall and drought with prolonged summer also promote the release of cyanotoxins, since the abundance of bloom is greater due to climate change (Paerl et al., 2011). Some filamentous cyanobacteria can form akinetes, which are spore-like dormant cells resistant to various unfavourable environmental fluctuations (Garg and Maldener, 2021). For example, activation of akinetes in species like *Aphanizomenon ovalisporum* is more temperature regulated (Cirés et al., 2012). Increases in ambient temperatures in the changing climate play an important role in the expansion of akinete-forming genera, including *Anabaena, Anabaenopsis, Nodularia*. Likewise, another filamentous toxin-producing bloom former is *Lyngbya* spp., a big concern in freshwater and marine ecosystems (Watkinson et al., 2005). As per literature, due to climatic and hydrologic changes like tropical cyclones, *L. wollei* can form colonies aggressively forming dense floating mats over surface water and compete with others for light (Paerl et al., 2012). *Cylindrospermopsis*, *Microcystis* and *Lyngbya* take advantage of climate change and flourish. If we analyse this critically, none of the environmental factors works independently. There exists a complex interaction between variables as well as anthropogenic interventions which contribute to the varied adaptation and expansion strategies of cyanobacteria.

3.5 CONCLUSION

Cyanobacteria are the most globally distributed ancient prokaryotes that can produce oxygen and form blooms. Due to their longest evolutionary history and journey over the years, they are equipped with various adaptations to face the ever-changing climate. Infestation of waterbodies brings many challenges to aquatic lives, water quality and human health. The freshwater ecosystem has been experiencing unprecedented climate change, which is coupled with eutrophication and subsequent bloom episodes. Climate change includes global warming, irregular precipitation and rainfall, cyclones and droughts, which strongly impact aquaculture, more specifically encouraging bloom expansion in the freshwater continuum. All these changes may act synergistically and amplify the consequences (Paerl et al., 2012). Hence, the interaction of these must be studied in depth in order to formulate an effective strategy for bloom control. Recent research and literature highlights bloom as a result of the combined effect of nutrient enrichment and climate change. In addition, continuous surveillance programmes should be conducted for analysing bloom composition and toxicity in freshwater ecosystems (Moreira et al., 2022). Growth of cyanobacteria is regulated by the N and P bank in the sediment, hydrological changes, the existence of grazers, and the magnitude of ingress of fertilizers, pesticides, xenobiotics and synthetics into the particular water body. All these factors are further influenced by climate change. Hence, while developing sustainable mitigation approaches, all these factors are to be critically analysed. Looking at the global concerns associated with cyano-bloom, various control measures are being developed in laboratories but are still limited to the bench. A few appear to be immediately effective but are also associated with environmental pollution (e.g., chemical algaecides). A few are environment friendly but cost intensive. Moreover, climate change is going to continue to threaten the ecosystem.

Many researchers have emphasised nutrient management and quenching, which needs critical attention due to the greater flexibility of bloom-forming cyanobacteria towards nutrient availability and further assimilation. However, dual nutrient reduction strategies are needed for substantial control of bloom (Paerl et al., 2016). The next step would probably be to develop tools that would break the synergy between various factors. Further, control of blooms in a live production pond in the presence of a culture of multiple species remains a big challenge and needs to be researched. In a nutshell, there needs to be brainstorming for a comprehensive strategy to reduce greenhouse gases. Of course, this is a hard task, but only this can provide a healthy environment for aquatic life, minimising the expansion of cyanobacterial bloom.

REFERENCES

Baldia, S. F., A. D. Evangelista, E. V. Aralar, and A. E. Santiago. "Nitrogen and phosphorus utilization in the cyanobacterium Microcystis aeruginosa isolated from Laguna de Bay, Philippines." *Journal of Applied Phycology* 19 (2007): 607–613.

Blomqvist, Peter, Annette Pettersson, and Per Hyenstrand. "Ammonium-nitrogen: A key regulatory factor causing dominance of non-nitrogen-fixing cyanobacteria in aquatic systems." *Archiv für Hydrobiologie* 132 (1994): 141–164.

Bridgeman, Thomas B., Justin D. Chaffin, and Jesse E. Filbrun. "A novel method for tracking western Lake Erie Microcystis blooms, 2002–2011." *Journal of Great Lakes Research* 39, no. 1 (2013): 83–89.

Canfield, D. E., E. Kristensen, and B. Thamdrup. "The sulfur cycle." In Donald E. Canfield, Erik Kristensen, Bo Thamdrup (Eds), *Advances in Marine Biology*, Academic Press, Volume 48, 2005, pp. 313–381, ISSN 0065-2881, ISBN 9780120261475, https://doi.org/10.1016/S0065-2881(05)48009-8.

Carmichael, Wayne W. "Health effects of toxin-producing cyanobacteria: 'The CyanoHABs'." *Human and Ecological Risk Assessment: An International Journal* 7, no. 5 (2001): 1393–1407.

Carr, N. G., and B. A. Whitten. *The biology of cyanobacteria*. Black well Scientific Publications, 1982.

Chen, Chao, Yiyao Wang, Kaining Chen, Xiaoli Shi, and Gang Yang. "Using hydrogen peroxide to control cyanobacterial blooms: A mesocosm study focused on the effects of algal density in Lake Chaohu, China." *Environmental Pollution* 272 (2021): 115923.

Cirés, Samuel, Lars Wörmer, Claudia Wiedner, and Antonio Quesada. "Temperature-dependent dispersal strategies of Aphanizomenon ovalisporum (Nostocales, Cyanobacteria): Implications for the annual life cycle." *Microbial Ecology* 65 (2013): 12–21.

Coles, James F., and R. Christian Jones. "Effect of temperature on photosynthesis-light response and growth of four phytoplankton species isolated from a tidal freshwater river." *Journal of Phycology* 36, no. 1 (2000): 7–16.

Davis, Timothy W., Dianna L. Berry, Gregory L. Boyer, and Christopher J. Gobler. "The effects of temperature and nutrients on the growth and dynamics of toxic and non-toxic strains of Microcystis during cyanobacteria blooms." *Harmful Algae* 8, no. 5 (2009): 715–725.

DeMott, William R., and Felix Moxter. "Foraging cyanobacteria by copepods: Responses to chemical defense and resource abundance." *Ecology* 72, no. 5 (1991): 1820–1834.

Elliott, J. A., I. D. Jones, and S. J. Thackeray. "Testing the sensitivity of phytoplankton communities to changes in water temperature and nutrient load, in a temperate lake." *Hydrobiologia* 559 (2006): 401–411.

Falkowski, Paul G., and John A. Raven. *Aquatic photosynthesis*. Princeton University Press, 2013.

Fernald, Sarah H., Nina F. Caraco, and Jonathan J. Cole. "Changes in cyanobacterial dominance following the invasion of the zebra mussel Dreissena polymorpha: Long-term results from the Hudson River Estuary." *Estuaries and Coasts* 30 (2007): 163–170.

Flöder, Sabine, Jotaro Urabe, and Zen-ichiro Kawabata. "The influence of fluctuating light intensities on species composition and diversity of natural phytoplankton communities." *Oecologia* 133 (2002): 395–401.

Food and Agriculture Organization (FAO). (2017). The State of World Fisheries and Aquaculture 2016 - Contributing to food security and nutrition for all. FAO.

Gallon, J. R. "Tansley review No. 44. Reconciling the incompatible: N2 fixation and O2." *New Phytologist* (1992): 571–609.

Galloway, James N., Frank J. Dentener, Douglas G. Capone, Elisabeth W. Boyer, Robert W. Howarth, Sybil P. Seitzinger, Gregory P. Asner et al. "Nitrogen cycles: past, present, and future." *Biogeochemistry* 70 (2004): 153–226.

Garg, Ritu, and Iris Maldener. "The formation of spore-like akinetes: A survival strategy of filamentous cyanobacteria." *Microbial Physiology* 31, no. 3 (2021): 296–305.

Gobler, Christopher J., JoAnn M. Burkholder, Timothy W. Davis, Matthew J. Harke, Tom Johengen, Craig A. Stow, and Dedmer B. Van de Waal. "The dual role of nitrogen supply in controlling the growth and toxicity of cyanobacterial blooms." *Harmful Algae* 54 (2016): 87–97.

Gobler, Christopher J., Owen M. Doherty, Theresa K. Hattenrath-Lehmann, Andrew W. Griffith, Yoonja Kang, and R. Wayne Litaker. "Ocean warming since 1982 has expanded the niche of toxic algal blooms in the North Atlantic and North Pacific oceans." *Proceedings of the National Academy of Sciences* 114, no. 19 (2017): 4975–4980.

Haney, James F., Don J. Forsyth, and Mark R. James. "Inhibition of zooplankton filtering rates by dissolved inhibitors produced by naturally occurring cyanobacteria." *Archiv für Hydrobiologie* (1994): 1–13.

Harke, Matthew J., and Christopher J. Gobler. "Global transcriptional responses of the toxic cyanobacterium, Microcystis aeruginosa, to nitrogen stress, phosphorus stress, and growth on organic matter." *PLoS One* 8, no. 7 (2013): e69834.

Healy, F. P. "Phosphate." In *The biology of cyanobacteria*, edited by N. G. Carr, Brian A. Whitton pp. 105–124, 1982, University of California Press, Berkley and Los Angeles

Herry-Allani, S. E., and N. Bouaïcha. "Cyanobacterial blooms in dams: Environmental factors, toxins, public health, and remedial measures." In *Dams: Structure, performance and safety management*, Ed: Slaheddine Khilifi, pp. 221–264, 2013 Nova Science Publishers, Inc. ISBN: 978-1-62417-702-6.

Hessen, Dag O., and Ellen Van Donk. "Morpholigical changes in Scenedesmus induced by substances released from Daphnia." *Archiv für Hydrobiologie* 127 (1993): 129–140.

Hudnell, H. Kenneth, Christopher Jones, Bo Labisi, Vic Lucero, Dennis R. Hill, and Joseph Eilers. "Freshwater harmful algal bloom (FHAB) suppression with solar powered circulation (SPC)." *Harmful Algae* 9, no. 2 (2010): 208–217.

Huisman, Jef, Jonathan Sharples, Jasper M. Stroom, Petra M. Visser, W. Edwin A. Kardinaal, Jolanda MH Verspagen, and Ben Sommeijer. "Changes in turbulent mixing shift competition for light between phytoplankton species." *Ecology* 85, no. 11 (2004): 2960–2970.

Jakobsen, Hans Henrik. "Escape of protists in predator-generated feeding currents." *Aquatic Microbial Ecology* 26, no. 3 (2002): 271–281.

Jang, Min-Ho, Kyong Ha, Gea-Jae Joo, and Noriko Takamura. "Toxin production of cyanobacteria is increased by exposure to zooplankton." *Freshwater Biology* 48, no. 9 (2003): 1540–1550.

Joehnk, K. D., Huisman, J. E. F., Sharples, J., Sommeijer, B. E. N., Visser, P. M., & Stroom, J. M. (2008). Summer heatwaves promote blooms of harmful cyanobacteria. *Global change biology, 14*(3), 495-512.

Kanoshina, Inga, Urmas Lips, and Juha-Markku Leppänen. "The influence of weather conditions (temperature and wind) on cyanobacterial bloom development in the Gulf of Finland (Baltic Sea)." *Harmful Algae* 2, no. 1 (2003): 29–41.

Ke, Zhi-Xin, Ping Xie, and Long-Gen Guo. "Impacts of two biomanipulation fishes stocked in a large pen on the plankton abundance and water quality during a period of phytoplankton seasonal succession." *Ecological Engineering* 35, no. 11 (2009): 1610–1618.

Kotak, Brian G., Angeline KY Lam, Ellie E. Prepas, and Steve E. Hrudey. "Role of chemical and physical variables in regulating microcystin-LR concentration in phytoplankton of eutrophic lakes." *Canadian Journal of Fisheries and Aquatic Sciences* 57, no. 8 (2000): 1584–1593.

Lean, D. R. S. "Movements of phosphorus between its biologically important forms in lake water." *Journal of the Fisheries Board of Canada* 30, no. 10 (1973): 1525–1536.

Liu, Lizhen, Qi Huang, and Boqiang Qin. "Characteristics and roles of Microcystis extracellular polymeric substances (EPS) in cyanobacterial blooms: A short review." *Journal of Freshwater Ecology* 33, no. 1 (2018): 183–193.

Maeda, Kaisei, Yukiko Okuda, Gen Enomoto, Satoru Watanabe, and Masahiko Ikeuchi. "Biosynthesis of a sulfated exopolysaccharide, synechan, and bloom formation in the model cyanobacterium Synechocystis sp. strain PCC 6803." *Elife* 10 (2021): e66538.

Martin-Creuzburg, Dominik, and Eric von Elert. "Ecological significance of sterols in aquatic food webs." In Kainz, M., Brett, M., Arts, M. (Eds), *Lipids in aquatic ecosystems*, pp. 43–64. New York, NY: Springer, 2009. https://doi.org/10.1007/978-0-387-89366-2_3

Masson-Delmotte, Valérie. "Global warming of 1.5° c: An IPCC Special Report on impacts of global warming of 1.5° c above pre-industrial levels and related global greenhouse gas emission pathways, in the context of strengthening the global response to the threat of climate change, sustainable development, and efforts to eradicate poverty." 2018.

Mat Nawi, Normi Izati, Nur Syakinah Abd Halim, Leong Chew Lee, Mohd Dzul Hakim Wirzal, Muhammad Roil Bilad, Nik Abdul Hadi Nordin, and Zulfan Adi Putra. "Improved nylon 6, 6 nanofiber membrane in a tilted panel filtration system for fouling control in microalgae harvesting." *Polymers* 12, no. 2 (2020): 252.

Matthijs, Hans CP, Petra M. Visser, Bart Reeze, Jeroen Meeuse, Pieter C. Slot, Geert Wijn, Renée Talens, and Jef Huisman. "Selective suppression of harmful cyanobacteria in an entire lake with hydrogen peroxide." *Water Research* 46, no. 5 (2012): 1460–1472.

Moreira, Cristiana, Vitor Vasconcelos, and Agostinho Antunes. "Cyanobacterial blooms: Current knowledge and new perspectives." *Earth* 3, no. 1 (2022): 127–135.

Nwankwegu, Amechi S., Yiping Li, Yanan Huang, Jin Wei, Eyram Norgbey, Linda Sarpong, Qiuying Lai, and Kai Wang. "Harmful algal blooms under changing climate and constantly increasing anthropogenic actions: The review of management implications." *3 Biotech* 9 (2019): 1–19.

Oliver, Roderick L., and George G. Ganf. "Freshwater blooms." In *The ecology of cyanobacteria: Their diversity in time and space*, pp. 149–194. Dordrecht: Springer Netherlands, 2000.

Paerl, H. W. "Combating the global proliferation of harmful cyanobacterial blooms by integrating conceptual and technological advances in an accessible water management toolbox." *Environmental Microbiology Reports* 5 (2013): 12–14.

Paerl, H. W., and R. S. Fulton III. "Ecology of harmful cyanobacteria." In *Ecology of harmful algae*, pp. 95–109. Berlin: Springer, 2006.

Paerl, Hans W. "Physiological ecology and regulation of N2 fixation in natural waters." In *Advances in microbial ecology*, pp. 305–344. Boston, MA: Springer US, 1990.

Paerl, Hans W., and Jef Huisman. "Blooms like it hot." *Science* 320, no. 5872 (2008): 57–58.

Paerl, Hans W., and Valerie J. Paul. "Climate change: Links to global expansion of harmful cyanobacteria." *Water Research* 46, no. 5 (2012): 1349–1363.

Paerl, Hans W., Hai Xu, Nathan S. Hall, Guangwei Zhu, Boqiang Qin, Yali Wu, Karen L. Rossignol, Linghan Dong, Mark J. McCarthy, and Alan R. Joyner. "Controlling cyanobacterial blooms in hypertrophic Lake Taihu, China: Will nitrogen reductions cause replacement of non-N2 fixing by N2 fixing taxa?." *PloS One* 9, no. 11 (2014): e113123.

Paerl, Hans W., Jane Tucker, and Patricia T. Bland. "Carotenoid enhancement and its role in maintaining blue-green algal (Microcystis aeruginosa) surface blooms." *Limnology and Oceanography* 28, no. 5 (1983): 847–857.

Paerl, Hans W., Nathan S. Hall, and Elizabeth S. Calandrino. "Controlling harmful cyanobacterial blooms in a world experiencing anthropogenic and climatic-induced change." *Science of the Total Environment* 409, no. 10 (2011): 1739–1745.

Paerl, Hans W., Wayne S. Gardner, Karl E. Havens, Alan R. Joyner, Mark J. McCarthy, Silvia E. Newell, Boqiang Qin, and J. Thad Scott. "Mitigating cyanobacterial harmful algal blooms in aquatic ecosystems impacted by climate change and anthropogenic nutrients." *Harmful Algae* 54 (2016): 213–222.

Passow, Uta. "Transparent exopolymer particles (TEP) in aquatic environments." *Progress in Oceanography* 55, no. 3–4 (2002): 287–333.

Peterson, Spencer A. "Lake restoration by sediment removal." *JAWRA Journal of the American Water Resources Association* 18, no. 3 (1982): 423–436.

Porter, Karen Glaus. "Enhancement of algal growth and productivity by grazing zooplankton." *Science* 192, no. 4246 (1976): 1332–1334.

Rajasekhar, Pradeep, Linhua Fan, Thang Nguyen, and Felicity A. Roddick. "A review of the use of sonication to control cyanobacterial blooms." *Water Research* 46, no. 14 (2012): 4319–4329.

Reichwaldt, Elke S., and Anas Ghadouani. "Effects of rainfall patterns on toxic cyanobacterial blooms in a changing climate: between simplistic scenarios and complex dynamics." *Water Research* 46, no. 5 (2012): 1372–1393.

Reynolds, Colin S. *The ecology of phytoplankton*. Cambridge University Press, 2006.

Robarts, Richard D., and Tamar Zohary. "Temperature effects on photosynthetic capacity, respiration, and growth rates of bloom-forming cyanobacteria." *New Zealand journal of Marine and Freshwater Research* 21, no. 3 (1987): 391–399.

Robb, Malcolm, Bruce Greenop, Zoe Goss, Grant Douglas, and John Adeney. "Application of Phoslock TM, an innovative phosphorus binding clay, to two Western Australian waterways: Preliminary findings." In *The interactions between sediments and water: Proceedings of the 9th international symposium on the interactions between sediments and water, held 5–10 May 2002 in Banff, Alberta, Canada*, pp. 237–243. Springer Netherlands, 2003.

Sandrini, Giovanni, Serena Cunsolo, J. Merijn Schuurmans, Hans CP Matthijs, and Jef Huisman. "Changes in gene expression, cell physiology and toxicity of the harmful cyanobacterium Microcystis aeruginosa at elevated CO2." *Frontiers in Microbiology* 6 (2015): 401.

Sandrini, Giovanni, Xing Ji, Jolanda MH Verspagen, Robert P. Tann, Pieter C. Slot, Veerle M. Luimstra, J. Merijn Schuurmans, Hans CP Matthijs, and Jef Huisman. "Rapid adaptation of harmful cyanobacteria to rising CO2." *Proceedings of the National Academy of Sciences* 113, no. 33 (2016): 9315–9320.

Saxton, Matthew A., Robert J. Arnold, Richard A. Bourbonniere, Robert Michael L. McKay, and Steven W. Wilhelm. "Plasticity of total and intracellular phosphorus quotas in Microcystis aeruginosa cultures and Lake Erie algal assemblages." *Frontiers in Microbiology* 3 (2012): 3.

Schrader, Kevin K., Craig S. Tucker, Terrill R. Hanson, Patrick D. Gerard, Susan K. Kingsbury, and Agnes M. Rimando. "Management of musty off-flavor in channel catfish from commercial ponds with weekly applications of copper sulfate." *North American Journal of Aquaculture* 67, no. 2 (2005): 138–147.

Scott, J. Thad, Mark J. McCarthy, Timothy G. Otten, Morgan M. Steffen, Bryant C. Baker, Erin M. Grantz, Steven W. Wilhelm, and Hans W. Paerl. "Comment: An alternative interpretation of the relationship between TN: TP and microcystins in Canadian lakes." *Canadian Journal of Fisheries and Aquatic Sciences* 70, no. 8 (2013): 1265–1268.

Šejnohová, Lenka, and Blahoslav Maršálek. "Microcystis." In Whitton, B. (eds), *Ecology of cyanobacteria II.* Springer, Dordrecht. https://doi.org/10.1007/978-94-007-3855-3_7

Selander, Erik, Tony Fagerberg, Sylke Wohlrab, and Henrik Pavia. "Fight and flight in dinoflagellates? Kinetics of simultaneous grazer-induced responses in Alexandrium tamarense." *Limnology and Oceanography* 57, no. 1 (2012): 58–64.

Sharpley, Andrew N., T. Daniel, T. Sims, J. Lemunyon et al. *Agricultural phosphorus and eutrophication*, 2nd edition. US Department of Agriculture, Agricultural Research Service, 2003, p. 44.

Smayda, Theodore J. "Complexity in the eutrophication–harmful algal bloom relationship, with comment on the importance of grazing." *Harmful Algae* 8, no. 1 (2008): 140–151.

Smith, Val H., and David W. Schindler. "Eutrophication science: Where do we go from here?." *Trends in Ecology & Evolution* 24, no. 4 (2009): 201–207.

Steffensen, Dennis A. "Economic cost of cyanobacterial blooms." In *Cyanobacterial harmful algal blooms: State of the science and research needs*, pp. 855–865. New York, NY: Springer, 2008.

Strom, Suzanne L., Elizabeth L. Harvey, Kerri A. Fredrickson, and Susanne Menden-Deuer. "Broad salinity tolerance as a refuge from predation in the harmful raphidophyte alga Heterosigma akashiwo (Raphidophyceae)." *Journal of Phycology* 49, no. 1 (2013): 20–31.

Tsai, Kuo-Pei, Habibullah Uzun, Huan Chen, Tanju Karanfil, and Alex T. Chow. "Control wildfire-induced Microcystis aeruginosa blooms by copper sulfate: Trade-offs between reducing algal organic matter and promoting disinfection byproduct formation." *Water Research* 158 (2019): 227–236.

Tucker, Craig S. "Off-flavor problems in aquaculture." *Reviews in Fisheries Science* 8, no. 1 (2000): 45–88.

Vanderploeg, Henry A., Gustav-Adolf Paffenhöfer, and James R. Liebig. "Diaptomus vs. net phytoplankton: Effects of algal size and morphology on selectivity of a behaviorally flexible, omnivorous copepod." *Bulletin of Marine Science* 43, no. 3 (1988): 377–394.

Verspagen, Jolanda MH, Dedmer B. Van de Waal, Jan F. Finke, Petra M. Visser, Ellen Van Donk, and Jef Huisman. "Rising CO2 levels will intensify phytoplankton blooms in eutrophic and hypertrophic lakes." *PloS One* 9, no. 8 (2014): e104325.

Visser, Petra Miranda. "Growth and vertical movement of the cyanobacterium Microcystis in stable and artificially mixed water columns." 1995 *Thesis, University of Amsterdam*, 147 pp

Wagner, Carola, and Rita Adrian. "Cyanobacteria dominance: Quantifying the effects of climate change." *Limnology and Oceanography* 54, no. 6 (2009): 2460–2468.

Walsby, A. E. "Gas vesicles." *Annual Review of Plant Physiology* 26, no. 1 (1975): 427–439.

Walsby, Anthony E., Paul K. Hayes, Rolf Boje, and Lucas J. Stal. "The selective advantage of buoyancy provided by gas vesicles for planktonic cyanobacteria in the Baltic Sea." *The New Phytologist* 136, no. 3 (1997): 407–417.

Watkinson, A. J., J. M. o'Neil, and W. C. Dennison. "Ecophysiology of the marine cyanobacterium, Lyngbya majuscula (Oscillatoriaceae) in Moreton Bay, Australia." *Harmful Algae* 4, no. 4 (2005): 697–715.

Weyhenmeyer, Gesa A. "Warmer winters: Are planktonic algal populations in Sweden's largest lakes affected?." *AMBIO: A Journal of the Human Environment* 30, no. 8 (2001): 565–571.

Wilhelm, Steven W., and Charles G. Trick. "Iron-limited growth of cyanobacteria: Multiple siderophore production is a common response." *Limnology and Oceanography* 39, no. 8 (1994): 1979–1984.

Xiao, Man, Ming Li, and Colin S. Reynolds. "Colony formation in the cyanobacterium Microcystis." *Biological Reviews* 93, no. 3 (2018): 1399–1420.

Zhang, Hui, Ge Meng, Feijian Mao, Wenxi Li, Yiliang He, Karina Yew-Hoong Gin, and Choon Nam Ong. "Use of an integrated metabolomics platform for mechanistic investigations of three commonly used algaecides on cyanobacterium, Microcystis aeruginosa." *Journal of Hazardous Materials* 367 (2019): 120–127.

Zhang, Xuezhen, Wei Ji, Huan Zhang, Wei Zhang, and Ping Xie. "Studies on the toxic effects of microcystin-LR on the zebrafish (Danio rerio) under different temperatures." *Journal of Applied Toxicology* 31, no. 6 (2011): 561–567.

Zhang, Xuezhi, Lan Wang, Milton Sommerfeld, and Qiang Hu. "Harvesting microalgal biomass using magnesium coagulation-dissolved air flotation." *Biomass and Bioenergy* 93 (2016): 43–49.

4 Benthic Microbial Community in the Aquatic Environment

Pramod Kumar Pandey, Richa Pathak,
Rameshori Yumnam, and Sumanta Kumar Mallik

4.1 INTRODUCTION

The benthic community, also known as benthos, refers to the ecological zone inhabited by organisms situated on, within, or near the seabed, riverbed, lakebed, or streambed. These communities are typically located in sedimentary environments, whether in marine or freshwater ecosystems, and can be found across a wide range of habitats, from intertidal zones along the shoreline to the continental shelf, and extending to the ocean's profound depths. The term "microbenthos" describes the tiniest living things, which are defined as those that are smaller than 0.1 mm and are found in aquatic environments, such as bacteria, diatoms, ciliates, fungi, and protozoans. According to Azovsky et al. (2013), this group has received less attention because of its smaller size, which makes it harder for the human eye to perceive, and the inherent challenges associated with identifying the microbial population.

Microorganisms residing in sedimentary environments play a pivotal role in material exchange and energy flow across diverse marine ecosystems, encompassing marine hypoxic zones and estuarine systems (Lozupone and Knight, 2007). They significantly contribute to the degradation and biogeochemical cycling of crucial elements such as nitrogen, sulfur, phosphorus, and carbon through metabolic processes involving uptake and decomposition. These sediment microorganisms act as essential agents in the cycling of biogenic elements, profoundly influencing the global biogeochemical cycle (Lozupone and Knight, 2007; Perkins et al., 2014; Zhu et al., 2013). Due to the heightened sensitivity of microorganisms to changes in aquatic ecosystem conditions, the characteristics of microbial diversity and community structure are closely intertwined with these environmental fluctuations. The abundance and diversity of microorganisms serve as pivotal parameters for evaluating the functionality of microbial communities within sedimentary settings (Lozupone and Knight, 2007; Perkins et al., 2014). The benthic zone constitutes the ecosystem where various organisms inhabit the bottom of aquatic bodies, including oceans, rivers, lakes, and streams. These communities are present in sedimentary marine and freshwater environments, ranging from the intertidal zones to the ocean's depths. Benthic communities encompass diverse organisms, including molluscs, sponges, and worms, that feed on other organisms or organic matter. They also include light-dependent species like algae, seagrass, mangroves, and corals that primarily rely on photosynthesis for energy.

There have been notable developments recently in the molecular technologies utilized to identify those groupings (Crespo et al., 2017). According to Lillebø et al. (1999), though they received less attention, they account for a sizable portion of benthos biomass and the majority of ecological processes that take place on sub-aquatic floors (up to 67% in a temperate estuary). Basic processes like the last phase of decomposition and nutrient recycling are carried out by this portion of the benthos. In the realm of benthic microbial ecology, a fundamental question revolves around whether similar environmental conditions give rise to similar microbial communities. Two contrasting hypotheses have emerged. The first hypothesis suggests that microbial assemblages in the

 DOI: 10.1201/9781003408543-4

oxic and oxidized zones surrounding burrows closely resemble those found in surface sediments. In contrast, the second hypothesis posits that burrow walls host unique microbial communities that substantially differ from those at the sediment surface (Kristensen and Kostka, 2005; Papaspyrou et al., 2005, 2006). These hypotheses raise inquiries regarding how microbial communities within burrows compare with those in the surrounding sediment and at the sediment surface.

4.2 CLASSIFICATION OF BENTHIC COMMUNITY

The categorization of benthos is mostly determined by the specific habitat they occupy, the way they accumulate nutrients, and their size. Within benthic ecosystems, we can identify distinct categories of communities. The phytobenthos and zoobenthos represent two primary benthic communities. Phytobenthos comprises various types of algae and aquatic plants, serving as the primary producers, while zoobenthos includes protozoans and benthic animals that function as predators. In addition, the benthic microflora, consisting of numerous protozoa, fungi, and bacteria, forms distinct communities, actively participating in energy and essential nutrient recycling. Macrozoobenthos, known as macroinvertebrates, includes larger benthic creatures that can be observed without a microscope. Microbenthos consists of tiny benthic organisms. Benthic life encompasses organisms inhabiting sediments, residing within or on top of the sediment. They primarily obtain oxygen for their diet from the surrounding sedimentary water. Some benthic species feed on food particles within the sediment, sifting through sand particles. Meiobenthos, or meiofauna, refers to small benthic invertebrates found in both saltwater and freshwater environments. Meiobenthos comprises organisms larger than microfauna but smaller than macrofauna. These organisms are commonly found in soft sediments but can also inhabit underwater algae, higher plants, and other solid substrates. Meiobenthic habitats and taxa are highly diverse. They play crucial roles as primary producers, decomposers, nutrient recyclers, and predators within benthic food chains. Bacteria primarily break down organic matter aerobically at the sediment's surface, where oxygen is abundant. However, due to oxygen consumption at this level, deeper sediment layers become oxygen-depleted, resulting in anaerobic conditions below the surface layer. Epibenthos are a group of benthic organisms that can be found on silt, rocks, tree fragments, or vegetation, or attached to other animals. The suffix added to their name identifies the substrate to which they are attached, such as ectoprocts and ectozoon. These epifaunal organisms, including zebra mussels, thrive in various benthic environments. The categorization of benthic communities based on their size is shown in Table 4.1.

TABLE 4.1
Categorization of Benthic Communities Based on Size

Size	Benthic Community	Organisms Inhibited
Organisms retained by a <0.1 mm mesh sieve	Microbenthos	Bacteria, diatoms, ciliates, amoebae, and flagellates
Organisms retained by a 0.1–1.0 mm mesh sieve	Meiobenthos	Foraminifers (unicellular protists), flatworms (turbellarians), and polychaetes (marine annelids)
>1 mm mesh sieve	Macrobenthos	Amphipods, crayfish, insects, snails, oysters, and aquatic insect larvae like blackflies, caddisflies, chironomids, midges, dragonflies, and stoneflies

4.3 ATTRIBUTES OF MICROBIAL BENTHIC COMMUNITIES AND THEIR CONSTITUENT ELEMENTS

Microbial benthic communities encompass diverse groupings of microorganisms that reside in the sediment or substrates of aquatic environments, including rivers, lakes, oceans, and estuaries. These communities represent a fundamental element of benthic ecosystems, playing pivotal roles in various biogeochemical cycles, the decomposition of organic matter, and the cycling of nutrients within the sedimentary milieu. These microbial benthic communities interact in complex ecological relationships and are interdependent. Their role in sustaining the health and functionality of aquatic ecosystems is essential, as they contribute to the recycling of nutrients, the decomposition of organic matter, and the broader biogeochemical cycling of elements within sediment layers.

These microbial benthic communities generally consist of the following components:

a) **Bacteria:** Within benthic communities, bacteria are prevalent and diverse, and they serve indispensable functions in the degradation of organic matter, nutrient cycling, and the transformation of various chemical compounds within sediment.
b) **Archaea:** Archaea are another group of microorganisms that inhabit benthic sediments. They partake in various metabolic processes, including methanogenesis and ammonia oxidation, contributing to the cycling of crucial elements such as carbon and nitrogen.
c) **Algae and microalgae:** Algae, including microalgae, can be present in benthic communities, especially in well-illuminated shallow aquatic environments. They contribute to primary production through photosynthesis and may serve as a food source for other benthic organisms.
d) **Protozoa:** Protozoa, which are single-celled eukaryotic microorganisms, graze on bacteria and algae in benthic sediments. They constitute an integral component of the microbial food web.
e) **Fungi:** Fungi participate in the decomposition of organic matter within sediments, aiding in the breakdown of complex organic compounds and contributing to nutrient cycling.
f) **Viruses:** Bacteriophages, or viruses that infect bacteria, are also encountered in benthic communities. They impact bacterial populations and nutrient cycling by regulating bacterial abundance.
g) **Small metazoans:** Some small benthic metazoans, such as nematodes and copepods, are associated with benthic sediments. They feed on microorganisms and organic matter, thus enhancing the overall biodiversity of the benthic community.

4.4 FACTORS INFLUENCING THESE COMMUNITIES

Microbial benthic communities in aquatic ecosystems are subject to the influence of a spectrum of abiotic (non-living) and biotic (living) factors. Comprehending the intricate interplay between these abiotic and biotic factors is pivotal for gaining insights into the dynamics and resilience of microbial benthic communities in aquatic ecosystems. Understanding of the broader ecological roles these communities play in processes such as nutrient cycling, organic matter decomposition, and biogeochemical transformations is very important. The majority of research on microbial diversity in marine environments has been conducted in the water column, with a focus on investigating biogeographical patterns exhibited by bacteria. Yet, the underlying processes governing these distribution patterns in the field remain inadequately understood (Suzuki, 2002; Castro-Gonzalez et al., 2005; Fuhrman et al., 2008). Although sediment systems harbour even higher microbial abundances and greater diversity, they have received relatively less attention (Jørgensen and Boetius, 2007). The relationship between habitat complexity and microbial diversity remains a subject of exploration (Fierer, 2008). Some of the primary abiotic and biotic elements that exert an impact on these communities are described in the following subsections.

4.4.1 Abiotic Factors

a) **Sediment characteristics:** The nature and composition of sediment, encompassing attributes like grain size, organic matter content, and mineral constitution, wield a substantial influence on microbial communities. These sediment properties impinge upon variables such as oxygen levels, moisture, and nutrient availability.
b) **Oxygen levels:** The presence of oxygen in sediments assumes a pivotal abiotic role. It dictates the kinds of microorganisms that can thrive, with some specially adapted to low-oxygen (anaerobic) environments and others reliant on oxygen (aerobic) for their metabolic processes.
c) **Temperature:** Temperature stands as a determinant of microbial metabolic rates. Elevated temperatures generally stimulate heightened microbial activity, while extremes of high or low temperatures can impose limitations.
d) **Salinity:** The salt concentration in water and sediment fluctuates across diverse aquatic environments. Microbial communities in benthic ecosystems must accommodate these shifts in salinity, with some microorganisms being specialized for freshwater habitats while others thrive in saltwater environments.
e) **pH Levels:** The pH of sediment and water contributes to the types of microorganisms that prosper. Some microorganisms are well suited to acidic conditions, while others exhibit a preference for alkaline settings.
f) **Nutrient Availability:** The availability of essential nutrients, such as nitrogen and phosphorus, plays a pivotal role in shaping microbial growth and community composition. Excessive nutrient inputs, often stemming from pollution sources, can lead to imbalances and alter the constitution of microbial communities.

4.4.1.1 Effect of Pollution

The establishment of benthic microbial communities in lakes hinges on their interactions with environmental factors. As microbial community complexity increases, external pollution inputs can significantly impact these communities (Brock, 2007). Various studies have explored the influence of single-source pollution on river microorganisms, but the effects of different pollution sources in lake ecosystems are diverse and dynamic (Kelsey et al., 2008; Liu et al., 2015; Ouyang et al., 2018). Therefore, it is valuable to investigate the combined effects of these diverse external pollution factors on river bacterial community composition. However, understanding community dynamics solely through non-biological factors is often incomplete. It is essential to consider the interactions among microorganisms, including interspecific interactions, which play a crucial role in community formation and ecosystem functioning. These microbial interactions are heavily influenced by abiotic environments and various external factors (Barberán et al., 2012; Faust and Raes, 2012). Furthermore, microbial interactions are pivotal in shaping the relationships between microorganisms and their unique functions within ecosystems (Braga et al., 2016). Co-occurrence network analysis has emerged as a powerful tool for investigating symbiotic patterns among different entities in a system and for understanding the role of bacteria in the ecological environment and the interactive effects between bacterial communities (Saunders et al., 2016; Chaffron et al., 2010; Fuhrman and Steele, 2008; Ruan et al., 2006).

4.4.2 Biotic Factors

a) **Predation:** Benthic microbial communities are not exempt from predation by larger organisms, and this factor significantly impacts their dynamics. Grazers, filter feeders, and various benthic organisms consume microorganisms, leading to fluctuations in microbial abundance and diversity.
b) **Competition:** Benthic microbial communities engage in resource competition, particularly for organic matter and nutrients. Competitive interactions are instrumental in moulding community structure, often favouring species with specific metabolic adaptations.

c) **Symbiotic relationships:** A number of microorganisms form mutualistic or symbiotic relationships with other benthic organisms. For instance, nitrogen-fixing bacteria in plant root nodules provide nitrogen in exchange for carbohydrates.
d) **Species interactions:** Microbial species participate in intricate ecological interactions, which encompass cooperative relationships (e.g., syntrophy) and competitive encounters. These interactions significantly influence community composition and function.
e) **Succession:** Over time, microbial communities undergo changes in species composition and diversity. Factors such as environmental disturbances, seasonal variations, and ecological succession contribute to this evolving landscape.
f) **Biological engineering:** Certain benthic organisms, notably burrowing macrofauna like worms and burrowing shrimp, actively modify sediment conditions, creating microenvironments with distinct abiotic properties. These physical alterations have a notable impact on the composition of microbial communities.

4.5 ENUMERATION OF MICROBIAL BENTHIC COMMUNITIES

The integration of the methods and technologies empowers researchers to develop a holistic comprehension of microbial benthic communities. This includes insights into their diversity, composition, metabolic processes, and ecological functions within aquatic ecosystems.

4.5.1 Methods of Sample Collection

a) **Sediment coring:** Sediment cores are collected from the benthic zone using coring devices. These cores capture vertical profiles of sediments, allowing researchers to study microbial communities at different depths (Jamieson et al., 2013).
b) **Grab samplers:** Grab samplers are used to collect sediment samples from specific locations. They are particularly useful for targeting areas of interest or for obtaining samples from different sediment layers (Danovaro, 2009).
c) **Sediment traps:** Sediment traps are deployed to collect settling particles, including microbial aggregates, from the water column. They are valuable for studying the vertical flux of organic matter (Joseph, 2013).
d) **Porewater samplers:** Porewater samplers collect water from within the sediments, enabling the analysis of microorganisms in the pore spaces (Wefer et al., 2012).

4.5.2 Laboratory-Based Analytical Techniques

a) **DNA sequencing:** High-throughput DNA sequencing techniques, such as metagenomics and amplicon sequencing (e.g., 16S rRNA or 18S rRNA gene sequencing), are used to identify and characterize microbial taxa within the samples (Rinke et al., 2013).
b) **Microscopy:** Microscopy methods, including epifluorescence microscopy and transmission electron microscopy (TEM), are employed to visualize and enumerate microorganisms in sediment samples (Bolhuis et al., 2013).
c) **Cultivation:** Microbial cultures are used to isolate and grow specific microorganisms from the samples. This method allows the study of individual species (Booth, 1971).
d) **Biogeochemical analysis:** Various biogeochemical analyses, such as nutrient measurements, enzymatic assays, and stable isotope analysis, provide insights into microbial metabolic activities and nutrient cycling processes (Landers et al., 1983)
e) **Exoenzyme activity assays:** These assays assess the activities of microbial extracellular enzymes involved in the degradation of organic matter. They offer information on the potential enzymatic capabilities of microbial communities (Arnosti, 2011; Findlay and Robert, 2003).
f) **Fluorescence In Situ Hybridization (FISH):** FISH involves the use of fluorescently labelled probes to target specific microbial groups within the samples. It allows the visualization and quantification of particular microorganisms (Amann, 1991).

4.5.3 Emerging Technologies for Analysis

a) **Next-Generation Sequencing (NGS):** Next-Generation Sequencing (NGS) technologies have revolutionized microbial community analysis by providing high-resolution genetic data. Metagenomic, metatranscriptomic, and metaproteomic approaches offer insights into functional capabilities and gene expression within microbial communities (Caporaso et al., 2012).
b) **Single-cell genomics:** This cutting-edge technology enables the genomic analysis of individual microbial cells within a community. It provides insights into the genetic diversity and functional potential of uncultured microorganisms (Kalisky and Quake, 2011).
c) **Nano-scale Secondary Ion Mass Spectrometry (NanoSIMS):** Nano-scale Secondary Ion Mass Spectrometry (NanoSIMS) allows the precise measurement of isotopic ratios in microbial cells. It is used to study nutrient uptake, metabolic activities, and interactions between microorganisms (Musat et al., 2008).
d) **Environmental DNA (eDNA) analysis:** eDNA analysis involves the extraction and sequencing of DNA from environmental samples. It can reveal the presence of specific microorganisms without the need for cultivation (Taberlet et al., 2018).
e) **High-resolution microscopy:** Advanced microscopy techniques, such as confocal laser scanning microscopy (CLSM) and cryo-electron microscopy, provide detailed visualizations of microbial structures and interactions (Joy and Pawley, 1992).
f) **Bioinformatics tools:** Bioinformatics tools and software are crucial for processing and analysing large datasets generated by NGS and other high-throughput techniques. They help researchers interpret microbial community data (Maltez et al., 2018).

4.6 BENTHIC BIOFILM BACTERIAL COMMUNITIES

The health of urban rivers and internal lakes has been progressively degraded due to human activities, with a major concern being the impact of wastewater treatment plants (WWTPs) (Kamjunke et al., 2019). As the volume of treated sewage continues to grow, the discharge from WWTPs has become a significant source of replenishment for these water bodies. WWTPs release a complex mixture of substances, including nutrients, dissolved organic matter (DOM), and micropollutants such as pharmaceuticals and personal care products (Burdon et al., 2020; Waiser et al., 2011). The influx of synthetic chemicals and nutrients from WWTPs triggers eutrophication and brings about substantial alterations in the biogeochemical processes within river ecosystems (Freixa et al., 2020). Water bodies receiving effluents are recognized as environmentally sensitive areas and have, therefore, become the focus of extensive research (Drury et al., 2013)

Effluent discharge has multifaceted impacts on ecosystems and the functioning of the receiving waters. It primarily affects the trophic level and composition of DOM molecules, leading to subsequent changes in pelagic algae and their communities. This shift in trophic levels has consequences for primary production. The effects of effluent discharge on other organisms, including benthic biofilm, macrophytes, and invertebrates, have also been documented. These impacts are often influenced by hydrological factors, such as alterations in wetting/drying patterns, seasonal variations, and the dilution of effluent discharge (Pereda et al., 2020). For example, a study investigated the effects of effluent and hydrological factors on river functioning, revealing that even highly diluted WWTP effluents can bring about changes in the structure of biofilm communities and the functioning of river ecosystems (Pereda et al., 2021). Benthic biofilms, which consist of living and deceased algae, microbes, and organic matter, form the cornerstone of the benthic food web. They are recognized as pioneering microbial aggregates that respond to effluent discharge (Battin et al., 2016). The structure and function of biofilms are in constant flux in response to variations in factors like dissolved oxygen, organic content, and hydrodynamics. Consequently, the total biomass, microbial composition, algal photosynthesis, and assimilation of organic matter are all adjusted as a result (Sabater et al., 2007). Research has shown that benthic biofilms can ameliorate pollutants, such as heavy metals, pharmaceuticals, and brominated flame retardants, and transfer these contaminants to higher trophic levels (Pu et al., 2019; Hobbs et al., 2019). As a result, biofilms are employed as indicators to evaluate ecological changes in aquatic environments by assessing parameters like respiratory rates and the capacity for soluble reactive phosphorus uptake (Pereda et al., 2021). Biofilms that have adapted to anthropogenic disturbances often exhibit greater resistance to effluent discharge.

Water-soluble organic matter (WSOM) is a highly active component of biofilm organic matter, comprising carbohydrates, amino acids, and organic acids. Effluent discharge has been shown to stimulate the activity of extracellular enzymes in benthic biofilms, such as leucine aminopeptidase and amino glucosidase, thereby enhancing the conversion of complex organic matter and its subsequent uptake and utilization by microbes (Qiu et al., 2016; Kamjunke et al., 2015). Various studies have demonstrated the complex and bidirectional relationship between organic carbon and benthic bacterial communities in rivers. River organic carbon serves as a carbon source and nutrient for heterotrophic bacteria and some algae, and can be metabolized by microorganisms in aquatic environments. Unlike organic content in other matrices, such as sediments, the organic compounds in biofilms can influence microbial community structure and alter food web dynamics and energy transfer efficiency (Battin et al., 2016). Chromophoric dissolved organic matter (CDOM), a light-absorbing component in the bulk DOM pool, undergoes significant changes in quantity and quality and affects diverse biogeochemical processes. Studies have shown the dynamic relationship between CDOM and the microbial community in rivers dominated by effluent discharge, impacting actinomycetes and protein components. Similarly, the connection between CDOM and bacterial community structure in the metabolism of specific bacterial communities has been observed in tropical and eutrophic lakes (Zhang et al., 2020). However, most research has focused on planktonic communities and DOM in streams, with limited attention given to bacterial community structure and organic compounds in benthic biofilms. Analysing the relationship between microbial communities and WSOM components in benthic biofilms can provide valuable insights into freshwater quality and ecosystem health. In recent years, advances in bioinformatics have provided tools to decipher how microbial community assembly and metabolic functions respond to environmental stresses. Some studies have investigated the resilience of microbial communities to effluent discharge, revealing the vulnerability of microbial communities to environmental fluctuations (Burdon et al., 2020). Co-occurrence network analysis has been utilized to elucidate interspecific interactions in microbial communities in various environments, such as suspended particulates, soils, and sediments, revealing non-random co-occurrence patterns and a modular structure in microbial communities (Zhao et al., 2021; Zhou et al., 2020; Dong et al., 2020). Nevertheless, no studies have explored microbial interactions in river biofilms in response to effluent discharge, and the co-occurrence patterns between bacterial communities and biofilm WSOM remain uncharted.

4.7 STABLE ISOTOPE ANALYSIS AND ECOLOGICAL MODELLING: UNRAVELLING THE FUNCTIONAL ROLES OF BENTHIC MICROORGANISMS

Stable isotope analysis and ecological modelling have emerged as powerful tools for unravelling the functional roles of benthic microorganisms in the intricate processes of energy flow and nutrient cycling within aquatic ecosystems. These methodologies have provided valuable insights into the dynamics of microbial communities and their interactions with the surrounding environment. Here, we explore how stable isotope analysis and ecological modelling have contributed to our understanding of the roles of benthic microorganisms in these critical ecological processes.

4.7.1 Stable Isotope Analysis

Stable isotope analysis involves the examination of isotopic ratios of elements such as carbon, nitrogen, and sulfur in biological samples. In the context of benthic microbial communities, it has proven to be a game-changer. Researchers can use this technique to trace the origin and fate of organic matter within aquatic ecosystems, shedding light on the contributions of benthic microorganisms to nutrient cycling and energy flow.

a) **Source tracking:** By analysing stable isotopes, scientists can differentiate between various sources of organic matter, distinguishing between autochthonous (locally produced) and allochthonous (external) carbon inputs. This helps in identifying the primary carbon sources supporting benthic microbial communities, whether they are derived from terrestrial, algal, or other aquatic production.
b) **Troop dynamics:** Stable isotope analysis can also reveal the trophic structure within benthic microbial communities. It helps identify the flow of carbon and nutrients between different microbial groups and their interactions with higher trophic levels in the food web. Isotope signatures can provide clues about which microorganisms are primary producers, consumers, or decomposers.

4.7.2 Ecological Modelling

Ecological modelling is another indispensable tool in studying the functional roles of benthic microorganisms. Models can simulate the complex interactions between these microorganisms, their environment, and other biota, enabling researchers to test hypotheses and make predictions.

a) **Carbon and nutrient flux models:** Ecological models can quantify the rates of carbon and nutrient fluxes associated with benthic microbial communities. They can help estimate the amount of organic matter processed by microbes, the release of nutrients, and the transfer of energy to higher trophic levels. These models provide a comprehensive view of the contributions of benthic microorganisms to ecosystem-level processes.
b) **Response to environmental changes:** Ecological models can also be used to simulate how benthic microbial communities respond to environmental changes such as temperature fluctuations, pollution, or shifts in organic matter inputs. This is crucial for predicting the impact of perturbations on nutrient cycling and energy flow within aquatic ecosystems.

By integrating stable isotope analysis and ecological modelling, scientists have gained a deeper understanding of the critical roles that benthic microorganisms play in aquatic ecosystems. These tools have not only provided insights into their contributions to nutrient cycling and energy flow but have also enhanced our ability to manage and conserve these ecosystems effectively. They offer a holistic perspective on the functions of benthic microorganisms, facilitating the development of more sustainable practices and policies for the protection of our vital aquatic environments.

4.8 MICROBIAL COMMUNITIES IN BENTHIC ECOSYSTEM: RESEARCH INPUTS

Benthic microbial communities are responsible for nutrient cycling, organic matter decomposition, and the formation of complex food webs that sustain higher trophic levels. They also help detoxify and remove pollutants from sediments, making them vital for water quality and environmental stability. As these microbial communities respond to changes in their environment, they serve as valuable indicators of ecosystem health and can provide insights into the impacts of pollution, climate change, and human activities on aquatic systems. Understanding and preserving the intricate relationships within benthic microbial communities is essential for the conservation and sustainable management of our aquatic ecosystems. Various research inputs on microbial communities identified in the benthic ecosystem are summarized in Table 4.2.

TABLE 4.2
Research Inputs on Microbial Communities of Benthic Ecosystem

S. No.	The Microbial Community Identified in the Benthic Ecosystem	Significance of the Study	References
1.	ε-Proteobacteria, *Vibrio* sp., δ-Proteobacteria	Examination of the effects of finfish cage aquaculture on the microbial community of sediments beneath fish cages in tropical marine ecosystems, alterations in microbial populations, and alterations in the composition of sedimentary bacterial communities serve as valuable indicators of organic disturbances in aquatic environments	Castine et al. (2009)
2.	Proteobacteria, Bacteroidetes, Firmicutes, and Flavobacteria	The microbial community distributions and their interactions with nutrient fluxes showed the most significant improvement in the constructed wetland, followed by the area under biofilm and sedimentation treatment	Nicholaus et al. (2021)
3.	Microeukaryotes (*Cryptophyta*), (Diatoms), Archaea	Impact on the sediment processes of photosynthesis and ammonia oxidation, defines the boundaries between clear freshwater and saline conditions and between sediment and water ecosystems within a connected coastal aquatic system, offers a structure for comprehending the varying significance of salinity, the differentiation between planktonic and benthic habitats, and the availability of nutrients in influencing microbial metabolic activities, especially in tidal lagoon systems	Tee et al. (2021)
4.	*Cylindrotheca* and *Cocconeis*	Indicates the immediate impact of Cu-based agents on microalgal community composition	Natali et al. (2023)
5.	*Licmophora* and *Synedra*	Shows the short-term effectiveness of Cu-based agents as antifouling measures	
6.	Helicobacteraceae (*Helicobacter*)	Recognized for their prevalence in highly affected coastal sediments, functioning as reliable indicators of pollution, especially in regions that surpass environmental standards, these microorganisms consist of both microaerophiles and anaerobes that play a role in sulfur-related processes and are linked to fecal contamination, detectable within a variety of gut microbiomes, including those of coastal invertebrates	92Menchaca et al. (2014); 4Aylagas et al. (2017); 55Mitchell et al. (2014); 12Buccheri et al. (2019); 81Unno et al. (2018); 58Nakagawa et al. (2017)

(*Continued*)

TABLE 4.2 (CONTINUED)
Research Inputs on Microbial Communities of Benthic Ecosystem

S. No.	The Microbial Community Identified in the Benthic Ecosystem	Significance of the Study	References
7.	Caldilineaceae, Desulfarculaceae, Desulfobacteraceae, Desulfobulbaceae, Flavobacteriaceae, Helicobacteraceae, Lachnospiraceae, Psychromonadaceae, Ruminococcaceae, Spirochaetaceae, Chromatiaceae, Opitutaceae and Thiotrichaceae, Acidaminobacteraceae, Vibrionaceae, Shewanellaceae, Ectothiorhodospiraceae, Flammeovirgaceae, Marinicellaceae, Nitrosomonadaceae, Nitrospiraceae, Pirellulaceae, Piscirickettsiaceae, Syntrophobacteraceae, Thermodesulfovibrionaceae, Alteromonadaceae, Desulfuromonadaceae, Fusobacteriaceae, Hyphomicrobiaceae, Nitrospinaceae, Planctomycetaceae, and Rhodospirillaceae	Act as markers to evaluate the extent of aquaculture-related influence on coastal ecosystems, even on a broad geographical scale, form a robust basis for the establishment of bioindicators to oversee and regulate aquaculture operations, offer potential focal points for designing specialized Polymerase Chain Reaction primers, facilitating quick and sequence-agnostic analysis of environmental samples in quantitative PCR assessments	Frühe et al. (2021)
8.	Caldilineaceae	Crucial constituents within networks of marine iron-oxidizing bacteria	Duchinski et al. (2019).
9.	Caldilineaceae	Indicating positive correlation of this family with increased organic enrichment and a negative correlation with oxygen concentration	Ape et al. (2019)
10.	*Psychromonas* spp.	Linked to a significant level of disturbance and an anaerobic carbohydrate metabolism	Miyazaki and Nogi (2014); Keeley et al. (2020)
11.	Desulfobulbaceae	Able to undergo sulfur disproportionation, converting sulfur with intermediate oxidation states into accessible sulfide	Finster et al. (1998); Frederiksen and Finster (2004)
12.	Desulfobacteraceae	Demonstrate a wide metabolic capability, including the utilization of various carbon sources and hydrogen. They play essential roles in processes related to the biodegradation of pollutants, organic matter decomposition, and the cycling of sulfur and carbon. Additionally, they have been linked to metal-contaminated coastal sediments	Isaksen and Teske (1996); Schwartz and Friedrich (2006); Zhang et al. (2008); Lanzen et al. (2021)
13.	*Nitrospira*	Chemoautotrophic nitrite-oxidizing bacteria, responsible for converting nitrite to nitrate, they have more recently been discovered to have the capacity for complete nitrification, effectively oxidizing ammonia to nitrate	Daims (2014); Daims et al. (2015)

(Continued)

TABLE 4.2 (CONTINUED)
Research Inputs on Microbial Communities of Benthic Ecosystem

S. No.	The Microbial Community Identified in the Benthic Ecosystem	Significance of the Study	References
14.	*Candidatus Nitrosopumilus*	Ammonia-oxidizing archaea (AOA) as aerobic nitrifiers with an exceptional affinity for ammonia, proficiently oxidize ammonia, yielding nitrate, nitrite, or even nitrous oxide as end products	Stahl and de la Torre (2012); Könneke et al. (2014); Martens-Habbena et al. (2015); Qin et al. (2016)
15.	*Candidatus Scalindua*	Anaerobic ammonia-oxidizing (anammox) bacteria employ autotrophic metabolism to convert ammonia into N_2, utilize various electron acceptors, including nitrite, nitrate, metal oxides, as well as oligopeptides and small organic molecules	Penton et al. (2006); Van de Vossenberg et al. (2013)
16.	Ulvibacter	Most prevalent genera to react to a phytoplankton bloom when algal cells break down	Teeling et al. (2012), (2016); Xing et al. (2015); Helbert (2017)
17.	Polaribacter	More abundant during the later stages of an algal bloom and has the capability to utilize sulfatases to break down highly sulfated algal material	Teeling et al. (2012), (2016); Xing et al. (2015); Helbert (2017)

4.9 BENTHIC ECOLOGY AND MICROBIAL COMMUNITY AS FOOD SOURCES

Benthic ecology focuses on the organisms and processes associated with the benthic zone, which is the ecological region at the bottom of aquatic environments, including oceans, lakes, rivers, and even some terrestrial ecosystems like wetlands. It explores the interactions between various benthic organisms, such as algae, bacteria, invertebrates, and sediment-dwelling creatures, as well as their relationships with their physical and chemical surroundings. Benthic ecosystems are vital components of aquatic environments, playing critical roles in nutrient cycling, sediment stability, and the overall health and functioning of these systems. Knowledge of benthic ecology, such as community structure, species diversity, adaptations to environmental conditions, and the ecological implications of human impacts, help us to understand and conserve aquatic habitats.

Benthic microbial communities also serve as essential food sources within aquatic ecosystems. They form the base of the benthic food web, and many organisms in the benthos, such as small invertebrates and filter-feeding species, rely on these microbial communities as a primary source of nutrition. The microorganisms, which include diatoms, algae, and bacteria, produce organic matter through photosynthesis and other metabolic processes, making them a critical energy source for various benthic and pelagic organisms. In turn, these primary consumers are preyed upon by higher trophic levels, creating a cascade of energy transfer that ultimately supports fish and other aquatic life that are crucial for human and ecosystem well-being. Thus, benthic microbial communities play a fundamental role in sustaining the productivity and diversity of aquatic ecosystems and contribute significantly to global food chains.

4.10 CONCLUSION

The study of benthic microbial communities in aquatic environments is a crucial area of research that provides valuable insights into the ecological dynamics and biogeochemical processes that shape our aquatic ecosystems. These communities, comprised of a diverse array of microorganisms, play a pivotal role in nutrient cycling, organic matter decomposition, and overall ecosystem health. By understanding the composition, diversity, and functional roles of benthic microbial communities, we gain essential knowledge for managing and preserving our aquatic ecosystems. Furthermore, benthic microbial communities are sensitive indicators of environmental changes and pollution, making them invaluable for monitoring water quality and ecosystem health. They can serve as early warning systems for the impacts of human activities on aquatic environments, helping us to develop strategies for their protection and restoration. Ongoing research in this field is likely to yield new discoveries about the intricate relationships between benthic microbes and their environments. This knowledge will be essential for addressing environmental challenges such as climate change, pollution, and habitat degradation in the context of aquatic ecosystems. As we continue to explore and comprehend the intricacies of benthic microbial communities, we will be better equipped to make informed decisions that support the sustainability and vitality of our aquatic environments.

REFERENCES

Amann, Rudolf, Nina Springer, Wolfgang Ludwig, Hans-Dieter Görtz, and Karl-Heinz Schleifer. "Identification in situ and phylogeny of uncultured bacterial endosymbionts." *Nature* 351, no. 6322 (1991): 161–164.

Ape, Francesca, Elena Manini, Grazia Marina Quero, Gian Marco Luna, Gianluca Sara, Paolo Vecchio, Pierlorenzo Brignoli, Sante Ansferri, and Simone Mirto. "Biostimulation of in situ microbial degradation processes in organically-enriched sediments mitigates the impact of aquaculture." *Chemosphere* 226 (2019): 715–725.

Arnosti, Carol. "Microbial extracellular enzymes and the marine carbon cycle." *Annual Review of Marine Science* 3 (2011): 401–425.

Aylagas, Eva, Ángel Borja, Michael Tangherlini, Antonio Dell'Anno, Cinzia Corinaldesi, Craig T. Michell, Xabier Irigoien, Roberto Danovaro, and Naiara Rodríguez-Ezpeleta. "A bacterial community-based index to assess the ecological status of estuarine and coastal environments." *Marine Pollution Bulletin* 114, no. 2 (2017): 679–688.

Azovsky, Andrey, Maria Saburova, Denis Tikhonenkov, Ksenya Khazanova, Anton Esaulov, and Yuri Mazei. "Composition, diversity and distribution of microbenthos across the intertidal zones of Ryazhkov Island (the White Sea)." *European Journal of Protistology* 49, no. 4 (2013): 500–515.

Barberán, Albert, Scott T. Bates, Emilio O. Casamayor, and Noah Fierer. "Using network analysis to explore co-occurrence patterns in soil microbial communities." *The ISME Journal* 6, no. 2 (2012): 343–351.

Battin, Tom J., Katharina Besemer, Mia M. Bengtsson, Anna M. Romani, and Aaron I. Packmann. "The ecology and biogeochemistry of stream biofilms." *Nature Reviews Microbiology* 14, no. 4 (2016): 251–263.

Bolhuis, Henk, Lucas Fillinger, and Lucas J. Stal. "Coastal microbial mat diversity along a natural salinity gradient." *PloS One* 8, no. 5 (2013): e63166.

Booth, Colin. *Methods in Microbiology*. Academic Press, 1971.

Braga, Raíssa Mesquita, Manuella Nóbrega Dourado, and Welington Luiz Araújo. "Microbial interactions: Ecology in a molecular perspective." *Brazilian Journal of Microbiology* 47 (2016): 86–98.

Brock, T. C. M. "Role of microbial communities in aquatic ecosystems: Lentic macrophyte-dominated freshwater ecosystems." In *New Improvements in the Aquatic Ecological Risk Assessment of Fungicidal Pesticides and Biocides*. SETAC Europe, 2007, 17–22.

Buccheri, Maria Antonietta, Eliana Salvo, Manuela Coci, Grazia M. Quero, Luca Zoccarato, Vittorio Privitera, and Giancarlo Rappazzo. "Investigating microbial indicators of anthropogenic marine pollution by 16S and 18S High-Throughput Sequencing (HTS) library analysis." *FEMS Microbiology Letters* 366, no. 14 (2019): fnz179.

Burdon, Francis J., Yaohui Bai, Marta Reyes, et al. "Stream microbial communities and ecosystem functioning show complex responses to multiple stressors in wastewater." *Global Change Biology* 26, no. 11 (2020): 6363–6382.

Caporaso, J. Gregory, Christian L. Lauber, William A. Walters, Donna Berg-Lyons, James Huntley, Noah Fierer, Sarah M. Owens, et al. "Ultra-high-throughput microbial community analysis on the Illumina HiSeq and MiSeq platforms." *The ISME Journal* 6, no. 8 (2012): 1621–1624.

Castine, Sarah A., David G. Bourne, Lindsay A. Trott, and David A. McKinnon. "Sediment microbial community analysis: Establishing impacts of aquaculture on a tropical mangrove ecosystem." *Aquaculture* 297, no. 1–4 (2009): 91–98.

Castro-González, Maribeb, Gesche Braker, Laura Farías, and Osvaldo Ulloa. "Communities of nirS-type denitrifiers in the water column of the oxygen minimum zone in the eastern South Pacific." *Environmental Microbiology* 7, no. 9 (2005): 1298–1306.

Chaffron, Samuel, Hubert Rehrauer, Jakob Pernthaler, and Christian Von Mering. "A global network of coexisting microbes from environmental and whole-genome sequence data." *Genome Research* 20, no. 7 (2010): 947–959.

Crespo, Daniel, Tiago Fernandes Grilo, Joana Baptista, João Pedro Coelho, Ana Isabel Lillebø, Fernanda Cássio, Isabel Fernandes, Cláudia Pascoal, Miguel Ângelo Pardal, and Marina Dolbeth. "New climatic targets against global warming: Will the maximum 2° C temperature rise affect estuarine benthic communities?." *Scientific Reports* 7, no. 1 (2017): 3918.

Daims, Holger, Elena V. Lebedeva, Petra Pjevac, Ping Han, Craig Herbold, Mads Albertsen, Nico Jehmlich, et al. "Complete nitrification by Nitrospira bacteria." *Nature* 528, no. 7583 (2015): 504–509.

Daims, Holger. "The family nitrospiraceae." *The Prokaryotes* (2014): 733–749.

Danovaro, Roberto, ed. *Methods for the Study of Deep-Sea Sediments, Their Functioning and Biodiversity.* CRC Press, 2009.

Dong, Yiyi, Jie Gao, Qingshan Wu, Yilang Ai, Yu Huang, Wenzhang Wei, Shiyu Sun, and Qingbei Weng. "Co-occurrence pattern and function prediction of bacterial community in Karst cave." *BMC Microbiology* 20, no. 1 (2020): 1–13.

Drury, Bradley, Emma Rosi-Marshall, and John J. Kelly. "Wastewater treatment effluent reduces the abundance and diversity of benthic bacterial communities in urban and suburban rivers." *Applied and Environmental Microbiology* 79, no. 6 (2013): 1897–1905.

Duchinski, Katherine, Craig L. Moyer, Kevin Hager, and Heather Fullerton. "Fine-Scale biogeography and the inference of ecological interactions among neutrophilic iron-oxidizing zetaproteobacteria as determined by a rule-based microbial network." *Frontiers in Microbiology* 10 (2019): 2389.

Faust, Karoline, and Jeroen Raes. "Microbial interactions: from networks to models." *Nature Reviews Microbiology* 10, no. 8 (2012): 538–550.

Fierer, Noah. "Microbial biogeography: Patterns in microbial diversity across space and time." In *Accessing Uncultivated Microorganisms: From the Environment to Organisms and Genomes and Back* (2008): Washington DC: ASM Press, 95–115.

Findlay, Stuart, and Robert L. Sinsabaugh, eds. *Aquatic Ecosystems: Interactivity of Dissolved Organic Matter.* Academic Press, 2003.

Finster, Kai, Werner Liesack, and B. O. Thamdrup. "Elemental sulfur and thiosulfate disproportionation by Desulfocapsasulfoexigens sp. nov., a new anaerobic bacterium isolated from marine surface sediment." *Applied and Environmental Microbiology* 64, no. 1 (1998): 119–125.

Frederiksen, Trine-Maria, and Kai Finster. "The transformation of inorganic sulfur compounds and the assimilation of organic and inorganic carbon by the sulphur disproportionating bacterium Desulfocapsa sulfoexigens." *Antonie van Leeuwenhoek* 85 (2004): 141–149.

Freixa, Anna, Núria Perujo, Silke Langenheder, and Anna M. Romaní. "River biofilms adapted to anthropogenic disturbances are more resistant to WWTP inputs." *FEMS Microbiology Ecology* 96, no. 9 (2020): fiaa152.

Frühe, Larissa, Verena Dully, Dominik Forster, Nigel B. Keeley, Olivier Laroche, Xavier Pochon, Shawn Robinson, Thomas A. Wilding, and Thorsten Stoeck. "Global trends of benthic bacterial diversity and community composition along organic enrichment gradients of salmon farms." *Frontiers in Microbiology* 12 (2021): 637811.

Fuhrman, Jed A., and Joshua A. Steele. "Community structure of marine bacterioplankton: Patterns, networks, and relationships to function." *Aquatic Microbial Ecology* 53, no. 1 (2008): 69–81.

Fuhrman, Jed A., Joshua A. Steele, Ian Hewson, et al. "A latitudinal diversity gradient in planktonic marine bacteria." *Proceedings of the National Academy of Sciences* 105, no. 22 (2008): 7774–7778.

Helbert, William. "Marine polysaccharide sulfatases." *Frontiers in Marine Science* 4 (2017): 6.

Hobbs, William O., Scott A. Collyard, Chad Larson, Andrea J. Carey, and Sandra M. O'Neill. "Toxic burdens of freshwater biofilms and use as a source tracking tool in rivers and streams." *Environmental Science & Technology* 53, no. 19 (2019): 11102–11111.

Isaksen, Mai Faurschou, and Andreas Teske. "Desulforhopalusvacuolatus gen. nov., sp. nov., a new moderately psychrophilic sulfate-reducing bacterium with gas vacuoles isolated from a temperate estuary." *Archives of Microbiology* 166 (1996): 160–168.

Jamieson, Alan J., Ben Boorman, and Daniel O. B. Jones. "Deep-sea benthic sampling." In A. Eleftheriou (Ed.), *Methods for the Study of Marine Benthos* (2013): 285–347.

Jørgensen, Bo Barker, and Antje Boetius. "Feast and famine—microbial life in the deep-sea bed." *Nature Reviews Microbiology* 5, no. 10 (2007): 770–781.

Joseph, Antony. *Measuring Ocean Currents: Tools, Technologies, and Data*. Newnes, 2013.

Joy, David C., and James B. Pawley. "High-resolution scanning electron microscopy." *Ultramicroscopy* 47, no. 1–3 (1992): 80–100.

Kalisky, Tomer, and Stephen R. Quake. "Single-cell genomics." *Nature Methods* 8, no. 4 (2011): 311–314.

Kamjunke, Norbert, Norbert Hertkorn, Mourad Harir, et al. "Molecular change of dissolved organic matter and patterns of bacterial activity in a stream along a land-use gradient." *Water Research* 164 (2019): 114919.

Kamjunke, Norbert, Peter Herzsprung, and Thomas R. Neu. "Quality of dissolved organic matter affects planktonic but not biofilm bacterial production in streams." *Science of the Total Environment* 506 (2015): 353–360.

Keeley, Nigel, Thomas Valdemarsen, Tore Strohmeier, Xavier Pochon, Thomas Dahlgren, and Raymond Bannister. "Mixed-habitat assimilation of organic waste in coastal environments–It's all about synergy!." *Science of the Total Environment* 699 (2020): 134281.

Kelsey, R. Heath, Laura F. Webster, David J. Kenny, Jill R. Stewart, and Geoffrey I. Scott. "Spatial and temporal variability of ribotyping results at a small watershed in South Carolina." *Water Research* 42, no. 8–9 (2008): 2220–2228.

Könneke, Martin, Daniel M. Schubert, Philip C. Brown, Michael Hügler, Sonja Standfest, Thomas Schwander, Lennart Schada von Borzyskowski, Tobias J. Erb, David A. Stahl, and Ivan A. Berg. "Ammonia-oxidizing archaea use the most energy-efficient aerobic pathway for CO2 fixation." *Proceedings of the National Academy of Sciences* 111, no. 22 (2014): 8239–8244.

Kristensen, Erik, and J. E. Kostka. "Macrofaunal burrows and irrigation in marine sediment: Microbiological and biogeochemical interactions." *Interactions between Macro-and Microorganisms in Marine Sediments* 60 (2005): 125–157.

Landers, D. H., M. B. David, and M. J. Mitchell. "Analysis of organic and inorganic sulfur constituents in sediments, soils and water." *International Journal of Environmental Analytical Chemistry* 14, no. 4 (1983): 245–256.

Lanzen, Anders, Inaki Mendibil, Angel Borja, and Laura Alonso-Sáez. "A microbial mandala for environmental monitoring: Predicting multiple impacts on estuarine prokaryote communities of the Bay of Biscay." *Molecular Ecology* 30, no. 13 (2021): 2969–2987.

Lillebø, Ana Isabel, Mogens R. Flindt, Miguel Ângelo Pardal, and João Carlos Marques. "The effect of macrofauna, meiofauna and microfauna on the degradation of Spartina maritima detritus from a salt marsh area." *Acta Oecologica* 20, no. 4 (1999): 249–258.

Liu, Hui, Gaboury Benoit, Tao Liu, Yong Liu, and Huaicheng Guo. "An integrated system dynamics model developed for managing lake water quality at the watershed scale." *Journal of Environmental Management* 155 (2015): 11–23.

Lozupone, Catherine A., and Rob Knight. "Global patterns in bacterial diversity." *Proceedings of the National Academy of Sciences* 104, no. 27 (2007): 11436–11440.

Maltez Thomas, Andrew, Felipe Prata Lima, Livia Maria Silva Moura, Aline Maria da Silva, Emmanuel Dias-Neto, and João C. Setubal. "Comparative metagenomics." In Setubal, J., Stoye, J., Stadler, P. (Eds), *Comparative Genomics: Methods and Protocols* (2018): vol 1704. Humana Press, New York, NY.

Martens-Habbena, Willm, Wei Qin, Rachel E. A. Horak, Hidetoshi Urakawa, Andrew J. Schauer, James W. Moffett, E. Virginia Armbrust, Anitra E. Ingalls, Allan H. Devol, and David A. Stahl. "The production of nitric oxide by marine ammonia-oxidizing archaea and inhibition of archaeal ammonia oxidation by a nitric oxide scavenger." *Environmental Microbiology* 17, no. 7 (2015): 2261–2274.

Menchaca, Iratxe, A. Borja, M. J. Belzunce-Segarra, J. Franco, J. M. Garmendia, J. Larreta, and J. G. Rodríguez. "Determination of PCB and PAH marine regional Sediment Quality Guidelines, within the European Water Framework Directive." *Chemical Ecology* 30 (2014): 693–700.

Mitchell, H. M., G. A. Rocha, N. O. Kaakoush, J. L. O'Rourke, and D. M. M. Queiroz. "The family Helicobacteraceae." In *The Prokaryotes*, eds E. Rosenberg, E. F. DeLong, S. Lory, E. Stackebrandt, and F. Thompson (2014): Springer, Berlin, Heidelberg.

Miyazaki, M., and Y. Nogi. "The family Psychromonadaceae." In *The Prokaryotes—Gammaproteobacteria*, eds E. Rosenberg, E. F. DeLong, S. Lory, E. Stackebrandt, and F. Thompson (Berlin: Springer) (2014): 583–590.

Musat, Niculina, Hannah Halm, Bärbel Winterholler, Peter Hoppe, Sandro Peduzzi, Francois Hillion, Francois Horreard, Rudolf Amann, Bo B. Jørgensen, and Marcel M. M. Kuypers. "A single-cell view on the eco-physiology of anaerobic phototrophic bacteria." *Proceedings of the National Academy of Sciences* 105, no. 46 (2008): 17861–17866.

Nakagawa, Satoshi, Hikari Saito, Akihiro Tame, Miho Hirai, Hideyuki Yamaguchi, Takashi Sunata, Masanori Aida, Hisashi Muto, Shigeki Sawayama, and Yoshihiro Takaki. "Microbiota in the coelomic fluid of two common coastal starfish species and characterization of an abundant Helicobacter-related taxon." *Scientific Reports* 7, no. 1 (2017): 8764.

Natali, Vanessa, Francesca Malfatti, and Tamara Cibic. "Ecological effect of differently treated wooden materials on microalgal biofilm formation in the Grado Lagoon (Northern Adriatic Sea)." *Microorganisms* 11, no. 9 (2023): 2196.

Nicholaus, Regan, Betina Lukwambe, Wen Yanga, and Zhongming Zhenga. "The relationship between benthic nutrient fluxes and bacterial community in Aquaculture Tail-water Treatment Systems." *bioRxiv* (2021): 2021–08.

Ouyang, Wei, Wanxin Yang, Mats Tysklind, Yixue Xu, Chunye Lin, Xiang Gao, and Zengchao Hao. "Using river sediments to analyze the driving force difference for non-point source pollution dynamics between two scales of watersheds." *Water Research* 139 (2018): 311–320.

Papaspyrou, Sokratis, Trine Gregersen, Erik Kristensen, Bjarne Christensen, and Raymond P. Cox. "Microbial reaction rates and bacterial communities in sediment surrounding burrows of two nereidid polychaetes (Nereis diversicolor and N. virens)." *Marine Biology* 148 (2006): 541–550.

Papaspyrou, Sokratis, Trine Gregersen, Raymond P. Cox, Maria Thessalou-Legaki, and Erik Kristensen. "Sediment properties and bacterial community in burrows of the ghost shrimp Pestarellatyrrhena (Decapoda: Thalassinidea)." *Aquatic Microbial Ecology* 38, no. 2 (2005): 181–190.

Penton, C. Ryan, Allan H. Devol, and James M. Tiedje. "Molecular evidence for the broad distribution of anaerobic ammonium-oxidizing bacteria in freshwater and marine sediments." *Applied and Environmental Microbiology* 72, no. 10 (2006): 6829–6832.

Pereda, Olatz, Daniel von Schiller, Gonzalo Garcia-Baquero, Jordi-Rene Mor, Vicenc Acuna, Sergi Sabater, and Arturo Elosegi. "Combined effects of urban pollution and hydrological stress on ecosystem functions of Mediterranean streams." *Science of the Total Environment* 753 (2021): 141971.

Pereda, Olatz, Libe Solagaistua, Miren Atristain, Ioar de Guzmán, Aitor Larrañaga, Daniel von Schiller, and Arturo Elosegi. "Impact of wastewater effluent pollution on stream functioning: A whole-ecosystem manipulation experiment." *Environmental Pollution* 258 (2020): 113719.

Perkins, Tracy L., Katie Clements, Jaco H. Baas, Colin F. Jago, Davey L. Jones, Shelagh K. Malham, and James E. McDonald. "Sediment composition influences spatial variation in the abundance of human pathogen indicator bacteria within an estuarine environment." *PloS One* 9, no. 11 (2014): e112951.

Pu, Yang, Wing Yui Ngan, Yuan Yao, and Olivier Habimana. "Could benthic biofilm analyses be used as a reliable proxy for freshwater environmental health?." *Environmental Pollution* 252 (2019): 440–449.

Qin, Wei, Willm Martens-Habbena, Julia N. Kobelt, and David A. Stahl. "Candidatus nitrosopumilus." In *Bergey's Manual of Systematics of Archaea and Bacteria* (2016). Bergey's Manual Trust. Published by John Wiley & Sons, Inc 8818–8827.

Qiu, Linlin, Hongyang Cui, Junqiu Wu, Baijie Wang, Yue Zhao, Jiming Li, Liming Jia, and Zimin Wei. "Snowmelt-driven changes in dissolved organic matter and bacterioplankton communities in the Heilongjiang watershed of China." *Science of the Total Environment* 556 (2016): 242–251.

Rinke, Jenny, Vivien Schäfer, Mathias Schmidt, Janine Ziermann, Alexander Kohlmann, Andreas Hochhaus, and Thomas Ernst. "Genotyping of 25 leukemia-associated genes in a single work flow by next-generation sequencing technology with low amounts of input template DNA." *Clinical Chemistry* 59, no. 8 (2013): 1238–1250.

Ruan, Quansong, Debojyoti Dutta, Michael S. Schwalbach, Joshua A. Steele, Jed A. Fuhrman, and Fengzhu Sun. "Local similarity analysis reveals unique associations among marine bacterioplankton species and environmental factors." *Bioinformatics* 22, no. 20 (2006): 2532–2538.

Sabater, Sergi, Helena Guasch, Marta Ricart, Anna Romaní, Gemma Vidal, Christina Klünder, and Mechthild Schmitt-Jansen. "Monitoring the effect of chemicals on biological communities. The biofilm as an interface." *Analytical and Bioanalytical Chemistry* 387 (2007): 1425–1434.

Saunders, Aaron M., Mads Albertsen, Jes Vollertsen, and Per H. Nielsen. "The activated sludge ecosystem contains a core community of abundant organisms." *The ISME Journal* 10, no. 1 (2016): 11–20.

Schwartz, E, and Bärbel Friedrich. "The H2-metabolizing prokaryotes." *The Prokaryotes* 7 (2006): 496–563.

Stahl, David A., and José R. De La Torre. "Physiology and diversity of ammonia-oxidizing archaea." *Annual Review of Microbiology* 66 (2012): 83–101.

Suzuki, Marcelino T. "Marine prokaryote diversity." In *Biodiversity of Microbial Life* (2002) Ed: James T. Staley, Anna-Louise Reysenbach, New York: Wiley, c2002.: 209–234.

Taberlet, Pierre, Aurélie Bonin, Lucie Zinger, and Eric Coissac. *Environmental DNA: For Biodiversity Research and Monitoring*. Oxford University Press, 2018.

Tee, Hwee Sze, David Waite, Gavin Lear, and Kim Marie Handley. "Microbial river-to-sea continuum: Gradients in benthic and planktonic diversity, osmoregulation and nutrient cycling." *Microbiome* 9, no. 1 (2021): 1–18.

Teeling, Hanno, Bernhard M. Fuchs, Christin M. Bennke, Karen Krüger, Meghan Chafee, Lennart Kappelmann, Greta Reintjes, et al. "Recurring patterns in bacterioplankton dynamics during coastal spring algae blooms." *elife* 5 (2016): e11888.

Teeling, Hanno, Bernhard M. Fuchs, Dörte Becher, Christine Klockow, Antje Gardebrecht, Christin M. Bennke, Mariette Kassabgy, et al. "Substrate-controlled succession of marine bacterioplankton populations induced by a phytoplankton bloom." *Science* 336, no. 6081 (2012): 608–611.

Unno, Tatsuya, Christopher Staley, Clairessa M. Brown, Dukki Han, Michael J. Sadowsky, and Hor-Gil Hur. "Fecal pollution: New trends and challenges in microbial source tracking using next-generation sequencing." *Environmental Microbiology* 20, no. 9 (2018): 3132–3140.

van de Vossenberg, Jack, Dagmar Woebken, Wouter J. Maalcke, Hans J. C. T. Wessels, Bas E. Dutilh, Boran Kartal, Eva M. Janssen-Megens, et al. "The metagenome of the marine anammox bacterium 'Candidatus Scalindua profunda' illustrates the versatility of this globally important nitrogen cycle bacterium." *Environmental Microbiology* 15, no. 5 (2013): 1275–1289.

Waiser, Marley J., Vijay Tumber, and Jennifer Holm. "Effluent-dominated streams. Part 1: Presence and effects of excess nitrogen and phosphorus in Wascana Creek, Saskatchewan, Canada." *Environmental Toxicology and Chemistry* 30, no. 2 (2011): 496–507.

Wefer, Gerold, Wolfgang H. Berger, Gerold Siedler, and David J. Webb. *The South Atlantic: Present and Past Circulation*. Springer Science & Business Media, 2012.

Xing, Peng, Richard L. Hahnke, Frank Unfried, Stephanie Markert, Sixing Huang, Tristan Barbeyron, Jens Harder, et al. "Niches of two polysaccharide-degrading Polaribacter isolates from the North Sea during a spring diatom bloom." *The ISME Journal* 9, no. 6 (2015): 1410–1422.

Zhang, Lei, Wangkai Fang, Xingchen Li, Wenxuan Lu, and Jing Li. "Strong linkages between dissolved organic matter and the aquatic bacterial community in an urban river." *Water Research* 184 (2020): 116089.

Zhang, Wen, Lin-sheng Song, Jang-Seu Ki, Chun-Kwan Lau, Xiang-Dong Li, and Pei-Yuan Qian. "Microbial diversity in polluted harbor sediments II: Sulfate-reducing bacterial community assessment using terminal restriction fragment length polymorphism and clone library of dsrAB gene." *Estuarine, Coastal and Shelf Science* 76, no. 3 (2008): 682–691.

Zhao, Qian, Shuangshuang Chu, Dan He, Daoming Wu, Qifeng Mo, and Shucai Zeng. "Sewage sludge application alters the composition and co-occurrence pattern of the soil bacterial community in southern China forestlands." *Applied Soil Ecology* 157 (2021): 103744.

Zhou, Hong, Ying Gao, Xiaohong Jia, Mengmeng Wang, Junjun Ding, Long Cheng, Fang Bao, and Bo Wu. "Network analysis reveals the strengthening of microbial interaction in biological soil crust development in the Mu Us Sandy Land, northwestern China." *Soil Biology and Biochemistry* 144 (2020): 107782.

Zhu, Jianyu, Jingxia Zhang, Qian Li, Tao Han, Jianping Xie, Yuehua Hu, and Liyuan Chai. "Phylogenetic analysis of bacterial community composition in sediment contaminated with multiple heavy metals from the Xiangjiang River in China." *Marine Pollution Bulletin* 70, no. 1–2 (2013): 134–139.

5 Sediment Microbiology in the Aquatic Environment

Sumanta Kumar Mallik, Richa Pathak, and Neetu Shahi

5.1 INTRODUCTION

The field of sediment microbiology in aquaculture refers to the study of microorganisms present in the sediment of aquaculture systems. Aquaculture involves the farming of aquatic organisms, such as fish, shellfish, and plants, in controlled environments such as ponds, tanks, or cages. The sediment in these aquaculture systems serves as a crucial component, as it can accumulate organic matter, nutrients, and microbial communities that interact with the surrounding water column (Naylor et al., 2000; Basaran et al., 2010; Martinez-Porchaz and Martinez-Cordova, 2012).

Research in sediment microbiology within aquacultures focuses on understanding the composition, diversity, and functional roles of microorganisms residing in the sediment. These microorganisms can include bacteria, archaea, fungi, and other microbial eukaryotes. They play essential roles in nutrient cycling, organic matter decomposition, disease dynamics, and overall ecosystem functioning within aquaculture systems (Stankevica et al., 2012). It also aims to unravel the microbial community structure, their interactions, and their responses to various environmental factors, such as water quality, organic loading, feed inputs, and management practices. This knowledge is crucial for optimizing aquaculture operations, improving water quality, preventing disease outbreaks, and ensuring sustainable and environmentally friendly aquaculture practices.

Researchers employ various techniques, including molecular biology tools, metagenomics, meta-transcriptomics, and microbial culturing, to characterize the microbial composition and functional potential of sediment communities in aquaculture systems. By understanding the sediment microbiology, the researchers can make informed decisions to enhance productivity, mitigate environmental impacts, and maintain the overall health and sustainability of the aquatic environment.

Sediment and its microbial communities offer a combination of biological, chemical, and physical attributes necessary for vital processes such as mineral cycling and the degradation of human-made substances. These qualities facilitate the day-to-day functioning of these essential ecological processes (Mooshammer et al., 2017). Sediment serves as an interface for various processes, including the transition between solid and liquid phases, aerobic and anaerobic conditions, as well as sorptive and desorptive activities (Nealson, 1997). It is inhabited by numerous microorganisms, some of which consider sediment their permanent home, while others seek temporary refuge, and yet for some, it is a hostile environment. The presence or absence of hydrodynamic conditions determines whether silt acts as a source or a sink for microorganisms in the surrounding environment (Moriarty, 1997).

5.2 MICROBIAL COMMUNITIES IN FRESHWATER SEDIMENTS

The microbial communities present in freshwater sediments are of significant interest and importance. These communities comprise a diverse array of microorganisms that play essential roles in the functioning and dynamics of freshwater ecosystems (Wilkes et al., 1999). Bacteria, archaea, fungi, and other microbial eukaryotes can be found in these sediments, forming intricate and interconnected networks of interactions. Freshwater sediments provide a unique and dynamic habitat

DOI: 10.1201/9781003408543-5

for microorganisms (Gilbert et al., 2009). They serve as a substrate for attachment and growth, offering a range of physical and chemical conditions. Sediments act as reservoirs for organic matter, nutrients, and various contaminants, making them hotspots for biogeochemical cycling processes. Microbial communities in freshwater sediments are involved in crucial ecological functions. They contribute to the decomposition and recycling of organic matter, nutrient cycling, and the transformation of various chemical compounds. These processes are vital for maintaining water quality, supporting primary production, and influencing the overall health and balance of freshwater ecosystems (Inagaki et al., 2003). Understanding the structure and dynamics of microbial communities in freshwater sediments is essential for gaining insights into ecosystem processes, predicting responses to environmental changes, and developing strategies for ecosystem management and restoration.

Freshwater ecosystems are vital resources, providing essential services such as drinking water, food production, animal habitats, and recreational opportunities. The sediment microbial communities within these ecosystems play a key role in important activities such as the carbon and nitrogen cycles, contributing to the overall stability and functioning of the ecosystem. However, freshwater sediment is also a primary site for the accumulation and attachment of bacteria and chemical pollutants, resulting in exposure of sediment microorganisms to various anthropogenic pollutants (Stankevica et al., 2015). These pollution events, known as "disturbances," can be categorized as either pulse disturbances, which are short-term and acute, or press disturbances, which are long-term and continuous. Disturbances have a significant impact on the structure and diversity of microbial communities. Two important factors used to assess the impact of disturbance on microbial ecosystems are resistance, which refers to the ability of a community to withstand and remain unchanged by disturbance, and resilience, which is the rate at which a community returns to its pre-disturbance composition.

By measuring changes in microbial community composition following anthropogenic disturbances, we can gain insights into ecosystem functioning and health, helping to predict and manage the impacts on freshwater ecosystems (Stankevica et al., 2015). These assessments are valuable for understanding the responses of microbial communities and their implications for the overall well-being of freshwater ecosystems (Figure 5.1).

Agriculture is the primary contributor to non-point source (NPS) pollution and the degradation of water quality in rivers and streams (Seiler and Berendonk, 2012; USEPA, 2017). Through agricultural runoff from farmland, livestock operations, and crop farming practices, substances such as nitrate, phosphorus, heavy metals, antibiotics, and non-native microorganisms are introduced into waterways. These pollutants have detrimental effects on water quality and ecosystem functioning. The transmission of microbial infections through manure is of significant concern. Livestock manure has been found to contain pathogens such as *Arcobacter*, *Acinetobacter*, methicillin-resistant *Staphylococcus aureus*, and toxic *Escherichia coli* (Sun et al., 2020a). Moreover, excessive amounts of nitrate and phosphate from agricultural fertilizers can impact the composition of sediment microbial communities. Studies have revealed that members of the Comamonadaceae family and the genera *Mucilaginibacter*, *Pseudospirillum*, and *Novosphingobium* strongly correlate with nitrate concentrations. Similarly, members of the Holophagaceae class, Gemmatimonadaceae family, and *Nitrospira* genus exhibit a strong correlation with phosphate concentrations (Wall et al., 2015). These findings highlight the significant role of agricultural practices in influencing microbial communities and the potential consequences for water quality and ecosystem health. Understanding these associations can aid in developing strategies to mitigate the impact of agricultural pollution and preserve the integrity of aquatic environments. Moreover, the occurrence of freshwater eutrophication episodes has been associated with an increase in Cyanobacteria, specifically *Microcystis*, *Anabaena*, *Planktothrix*, and *Aphanizomenon* (Havens, 2008; Graham et al., 2012; Dolman et al., 2012). Various parameters, including the type, number, duration, and intensity of disturbances, influence the responses of microbial communities to such disturbances. Several studies have demonstrated the high sensitivity of microbial communities to anthropogenic disruptions, and the Microbiome Stress Project offers a comprehensive database for comparing significant findings

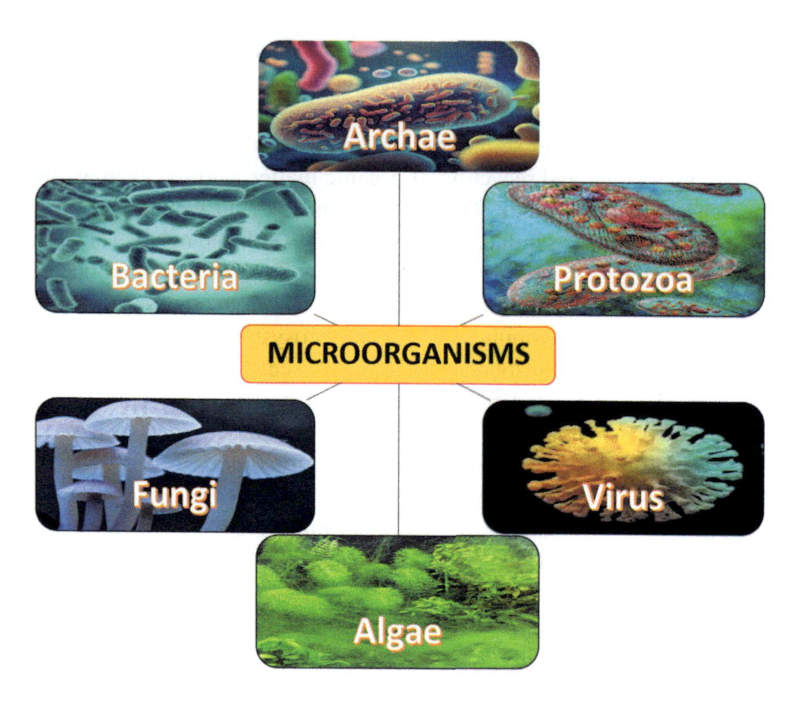

FIGURE 5.1 Microorganisms in a freshwater ecosystem.

TABLE 5.1
Microbial Communities in Freshwater Habitats – Physical and Biological Factors

S. No.	Habitat	Major Microbial Community	Major Physical Characteristic	Major Biological Constraints
1.	Lake	Plankton	Stratification, wind-generated turbulence	Nutrient competition, grazing, parasitism
2.	Flood plain	Plankton, biofilms attached to plants	Periodic desiccation	Competition between algae and macrophytes
3.	Rivers and streams	Benthic microbes	Flow-generated turbulence	Colonization competition, biofilm grazing
4.	Estuary – mud flats	Biofilms	Desiccation, high light, salt exposure	Competition and grazing at mud surface
5.	Estuary – outflow	Plankton	Mixing with saltwater, turbidity	Grazing in water column

(Pundyte et al., 2011). However, the resilience of native microbial communities following environmental perturbations remains poorly understood, and most of the research findings have relied on controlled laboratory experiments (Lourenço et al., 2018; Allison and Martiny, 2008). It is crucial to comprehend the impact of prolonged manure disturbances on the resistance and resilience of microbial community composition in order to assess risks to ecological and human health. Resistant and resilient microbial communities, known for their diversity and functional redundancy, are better equipped to withstand the influx of pathogenic organisms and chemical pollutants (Allison and Martiny, 2008) (Table 5.1).

5.2.1 Microbial Communities in Water and Sediment of a Lentic Ecosystem

The role of the intestinal microbiome in regulating human physiology and health has gained increasing recognition (Clemente et al., 2012; Le Chatelier et al., 2013; Jie et al., 2017). Both host genetics and environmental factors have been found to influence the composition of the human intestinal microbiota, with environmental factors appearing to have a greater impact (Benson et al., 2010; Spor et al., 2011; Turpin et al., 2016). Understanding the interplay between the host genetics and environment in shaping the gut microbial population is crucial. Microorganisms provide numerous environmental services, including maintaining aquatic ecosystem quality, impacting animal health and disease control, and influencing element cycling and water quality (Rothschild et al., 2018; Sun et al., 2020).

In aquaculture systems, microbial communities of various habitats (Rungrassamee et al., 2014; Fan et al., 2016; Hou et al., 2018), such as the surrounding water, animal intestines, and sediment, are closely associated with the occurrence of aquatic animal diseases, posing challenges in comprehending the structure, function, and interactions within these complex ecosystems (Yan et al., 2016; Hou et al., 2017; Li et al., 2017; Zeng et al., 2017). The global demand for animal proteins has led to a significant expansion of aquaculture production, making it the third-largest source of animal proteins worldwide, with China accounting for over 60% of the total output (Yan et al., 2017). However, the aquaculture industry faces the risk of frequent disease outbreaks. Bacterial diseases, such as white feces syndrome and various hepatopancreatic necrosis diseases, have caused substantial losses in shrimp production globally (Sriurairatana et al., 2014; Huang et al., 2020). Dysbiosis in gut microbiota has been associated with disorders in both humans and animals. The shrimp culture pond ecosystem (SCPE) is a unique anthropogenic aquaculture ecosystem that undergoes artificial manipulation and management, comprising multiple habitats, including water, shrimp intestines, and sediment (Huang et al., 2016; Hou et al., 2018; Xiong et al., 2017). Understanding the microbial ecology of the SCPE metacommunity, where aquatic animals reside, is crucial for the development of sustainable aquaculture practices. Though microbial communities in aquaculture ecosystems exhibit high diversity and are influenced by various environmental and geographic factors, such as host developmental stages, environmental conditions, and geographical distance, the mechanisms driving microbial assembly in these ecosystems remain poorly understood (Hou et al., 2017). It is generally believed that ecological processes play a significant role in shaping microbial diversity among species (Figure 5.2).

5.2.2 Influence of Ecological Processes on Microbial Communities in Lotic Ecosystems

Recent studies have greatly expanded our understanding of the role of ecological processes in shaping microbial communities. Deterministic mechanisms have been found to drive the structure of microbial communities in certain aquatic ecosystems, such as lakes, in both water and sediment habitats. In contrast, stochastic mechanisms play a dominant role in shaping the assemblage of microbial communities in the intestines of aquatic animals (Burns et al., 2016).

In lotic ecosystems, the microbial communities in water, animal intestines, sediment, and other associated habitats form a metacommunity, where interactions between these communities play a crucial role in aquatic animal health and productivity (Clemente et al., 2012; Le Chatelier et al., 2013; Jie et al., 2017). Understanding the relationship between microbial communities in animal intestines and their surrounding habitats is of utmost importance in the context of aquaculture. However, the contribution of microbial communities from diverse habitats, including water, animal intestines, and sediment, to the metacommunity, as well as the ecological interplay between environmental populations and intestinal microbiota in lotic settings, remains poorly understood (Rothschild et al., 2018; Sun et al., 2020).

5.2.3 Sediment Microbial Communities and Environmental Pollution

Microbial communities residing in sediment play a crucial role in environmental pollution, and their significance is increasingly recognized. These communities have the potential to influence

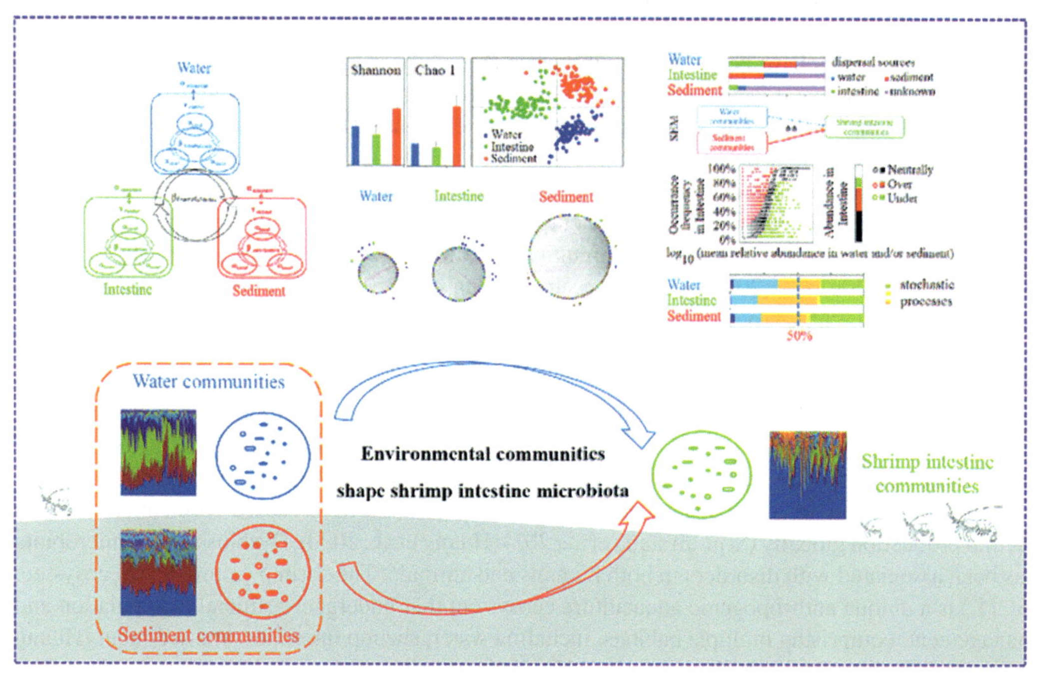

FIGURE 5.2 Schematic representation of microbial assembly procedures and interactions in the SCPE metacommunity's microbial communities of water, shrimp intestine, and sediment habitats. (From Huang, Z. et al., *Frontiers in Microbiology* 12, 772149, 2021.)

various ecological functions and processes related to pollution in the environment. Sediment microbial communities contribute to the degradation and transformation of pollutants, including organic compounds and nutrients. They can break down complex organic matter and participate in cycles of nitrogen, phosphorus, and sulfur, influencing the overall biogeochemical processes in the ecosystem. Additionally, sediment serves as a primary location for the accumulation and attachment of bacterial and chemical pollutants. Microorganisms in the sediment are exposed to a wide range of anthropogenic pollutants, such as heavy metals, antibiotics, and non-native microorganisms, which are introduced through activities like agriculture and urban runoff (Stankevica et al., 2015; Pundyte et al., 2011). Understanding the composition and dynamics of sediment microbial communities is crucial for assessing the impact of environmental pollution. By studying these communities, researchers can gain insights into the mechanisms underlying pollutant degradation, bioaccumulation, and the overall health of aquatic ecosystems. Investigating the structure and function of sediment microbial communities and their interactions with pollutants can aid in the development of strategies for pollution prevention, control, and remediation. Furthermore, studying the responses and resilience of these communities to environmental disturbances can provide valuable information for managing and mitigating pollution impacts in aquatic ecosystems.

5.3 MARINE SEDIMENTS AND THEIR MICROBIAL COMMUNITIES

Sediments play a crucial role in the marine environment, serving as the largest reservoir of organic carbon worldwide (Parkes et al., 2000). These sediments are primarily formed through the transportation and deposition of biological materials from the continents, making them nutrient-rich and providing important habitats for microbial communities (Kallmeyer et al., 2012; Parkes et al., 2014).

Microorganisms residing in marine sediments have vital functions in regulating major geochemical and environmental processes in marine ecosystems. They contribute to nutrient dynamics and participate in essential biogeochemical cycles that drive ecosystem functioning (Fuhrman, 2009; Nogales et al., 2011; Graham et al., 2012). The structure and diversity of microbial communities in marine sediments are influenced by various environmental factors. Factors such as pH, water temperature, silicate concentrations, and ocean currents contribute to the differences, observed in the composition and abundance of microorganisms (Gilbert et al., 2009; Hollister et al., 2010; Kirchman et al., 2010). Hydrography, including ocean currents and tides, also plays a significant role in shaping the distribution and migration patterns of microorganisms in marine sediments (Hamdan et al., 2013). Sediment grain-size distribution provides valuable information about hydrodynamic conditions and reflects the complex forces at play (Anthony and Héquette, 2007; Hu et al., 2012). Understanding the microbial communities and their interactions within marine sediments is crucial for comprehending the functioning and resilience of marine ecosystems. Marine sediments, the accumulations of particles, organic matter, and minerals settled at the bottom of oceans, represent one of Earth's largest ecosystems. Beneath the tranquil waves, an intricate world of microbial communities thrives in these seemingly inhospitable environments and plays a pivotal role in shaping the biogeochemical cycles, nutrient dynamics, and carbon sequestration processes that are fundamental to the health of our oceans and the global environment (Figure 5.3).

5.3.1 Diversity and Adaptation

The microbial communities within marine sediments exhibit remarkable diversity, mirroring the heterogeneity of their habitats. From shallow coastal regions to the abyssal plains, sediments vary in terms of grain size, organic content, and oxygen availability, creating unique niches for different microorganisms. Some are adapted to life in the oxygenated upper layers, while others thrive in the anoxic zones deeper within the sediments. This diversity has been fostered through evolutionary processes, allowing microbes to adapt to extreme conditions such as high hydrostatic pressures, varying temperatures, and nutrient limitations.

5.3.2 Carbon Cycling

One of the primary roles of marine sediment microbial communities is their involvement in carbon cycling. As organic matter from phytoplankton, detritus, and other sources sinks to the ocean floor,

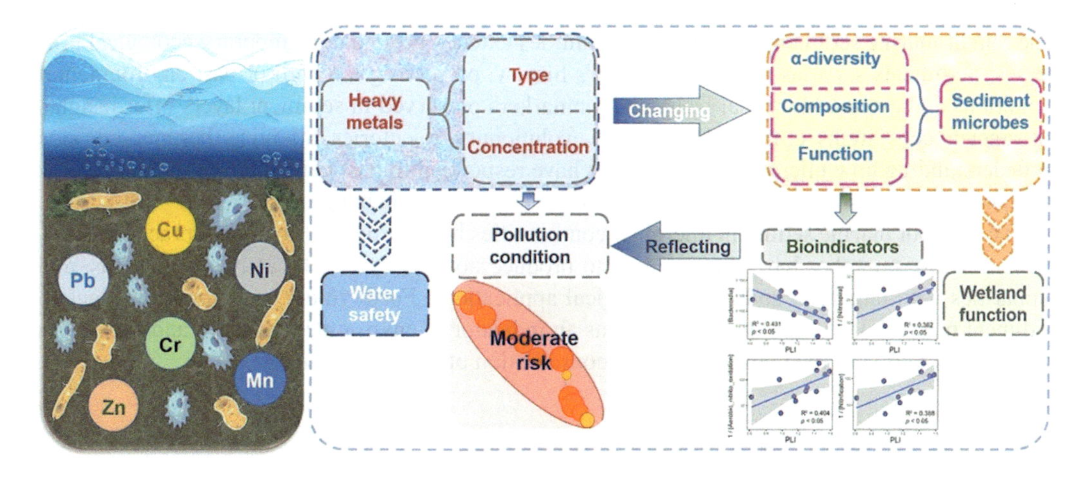

FIGURE 5.3 Heavy metal effects on microbial populations in sediments and the development of bioindicators. (From Li, C. et al., *Science of the Total Environment* 749, 141555, 2020.)

microbial degradation becomes a key process in remineralizing and recycling carbon. In the surface sediments, aerobic bacteria and fungi break down organic matter through respiration, releasing carbon dioxide back into the water column. In the deeper, oxygen-depleted layers, anaerobic bacteria and archaea utilize alternative electron acceptors such as nitrate and sulfate to degrade organic matter, producing methane as a byproduct. This process, known as methanogenesis, is a significant contributor to the global carbon cycle.

5.3.3 Nitrogen, Phosphorus, and Sulfur Cycling

Microbial communities in marine sediments also contribute significantly to nutrient cycling. Nitrogen and phosphorus, essential elements for marine life, are transformed and recycled by these microbes (Deming and Baross, 1993; Gooday, 2002; Vezzulli et al., 2002). Denitrifying bacteria play a crucial role in converting nitrate to nitrogen gas, returning it to the atmosphere. Similarly, sulfate-reducing bacteria participate in sulfur cycling, producing hydrogen sulfide as they reduce sulfate. These processes are essential for maintaining nutrient balance and preventing excessive accumulation of these compounds in the sediment.

5.3.4 Role in Elemental Cycling

Beyond carbon, nitrogen, and sulfur cycling, marine sediment microbes are involved in the cycling of various other elements. Iron and manganese cycling, for instance, is facilitated by iron-reducing and manganese-oxidizing bacteria, impacting the mobility and availability of these metals in the sediments. Additionally, microbial communities influence the cycling of trace elements, such as arsenic, selenium, and mercury, which can have toxic effects on marine life when their concentrations increase.

5.3.5 Biogeochemical Interactions

The microbial interactions within marine sediments are complex and interconnected. Synergistic and competitive relationships between different microorganisms drive the ecosystem dynamics. For instance, sulfate-reducing bacteria and methanogenic archaea often compete for organic substrates, affecting the balance between methane and sulfate production. These interactions shape the overall biogeochemical processes within the sediments and influence the fluxes of gases between the sediments and the overlying water column.

5.3.6 Climate, Earth's History, and Applied Implications

The role of marine sediment microbial communities extends beyond contemporary ecological processes. Sediments serve as archives of Earth's history, preserving a record of past environmental conditions and events. Microbial DNA and biomarkers preserved in sediment layers offer insights into ancient ecosystems, climate changes, and evolutionary trajectories. Studying these records aids in understanding how microbial communities have responded to past perturbations, helping scientists predict potential future responses to ongoing environmental changes.

The study of marine sediment microbial communities holds promise for various practical applications. These microbes have the potential to produce novel enzymes, bioactive compounds, and bioplastics with industrial and biotechnological applications. Furthermore, monitoring changes in sediment microbial communities can serve as an indicator of environmental health and ecosystem disturbance, aiding in the management and conservation of marine resources.

5.4 GEOGRAPHIC VARIATION IN MICROBIAL COMMUNITIES IN SEDIMENT SAMPLES

Studies conducted in various regions, including the Sea of Japan, the South China Sea (SCS) Trough, the Sea of Okhotsk, and the Peruvian Marginal Sea, have revealed differences in the

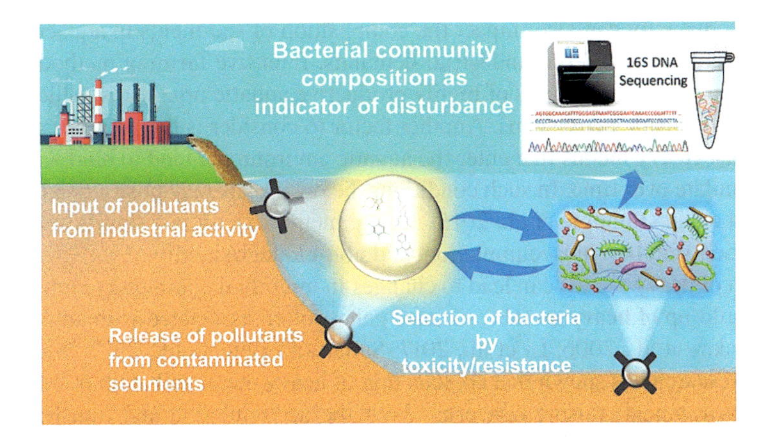

FIGURE 5.4 Bacterial colonies as markers of contamination in the environment. (From Rodríguez, J. et al., *Environmental Pollution* 268, 115690, 2021.)

composition of microbial communities in sediment samples. These variations reflect the diverse environmental conditions across these regions (Rochelle et al., 1994; Inagaki et al., 2003; Newberry et al., 2004; Webster et al., 2006). In sediments heavily impacted by human activities, such as Hangzhou Bay and Bohai Bay, there is an increase in the abundance of certain genera belonging to the phyla *Proteobacteria*, *Firmicutes*, *Actinobacteria*, and *Bacteroides*. These genera have the ability to utilize petroleum as a carbon source, indicating their adaptation to the presence of pollutants. Additionally, classes such as δ-*Proteobacteria* and γ-*Proteobacteria* have been identified as dominant in these polluted coastal areas, making them potential indicators of environmental contamination (Kochling et al., 2011; Wang et al., 2013; Guan et al., 2020; Lu et al., 2016; Su et al., 2018; Xiong et al., 2014). Prevalence of the Firmicutes phylum and Bacilli class has been observed in locations extensively affected by human disturbances and contamination (Zhang et al., 2008; Lu et al., 2016). These microorganisms are associated with altered and disturbed environments, reflecting their potential as indicators of human activity impacts on sediment ecosystems. Spatial distribution and environmental conditions play a significant role in shaping microbial communities in marine sediments. The identification of certain microorganisms as environmental indicators can provide valuable insights into the ecological status and the impact of human activities on marine sediment ecosystems (Gillan et al., 2005; Glasl et al., 2019) (Figure 5.4).

5.5 INFLUENCE OF INTENSIVE AQUACULTURE ON MICROBIAL COMMUNITIES IN POND SEDIMENTS

According to a 2020 report by the Food and Agriculture Organization of the United Nations, aquaculture has become a vital sector in the food industry, with an annual growth rate of 3.1% between 1961 and 2017. This growth surpasses the global population increase (1.6%) by more than twice. Fish ponds, due to extensive feeding and fecal discharge, contain high levels of dissolved organic compounds. Over time, these ponds accumulate sediments through the water regime processes of filling and discharging. These sediments comprise biological remnants from the ponds and their surrounding areas, along with soil particles and other non-biological substances, carried into the pond. The prevailing type of material found in these sediments is organic matter (Stankevica et al., 2012). The intensity and combination of these processes vary significantly, based on geological and geomorphological contexts, hydrological regimes, climatic conditions, and human activities (Wilkes et al., 1999; Stankevica et al., 2015).

Changes in biodiversity can also impact the composition of sediments in aquaculture ponds, as the focus of aquaculture is meeting industrial demands. Intensive farming methods are employed to establish and sustain monocultures of highly productive aquatic organisms, which differ significantly from natural aquatic ecosystems. Another factor to consider is the presence of heavy metals in pond sediments. Heavy metals that enter the aquatic environment tend to bind with bottom sediments and accumulate over time. In such conditions, heavy metals may pose a risk to human health through the food chain (Pundyte et al., 2011). Heavy metal toxicity is a significant environmental concern due to their stability, bioaccumulation, and non-biodegradability.

Accumulation of heavy metals can lead to changes in microbial community composition and the activation and build-up of heavy metal resistance genes, often associated with antibiotic resistance genes (Stepanauskas et al., 2006; Li et al., 2017; Seiler and Berendonk, 2012; Wales and Davies, 2015). Previous research has shown that co-selection of heavy metal and antibiotic resistance genes occurs in the environment, raising concerns about the accumulation and transfer of potentially harmful antibiotic resistance genes from the environment to humans (Seiler and Berendonk, 2012).

The use of antimicrobial drugs in aquaculture also directly affects the microbial community composition and causes the accumulation and spread of antibiotic-resistant microbial communities. In the Eurozone, the use of veterinary pharmaceuticals in aquaculture is regulated by EU Council Regulations, which establish maximum residual limits for veterinary medicinal products in animal-derived food. Only two broad-spectrum antibiotics, florfenicol and oxytetracycline, are approved for use in aquaculture in Lithuania. Florfenicol is a chloramphenicol structural analogue that is more effective against certain bacteria than chloramphenicol (Cannon et al., 1990). Oxytetracycline is a broad-spectrum tetracycline antibiotic with bacteriostatic action, used to treat systemic bacterial infections in fish (Jerbi et al., 2011a). Approximately 73% of the 11 major aquaculture-producing countries utilize oxytetracycline and florfenicol (Lulijwa et al., 2020). Antibiotics that are unconsumed or expelled by fed animals can reach the water and persist or concentrate in sediments. Antibiotic residues in the environment can have significant impacts on human health and ecosystems (Daughton, 2004). However, these antibiotics are not currently included in the Water Framework Directive's Watch List (Loos et al., 2018). The issue of antibiotic pollution worsens when animal feces from farms are used to enhance fish pond production (Xiong et al., 2015). Understanding the composition of sediments in long-established aquaculture farms is crucial for assessing the impact of human activities and the dynamics of pond ecosystems. Comparing sediments in fish ponds, both upstream and downstream, can unveil the influence of intensive aquaculture on adjacent water ecosystems and the likelihood of antibiotic resistance gene transmission to human hosts (Lastauskiene et al., 2021).

5.6 MAINTAINING ADEQUATE SEDIMENT QUALITY IN POND AQUACULTURE

During the construction of new ponds, the removal of native topsoil results in low concentrations of organic matter in the pond bottoms. However, through sedimentation, the organic matter concentration in the bottom soil increases during aquaculture. Particles originating from mineral soil and organic waste, including uneaten feed, excrement, and remains of organisms, which do not disperse or dissolve in pond water, settle at the bottom. While most fish and shrimp farmers understand the importance of good sediment quality in pond aquaculture, they often lack a comprehensive understanding of the biogeochemical processes occurring in the sediment (Figure 5.5).

5.6.1 Composition of Organic Matter

Bacteria and other organisms present in sediment break down organic matter, although initially, the input rate exceeds the decomposition rate. In pond soil that receives a consistent supply of organic matter, the concentration of organic matter increases over a few years until a balance is reached between input and decay rates. Typically, the sediment contains no more than 2–3% organic carbon

FIGURE 5.5 The surface sediment of the pond appears brown due to oxidation. Removing the thin surface layer exposes the underlying black sediment, which is anaerobic. (From Boyd, C.E., Sediment microbiology, management. Global Seafood Alliance, 2004. www.globalseafood.org/advocate/sediment-microbiology-management/)

or 4–6% organic matter when equilibrium is achieved. Organic matter consists of substances that decompose at different rates, including labile (quick decomposition) and refractory (slow decomposition) compounds. Labile compounds degrade within days or weeks, while refractory compounds take months or even years to break down. Freshly deposited organic waste from uneaten feed, excrement, and deceased plankton in aquaculture ponds is primarily composed of labile compounds, but it also contains a refractory component. The chemical oxygen demand (COD) of the sediment is mainly driven by the labile portion of newly deposited organic materials. It is important to note that even a small input of labile organic matter to bottom soil with low organic content can lead to anaerobic conditions at the sediment surface. For instance, uneaten feed may cause localized issues with sediment quality in a new pond that has not been previously used for aquaculture (Boyd, 2004).

5.6.2 Consumption of Oxygen

To better understand the decomposition of organic matter in pond bottoms, let's consider a small, sealed container filled with fresh aerobic sediment. Initially, the water within the sediment pores contains dissolved oxygen, which aerobic bacteria utilize to convert organic matter into carbon dioxide and water. Eventually, the microbes deplete the dissolved oxygen. In the absence of molecular oxygen, fermentation begins. Fermenting microorganisms break down complex organic materials into carbon dioxide and simpler carbon molecules such as alcohols, aldehydes, and ketones. However, fermentation is not as efficient as aerobic respiration in fully mineralizing organic matter into carbon dioxide. Anaerobic chemotrophic bacteria degrade organic materials by extracting oxygen from inorganic molecules instead of using molecular oxygen. Denitrifying bacteria utilize oxygen from nitrate and nitrite for respiration after the oxygen is reduced in sediment pore water, resulting in the emission of nitrogen gas as a byproduct. Once nitrate and nitrite are depleted through denitrification, other bacteria utilize oxygen to respire iron and manganese oxides, producing ferrous and manganous compounds as metabolic byproducts. Following the depletion of these compounds, some bacteria can extract oxygen from sulfate for respiration, converting sulfate to sulfide in the process. Instead of complex organic materials, chemotrophic bacteria often utilize the simple carbon molecules produced during fermentation as carbon sources. Lastly, the accumulation of carbon dioxide in the sediment, resulting from the decomposition of organic matter, can serve as a source of oxygen for methane-producing bacteria (Boyd, 2004).

5.6.3 Pond Sediment Regulation

To prevent the entry of toxic metabolic wastes, such as hydrogen sulfide, into the water and protect aquaculture species, it is advisable to maintain an oxidized layer at the sediment–water interface. The most effective approach to ensure an aerobic sediment surface is to manage ponds in a way that maintains dissolved oxygen concentrations above 3 mg L^{-1} and promotes strong water circulation across the pond bottoms (Boyd, 2004). Additionally, application of nitrate to the pond water can help prevent low redox potential at the sediment–water interface by stimulating denitrification. When ponds are drained for harvest, it is beneficial to allow the bottom soils to dry, crack, and naturally aerate. This process facilitates the breakdown of labile organic matter from the previous crop and the oxidation of reduced chemicals in the soil. During the subsequent crop, ferrous iron is oxidized to ferric iron, and sulfur is transformed into sulfate, providing oxygen sources for microbial activity. Microorganisms decompose organic matter most efficiently at pH 7 to 8, so it is recommended to lime acidic soils. Tilling the pond bottoms also enhances aeration. In cases where soils cannot be completely dried, the use of nitrate fertilizer can serve as a soil oxidant.

5.6.4 Potential for Oxidation

The oxidation-reduction (redox) potential indicates the degree of oxidation or reduction of a substance. Water saturated with dissolved oxygen has a redox potential of approximately 560 mV, while water containing only 1 mg L^{-1} has a redox potential above 500 mV. As oxygen is consumed, the redox potential decreases. In anaerobic pond sediment, the redox potential declines significantly with increasing depth. In highly reduced sediment, the redox potential can reach 0 mV or even become negative. Microbial activities responsible for organic matter decomposition occur within specific ranges of redox potential. As long as the dissolved oxygen level remains at or above 1 mg L^{-1}, the redox potential remains high, and organic matter is decomposed aerobically. However, once oxygen becomes depleted, the redox potential decreases, and aerobic bacteria can no longer decompose organic matter (Boyd, 2004) (Figure 5.6).

5.7 SEDIMENTARY MICROBES AND PROCESSES ASSOCIATED WITH THEM

The microbial population and their mediated processes in aquatic sediments are influenced by various factors. These include the factors described in the following subsections.

5.7.1 The Nature and Source of Organic Matter

Factors such as whether the organic matter is freshly deposited algal material or previously processed soil organic matter from the surrounding area can affect its decomposition and modification in the sediment.

5.7.2 Temperature

The temperature of the sediment plays an important role in microbial activity and the rate of organic matter decomposition.

5.7.3 Availability of Electron Acceptors

The presence of electron acceptors, such as oxygen or other compounds that microbes can use during respiration, impacts the microbial processes involved in organic matter decomposition.

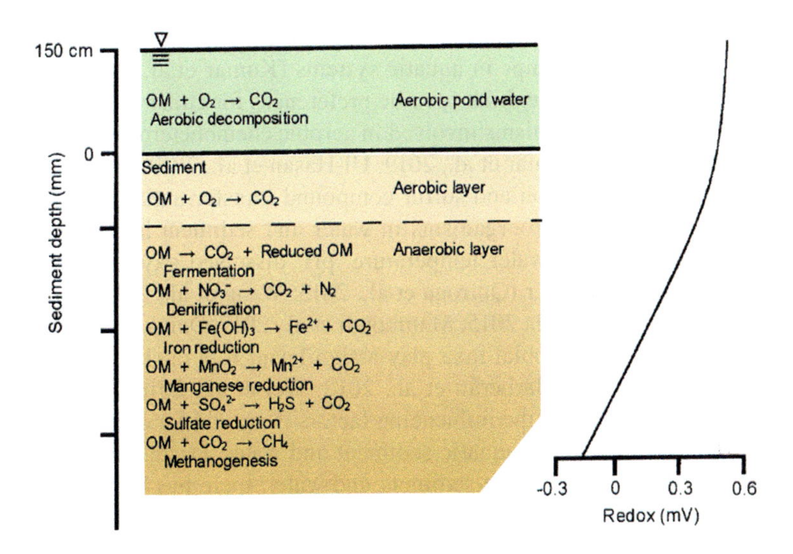

FIGURE 5.6 Schematic illustration of variations in microbial respiration in pond sediment as redox potential decreases. (From Boyd, C.E., Sediment microbiology, management. Global Seafood Alliance, 2004. www.globalseafood.org/advocate/sediment-microbiology-management/)

5.7.4 Bioturbation

Sediment mixing, caused by animals and known as bioturbation, can affect microbial activity and the distribution of organic matter in the sediment.

5.7.5 Presence of Specific Decomposer Microbes

The composition and presence of particular microbial species in the sediment contribute to the decomposition and modification of organic matter.

5.8 DISTRIBUTION PATTERN OF MICROORGANISMS IN THE AQUATIC ENVIRONMENT

Microorganisms are crucial in the sediment and water of natural water bodies, playing important roles in various ecological functions (Massana and Logares, 2013; Segovia et al., 2015; Wörmer et al., 2019). Bacteria, for instance, contribute to organic matter degradation and influence its preservation in sediments (Arnosti, 2011; Koho et al., 2013), as well as the nitrogen (Kuypers et al., 2018), phosphorus (McMahon and Read, 2013; Figueroa and Coates, 2017), and sulfur cycles (Jörgensen et al., 2019). Eukaryotic microorganisms, including protists, algae, unicellular fungi, and small invertebrates, also contribute to biogeochemical processes (Edgcomb et al., 2011; Anderson et al., 2013) and food network interactions (Massana, 2011; Grujcic et al., 2018). The aquatic system comprises two interconnected components: the water column and sediment, which are crucial to understanding the system. Sediment serves as a source or sink of nutrients to or from the water column, indicating a potential exchange of microorganisms between the two habitats. Consequently, similar microbiological communities can be expected between sediment and water (Powers et al., 2016; Pearce et al., 2017; Shaughnessy et al., 2019). Indeed, studies have identified shared bacterial (Ekwanzala et al., 2017; Kumar et al., 2019) and eukaryotic (Kosolapov et al., 2017; Shi et al., 2020) taxa between sediment and water samples, suggesting the sediment's contribution as a source of bacteria (Fries et al., 2008) and invertebrates (Wang et al., 2020) to the water column. However, these studies, along with others, have also revealed distinct taxa and distribution patterns of microorganisms between sediment and the water column, indicating complex microbiological relationships in aquatic systems (Kumar et al., 2019; Shi et al., 2020; Wang et al., 2020). Microbial taxa exhibit specific preferences for either sediment or water environments. For instance, microorganisms involved in aerobic chemoheterotrophy and phototrophy are commonly found in water (Kumar et al., 2019; Ul-Hasan et al., 2019), while those participating in organic matter decomposition and sulfur compound transformation are often detected in sediment. The distribution of microorganisms in water and sediment is influenced by various environmental factors, including water temperature, pH, dissolved oxygen (DO), nutrient levels, and sedimentary organic matter (Quiroga et al., 2013; Wang et al., 2015; Merlo et al., 2014; Fodelianakis et al., 2014; Zhao et al., 2015; Mahmoudi et al., 2017; Dai et al., 2018). Furthermore, interactions among different microbial taxa play a significant role in shaping their distribution patterns in sediment and water (Barberán et al., 2012; Wei et al., 2016). Understanding these microbial community patterns and the influencing factors is crucial for comprehending the relationships among microorganisms in aquatic sediment and water. Despite the significant importance of microorganisms in aquaculture sediment and water, there has been limited analysis of their distributions and differences between the two habitats. While some studies have examined microorganisms in aquaculture systems, the specific characteristics and distribution patterns remain largely unknown.

5.9 TOOLS AND TECHNIQUES USED IN SEDIMENT MICROBIOLOGY

In aquaculture, various tools and techniques are utilized to study sediment microbiology. The use of these tools and techniques in aquaculture facilitates the exploration of sediment microbiology, leading to a proper understanding of microbial communities and their roles in aquaculture systems. This knowledge helps in implementing effective management strategies and promoting sustainable practices in aquaculture. These tools include the following.

5.9.1 DNA Sequencing

High-throughput DNA sequencing methods, such as metagenomics and amplicon sequencing, enable the identification and characterization of microbial communities in sediment samples. This helps in understanding the diversity and functional potential of sediment microbes.

5.9.2 Microbial Isolation and Cultivation

Isolating and culturing specific microbial strains from sediment samples enables studying their metabolic capabilities, interactions, and physiological characteristics. This approach provides insights into the roles of individual microbes in sediment processes.

5.9.3 Microscopic Analysis

Microscopy techniques, such as light microscopy and electron microscopy, are employed to visualize sediment microbes at different magnifications. This aids in identifying microbial cells, studying their morphology, and examining their distribution within the sediment.

5.9.4 Functional Assays

Functional assays measure specific activities or processes mediated by sediment microbes. For example, enzyme assays can quantify enzymatic activities related to organic matter degradation or nutrient cycling. These assays provide information about the functional potential of sediment microbiota.

5.9.5 Stable Isotope Probing

Stable isotope probing (SIP) is used to track the incorporation of isotopically labelled substrates into specific microbial groups within sediment samples. This helps identify key microbial players involved in particular metabolic processes, such as organic matter degradation or nutrient transformations.

5.9.6 Molecular Techniques

Molecular techniques like quantitative PCR (qPCR) and fluorescence in situ hybridization (FISH) based on polymerase chain reaction (PCR) enable the quantification and detection of specific microbial groups or target genes in sediment samples. These techniques provide insights into the abundance and activity of the microbial populations of interest.

5.9.7 Bioinformatics Analysis

Advanced bioinformatics tools and software are employed to analyse large-scale sequencing data obtained from sediment microbiome studies. This involves taxonomic classification, functional annotation, and statistical comparisons to unravel the complex microbial communities and their functional potential.

5.10 USE OF SEDIMENT MICROBIOLOGY IN AQUACULTURE

Sediment microbiology plays a crucial role in aquaculture by providing insights and applications in various aspects of the industry (Nealson, 1997). It provides valuable knowledge and tools to optimize aquaculture operations, improve water quality, prevent diseases, and ensure the long-term sustainability of the industry. The following subsections give some examples.

5.9.1 Nutrient Cycling

Sediment microbes are involved in the breakdown and recycling of organic matter, playing a key role in nutrient cycling within aquaculture systems. They help converting organic waste, uneaten feed, and other organic materials into forms that can be utilized by aquatic organisms as nutrients.

5.9.2 Water Quality Management

Sediment microbiology is essential for maintaining good water quality in aquaculture systems. Certain microbes in the sediment can help remove or transform pollutants, such as ammonia and nitrate, which are byproducts of aquaculture activities. They contribute to the biological filtration and removal of harmful substances, thereby improving water quality for the cultured organisms.

5.9.3 Disease Control

Sediment microbiology can be used in the control and prevention of diseases in aquaculture. Beneficial microbes, such as probiotics, can be introduced into the sediment to promote a healthy microbial community and suppress the growth of harmful pathogens. This can help enhance the immune response and overall health of the cultured organisms, reducing the incidence of diseases.

5.9.4 Bioremediation

Sediment microbiology has the potential for bioremediation of aquaculture systems. Certain microbial species have the ability to degrade and detoxify pollutants present in sediments, such as heavy metals or organic contaminants. By harnessing the natural abilities of sediment microbes, bioremediation techniques can be employed to mitigate the impacts of pollutants and restore the ecological balance of aquaculture environments.

5.9.5 Ecological Balance

Understanding sediment microbiology is crucial for maintaining the ecological balance of aquaculture systems. By studying the microbial community composition and dynamics in sediments, aquaculturists can make informed decisions regarding feed management, waste treatment, and habitat restoration, thereby promoting sustainable and environmentally friendly aquaculture practices.

5.10 ROLE OF FISH IN SEDIMENT MICROBIOLOGY

Fish play a vital role in sediment microbiology studies within aquaculture. Their involvement provides researchers with a holistic understanding of the interactions between fish, sediment microbes, and the overall dynamics of the ecosystem. This knowledge is instrumental in improving management practices, preventing diseases, and promoting sustainable operations in aquaculture. There are several ways in which fish contribute to sediment microbiology.

5.10.1 Influence on Sediment Microbial Communities

Fish excrete organic waste, such as feces and uneaten feed, which serve as a source of nutrients for sediment microbes. The presence of fish in aquaculture systems can influence the composition, abundance, and activity of microbial communities in sediments.

5.10.2 Nutrient Cycling and Bioturbation

Fish activities, such as burrowing and feeding, can lead to bioturbation of sediment. This physical disturbance enhances nutrient cycling and mixing of sediment layers, affecting microbial processes and community structure. Fish contribute to the redistribution of organic matter and nutrients within sediments, influencing microbial decomposition and nutrient transformations.

5.10.3 Fish–Microbe Interactions

Fish harbour microbiota on their surfaces, including the skin, gills, and digestive tract. These microbiotas can interact with sediment microbes when fish deposit their waste or come into contact with sediments. The interactions between fish-associated microbiota and sediment microbes can influence sediment processes and microbial community dynamics.

5.10.4 Indicators of Sediment Quality

The health and behaviour of fish in aquaculture systems can serve as indicators of sediment quality. Poor sediment conditions, such as high levels of toxic compounds or oxygen depletion, can negatively impact fish health and performance. Monitoring fish health and responses can provide insights into the state of sediment microbiology and potential issues within the system.

5.10.5 Experimental Studies

Fish can be included in controlled laboratory or mesocosm experiments to investigate the effects of sediment microbiota on fish health and performance. By manipulating sediment conditions and microbial communities, researchers can examine the direct and indirect impacts of sediment microbiology on fish physiology, immunity, and overall well-being.

5.11 CONCLUSION

The fascinating realm of sediment microbiology unravels a complex ecosystem that plays a pivotal role in the Earth's biogeochemical cycles. Microorganisms inhabiting sediments demonstrate remarkable adaptations to extreme environments, contributing to nutrient cycling, organic matter degradation, and elemental transformation. As research continues to delve deeper into the interactions among sediment-dwelling microbes and their environment, we gain profound insights into the intricate web of life beneath our feet. The study of sediment microbiology not only expands our understanding of microbial ecology but also underscores the interconnectedness of all living systems on our planet.

REFERENCES

Allison, Steven D., and Jennifer B. H. Martiny. "Resistance, resilience, and redundancy in microbial communities." *Proceedings of the National Academy of Sciences* 105, no. supplement_1 (2008): 11512–11519.

Anderson, Ruth, Claudia Wylezich, Sabine Glaubitz, Matthias Labrenz, and Klaus Jürgens. "Impact of protist grazing on a key bacterial group for biogeochemical cycling in Baltic Sea pelagic oxic/anoxic interfaces." *Environmental Microbiology* 15, no. 5 (2013): 1580–1594.

Anthony, Edward J., and Arnaud Héquette. "The grain-size characterisation of coastal sand from the Somme estuary to Belgium: Sediment sorting processes and mixing in a tide-and storm-dominated setting." *Sedimentary Geology* 202, no. 3 (2007): 369–382.

Arnosti, Carol. "Microbial extracellular enzymes and the marine carbon cycle." *Annual Review of Marine Science* 3 (2011): 401–425.

Barberán, Albert, Scott T. Bates, Emilio O. Casamayor, and Noah Fierer. "Using network analysis to explore co-occurrence patterns in soil microbial communities." *The ISME Journal* 6, no. 2 (2012): 343–351.

Basaran, Asli Kaymakci, Mehmet Aksu, and Ozdemir Egemen. "Impacts of the fish farms on the water column nutrient concentrations and accumulation of heavy metals in the sediments in the eastern Aegean Sea (Turkey)." *Environmental Monitoring and Assessment* 162 (2010): 439–451.

Benson, Andrew K., Scott A. Kelly, Ryan Legge, Fangrui Ma, Soo Jen Low, Jaehyoung Kim, Min Zhang et al. "Individuality in gut microbiota composition is a complex polygenic trait shaped by multiple environmental and host genetic factors." *Proceedings of the National Academy of Sciences* 107, no. 44 (2010): 18933–18938.

Boyd, Claude E. "Maintaining good sediment quality is important in pond aquaculture." Sediment-microbiology-management. Global Seafood Alliance's. (2004). (https://www.globalseafood.org/advocate/sediment-microbiology-management/).

Burns, Adam R., W. Zac Stephens, Keaton Stagaman, Sandi Wong, John F. Rawls, Karen Guillemin, and Brendan J. M. Bohannan. "Contribution of neutral processes to the assembly of gut microbial communities in the zebrafish over host development." *The ISME Journal* 10, no. 3 (2016): 655–664.

Cannon, M., S. Harford, and J. Davies. "A comparative study on the inhibitory actions of chloramphenicol, thiamphenicol and some fluorinated derivatives." *Journal of Antimicrobial Chemotherapy* 26, no. 3 (1990): 307–317.

Clemente, Jose C., Luke K. Ursell, Laura Wegener Parfrey, and Rob Knight. "The impact of the gut microbiota on human health: An integrative view." *Cell* 148, no. 6 (2012): 1258–1270.

Dai, Lili, Chengqing Liu, Liqin Yu, Chaofeng Song, Liang Peng, Xiaoli Li, Ling Tao, and Gu Li. "Organic matter regulates ammonia-oxidizing bacterial and archaeal communities in the surface sediments of Ctenopharyngodon idellus aquaculture ponds." *Frontiers in Microbiology* 9 (2018): 2290.

Daughton, Christian G. "Non-regulated water contaminants: Emerging research." *Environmental Impact Assessment Review* 24, no. 7–8 (2004): 711–732.

Deming, Jody W., and John A. Baross. "The early diagenesis of organic matter: Bacterial activity." In *Organic Geochemistry: Principles and Applications*, pp. 119–144. Boston, MA: Springer US, 1993.

Dolman, Andrew M., Jacqueline Rücker, Frances R. Pick, Jutta Fastner, Thomas Rohrlack, Ute Mischke, and Claudia Wiedner. "Cyanobacteria and cyanotoxins: The influence of nitrogen versus phosphorus." *PloS One* 7, no. 6 (2012): e38757.

Edgcomb, Virginia P., David Beaudoin, Rebecca Gast, Jennifer F. Biddle, and Andreas Teske. "Marine subsurface eukaryotes: The fungal majority." *Environmental Microbiology* 13, no. 1 (2011): 172–183.

Ekwanzala, Mutshiene Deogratias, Akebe Luther King Abia, Eunice Ubomba-Jaswa, Jitendra Keshri, and Ndombo Benteke Maggy Momba. "Genetic relatedness of faecal coliforms and enterococci bacteria isolated from water and sediments of the Apies River, Gauteng, South Africa." *AMB Express* 7, no. 1 (2017): 1–10.

Fan, Li Min, Kamira Barry, Geng Dong Hu, Shun Long Meng, Chao Song, Wei Wu, Jia Zhang Chen, and Pao Xu. "Bacterioplankton community analysis in tilapia ponds by Illumina high-throughput sequencing." *World Journal of Microbiology and Biotechnology* 32 (2016): 1–11.

Figueroa, I. A., and J. D. Coates. "Microbial phosphite oxidation and its potential role in the global phosphorus and carbon cycles." *Advances in Applied Microbiology* 98 (2017): 93–117.

Fodelianakis, Stilianos, Nafsika Papageorgiou, Paraskevi Pitta, Panagiotis Kasapidis, Ioannis Karakassis, and Emmanuel D. Ladoukakis. "The pattern of change in the abundances of specific bacterioplankton groups is consistent across different nutrient-enriched habitats in Crete." *Applied and Environmental Microbiology* 80, no. 13 (2014): 3784–3792.

Fries, J. Stephen, Gregory W. Characklis, and Rachel T. Noble. "Sediment–water exchange of Vibrio sp. and fecal indicator bacteria: Implications for persistence and transport in the Neuse River Estuary, North Carolina, USA." *Water Research* 42, no. 4–5 (2008): 941–950.

Fuhrman, Jed A. "Microbial community structure and its functional implications." *Nature* 459, no. 7244 (2009): 193–199.

Gilbert, Jack A., Dawn Field, Paul Swift, Lindsay Newbold, Anna Oliver, Tim Smyth, Paul J. Somerfield, Sue Huse, and Ian Joint. "The seasonal structure of microbial communities in the Western English Channel." *Environmental Microbiology* 11, no. 12 (2009): 3132–3139.

Gillan, David C., Bruno Danis, Philippe Pernet, Guillemette Joly, and Philippe Dubois. "Structure of sediment-associated microbial communities along a heavy-metal contamination gradient in the marine environment." *Applied and Environmental Microbiology* 71, no. 2 (2005): 679–690.

Glasl, Bettina, David G. Bourne, Pedro R. Frade, Torsten Thomas, Britta Schaffelke, and Nicole S. Webster. "Microbial indicators of environmental perturbations in coral reef ecosystems." *Microbiome* 7, no. 1 (2019): 1–13.

Gooday, Andrew J. "Biological responses to seasonally varying fluxes of organic matter to the ocean floor: A review." *Journal of Oceanography* 58 (2002): 305–332.

Graham, David E., Matthew D. Wallenstein, Tatiana A. Vishnivetskaya, Mark P. Waldrop, Tommy J. Phelps, Susan M. Pfiffner, Tullis C. Onstott et al. "Microbes in thawing permafrost: The unknown variable in the climate change equation." *The ISME Journal* 6, no. 4 (2012): 709–712.

Grujcic, Vesna, Julia K. Nuy, Michaela M. Salcher, Tanja Shabarova, Vojtech Kasalicky, Jens Boenigk, Manfred Jensen, and Karel Simek. "Cryptophyta as major bacterivores in freshwater summer plankton." *The ISME Journal* 12, no. 7 (2018): 1668–1681.

Guan, Xiaoyan, Bai Wang, Ping Duan, Jiashen Tian, Ying Dong, Jingwei Jiang, Bing Sun, and Zunchun Zhou. "The dynamics of bacterial community in a polyculture aquaculture system of Penaeus chinensis, Rhopilema esculenta and Sinonovacula constricta." *Aquaculture Research* 51, no. 5 (2020): 1789–1800.

Hamdan, Leila J., Richard B. Coffin, Masoumeh Sikaroodi, Jens Greinert, Tina Treude, and Patrick M. Gillevet. "Ocean currents shape the microbiome of Arctic marine sediments." *The ISME Journal* 7, no. 4 (2013): 685–696.

Havens, Karl E. "Cyanobacteria blooms: Effects on aquatic ecosystems." In *Cyanobacterial Harmful Algal Blooms: State of the Science and Research Needs* (ed. Hudnell, H.K.), Springer (2008): 733–747.

Hollister, Emily B., Amanda S. Engledow, Amy Jo M. Hammett, Tony L. Provin, Heather H. Wilkinson, and Terry J. Gentry. "Shifts in microbial community structure along an ecological gradient of hypersaline soils and sediments." *The ISME Journal* 4, no. 6 (2010): 829–838.

Hou, D., Z. Huang, S. Zeng, J. Liu, S. Weng, and J. He. "Comparative analysis of the bacterial community compositions of the shrimp intestine, surrounding water and sediment." *Journal of Applied Microbiology* 125, no. 3 (2018): 792–799.

Hou, Dongwei, Zhijian Huang, Shenzheng Zeng, Jian Liu, Dongdong Wei, Xisha Deng, Shaoping Weng, Zhili He, and Jianguo He. "Environmental factors shape water microbial community structure and function in shrimp cultural enclosure ecosystems." *Frontiers in Microbiology* 8 (2017): 2359.

Hu, Limin, Xuefa Shi, Zhigang Yu, Tian Lin, Houjie Wang, Deyi Ma, Zhigang Guo, and Zuosheng Yang. "Distribution of sedimentary organic matter in estuarine–inner shelf regions of the East China Sea: Implications for hydrodynamic forces and anthropogenic impact." *Marine Chemistry* 142 (2012): 29–40.

Huang, Z., Y. Chen, S. Weng, X. Lu, L. Zhong, W. Fan, X. Chen, H. Zhang, and J. He. "Multiple bacteria species were involved in hepatopancreas necrosis syndrome (HPNS) of Litopenaeus vannamei." *Acta Scientiarum Naturalium Universitatis SunYatseni* 55, no. 1 (2016): 1–11.

Huang, Zhijian, Dongwei Hou, Renjun Zhou, Shenzheng Zeng, Chengguang Xing, Dongdong Wei, Xisha Deng et al. "Environmental water and sediment microbial communities shape intestine microbiota for host health: The central dogma in an anthropogenic aquaculture ecosystem." *Frontiers in Microbiology* 12 (2021): 772149.

Huang, Zhijian, Shenzheng Zeng, Jinbo Xiong, Dongwei Hou, Renjun Zhou, Chengguang Xing, Dongdong Wei et al. "Microecological Koch's postulates reveal that intestinal microbiota dysbiosis contributes to shrimp white feces syndrome." *Microbiome* 8 (2020): 1–13.

Inagaki, Fumio, Masae Suzuki, Ken Takai, Hanako Oida, Tatsuhiko Sakamoto, Kaori Aoki, Kenneth H. Nealson, and Koki Horikoshi. "Microbial communities associated with geological horizons in coastal subseafloor sediments from the Sea of Okhotsk." *Applied and Environmental Microbiology* 69, no. 12 (2003): 7224–7235.

Jerbi, M. A., Z. Ouanes, Z. Haouas, L. Achour, and A. Kacem. "Single and combined effects associated with two xenobiotics widely used in intensive aquaculture on European sea bass (Dicentrarchus labrax)." *Toxicology Letters* 205 (2011a): S119.

Jerbi, Mohamed Ali, Zouhour Ouanes, Raouf Besbes, Lotfi Achour, and Adnen Kacem. "Single and combined genotoxic and cytotoxic effects of two xenobiotics widely used in intensive aquaculture." *Mutation Research/Genetic Toxicology and Environmental Mutagenesis* 724, no. 1–2 (2011b): 22–27.

Jie, Zhuye, Huihua Xia, Shi-Long Zhong, Qiang Feng, Shenghui Li, Suisha Liang, Huanzi Zhong et al. "The gut microbiome in atherosclerotic cardiovascular disease." *Nature Communications* 8, no. 1 (2017): 845.

Jørgensen, Bo Barker, Alyssa J. Findlay, and André Pellerin. "The biogeochemical sulfur cycle of marine sediments." *Frontiers in Microbiology* 10 (2019): 849.

Kallmeyer, Jens, Robert Pockalny, Rishi Ram Adhikari, David C. Smith, and Steven D'Hondt. "Global distribution of microbial abundance and biomass in subseafloor sediment." *Proceedings of the National Academy of Sciences* 109, no. 40 (2012): 16213–16216.

Kirchman, David L., Matthew T. Cottrell, and Connie Lovejoy. "The structure of bacterial communities in the western Arctic Ocean as revealed by pyrosequencing of 16S rRNA genes." *Environmental Microbiology* 12, no. 5 (2010): 1132–1143.

Köchling, Thorsten, Pablo Lara-Martín, Eduardo González-Mazo, Ricardo Amils, and José Luis Sanz. "Microbial community composition of anoxic marine sediments in the Bay of Cádiz (Spain)." *International Microbiology* 14, no. 3 (2011): 143–154.

Koho, K. A., K. G. J. Nierop, Leon Moodley, J. J. Middelburg, Lara Pozzato, Karline Soetaert, J. Van der Plicht, and Gert-Jan Reichart. "Microbial bioavailability regulates organic matter preservation in marine sediments." *Biogeosciences* 10, no. 2 (2013): 1131–1141.

Kosolapov, D. B., A. I. Kopylov, and N. G. Kosolapova. "Heterotrophic nanoflagellates in water column and bottom sediments of the Rybinsk Reservoir: Species composition, abundance, biomass and their grazing impact on bacteria." *Inland Water Biology* 10 (2017): 192–202.

Kumar, Amit, Daphne H. P. Ng, Yichao Wu, and Bin Cao. "Microbial community composition and putative biogeochemical functions in the sediment and water of tropical granite quarry lakes." *Microbial Ecology* 77 (2019): 1–11.

Kuypers, Marcel M. M., Hannah K. Marchant, and Boran Kartal. "The microbial nitrogen-cycling network." *Nature Reviews Microbiology* 16, no. 5 (2018): 263–276.

Lastauskienė, Eglė, Vaidotas Valskys, Jonita Stankevičiūtė, Virginija Kalcienė, Vilmantas Gėgžna, Justinas Kavoliūnas, Modestas Ružauskas, and Julija Armalytė. "The impact of intensive fish farming on pond sediment microbiome and antibiotic resistance gene composition." *Frontiers in Veterinary Science* 8 (2021): 673756.

Le Chatelier, Emmanuelle, Trine Nielsen, Junjie Qin, Edi Prifti, Falk Hildebrand, Gwen Falony, Mathieu Almeida et al. "Richness of human gut microbiome correlates with metabolic markers." *Nature* 500, no. 7464 (2013): 541–546.

Li, Changchao, Quan Quan, Yandong Gan, Junyu Dong, Jiaohui Fang, Lifei Wang, and Jian Liu. "Effects of heavy metals on microbial communities in sediments and establishment of bioindicators based on microbial taxa and function for environmental monitoring and management." *Science of the Total Environment* 749 (2020): 141555.

Li, Tongtong, Huan Li, François-Joël Gatesoupe, Rong She, Qiang Lin, Xuefeng Yan, Jiabao Li, and Xiangzhen Li. "Bacterial signatures of 'Red-Operculum' disease in the gut of crucian carp (Carassius auratus)." *Microbial Ecology* 74 (2017): 510–521.

Loos, Robert, Dimitar Marinov, Isabella Sanseverino, Dorota Napierska, and Teresa Lettieri. "Review of the 1st watch list under the water framework directive and recommendations for the 2nd Watch List." Publications Office of the European Union, Luxembourg (2018).

Lourenço, Késia Silva, Afnan K. A. Suleiman, A. Pijl, J. A. Van Veen, H. Cantarella, and E. E. Kuramae. "Resilience of the resident soil microbiome to organic and inorganic amendment disturbances and to temporary bacterial invasion." *Microbiome* 6, no. 1 (2018): 1–12.

Lu, Xiao-Ming, Chang Chen, Tian-Ling Zheng, and Jian-Jun Chen. "Temporal–spatial variation of bacterial diversity in estuary sediments in the south of Zhejiang Province, China." *Applied Microbiology and Biotechnology* 100 (2016): 2817–2828.

Lulijwa, Ronald, Emmanuel Joseph Rupia, and Andrea C. Alfaro. "Antibiotic use in aquaculture, policies and regulation, health and environmental risks: A review of the top 15 major producers." *Reviews in Aquaculture* 12, no. 2 (2020): 640–663.

Mahmoudi, Nagissa, Steven R. Beaupré, Andrew D. Steen, and Ann Pearson. "Sequential bioavailability of sedimentary organic matter to heterotrophic bacteria." *Environmental Microbiology* 19, no. 7 (2017): 2629–2644.

Martinez-Porchas, Marcel, and Luis R. Martinez-Cordova. "World aquaculture: Environmental impacts and troubleshooting alternatives." *The Scientific World Journal* 2012 (2012): 1–9.

Massana, Ramon, and Ramiro Logares. "Eukaryotic versus prokaryotic marine picoplankton ecology." *Environmental Microbiology* 15, no. 5 (2013): 1254–1261.

Massana, Ramon. "Eukaryotic picoplankton in surface oceans." *Annual Review of Microbiology* 65 (2011): 91–110.

McMahon, Katherine D., and Emily K. Read. "Microbial contributions to phosphorus cycling in eutrophic lakes and wastewater." *Annual Review of Microbiology* 67 (2013): 199–219.

Merlo, Carolina, Luciana Reyna, Adriana Abril, María Valeria Amé, and Susana Genti-Raimondi. "Environmental factors associated with heterotrophic nitrogen-fixing bacteria in water, sediment, and riparian soil of Suquía River." *Limnologica* 48 (2014): 71–79.

Mooshammer, Maria, Florian Hofhansl, Alexander H. Frank, Wolfgang Wanek, Ieda Hämmerle, Sonja Leitner, Jörg Schnecker et al. "Decoupling of microbial carbon, nitrogen, and phosphorus cycling in response to extreme temperature events." *Science Advances* 3, no. 5 (2017): e1602781.

Moriarty, David J. W. "The role of microorganisms in aquaculture ponds." *Aquaculture* 151, no. 1–4 (1997): 333–349.

Naylor, Rosamond L., Rebecca J. Goldburg, Jurgenne H. Primavera, Nils Kautsky, Malcolm C. M. Beveridge, Jason Clay, Carl Folke, Jane Lubchenco, Harold Mooney, and Max Troell. "Effect of aquaculture on world fish supplies." *Nature* 405, no. 6790 (2000): 1017–1024.

Nealson, Kenneth H. "Sediment bacteria: Who's there, what are they doing, and what's new?." *Annual Review of Earth and Planetary Sciences* 25, no. 1 (1997): 403–434.

Newberry, Carole J., Gordon Webster, Barry A. Cragg, R. John Parkes, Andrew J. Weightman, and John C. Fry. "Diversity of prokaryotes and methanogenesis in deep subsurface sediments from the Nankai Trough, Ocean Drilling Program Leg 190." *Environmental Microbiology* 6, no. 3 (2004): 274–287.

Nogales, Balbina, Mariana P. Lanfranconi, Juana M. Piña-Villalonga, and Rafael Bosch. "Anthropogenic perturbations in marine microbial communities." *FEMS Microbiology Reviews* 35, no. 2 (2011): 275–298.

Parkes, R. John, Barry A. Cragg, and Peter Wellsbury. "Recent studies on bacterial populations and processes in subseafloor sediments: A review." *Hydrogeology Journal* 8 (2000): 11–28.

Parkes, R. John, Barry Cragg, Erwan Roussel, Gordon Webster, Andrew Weightman, and Henrik Sass. "A review of prokaryotic populations and processes in sub-seafloor sediments, including biosphere: Geosphere interactions." *Marine Geology* 352 (2014): 409–425.

Pearce, Alexandra R., Lisa G. Chambers, and Elizabeth A. Hasenmueller. "Characterizing nutrient distributions and fluxes in a eutrophic reservoir, Midwestern United States." *Science of the Total Environment* 581 (2017): 589–600.

Powers, Stephen M., Thomas W. Bruulsema, Tim P. Burt, Neng Iong Chan, James J. Elser, Philip M. Haygarth, Nicholas J. K. Howden et al. "Long-term accumulation and transport of anthropogenic phosphorus in three river basins." *Nature Geoscience* 9, no. 5 (2016): 353–356.

Pundytė, Neringa, Edita Baltrėnaitė, Paulo Pereira, and Dainius Paliulis. "Anthropogenic effects on heavy metals and macronutrients accumulation in soil and wood of Pinus sylvestris L." *Journal of Environmental Engineering and Landscape Management* 19, no. 1 (2011): 34–43.

Quiroga, M. Victoria, Fernando Unrein, Gabriela González Garraza, Gabriela Kueppers, Rubén Lombardo, M. Cristina Marinone, Silvina Menu Marque, Alicia Vinocur, and Gabriela Mataloni. "The plankton communities from peat bog pools: Structure, temporal variation and environmental factors." *Journal of Plankton Research* 35, no. 6 (2013): 1234–1253.

Rochelle, Paul A., Barry A. Cragg, John C. Fry, R. John Parkes, and Andrew J. Weightman. "Effect of sample handling on estimation of bacterial diversity in marine sediments by 16S rRNA gene sequence analysis." *FEMS Microbiology Ecology* 15, no. 1–2 (1994): 215–225.

Rodríguez, Juanjo, Christine M. J. Gallampois, Peter Haglund, Sari Timonen, and Owen Rowe. "Bacterial communities as indicators of environmental pollution by POPs in marine sediments." *Environmental Pollution* 268 (2021): 115690.

Rothschild, Daphna, Omer Weissbrod, Elad Barkan, Alexander Kurilshikov, Tal Korem, David Zeevi, Paul I. Costea et al. "Environment dominates over host genetics in shaping human gut microbiota." *Nature* 555, no. 7695 (2018): 210–215.

Rungrassamee, Wanilada, Amornpan Klanchui, Sawarot Maibunkaew, Sage Chaiyapechara, Pikul Jiravanichpaisal, and Nitsara Karoonuthaisiri. "Characterization of intestinal bacteria in wild and domesticated adult black tiger shrimp (Penaeus monodon)." *PloS One* 9, no. 3 (2014): e91853.

Segovia, Bianca Trevizan, Danielle Goeldner Pereira, Luis Mauricio Bini, Bianca Ramos de Meira, Verônica Sayuri Nishida, Fabio Amodêo Lansac-Tôha, and Luiz Felipe Machado Velho. "The role of microorganisms in a planktonic food web of a floodplain lake." *Microbial Ecology* 69 (2015): 225–233.

Seiler, Claudia, and Thomas U. Berendonk. "Heavy metal driven co-selection of antibiotic resistance in soil and water bodies impacted by agriculture and aquaculture." *Frontiers in Microbiology* 3 (2012): 399.

Shaughnessy, A. R., J. J. Sloan, M. J. Corcoran, and E. A. Hasenmueller. "Sediments in agricultural reservoirs act as sinks and sources for nutrients over various timescales." *Water Resources Research* 55, no. 7 (2019): 5985–6000.

Shi, Tian, Mingcong Li, Guangshan Wei, Jiai Liu, and Zheng Gao. "Distribution patterns of microeukaryotic community between sediment and water of the Yellow River estuary." *Current Microbiology* 77 (2020): 1496–1505.

Spor, Aymé, Omry Koren, and Ruth Ley. "Unravelling the effects of the environment and host genotype on the gut microbiome." *Nature Reviews Microbiology* 9, no. 4 (2011): 279–290.

Sriurairatana, Siriporn, Visanu Boonyawiwat, Warachin Gangnonngiw, Chaowanee Laosutthipong, Jindanan Hiranchan, and Timothy W. Flegel. "White feces syndrome of shrimp arises from transformation, sloughing and aggregation of hepatopancreatic microvilli into vermiform bodies superficially resembling gregarines." *PloS One* 9, no. 6 (2014): e99170.

Stankevica, Karina, Maris Klavins, Liga Rutina, and Aija Cerina. "Lake sapropel: A valuable resource and indicator of lake development." In *Advances in Environment, Computational Chemistry and Bioscience* (eds Oprisan, S., Zaharim, A., Eslamian. S., Jian. M.S., Aiub, C.A.F. & Azami, A.), Riga (2012): 247–252.

Stankevica, Karina, Agnese Pujate, Laimdota Kalnina, Maris Klavins, Aija Cerina, and Anda Drucka. "Records of the anthropogenic influence on different origin small lake sediments of Latvia." *Baltica* 28, no. 2 (2015): 135–150.

Stepanauskas, Ramunas, Travis C. Glenn, Charles H. Jagoe, R. Cary Tuckfield, Angela H. Lindell, Catherine J. King, and J. V. McArthur. "Coselection for microbial resistance to metals and antibiotics in freshwater microcosms." *Environmental Microbiology* 8, no. 9 (2006): 1510–1514.

Su, Lei, Huiwen Cai, Prabhu Kolandhasamy, Chenxi Wu, Chelsea M. Rochman, and Huahong Shi. "Using the Asian clam as an indicator of microplastic pollution in freshwater ecosystems." *Environmental Pollution* 234 (2018): 347–355.

Sun, Jian, Xiao-Ping Liao, Alaric W. D'Souza, Manish Boolchandani, Sheng-Hui Li, Ke Cheng, José Luis Martínez et al. "Environmental remodeling of human gut microbiota and antibiotic resistome in livestock farms." *Nature Communications* 11, no. 1 (2020a): 1427.

Sun, Yunfei, Wenfeng Han, Jian Liu, Xiaoshuai Huang, Wenquan Zhou, Jinbiao Zhang, and Yongxu Cheng. "Bacterial community compositions of crab intestine, surrounding water, and sediment in two different feeding modes of Eriocheir sinensis." *Aquaculture Reports* 16 (2020b): 100236.

Turpin, Williams, Osvaldo Espin-Garcia, Wei Xu, Mark S. Silverberg, David Kevans, Michelle I. Smith, David S. Guttman et al. "Association of host genome with intestinal microbial composition in a large healthy cohort." *Nature Genetics* 48, no. 11 (2016): 1413–1417.

Ul-Hasan, Sabah, Robert M. Bowers, Andrea Figueroa-Montiel, Alexei F. Licea-Navarro, J. Michael Beman, Tanja Woyke, and Clarissa J. Nobile. "Community ecology across bacteria, archaea and microbial eukaryotes in the sediment and seawater of coastal Puerto Nuevo, Baja California." *PloS One* 14, no. 2 (2019): e0212355.

USEPA. 2017. Nonpoint source: Agriculture. Avaliable at: https://www.epa.gov/nps/nonpoint-source-agriculture (accessed November 18, 2019).

Vezzulli, Luigi, Paolo Povero, and Mauro Fabiano. "The distribution and biochemical composition of biogenic particles across the subtropical Front in June 1993 (Azores-Madeira region, Northeast Atlantic)." *Scientia Marina* 66, no. 3 (2002): 205–214.

Wales, Andrew D., and Robert H. Davies. "Co-selection of resistance to antibiotics, biocides and heavy metals, and its relevance to foodborne pathogens." *Antibiotics* 4, no. 4 (2015): 567–604.

Wall, Diana H., Uffe N. Nielsen, and Johan Six. "Soil biodiversity and human health." *Nature* 528, no. 7580 (2015): 69–76.

Wang, Binhao, Xiafei Zheng, Hangjun Zhang, Fanshu Xiao, Hang Gu, Keke Zhang, Zhili He, Xiang Liu, and Qingyun Yan. "Bacterial community responses to tourism development in the Xixi National Wetland Park, China." *Science of the Total Environment* 720 (2020): 137570.

Wang, Jianjun, J. I. Shen, Yucheng Wu, Chen Tu, Janne Soininen, James C. Stegen, Jizheng He, Xingqi Liu, Lu Zhang, and Enlou Zhang. "Phylogenetic beta diversity in bacterial assemblages across ecosystems: Deterministic versus stochastic processes." *The ISME Journal* 7, no. 7 (2013): 1310–1321.

Wang, Kai, Xiansen Ye, Heping Chen, Qunfen Zhao, Changju Hu, Jiaying He, Yunxia Qian, Jinbo Xiong, Jianlin Zhu, and Demin Zhang. "Bacterial biogeography in the coastal waters of northern Z hejiang, East China Sea is highly controlled by spatially structured environmental gradients." *Environmental Microbiology* 17, no. 10 (2015): 3898–3913.

Webster, Gordon, R. John Parkes, Barry A. Cragg, Carole J. Newberry, Andrew J. Weightman, and John C. Fry. "Prokaryotic community composition and biogeochemical processes in deep subseafloor sediments from the Peru Margin." *FEMS Microbiology Ecology* 58, no. 1 (2006): 65–85.

Wei, Guangshan, Mingcong Li, Fenge Li, Han Li, and Zheng Gao. "Distinct distribution patterns of prokaryotes between sediment and water in the Yellow River estuary." *Applied Microbiology and Biotechnology* 100 (2016): 9683–9697.

Wilkes, Heinz, Antje Ramrath, and Jörg F. W. Negendank. "Organic geochemical evidence for environmental changes since 34,000 yrs BP from Lago di Mezzano, central Italy." *Journal of Paleolimnology* 22 (1999): 349–365.

Wörmer, Lars, Tatsuhiko Hoshino, Marshall W. Bowles, Bernhard Viehweger, Rishi R. Adhikari, Nan Xiao, Go-ichiro Uramoto et al. "Microbial dormancy in the marine subsurface: Global endospore abundance and response to burial." *Science Advances* 5, no. 2 (2019): eaav1024.

Xiong, Jinbo, Jinyong Zhu, Wenfang Dai, Chunming Dong, Qiongfen Qiu, and Chenghua Li. "Integrating gut microbiota immaturity and disease-discriminatory taxa to diagnose the initiation and severity of shrimp disease." *Environmental Microbiology* 19, no. 4 (2017): 1490–1501.

Xiong, Jinbo, Xiansen Ye, Kai Wang, Heping Chen, Changju Hu, Jianlin Zhu, and Demin Zhang. "Biogeography of the sediment bacterial community responds to a nitrogen pollution gradient in the East China Sea." *Applied and Environmental Microbiology* 80, no. 6 (2014): 1919–1925.

Xiong, Wenguang, Yongxue Sun, Tong Zhang, Xueyao Ding, Yafei Li, Mianzhi Wang, and Zhenling Zeng. "Antibiotics, antibiotic resistance genes, and bacterial community composition in fresh water aquaculture environment in China." *Microbial Ecology* 70 (2015): 425–432.

Yan, Qingyun, James C. Stegen, Yuhe Yu, Ye Deng, Xinghao Li, Shu Wu, Lili Dai et al. "Nearly a decade-long repeatable seasonal diversity patterns of bacterioplankton communities in the eutrophic Lake Donghu (Wuhan, China)." *Molecular Ecology* 26, no. 14 (2017): 3839–3850.

Yan, Qingyun, Jinjin Li, Yuhe Yu, Jianjun Wang, Zhili He, Joy D. Van Nostrand, Megan L. Kempher et al. "Environmental filtering decreases with fish development for the assembly of gut microbiota." *Environmental Microbiology* 18, no. 12 (2016): 4739–4754.

Zeng, Shenzheng, Zhijian Huang, Dongwei Hou, Jian Liu, Shaoping Weng, and Jianguo He. "Composition, diversity and function of intestinal microbiota in pacific white shrimp (Litopenaeus vannamei) at different culture stages." *PeerJ* 5 (2017): e3986.

Zhang, Wen, Jang-Seu Ki, and Pei-Yuan Qian. "Microbial diversity in polluted harbor sediments I: Bacterial community assessment based on four clone libraries of 16S rDNA." *Estuarine, Coastal and Shelf Science* 76, no. 3 (2008): 668–681.

Zhao, XinYu, Zimin Wei, Yue Zhao, Beidou Xi, Xueqin Wang, Taozhi Zhao, Xu Zhang, and Yuquan Wei. "Environmental factors influencing the distribution of ammonifying and denitrifying bacteria and water qualities in 10 lakes and reservoirs of the Northeast, China." *Microbial Biotechnology* 8, no. 3 (2015): 541–548.

6 Biogeochemical Cycles and Their Significance in Nutrient Recycling

Abhay Kumar Giri, Sumanta Kumar Mallik, Parvaiz Ahmad Ganie, and Suresh Chandra

6.1 INTRODUCTION

The cyclic movement as well as transformation of various materials between living forms and their surrounding environments, such as the atmosphere, the hydrosphere and the pedosphere/lithosphere, including the Earth's crust, is termed the biogeochemical cycle. In the biogeochemical cycle, chemical substances are channelled through the biotic as well as abiotic segments of the Earth. The biosphere is recognized as the biotic unit, whereas the abiotic sections are the atmosphere, the lithosphere and the hydrosphere. In the hydrosphere, the biogeochemical cycles, i.e., the conversion and passage of chemical compounds with the help of living beings, take place in various forms of water bodies, like lakes, rivers, streams, reservoirs, ponds and oceans, with the major share being in the oceans and seas. The most productive marine areas are estuarine and coastal ecosystems, which comprise a minor share in terms of surface area of the whole oceanic system. But, these ecosystems have a huge impact on global biogeochemical cycles and are driven by diverse groups of microbial communities, representing about 90% of the ocean's biomass (Alexander et al., 2011). The considerable anthropogenic pressure on these marine ecosystems not only imparts significant impact on the inhabited flora and fauna of the ecosystem but also creates hindrances to the recycling of both energy and nutrients (Galton, 1884; Hasler, 1969 and Jickells et al., 2017). One of the most important cases of human interference is cultural eutrophication, where the refuse from intensive farming systems, such as agriculture, aquaculture, piggery, poultry, dairy, etc., is discarded into coastal aquatic ecosystems. The outflows of these farming units cause eutrophication in the coastal ecosystems by increasing the level of various nutrients, especially the content of nitrogen and phosphorus, causing harmful algal blooms with prevalence of anoxic conditions and increased emissions of greenhouse gases as well (Bouwman et al., 2005). As a result of this, the smooth functioning of various biogeochemical cycles is affected, particularly the cyclic movement of nitrogen and carbon. Apart from cultural eutrophication, ongoing climate change also has a negative impact on the cryosphere of the globe, resulting in an intensified marine stratification, including qualitative and quantitative alterations of microbial assemblages at an unprecedented rate (Murillo et al., 2019; Altieri and Gedan, 2015; Breitburg et al., 2018; Cavicchioli et al., 2019 and Hutchins et al., 2019). Besides these phenomena, the oceanic hydrosphere is also subject to an intensive acidification process, with a hike of about ~0.1 pH units between the pre-industrial period and today. The acidification process has a significant impact on the buffer chemistry of carbonate and bicarbonate, which in turn leaves an adverse impact on the calcifying taxa of the planktonic biomass (Stillman and Paganini, 2015). As consequences of increase in global temperature, ocean stratification and deoxygenation, the microbial processes are being hindered, and hence, the production of toxic volatile materials such as greenhouse gases (e.g., N_2O, CH_4, etc.) (Breitburg et al., 2018), oxygen minimum zones (Bertagnolli and Stewart, 2018) and anoxic marine zones (Ulloa et al., 2012) are trending upwards, resulting in the loss of 25–50% of nitrogen from the

DOI: 10.1201/9781003408543-6

ocean to the atmosphere. In addition to the microbial and planktonic biomass, nekton communities are also adversely affected by the reduced form of vital elements, especially by reduced sulfur (e.g., H_2S). As the biogeochemical cycling of various important elements on the Earth is primarily driven by several living forms, more precisely by microbial entities (Falkowski et al., 2008 and Zakem et al., 2020), their wellbeing is of utmost priority for better functioning of the ecosystems (Murillo et al., 2019 and Cavicchioli et al., 2019) with respect to both fisheries and aquaculture.

6.2 PRINCIPLES OF BIOGEOCHEMICAL CYCLES

The biogeochemical cycles of the globe follow the principle of the 'Law of Conservation of Matter'. According to the principle, the elements or materials present on the Earth cannot be created or destroyed but can be transformed from one form to another. Various elements and compounds such as water, carbon, oxygen, nitrogen, sulfur, and phosphorus, including the metals and metalloids, are converted to their different forms for the wellbeing and prosperity of both organisms and their ecosystems. The elements are transformed between the abiotic and biotic compartments of ecosystems either as bulk materials or as different ionic forms (Hartland et al., 2013). All life forms, from unicellular microscopic organisms to the higher vertebrates and plants, are involved in the biogeochemical cycling processes for the stability of their ecosystems.

6.3 DRIVING PROCESSES OF BIOGEOCHEMICAL CYCLING

The transformation and cycling of various elements are possible with the involvement of and interaction among different processes, such as biological, chemical and geological. The biological processes are critically driven by several microbial consortia performing wide ranges of metabolic activities (Madsen, 2011). Without the involvement of microorganisms, the biological processes and the recycling of the nutrients and chemicals throughout global ecosystems could not be operational. All the cycles that occur in an ecosystem are interlinked to each other, which in turn facilitates and supports the flourishing of the living forms that inhabit it, like beneficial microbes, plants, plankton and other organisms for maintaining a healthy ecosystem. But, any sorts of anthropogenic interventions, like the burning of fossil fuels and unregulated disposal of chemicals and fertilizers in agricultural and other allied farming systems, alters the nutrient cycling processes and thereby, facilitates a negative impact on human health due to outcomes like pollution and climate change.

The microorganisms that are involved in the healthy functioning of different biogeochemical cycles derive energy from different sources for their performance as well as growth, e.g., organotrophs extract energy from sunlight, while inorganic molecules are the main energy sources for chemotrophs. The fungi and cyanogenic microbes are the major groups of organotrophs that help in the transformation of different organic materials through processes like bioleaching (Natarajan and Ting, 2014). Chemical cycles are particularly active and effective within hydrological, sedimentary, and gaseous processes. Among the various chemical cycling processes, some are the basic sources of renewable energy. In addition to the production of renewable energy, the production of several organic compounds and prebiotic chemistry, including complex chemical reactions, are the outcomes of other chemical cycles. Besides the biological and chemical cycles, the transformation and cycling of various nutrient elements in the biogeochemical cycles are also regulated by several geological processes, like weathering, surface runoffs, cyclones, tornadoes, earthquakes and tectonic depressions, including the subduction of the continental plates. Through these geological processes, inorganic materials are recycled among living biota and their non-living environments for the balance and stability of the ecosystem.

6.4 SIGNIFICANCE OF DIFFERENT BIOGEOCHEMICAL CYCLES

In general, the biogeochemical cycles (Figure 6.1) are operated through interrelation among various spheres of the Earth, including the Earth's crust, outer space and the solar system. The key

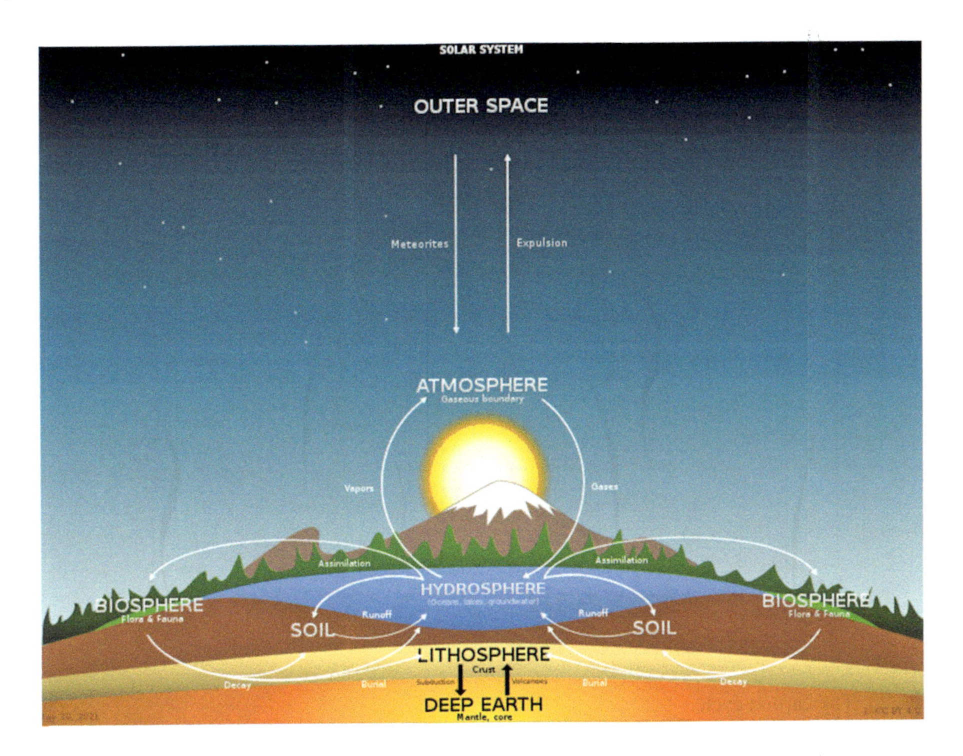

FIGURE 6.1 Overview of biogeochemical cycling. (From https://en.wikipedia.org/wiki/Biogeochemical_cycle#/media/File:BIOGEOCHEMICAL_CYCLING_OF_ELEMENTS.svg)

biogeochemical cycles are the carbon, nitrogen, oxygen, phosphorus and sulfur cycles, which involve substances like carbon, nitrogen, hydrogen, oxygen, phosphorus and sulfur along with their various forms. These cycles are in operation at various sites, such as the atmosphere, land and water, including the Earth's crust for stabilizing the ecosystem in equilibrium. There are various other elements, such as hydrogen, calcium, iron, mercury, selenium and silica, including some other elements that also undergo biogeochemical processes. The minerals are channelled between different living forms that are present in the biotic and abiotic components of the biosphere (Fisher, 2019). The rock cycle and the human-induced cycles for synthetic compounds, such as for polychlorinated biphenyls (PCBs), are not omitted from the cycling processes and are included among macroscopic cycles. Among some of the above-mentioned cycles, geological reservoirs come into the picture, where the elements or materials are stored for relatively longer periods of time.

For the occurrence of any biogeochemical cycle, a source of energy is the most important, and for this, sunlight and inorganic molecules act as energy sources for the functioning of organotrophs and chemotrophs, respectively. In general, sunlight combines carbon, hydrogen and oxygen for the source of energy, but organisms in the deep sea (where sunlight is lacking) obtain energy from sulfur through redox processes (sulfur to sulfite and then to sulfate). The organisms that are present near hydrothermal vents, such as the giant tube worms, obtain their energy from H_2S. In general, different biogeochemical cycles are responsible for the transformation and movement of various substances around the entire globe.

6.4.1 The Carbon Cycle

The carbon content of the Earth is transformed into its various forms and cycles (Figure 6.2) through the atmosphere, hydrosphere, biosphere and lithosphere, including fossil carbon, which act as major reservoirs (Prentice et al., 2001; Noguer et al., 2001 and Solomon, 2007). Among the

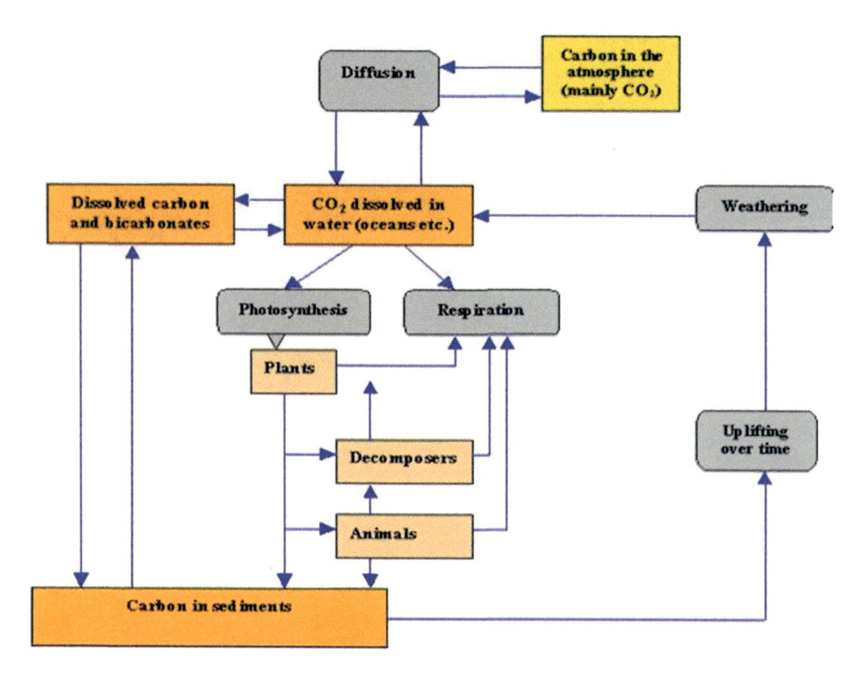

FIGURE 6.2 The carbon cycle.

various reservoirs of carbon, oceans are the major reservoirs, containing 37,000 Pg of inorganic carbon and 1,000 Pg of organic carbon, which is about 50 and 70 times the carbon present in the atmosphere and terrestrial vegetation, respectively. Among the different zones of the ocean, the mesopelagic and bathypelagic regions contribute the maximum share, while the sediments contain a significantly lower amount (6,000 Pg) with a very slow renewal or turnover time.

Carbon in the form of CO_2 is utilized by plants for their photosynthesis, and then subsequently, O_2 is released from the plants, and is in turn taken up by plants and animals for their respiration. CO_2 also contributes to the atmosphere through artificially induced human interventions like the burning of fossil fuels. Algae, phytoplankton and plants of the aquatic ecosystem also utilize CO_2 and produce O_2 in a similar manner. In addition to this, aquatic environments also receive carbon due to surface runoff. Excretion by animals as well as the death and decay of all living beings contributes to the inflow of carbon in land as well as aqueous systems. In water, CO_2 is either stored in rocks, sediments and the skeletons of living animals (molluscs and corals) or remains in aqueous solution in the form of carbonates and bicarbonates, acting as a buffer offering resistance to any changes in water pH. These carbonates and bicarbonates are returned to the atmosphere in the form of CO_2 with the heating of water by the Sun.

6.4.2 Oxygen Cycle

The oxygen cycle is among the simplest kinds of biogeochemical cycle (Figure 6.3). The atmosphere serves as the major pool for oxygen, and from there, the gas is passed to the aquatic ecosystems through the process of diffusion. In spite of the thorough mixing of oxygen in the atmosphere, seasonal variation has been noticed with respect to the latitudes, with higher latitudes registering more seasonal variation (Keeling and Shertz, 1992). Plants and animals utilize oxygen for their respiration and release carbon dioxide and water to their surrounding environment. Carbon dioxide and water are taken up by the plants for photosynthesis to produce oxygen and organic compounds, which in turn are used by plants and animals for their metabolic activities and respiration to complete the cycling process.

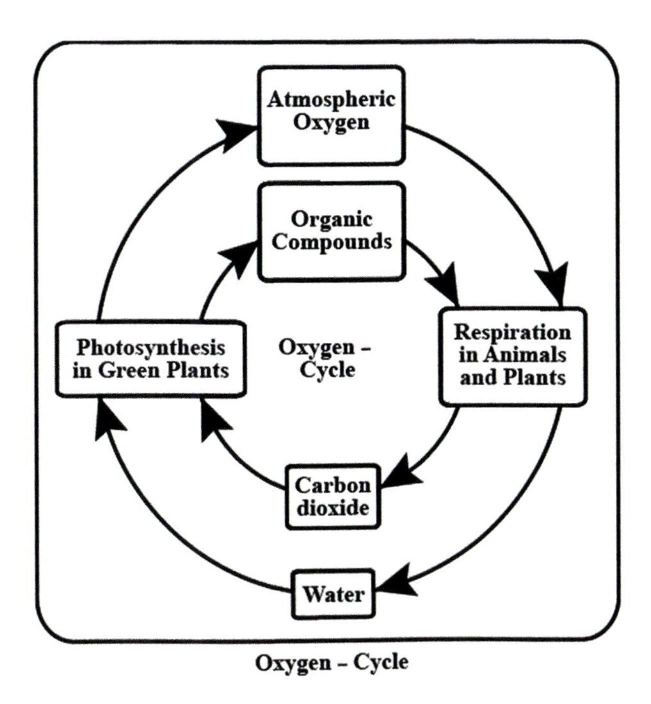

FIGURE 6.3 The oxygen cycle. (From www.toppr.com/ask/question/explain-oxygen-cycle-in-details/)

6.4.3 Nitrogen Cycle

The atmosphere is the major reservoir for nitrogen, containing about >78% of N_2 gas. In the nitrogen cycle, the atmospheric N_2 is moved to the land and aquatic ecosystems by both natural as well as artificially induced human activities (Galloway et al., 2004). Lightning is the natural process by which the atmospheric N_2 is directly fixed to the ecosystems. Volcanic eruptions (a natural phenomenon) and the emissions of fossil fuels (an artificial act) are the processes by which the nitrogen is transferred from the atmosphere to both aquatic and terrestrial environments through precipitation. Nitrogen is also released by excretion (animals) and the death and decomposition of both animals and plants. In addition to these, some of the bacteria and cyanobacteria, including the archaea, are capable of fixing nitrogen in both ecosystems. Free-living microbes like *Azotobacter* and *Azospirillium* and symbiotic bacteria like *Rhizobium* (Rhizobiaceae family) are capable of fixing nitrogen, and are called diazotrophs. The use of nitrogen-based chemical fertilizers in agriculture and allied farming systems also helps in enriching the nitrogen content of the soil as well as water.

The ammoniacal forms of nitrogen are converted to nitrite and then to nitrate by ammonia-oxidizing bacteria like *Nitrosomonas* and nitrite-oxidizing bacteria like *Nitrobacter*, respectively. The nitrates are then acted upon by denitrifying bacteria like *Pseudomonas* to release nitrogen gas back to the atmosphere (Figure 6.4). As well as these, there are other biological processes by which ammonia and nitrite are directly converted to N_2, called the anammox (Mulder et al., 1995). The anammox is carried out by chemolitho-autotrophic microbes called planctomycetes.

6.4.4 The Phosphorus Cycle

Phosphorus is the most important critical element for the survival and growth of both terrestrial and aquatic animals and plants. This element is unavailable in the atmosphere, and the main reservoir is phosphate-bearing rocks. It is a limiting element in the biological productivity of the land ecosystem (Lajtha and Schlesinger, 1988). Cycling of phosphorus involves entities like water and living organisms, including the Earth's crust, and phosphorus enters the food chain through the weathering of

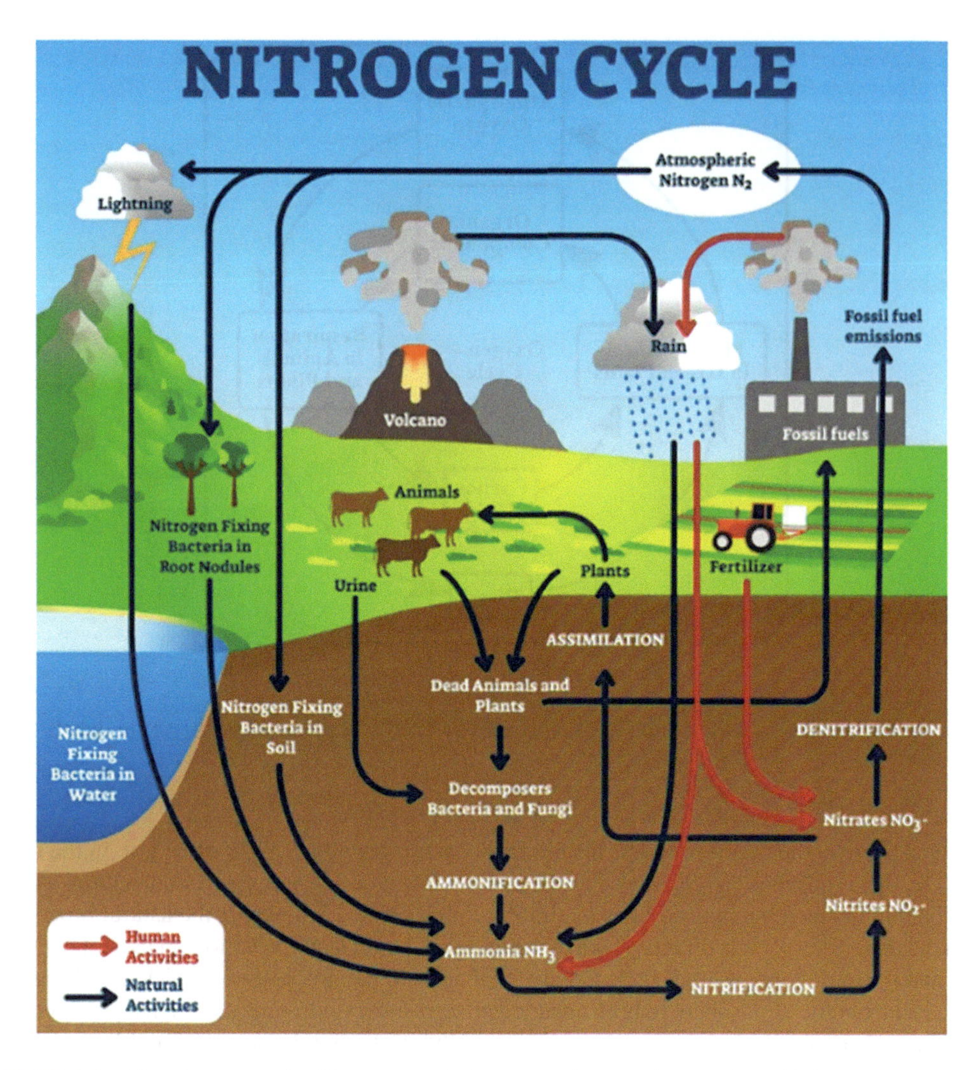

FIGURE 6.4 The nitrogen cycle. (From www.shalom-education.com/courses/gcse-biology/lessons/ecosystems/topic/the-nitrogen-cycle/)

phosphate rocks, which is a relatively slow process (Figure 6.5). Land and aquatic environments are also enriched with phosphorus due to the death and decay of both plants and animals in their respective environments. The animal metabolic outputs and the soil runoffs (for aquatic environment) also contribute to regaining of phosphorus by the ecosystem. Phosphates in the soil are taken up by plants, and are then passed to animals due to their consumption of plants and/or herbivores.

Phosphorus is also made available to the ecosystem by a group of bacteria called the phosphate solubilizing bacteria. Bacteria such as *Pantoea agglomerans*, *Microbacterium laevaniformans*, *Pseudomomas putida*, and *Acetobacter diazotropicus*, including some *Bacillus* and *Micrococcus* species, act on the inorganic phosphates of some insoluble compounds and release a usable form of phosphorus to the surrounding environment.

6.4.5 Sulfur Cycle

The major natural reservoirs of sulfur are rocks or sediments, where the element exists in oxidized forms (e.g., gypsum) or in reduced forms such as pyrite, cinnabar and galena, including

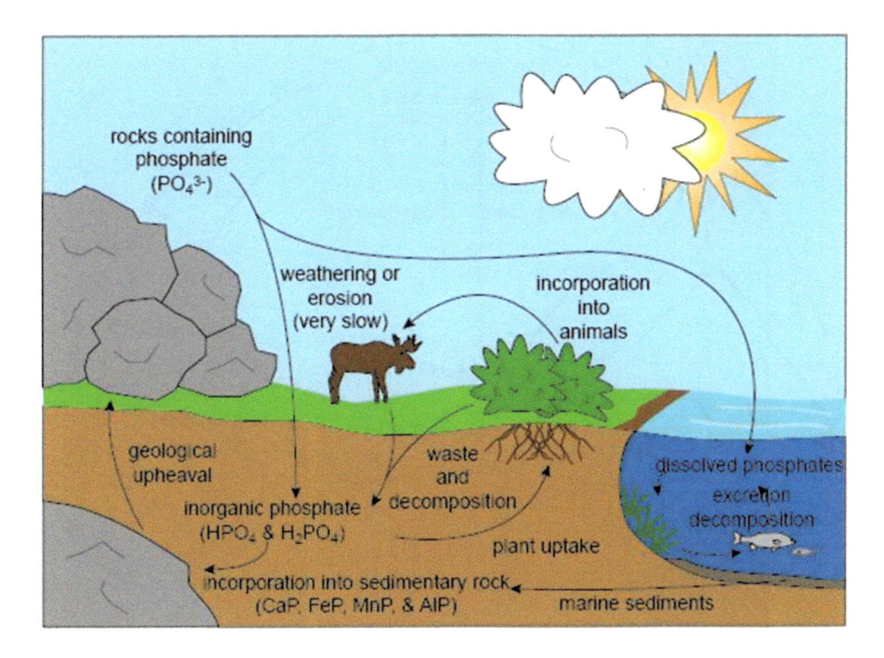

FIGURE 6.5 The phosphorus cycle.

some other metallic compounds. Oceans also serve as potent reservoirs for sulfur, containing assimilable sulfur in the form of soluble sulfate, while living materials contain only 1% of sulfur on a dry weight basis. Natural phenomena like volcanic emissions, bacterial activities and decomposition of living materials act as resources for sulfur. The sulfur cycle is mediated through diverse metabolic processes coupled with distinctive stable isotopic fractionations of sulfur species (Leavitt et al., 2013). In addition to these, artificial human interventions like industrial emissions with the burning of fossil fuels also release large amounts of sulfur dioxide and hydrogen sulfide to the atmosphere. The sulfur dioxide in the atmosphere interacts either with oxygen to produce sulfur trioxide gas and then sulfate (contributing to acid rain) or with various chemicals to produce different salts of sulfur. But in aquatic environments, the reactions between sulfur dioxide and water give rise to sulfuric acid, which can also be produced from dimethyl sulfide (released by plankton to the atmosphere). The sulfur compounds of soil as well as water release various gases containing sulfur to the atmosphere through the process of gasification to complete the cycling of sulfur (Figure 6.6).

6.4.6 Cycling of Metals and Metalloids

In addition to the cycling of major elements, the biogeochemical cycles of metals and metalloids are also critical processes for the existence of life on Earth. Microorganisms drive the cycling of these substances by altering their redox properties or chemical forms. Many prokaryotes, such as archaea and bacteria, play a significant role in the transformation of many metallic substances (Kim and Gadd, 2008). Some of the metals, such as iron, manganese, cobalt, etc., are essential for any life forms, while others, like mercury, cadmium, arsenic, etc., are toxic even in trace amounts. During this process, the microbes adopt various operational mechanisms, like resistance, detoxification, and donating or accepting electrons, including some incidental or indirect mechanisms. Besides these, the metals and metalloids are also recycled through physico-chemical processes mediated by non-enzymatic redox reactions. Both the biotic as well as the abiotic processes depend on several factors, like the surrounding environment, oxygen concentration and pH of the medium (Figure 6.7).

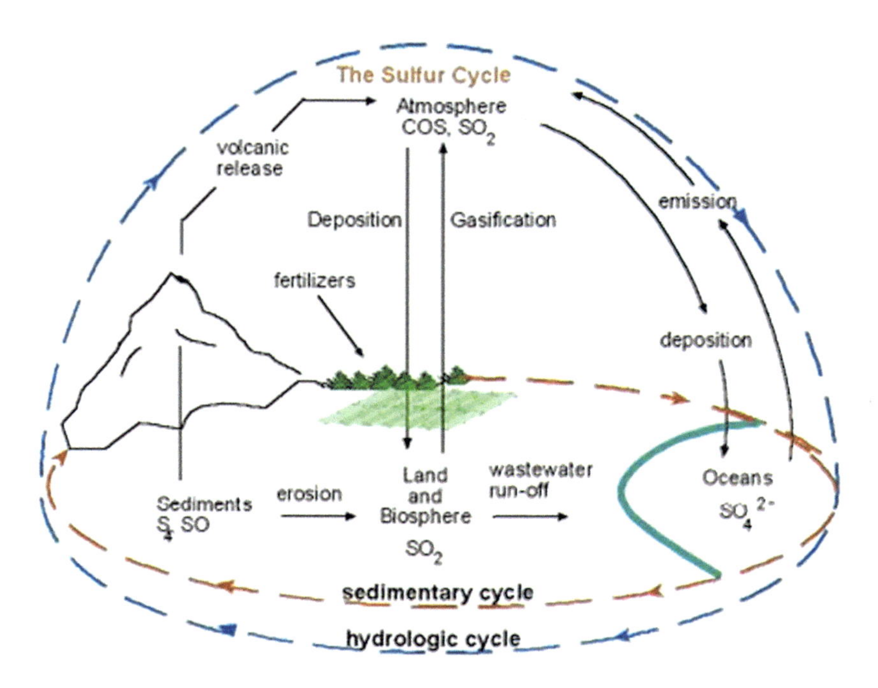

FIGURE 6.6 The sulfur cycle.

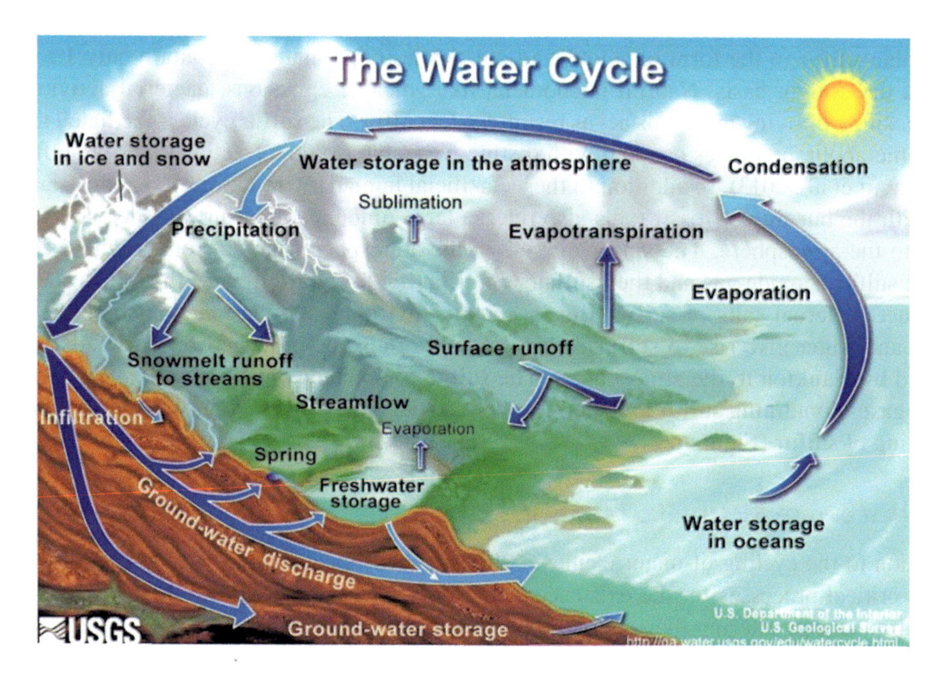

FIGURE 6.7 The water cycle. (From https://education.nationalgeographic.org/resource/800px-water-cycle/)

6.4.7 Cycling of Nutrients in the Marine Environment

Marine ecosystems such as oceans, seas and estuaries, including the coastal areas, are one of the major hubs for the movement of nutrients through biogeochemical cycles. During this process, various nutrient elements pass through different ecological niches of the marine environment. Several nutrient cycling processes are common in both freshwater as well as marine ecosystems for the decomposition of organic matter and regeneration of nutrients as well (Anderson and Macfadyen, 1976). Both photic and aphotic zones are involved in the processes of nutrient cycling (Figure 6.8). The plants present in the photic zone are the primary producers of the ecosystem, which derive energy and nutrients both from sunlight as well as from the inorganic matter present in the aphotic zone. Animals then consume the plants in both living as well as dead forms for their growth and survival. Microscopic organisms such as bacteria of the aphotic zones basically act upon the organic matter of the dead animals and plants to release inorganic nutrients, which are then transfected to the photic zone through upwelling and mixing of water. The cycling of materials and nutrients in the marine environment operates continuously with the involvement of these processes.

6.4.8 Types of Biogeochemical Cycles

Biogeochemical cycles are categorized as fast and slow based on their rate of progress. Deep biogeochemical cycles are the processes that occur beneath the terrestrial surface. The biosphere is the prime site for the operation of fast cycles, while the sites of occurrence for the slow cycles are the rocks. The fast cycles are the biological cycles, which are generally completed within some years, whereas geological cycles are categorized as slow cycles, as these processes take millions to billions of years for completion. In the fast biological cycles, materials are exchanged between biosphere and atmosphere, and vice versa, while the materials are moved through the Earth's crust between rocks, soil, ocean and atmosphere in the case of the slow geological cycles (Libes, 2015). The fast cycles may be annual, such as the photosynthesis of plant-like organisms, or may be the decadal, involving vegetative growth and decomposition. The increased level of unwanted human interventions is the main driving force for increasing the speed of some cycling processes, especially for carbon, which leads to immediate impacts like climate change (Bush, 2019; Rothman, 2002; Carpinteri and Niccolini, 2019 and Rothman, 2015). The mountain-building processes as well as the configuration of Earth's mantle through the formation of sedimentary rocks are critically driven by slow geochemical processes with the prime involvement of carbon. However, there are some processes by which carbon is returned to the atmosphere, e.g., weathering of rocks, degassing of oceans by rivers, hydrothermal emissions of calcium ions, and volcanic eruptions. In a volcanic eruption, the atmosphere receives the geological carbon directly in the form of CO_2, but this is <1% of the CO_2 that is released to the atmosphere by the burning of fossil fuels (coal, oil and natural gases) (Libes, 2015 and Bush, 2019).

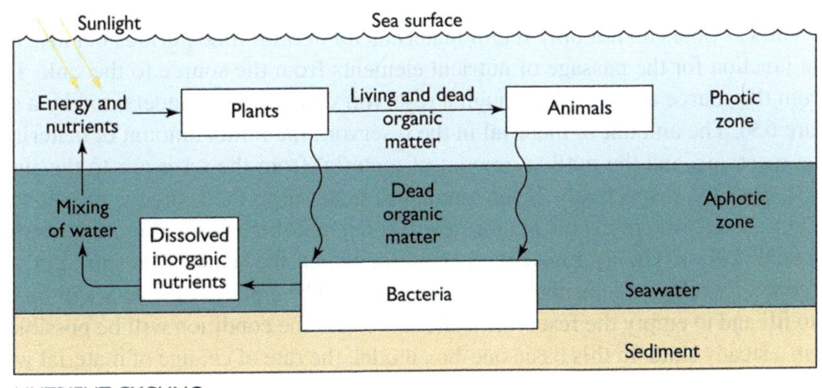

FIGURE 6.8 Cycling of nutrients in marine environments.

The deep cycles are basically functional at the terrestrial subsurface, which is the largest reservoir of carbon on Earth, containing about 14–135 Pg of carbon (McMahon and Parnell, 2014). The transformation and movement of its organic as well as inorganic forms are largely carried out by different groups of microbes. But, the qualitative as well as the quantitative diversity of the terrestrial subsurface microorganisms are still not fully explored, as their 16S rRNA sequences contribute only about <8% to the public databases (Schloss et al., 2016), and among these, only a few are presented in the form of genomes or isolates. Hence, there is a dearth of information about their metabolic activities, including the interlinking and interacting mechanisms among them in the subsurface areas. However, some studies on the syntrophic (cultivation-based) (Abreu and Taga, 2016; Bosse et al., 2015 and Braker et al., 2012) as well as on the natural (small-scale metagenomics) (Hug et al., 2016; McCarren et al., 2010 and Embree et al., 2015) consortia reveal that these microbes are interconnected with each other through metabolic handoffs, i.e., the exchange of redox reaction outputs with each other for their wellbeing and growth. But, no intensive studies have been carried out to date to fully understand the operational principles of these microbes, so the cycling of carbon and other nutrients in the deeper subsurface regions is still a mystery (Long et al., 2016). Therefore, advanced research programmes may be encouraged in the line of metagenomics or even complete genome sequencing to fully unfurl and understand the biogeochemical processes existing in deeper terrestrial subsurfaces (Hug et al., 2016; Eren et al., 2015; Alneberg et al., 2014 and Anantharaman et al., 2016).

6.5 ROLE OF RESERVOIRS/POOLS IN BIOGEOCHEMICAL CYCLING

During the processes of biogeochemical cycling, sometimes some of the materials or elements are held in one place for longer durations, depending on the nature of the materials as well as their surrounding environments. A place that holds the material for a longer time period is called a 'reservoir'. One example of a reservoir is the coal deposits, which store carbon for much longer time periods (Baedke and Fichter, 2017). On the other hand, the 'exchange pool' is a kind of temporary place where the materials are stored for relatively shorter durations. Plants and animals are the best examples of exchange pools (Baedke and Fichter, 2017), as these living forms store carbon to synthesize different biomolecules like proteins, lipids and carbohydrates. These biomolecules are then utilized by them either for their internal construction or to provide energy to carry out their various metabolic processes. After the completion of their internal processes, carbon is released back to the surrounding environment. The holding period for carbon in plants and animals is comparatively much shorter than the storing time of this element in the coal deposits. The reservoirs are generally abiotic entities, while the exchange pools are biotic ones. Whether it is a reservoir or an exchange pool, the time period for which the material is kept at one place is called the 'residence time or turnover time or renewal time or exit age' (Baedke and Fichter, 2017).

The reservoirs or pools do not only retain materials for certain time periods but also serve as an intermediate junction for the passage of nutrient elements from the source to the sink. The flow of materials from the source to the sink through a reservoir can easily be understood by a simple box model (Figure 6.9). The amount of material in the reservoir, the influx amount of material from the source to the reservoir, and the outflux amount of material from the reservoir to the sink are represented as M, Q and S, respectively. If the amount of material in the source is in balance with the sink, i.e., if $Q = S$, then the reservoir is in a steady state, and there is no alteration over the period of time (Bianchi, 2007). The average time period for which the material is stored or held in the reservoir is generally referred to as the 'turnover time (τ)'. The equation $\tau = M/S$ will be valid if the time taken to fill and to empty the reservoir is the same, and the condition will be possible when the reservoir is in a steady state. In this basic one-box model, the rate of change of material with respect to time is generally represented by the equation:

$$\frac{dM}{dt} = Q - S = A - \frac{M}{\tau}$$

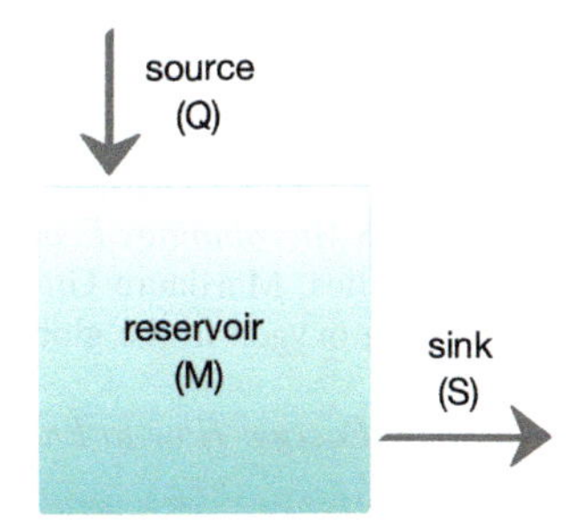

FIGURE 6.9 Simple box model. (From https://en.wikipedia.org/wiki/Biogeochemical_cycle#:~:text=A biogeochemical cycle%2C or more, cycle and the water cycle.)

6.6 CONCLUSION

Biogeochemical cycles are vital factors for the existence of any life forms on the planet Earth. The cycles also keep the ecosystems in equilibrium and contribute to the growth and wellbeing of both plants and animals. All the major nutrients, such as carbon, nitrogen, hydrogen, oxygen, phosphorus and sulfur, including metals and metalloids, and different synthetic compounds are cycled to plants, animals and microbes through various biogeochemical processes. These substances are channelled through biotic and abiotic compartments like the biosphere, hydrosphere, atmosphere, lithosphere, etc., with the involvement of intermediate sites such as reservoirs and exchange pools. The cycling of some chemicals or compounds is mediated very fast, while some are cycled at slow rates. The cycles that operate in the deep ocean and the deepest part of Earth's crust are quite different from the conventional biogeochemical cycling processes, and their exact mechanisms need to be unravelled and clearly understood. Extensive in-depth studies at genome level may be undertaken to gain technical knowhow regarding the diversity and role of microbes, especially for deep and rock cycles.

REFERENCES

Abreu, Nicole A., and Michiko E. Taga. "Decoding molecular interactions in microbial communities." *FEMS Microbiology Reviews* 40, no. 5 (2016): 648–663.

Alexander, Vera, Patricia Miloslavich, and Kristen Yarincik. "The census of marine life—Evolution of worldwide marine biodiversity research." *Marine Biodiversity* 41 (2011): 545–554.

Alneberg, Johannes, Brynjar Smári Bjarnason, Ino De Bruijn, Melanie Schirmer, Joshua Quick, Umer Z. Ijaz, Leo Lahti, Nicholas J. Loman, Anders F. Andersson, and Christopher Quince. "Binning metagenomic contigs by coverage and composition." *Nature Methods* 11, no. 11 (2014): 1144–1146.

Altieri, Andrew H., and Keryn B. Gedan. "Climate change and dead zones." *Global Change Biology* 21, no. 4 (2015): 1395–1406.

Anantharaman, Karthik, Christopher T. Brown, Laura A. Hug, Itai Sharon, Cindy J. Castelle, Alexander J. Probst, Brian C. Thomas et al. "Thousands of microbial genomes shed light on interconnected biogeochemical processes in an aquifer system." *Nature Communications* 7, no. 1 (2016): 13219.

Anderson, J. M., and Macfadyen, A. (Eds.). (1976). *The role of terrestrial and aquatic organisms in decomposition processes. The 17th symposium of the British Ecological Society, 15-18 April 1975* (pp. xi+-474).

Baedke, Steve J., and Lynn S. Fichter. "Biogeochemical cycles: Carbon cycle." *Supplimental Lecture Notes for Geol* 398 (2017).

Bertagnolli, Anthony D., and Frank J. Stewart. "Microbial niches in marine oxygen minimum zones." *Nature Reviews Microbiology* 16, no. 12 (2018): 723–729.

Bianchi, Thomas S. *Biogeochemistry of Estuaries*. Oxford University Press, 2007. ISBN 9780195160826.

Bosse, Magnus, Alexander Heuwieser, Andreas Heinzel, Ivan Nancucheo, Hivana Melo Barbosa Dall'Agnol, Arno Lukas, George Tzotzos, and Bernd Mayer. "Interaction networks for identifying coupled molecular processes in microbial communities." *BioData Mining* 8 (2015): 1–17.

Bouwman, A. F., G. Van Drecht, J. M. Knoop, A. H. W. Beusen, and C. R. Meinardi. "Exploring changes in river nitrogen export to the world's oceans." *Global Biogeochemical Cycles* 19, no. 1 (2005). GB1002, doi:10.1029/2004GB002314.

Braker, Gesche, Peter Dörsch, and Lars R. Bakken. "Genetic characterization of denitrifier communities with contrasting intrinsic functional traits." *FEMS Microbiology Ecology* 79, no. 2 (2012): 542–554.

Breitburg, Denise, Lisa A. Levin, Andreas Oschlies, Marilaure Grégoire, Francisco P. Chavez, Daniel J. Conley, Véronique Garçon et al. "Declining oxygen in the global ocean and coastal waters." *Science* 359, no. 6371 (2018): eaam7240.

Bush, Martin J. *Climate Change and Renewable Energy: How to End the Climate Crisis*. Springer Nature, 2019. ISBN 978-3-030-15423-3.

Carpinteri, Alberto, and Gianni Niccolini. "Correlation between the fluctuations in worldwide seismicity and atmospheric carbon pollution." *Science* 1, no. 1 (2019): 17.

Cavicchioli, Ricardo, William J. Ripple, Kenneth N. Timmis, Farooq Azam, Lars R. Bakken, Matthew Baylis, Michael J. Behrenfeld et al. "Scientists' warning to humanity: Microorganisms and climate change." *Nature Reviews Microbiology* 17, no. 9 (2019): 569–586.

Embree, Mallory, Joanne K. Liu, Mahmoud M. Al-Bassam, and Karsten Zengler. "Networks of energetic and metabolic interactions define dynamics in microbial communities." *Proceedings of the National Academy of Sciences* 112, no. 50 (2015): 15450–15455.

Eren, A. Murat, Özcan C. Esen, Christopher Quince, Joseph H. Vineis, Hilary G. Morrison, Mitchell L. Sogin, and Tom O. Delmont. "Anvi'o: An advanced analysis and visualization platform for 'omics data." *PeerJ* 3 (2015): e1319.

Falkowski, Paul G., Tom Fenchel, and Edward F. Delong. "The microbial engines that drive Earth's biogeochemical cycles." *Science* 320, no. 5879 (2008): 1034–1039.

Fisher, M. R. (Ed.). *Environmental Biology*. 3.2 Biogeochemical Cycles. (2019). (https://web.archive.org/web/20210927040314/https://openoregon.pressbooks.pub/envirobiology/chapter/3-2-biogeochemical-cycles/) 2021-09-27 at the Wayback Machine, OpenStax.

Galloway, J. N., Dentener, F. J., Capone, D. G., Boyer, E. W., Howarth, R. W., Seitzinger, S. P., ... and Vöosmarty, C. J. "Nitrogen cycles: Past, present, and future." *Biogeochemistry* 70, (2004): 153–226.

Galton, Douglas. "10th meeting: - 'Report of the royal commission on metropolitan sewage'." *RSA Journal* 33 (1884): 290.

Hartland, A., Lead, J. R., Slaveykova, V. I., O'Carroll, D., and Valsami-Jones, E. "The environmental significance of natural nanoparticles." *Nature education knowledge* 4, no. 8 (2013): 7.

Hasler, Arthur D. "Cultural eutrophication is reversible." *BioScience* 19, no. 5 (1969): 425–431.

Hug, Laura A., Brian C. Thomas, Itai Sharon, Christopher T. Brown, Ritin Sharma, Robert L. Hettich, Michael J. Wilkins, Kenneth H. Williams, Andrea Singh, and Jillian F. Banfield. "Critical biogeochemical functions in the subsurface are associated with bacteria from new phyla and little studied lineages." *Environmental Microbiology* 18, no. 1 (2016): 159–173.

Hutchins, David A., Janet K. Jansson, Justin V. Remais, Virginia I. Rich, Brajesh K. Singh, and Pankaj Trivedi. "Climate change microbiology—Problems and perspectives." *Nature Reviews Microbiology* 17, no. 6 (2019): 391–396.

Jickells, T. D., Erik Buitenhuis, Katye Altieri, A. R. Baker, D. Capone, Robert A. Duce, Frank Dentener et al. "A reevaluation of the magnitude and impacts of anthropogenic atmospheric nitrogen inputs on the ocean." *Global Biogeochemical Cycles* 31, no. 2 (2017): 289–305.

Keeling, R. F., and Shertz, S. R. "Seasonal and interannual variations in atmospheric oxygen and implications for the global carbon cycle." *Nature* 358, no. 6389 (1992): 723–727.

Kim, B. H., and Gadd, G. M. *Bacterial physiology and metabolism*. (2008). Cambridge University Press. New York, NY.

Lajtha, K., and Schlesinger, W. H. "The biogeochemistry of phosphorus cycling and phosphorus availability along a desert soil chronosequence." *Ecology* 69, no. 1 (1988): 24–39.

Leavitt, W. D., Halevy, I., Bradley, A. S., and Johnston, D. T. Influence of sulfate reduction rates on the Phanerozoic sulfur isotope record. *Proceedings of the National Academy of Sciences* 110, no. 28 (2013): 11244–11249.

Libes, Susan M. "Blue planet: The role of the oceans in nutrient cycling, maintaining the atmospheric system, and modulating climate change." In *Routledge Handbook of Ocean Resources and Management*, pp. 89–107. Routledge, 2015. ISBN 9781136294822.

Long, Philip E., Kenneth H. Williams, Susan S. Hubbard, and Jillian F. Banfield. "Microbial metagenomics reveals climate-relevant subsurface biogeochemical processes." *Trends in Microbiology* 24, no. 8 (2016): 600–610.

Madsen, Eugene L. "Microorganisms and their roles in fundamental biogeochemical cycles." *Current Opinion in Biotechnology* 22, no. 3 (2011): 456–464.

McCarren, Jay, Jamie W. Becker, Daniel J. Repeta, Yanmei Shi, Curtis R. Young, Rex R. Malmstrom, Sallie W. Chisholm, and Edward F. DeLong. "Microbial community transcriptomes reveal microbes and metabolic pathways associated with dissolved organic matter turnover in the sea." *Proceedings of the National Academy of Sciences* 107, no. 38 (2010): 16420–16427.

McMahon, Sean, and John Parnell. "Weighing the deep continental biosphere." *FEMS Microbiology Ecology* 87, no. 1 (2014): 113–120.

Mulder, Al, Astfid A. Van de Graaf, L. A. Robertson, and J. G. Kuenen. "Anaerobic ammonium oxidation discovered in a denitrifying fluidized bed reactor." *FEMS Microbiology Ecology* 16, no. 3 (1995): 177–183.

Murillo, Alejandro A., Verónica Molina, Julio Salcedo-Castro, and Chris Harrod. "Marine microbiome and biogeochemical cycles in marine productive areas." *Frontiers in Marine Science* 6 (2019): 657.

Natarajan, G., and Ting, Y. P. "Pretreatment of e-waste and mutation of alkali-tolerant cyanogenic bacteria promote gold biorecovery." *Bioresource technology* 152, (2014): 80–85.

Prentice, Iain Colin, G. D. Farquhar, M. J. R. Fasham, Michael L. Goulden, Martin Heimann, V. J. Jaramillo, H. S. Kheshgi et al. "The carbon cycle and atmospheric carbon dioxide." *Climate change 2001: the scientific basis, Intergovernmental panel on climate change*, (2001).

Rothman, Daniel H. "Atmospheric carbon dioxide levels for the last 500 million years." *Proceedings of the National Academy of Sciences* 99, no. 7 (2002): 4167–4171.

Rothman, Daniel. "Earth's carbon cycle: A mathematical perspective." *Bulletin of the American Mathematical Society* 52, no. 1 (2015): 47–64.

Schloss, Patrick D., Rene A. Girard, Thomas Martin, Joshua Edwards, and J. Cameron Thrash. "Status of the archaeal and bacterial census: An update." *MBio* 7, no. 3 (2016): 10–1128.

Solomon, Susan, ed. *Climate Change 2007–The Physical Science Basis: Working Group I Contribution to the Fourth Assessment Report of the IPCC*. Vol. 4. Cambridge University Press, 2007.

Stillman, Jonathon H., and Adam W. Paganini. "Biochemical adaptation to ocean acidification." *The Journal of Experimental Biology* 218, no. 12 (2015): 1946–1955.

Ulloa, Osvaldo, Donald E. Canfield, Edward F. DeLong, Ricardo M. Letelier, and Frank J. Stewart. "Microbial oceanography of anoxic oxygen minimum zones." *Proceedings of the National Academy of Sciences* 109, no. 40 (2012): 15996–16003.

Zakem, Emily J., Martin F. Polz, and Michael J. Follows. "Redox-informed models of global biogeochemical cycles." *Nature Communications* 11, no. 1 (2020): 5680.

7 Probiotics in Aquaculture

Suresh Chandra and Nupur Joshi

7.1 INTRODUCTION

Aquaculture, since its inception, has played a pivotal role in economic development and food security across the globe. China, India, Indonesia, Japan, and the Philippines were the top five producers of fish globally in 2020, with a total output of 178.5 MMT. By the year 2030, fish production meant for human consumption is estimated to reach 89%, aided collectively by an increase in urbanization, dietary trends focusing on enhanced nutrition, and advancements in post-harvest technologies (FAO 2020). However, intensive farming practices, along with stringent environmental regulations, limited natural resources, including land and water, and an increase in disease outbreaks, have resulted in a slowdown in terms of production.

An increase in disease outbreaks in recent years has led to a surge in the consumption of chemotherapeutics, which include drugs and antibiotics. These antimicrobial compounds are deployed as therapeutics, metaphylactics, prophylactics, growth promoters, and for the control of algal blooms (Lazado et al., 2015) in aquaculture practices. Despite their multitude of applications, antimicrobial agents pose collateral damage, including adverse ecological implications and promoting both co- and cross-resistance. Emergence of antimicrobial resistance (AMR) due to the indiscreet use of antibiotics is a major challenge in sustainable aquaculture, resulting in inferior quality of produce and economic losses (Schar et al., 2021 and Masoomi et al., 2019). Thus, to address this important issue, it becomes imperative to envisage alternative therapeutics, which include prebiotics, probiotics, alternative fish feeding strategies, enhanced fish health, and good management practices.

7.2 PROBIOTICS

The term "probiotic" is derived from the Greek words *pro* and *bios*, which mean "prolife." A probiotic is essentially a non-pathogenic microbial supplement that modulates the host immune system by improving microbial balance, stimulating the immune response, and deterring the growth of pathogenic organisms. Probiotics have been used in lieu of antimicrobial substances and incorporated into feed, which consequently leads to improved nutrition, aiding intestinal microflora, and stimulating the immune response against pathogenic bacteria (Feckaninova et al., 2017). In the year 2021, probiotics in the animal feed market were valued at 4.9 billion US dollars and are projected to reach 7.7 billion US dollars by 2030 (Spherical Insights, 2023). Probiotics may contain individual microbial species or consortia. Consortia are preferred over individual strains, as studies have reported that mixed probiotics are more efficient and provide a broad range of protection against various opportunistic pathogens. Commercial preparations of probiotics can be in either dry or liquid forms. Dry forms can be mixed with feed or applied directly to clean, disinfected water by blending them uniformly. Liquid forms are preferentially utilized by hatcheries, where they are added to hatchery tanks or blended with feed (Cruz et al., 2012).

7.3 TYPES OF PROBIOTICS

Based on their utility, probiotics can be segmented into two types.

 DOI: 10.1201/9781003408543-7

7.3.1 Feed Probiotics

This category of probiotics is administered orally to the host, either via blending or via encapsulation in the feed. These, when ingested by the host, enter the host's gut and colonize themselves, ultimately establishing and promoting the health of indigenous microflora and aiding in health supplementation.

7.3.2 Water Probiotics

These probiotics are directly introduced into aquatic systems, where they reduce organic pollution by assimilating organic waste and converting it into smaller fragments, assisting in water quality improvement. They also compete with other pathogenic microbial strains for nutrients, ultimately killing them due to starvation (Pandiyan et al., 2013).

7.4 SOURCES OF PROBIOTICS

Probiotics deployed in aquaculture can be derived from the host or non-host sources. The bacteria can be obtained from aquatic as well as terrestrial environments; the differences in the environments of both ecological niches play a crucial role in determining the efficacy of the probiotics (Pandiyan et al., 2013). The recent advent of aquatic-derived probiotics harnesses the potential of indigenous bacterial species to survive in their native environment. The commonly used probiotics utilized in aquaculture include the following bacterial species: *Bacillus*, *Clostridium*, *Enterococcus*, *Carnobacterium*, and yeast (Mao et al., 2020 and Van Doan et al., 2020).

7.5 SELECTION OF PROBIOTICS

Probiotics, because of their benefits, are utilized to maintain a homeostasis between the autochthonous beneficial and harmful bacterial populations present in fish guts. A probable candidate is initially screened for various parameters, which confirm its utility as a probiotic. Various in-vitro analyses that explore both intrinsic and extrinsic properties of probiotics are utilized in routine selection. Pathogen antagonism tests are performed, which encompass disc diffusion assays to check the production of antimicrobial compounds by probiotics; blood hemolysis, a characteristic property of pathogenic organisms, to identify non-pathogenic probiotics; and assays for assessing the tolerance of probiotics to the extreme microenvironment in fish gut and intestine (Hasan and Banerjee, 2020). Another critical test for potential probiotics is the adhesion assay, which includes adherence to solvents, hydrophobicity, and biofilm formation (Gomez-Gill et al., 2000).

An efficient probiotic should have a series of salient features that make it both effective and commercially viable:

a) It should support the growth and development of aquaculture species, conferring on them immunological advances.
b) It should be benign to the host, along with the capacity to maintain hereditary traits and the inability to develop drug resistance.
c) If incorporated as feed, it should be resistant to gastric juices, acid and bile tolerant, adhere to the surface, and promote fish gut health, aiding in the elevated immune system.
d) Apart from stimulating the fish immune response, it needs to have user-friendly application, tolerance toward drying processes, and viability during storage and packaging.

A microbial species fulfilling all these criteria forms a paradigmatic probiotic for use in sustainable aquaculture (Pereira et al., 2022).

7.6 MODE OF ACTION

The general mode of action of probiotics in humans and agriculture has been extensively studied and well documented. But aquatic probiotics and their mode of action are divided across different research publications and require more structured and precise elucidation. The general mode of action involves a range of mechanisms that contribute to improved health and performance of the cultured organisms.

7.6.1 Digestive Enzymes

Probiotics, when assimilated into the host, can result in the production of various exo-enzymes, which include carbohydrases, phosphatases, esterases, lipases, and peptidases, stimulating the intestinal health of the host system by facilitating the digestion process (Sotlani et al., 2019). Dietary non-starch polysaccharides (NSPs) form a major portion of fish feed and are comprised of lignin, cellulose, hemicellulose, and pectin. These fractions of NSPs have a depleted nutritional quotient due to their antinutritional properties and narrow range of digestibility (Kabir et al., 2020). The endogenous enzymes that degrade polysaccharides, especially cellulose, by hydrolyzing the glycosidic linkage and yielding glucose also contribute to enhanced digestibility. Similarly, various symbiotic gut bacteria degrade cellulose and hemicellulose through fermentation in the fish (Soltani et al., 2019). Additionally, gut bacteria also facilitate the breakdown of certain triglycerides, either directly or by modifying the activity of pancreatic lipase. The activity of digestive enzymes is an indicator of fish capacity to utilize and digest food, ultimately enhancing growth and resilience against pathogens. Probiotics are also efficient in remediating the adverse effects of the common feed additive soybean meal by relieving inflammation and modulating the gut microbiota of the host species (Liu et al., 2023).

7.6.2 Inhibitory Substances

The probiotic bacteria secrete various chemical compounds, like hydrogen peroxide, bacteriocins, lysozymes, siderophores, proteases, etc., which can have bacteriostatic or bactericidal effects on various pathogenic microbial populations (Kesarcodi et al., 2008). Lipopeptides, produced by certain microbial species, exert their antimicrobial activity by forming ion-conducting channels in bacterial cell membranes. Certain probiotic bacteria operate by altering the pH of the gastrointestinal (GI) tract by producing volatile fatty acids, organic acids, ammonia, and diacetyl, which deter the growth of other unwanted microbes. Recently, a potential antimicrobial compound, produced by certain probiotics including indole (2,3-benzopyrrole), has been utilized; it works by hampering the growth of opportunistic pathogens.

Some bacterial strains with potential probiotic use can exert antiviral effects; experimental data indicates that extracts of different bacterial strains can inactivate the virus. The exact mechanism that results in this antiviral action is still unknown and needs further exploration to be utilized to its full potential. Probiotics intrinsically inhibit the growth of other microbial species, which may inhibit the growth of viruses, as they need a live vehicle to cause infection. Probiotic candidates such as *Pseudomonas* sp. and *Vibrio* sp. are reported to be effective against infectious hematopoietic necrosis virus (IHNV), and *Lactobacillus* can be supplemented to prevent infection by lymphocystis disease virus (LCDV) (Hasan and Banerjee, 2020).

7.6.3 Immune Stimulation

Recent years have focused on improving the immune system in aquaculture species, due to which emphasis has been laid on different probiotics that can efficiently modulate the immune system of the host to confer protection against various pathogenic microbial strains. The immune response can be upregulated by administering probiotics. Several immunostimulants, such as lipopolysaccharides, glucans, chitin, teichoic acid, lipoteichoic acid, flagella, and microbial nucleic acids, are termed "microbial-associated molecular patterns" (MAMPs) that bind to Toll-like receptors (TLRs), which modulate expression of the immune system. Interaction of pattern recognition receptors (PRRs) with MAMPs translates to increased phagocytic activity through opsonization, resulting in the release of cytokines (like interleukin [IL]-6 and IL-4), which induce the production of tumor necrosis factor (TNF) and interferons (INF) from macrophages and dendritic cells (Dawood et al., 2020). Cytokines are protein molecules with an approximate size of 5–25 kDa that form an integral part of the adaptive and innate immune systems; these include ILs, interferon, TNFs, and transforming growth factors (Yang et al., 2021; Lazado et al., 2015).

Several *Bacillus* probiotics have been reported to prompt modifications in the cell physiology of the host, which may include enhanced neutrophil binding capacity and migration along with plasma bactericidal activity (Kuebutornye et al., 2019).

7.6.4 Antioxidant Production

The presence of an efficient antioxidant system in any host system is a measure of its immune efficiency. The events of phagocytosis lead to the generation of various reactive oxygen species (ROS), such as hydrogen peroxide, superoxide anion, and hydroxyl radical. The oxidative stress generated by these species is forestalled by the production of various antioxidant enzymes, including catalase, superoxide dismutase (SOD), glutathione reductase, and myeloperoxidase, which are upregulated by the consumption of probiotics, leading to the scavenging of ROS.

7.6.5 Interference with Quorum Sensing

Quorum sensing (QS) is a bacterial communication process that involves the production of a small signaling molecule called auto-inducer in response to fluctuations in the bacterial cell population. QS consists of two types of auto-inducer: the first group comprises N-acylhomoserine lactone (AHL) as a signaling molecule, and the second group contains a multichannel QS system. Disruption in the process of QS can be exploited to develop remedial processes to counteract opportunistic pathogens. Some probiotic strains, including *Bacillus* sp., produce quorum quenching (QQ) enzymes, such as lactonase, which degrades AHL, thereby interrupting the process of cell communication. This property of microbial species to produce QQ enzymes can be explored as an alternative biocontrol agent (Defoirdt et al., 2011).

7.6.6 Competition

Bacterial adhesion is a crucial step for the establishment of pathogenesis. Probiotics produce certain mucous-binding proteins that accelerate the adherence of probiotics to colonic cell lines. Bacterial antagonism is another mechanism in which probiotics produce several agents, such as bacteriocins, siderophores, lysozymes, and proteases, that prevent the adhesion of pathogens to the colonic epithelial cells. Alternatively, the probiotics compete with other unwanted bacterial strains for energy (Zorriehzahra et al., 2016). Assimilation of iron is an important parameter for the growth of marine microbes, so they produce siderophores to chelate iron from the surrounding environment. Probiotics present in the host system also produce siderophores, compete for available iron, and deprive pathogens of the same (Figure 7.1).

7.7 APPLICATIONS OF PROBIOTICS IN AQUACULTURE

Probiotics aim to improve the health, growth, and overall well-being of the aquaculture system. The application of probiotics is an environmentally sustainable process where the use of biotic agents mitigates several issues that arise due to the use of chemotherapeutants. They enhance the quality and quantity of produce, generating sustainable livelihoods.

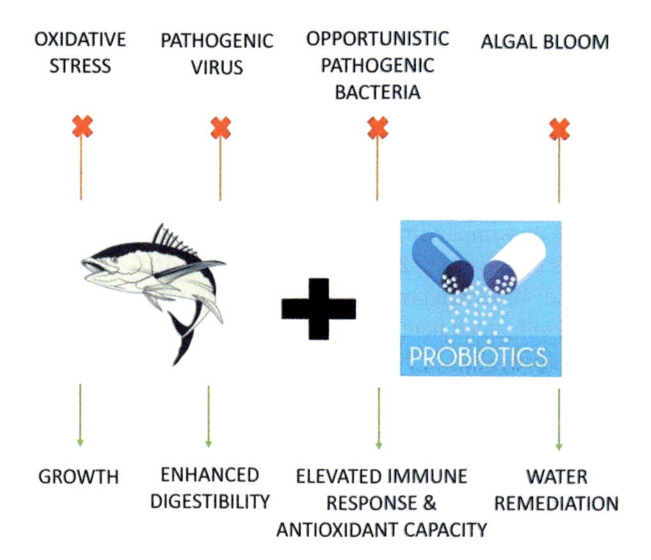

FIGURE 7.1 Role of probiotics in aquaculture.

7.7.1 Growth Promoter

The application of probiotics to an aquaculture system enhances the digestibility of the feed and elevates the general well-being of the host (Zhou et al., 2019). The use of autochthonous lactic acid bacteria (LAB) and *Bacillus* as probiotics in tilapia exhibited improved growth when supplemented with low-protein diets, consequently reducing the cost of farming (El Saadony et al., 2021). Dietary supplementation with heat-killed *Bacillus* also increases the digestibility of lipids, carbohydrates, and proteins through the release of certain digestive enzymes, which ultimately results in increased growth and feed utilization (Hasan et al., 2019).

Wu et al. (2020) supplemented sturgeon with the paraprobiotic HWFTM for three weeks. After three weeks, the sturgeon had an increase in weight as compared with the control. The abundance of *firmicutes*, *Lactococcus*, and *Clostridium* in probiotic-fed sturgeon indicated a healthy microbiota. Further, germ-free zebrafish were colonized, and with the gut microbiota of HWF TM-supplemented sturgeon, upregulation in the expression of genes related to growth, inflammation, and non-specific immunity was observed. Similarly, Choi et al. (2022) concluded that supplementation with probiotics resulted in a significantly higher growth rate in olive flounder in comparison to the control and antibiotic-treated fish populations. *C. butyricum* has a substantial probiotic benefit in aquaculture, as it promotes growth, immune response, and feed utilization (Tran et al., 2020, 2022). Similarly, probiotics have also been extensively used for the removal of heavy metals from the fish gut, a process defined as "gut remediation." These probiotics reduce the absorption of heavy metals through enhanced sequestration of heavy metals in the intestine, biotransformation of heavy metals, modulating the activity of metal transporter proteins, and protecting the gut barrier function (Kakade et al., 2022).

7.7.2 Water Quality Improvement

Elevated levels of ammonia and nitrite in aquaculture have deleterious consequences, leading to poor water quality and breeding grounds for various opportunistic pathogens. Probiotics in high concentrations can efficiently remediate the total ammonia, stabilizing the water quality (Zhang et al., 2022; Vershuere et al., 2000; Caipang and Lazado, 2015; Hasan and Banerjee, 2020). Several studies have elucidated the role of *Bacillus* sp. as a probable probiotic candidate with the potential to transform water quality parameters, where their performance is governed by various factors like mode of application, dissolved oxygen, pH, temperature, and nutrients (Hlordzi et al., 2020; Chauhan and Singh, 2019).

Li et al (2022), in their study, utilized two groups of mixed *Bacillus* that were introduced in crucian carp. The consortia of *Bacillus megaterium* and *subtilis* exerted a positive effect on water quality, reducing the levels of ammonia to 46.3% and total phosphorus to 80.3%. In addition to water remediation, the mixed probiotic was able to increase the microbial diversity of bacteria related to nitrogen and phosphorus removal (Rohani et al., 2022).

7.7.3 Enhanced Reproductive Health

Probiotics increase the digestibility of feed, which leads to enhanced nutrition, which ultimately influences growth, reproduction, and general fish health (Lall and Tibbets, 2009; Banerjee and Ray, 2017). During the reproduction period, female fish require more energy, and probiotics have been reported to increase fecundity. Several studies reveal that supplementation with probiotics resulted in a higher number of vitellogenic follicles along with improved oocyte competence and maturation (Carnevali et al., 2017).

Nargeshi et al. (2020) demonstrated the effects of dietary probiotics on the reproductive health of female rainbow trout (*Oncorhynchus mykiss*) brood stock. High levels of dietary probiotics (4×10^9) resulted in increased egg diameter, fertilization, and hatching rates. Similarly, in another study, supplementing the feed with *Bacillus* improved the gonad somatic index (GSI), and larvae from the same fish exhibited an increase in length and a higher survivability rate (Gioacchini et al., 2014). Probiotics can also exert a positive effect on spermatogenesis by increasing the levels of genes associated with it.

7.7.4 Effect on Phytoplankton

Supplementation of an aquatic ecosystem with probiotics helps to improve water quality and species composition. Probiotics work by controlling the processes of nitrification and denitrification, resulting in reduced ammonia levels, which ultimately hamper the growth of undesired algae. Several probiotics have also been reported to have an algicidal effect (Vershuere et al., 2000). Pant et al. (2019) evaluated a commercial probiotic in a set-up consisting of carp fingerlings for 90 days. At the end of the study, it was concluded that probiotics not only enhanced water quality and plankton concentration but also stimulated breakdown of the unused feed and checked algal bloom.

7.8 PREBIOTICS IN AQUACULTURE

Prebiotics are defined as non-digestible functional polysaccharides that stimulate the growth of indigenous bacteria which are responsible for production of short chain fatty acid (SCFA), ultimately leading to changes in blood lipids, immunomodulation, and promoting the functionality of the innate immune system. Prebiotics have also been reported to enhance the activity of various antioxidant enzymes, which indicates an efficient immune response.

Several prebiotics that are being widely used in aquaculture include insulin, oligo-fructose, xylo-oligosaccharide, fructo-oligosaccharide (FOS), mannan-oligosaccharide, galacto-oligosaccharide (GOS), and beta glucans (Dawood et al., 2020; Rohani et al., 2022).

7.9 SAFETY ONSIDERATIONS

For decades, probiotics have been used in the food industry, and to date, there has been no report of the deleterious effects of probiotics on human health. But the field of aquaculture in terms of probiotics is in its nascent stage, and continuous monitoring is imperative for effective functionality. Various theoretical implications for health have been hypothesized in suspected hosts, which

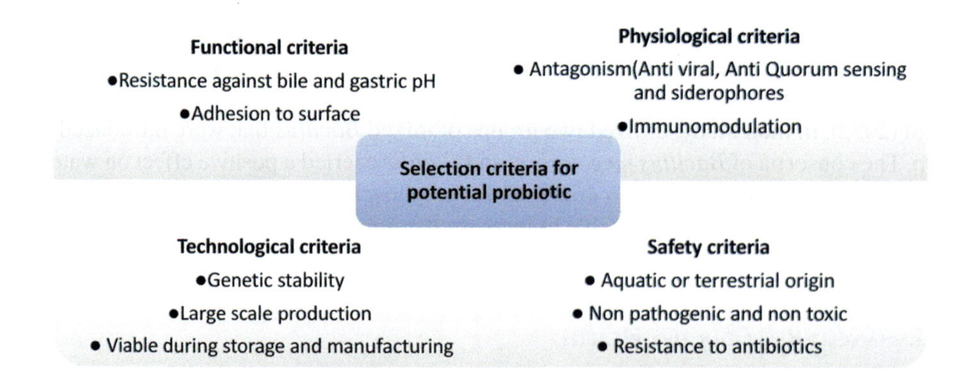

FIGURE 7.2 Various criteria for selection of potential probiotics in aquaculture.

include system infection, harmful metabolic activities, and overstimulation of the immune system. Nonetheless, there is a requirement to continuously monitor and evaluate the intrinsic properties of strains being utilized as probiotics, their pharmacokinetics, and their interactions with macro- and microenvironments (Wang et al., 2019) (Figure 7.2).

7.9.1 Quality Control

With the emergence of antimicrobial resistance (AMR), the Food and Agriculture Organization (FAO) has placed emphasis on the use of probiotics to enhance quality and improve production rates in aquaculture (Zaman and Cho, 2023). A candidate microbial strain, once identified as a potential probiotic, needs to be evaluated for various features. Before large-scale manufacturing, it becomes mandatory to develop monitoring tools that can rationalize its production and application (Verschuere et al., 2000). The bacterial strain being used as a probiotic should maintain its genetic stability and remain unaltered during its course of action.

The problem of inactivation of probiotics during their storage and processing is another bottleneck that is presented to aquaculture; hence, there is a need to develop better technology, stringent quality control, and good manufacturing practices so that the quality of probiotics during production and packaging is conserved. Several techniques are being used to assess the composition of probiotics, which include polymerase chain reaction-denaturing gradient gel electrophoresis/temperature gradient gel electrophoresis (PCR-DGGE/TGEE) and fluorescent in-situ hybridization (FISH) (Qi et al., 2009).

7.9.2 Novel Delivery Methods

To preserve the characteristics of probiotics, systems are being developed for immobilization, especially using encapsulation. Encapsulation increases the bioavailability of active compounds by increasing their uptake by the host system. Encapsulation can be done in the form of either microcapsules or microspheres using various methods such as emulsification, extrusion, spray drying, and freeze drying (Cruz et al., 2012).

Microbial cells with high density are entrapped using a colloidal matrix, which can be alginate, chitosan, carboxymethyl cellulose (CMC), pectin, gellan gum, gelatin, or whey protein. Encapsulation using alginate matrices has demonstrated resistance against the acidic pH and digestive enzymes in the gut, maintaining the integrity of probiotics. In recent advancements, viral proteins like cages have been utilized as biomaterials for nano encapsulation (Masoomi et al., 2019).

7.10 CONCLUSION AND FUTURE PROSPECTS

Probiotics in aquaculture practices have enormous potential to enhance the health of the environment and fish. Probiotics serve as essential intermediaries in the aquatic environment, offering a variety of advantages that promote long-term aquaculture productivity. With the increase in demand and preference for safe food fish, there is an urgent need to explore biogenic entities for improved fish health. The use of probiotics in aquaculture unveils new dimensions for improving the quality of farmed products. Fish health and welfare are integral parts of fish farming. The use of probiotics results in enhanced growth, reproduction, water quality, and digestibility of feed. It also minimizes the dependency on chemotherapeutants, ultimately leading to the generation of a sustainable farming system with no deleterious effects. The development of an efficient probiotic presents some key issues that need to be addressed, including genetic stability, shelf life, interaction with the ambient environment, and adaptability in the host system. There is a need to establish a multi-perspective approach through the development of control or in-vitro experiments to quantify the benefits conferred by probiotics on the host system. Additionally, more research is needed in the area of the role of probiotics in reproductive health and the elucidation of their mode of action in more depth for a comprehensive study. Investigating particular strains, doses, and application methods is essential to maximizing the benefits of probiotics. Sustainable production can be achieved by using probiotics as a key instrument in aquaculture management.

REFERENCES

Banerjee, Goutam, and Arun Kumar Ray. "The advancement of probiotics research and its application in fish farming industries." *Research in Veterinary Science* 115 (2017): 66–77.

Caipang, Christopher Marlowe A., and Carlo C. Lazado. "Nutritional impacts on fish mucosa: Immunostimulants, pre-and probiotics." In *Mucosal Health in Aquaculture*, pp. 211–272. Academic Press, 2015.

Carnevali, Oliana, Francesca Maradonna, and Giorgia Gioacchini. "Integrated control of fish metabolism, wellbeing and reproduction: The role of probiotic." *Aquaculture* 472 (2017): 144–155.

Chauhan, Arun, and Rahul Singh. "Probiotics in aquaculture: A promising emerging alternative approach." *Symbiosis* 77, no. 2 (2019): 99–113.

Choi, Wonsuk, Mohammad Moniruzzaman, Jinho Bae, Ali Hamidoghli, Seunghan Lee, Youn-Hee Choi, Taesun Min, and Sungchul C. Bai. "Evaluation of dietary probiotic bacteria and processed yeast (Gropro-aqua) as the alternative of antibiotics in juvenile olive flounder Paralichthys olivaceus." *Antibiotics* 11, no. 2 (2022): 129.

Dawood, Mahmoud A. O., Haitham G. Abo-Al-Ela, and Md Tawheed Hasan. "Modulation of transcriptomic profile in aquatic animals: Probiotics, prebiotics and synbiotics scenarios." *Fish & Shellfish Immunology* 97 (2020): 268–282.

Defoirdt, Tom, Patrick Sorgeloos, and Peter Bossier. "Alternatives to antibiotics for the control of bacterial disease in aquaculture." *Current Opinion in Microbiology* 14, no. 3 (2011): 251–258.

El-Saadony, Mohamed T., Mahmoud Alagawany, Amlan K. Patra, Indrajit Kar, Ruchi Tiwari, Mahmoud A. O. Dawood, Kuldeep Dhama, and Hany M. R. Abdel-Latif. "The functionality of probiotics in aquaculture: An overview." *Fish & Shellfish Immunology* 117 (2021): 36–52.

FAO. 2020. "The State of World Fisheries and Aquaculture 2020. Sustainability in action". Rome. https://doi.org/10.4060/ca9229en.

Fečkaninová, Adriána, Jana Koščová, Dagmar Mudroňová, Peter Popelka, and Julia Toropilova. "The use of probiotic bacteria against Aeromonas infections in salmonid aquaculture." *Aquaculture* 469 (2017): 1–8.

Gioacchini, Giorgia, Elisabetta Giorgini, Lisa Vaccari, and Oliana Carnevali. "Can probiotics affect reproductive processes of aquatic animals?." *Aquaculture Nutrition: Gut Health, Probiotics and Prebiotics* (2014): 328–346.

Gomez-Gil, Bruno, Ana Roque, and James F. Turnbull. "The use and selection of probiotic bacteria for use in the culture of larval aquatic organisms." *Aquaculture* 191, no. 1–3 (2000): 259–270.

Hasan, Kazi Nurul, and Goutam Banerjee. "Recent studies on probiotics as beneficial mediator in aquaculture: A review." *The Journal of Basic and Applied Zoology* 81, no. 1 (2020): 1–16.

Hasan, Md Tawheed, Won Je Jang, Bong-Joo Lee, Kang Woong Kim, Sang Woo Hur, Sang Gu Lim, Sungchul C. Bai, and In-Soo Kong. "Heat-killed *Bacillus* sp. SJ-10 probiotic acts as a growth and humoral innate immunity response enhancer in olive flounder (Paralichthys olivaceus)." *Fish & Shellfish Immunology* 88 (2019): 424–431.

Hlordzi, Vivian, Felix K. A. Kuebutornye, Gyamfua Afriyie, Emmanuel Delwin Abarike, Yishan Lu, Shuyan Chi, and Melody A. Anokyewaa. "The use of *Bacillus* species in maintenance of water quality in aquaculture: A review." *Aquaculture Reports* 18 (2020): 100503.

Kabir, K. A., M. C. J. Verdegem, J. A. J. Verreth, M. J. Phillips, and J. W. Schrama. "Dietary non-starch polysaccharides influenced natural food web and fish production in semi-intensive pond culture of Nile tilapia." *Aquaculture* 528 (2020): 735506.

Kakade, Apurva, Monika Sharma, El-Sayed Salama, Peng Zhang, Lihong Zhang, Xiaohong Xing, Jianwei Yue et al. "Heavy metals (HMs) pollution in the aquatic environment: Role of probiotics and gut microbiota in HMs remediation." *Environmental Research* (2022): 115186.

Kesarcodi-Watson, Aditya, Heinrich Kaspar, M. Josie Lategan, and Lewis Gibson. "Probiotics in aquaculture: The need, principles and mechanisms of action and screening processes." *Aquaculture* 274, no. 1 (2008): 1–14.

Kuebutornye, Felix K. A., Emmanuel Delwin Abarike, and Yishan Lu. "A review on the application of *Bacillus* as probiotics in aquaculture." *Fish & Shellfish Immunology* 87 (2019): 820–828.

Lall, Santosh P., and Sean M. Tibbetts. "Nutrition, feeding, and behavior of fish." *Veterinary Clinics of North America: Exotic Animal Practice* 12, no. 2 (2009): 361–372.

Lauzon, Hélène L., Arkadios Dimitroglou, Daniel L. Merrifield, Einar Ringø, and Simon J. Davies. "Probiotics and prebiotics: Concepts, definitions and history." In *Aquaculture Nutrition: Gut Health, Probiotics and Prebiotics* Ed(s) Daniel L. Merrifield and Einar Ringo. Published by @2014 John Willey & Sons Ltd (2014): 169–184.

Lazado, Carlo C., Christopher Marlowe A. Caipang, and Erish G. Estante. "Prospects of host-associated microorganisms in fish and penaeids as probiotics with immunomodulatory functions." *Fish & Shellfish Immunology* 45, no. 1 (2015): 2–12.

Li, Xue, Tianjie Wang, Baorong Fu, and Xiyan Mu. "Improvement of aquaculture water quality by mixed *Bacillus* and its effects on microbial community structure." *Environmental Science and Pollution Research* 29, no. 46 (2022): 69731–69742.

Liu, Zi-Yan, Hong-Ling Yang, Xi-Yue Ding, Sha Li, Guo-He Cai, Ji-Dan Ye, Chun-Xiao Zhang, and Yun-Zhang Sun. "Commensal *Bacillus* siamensis LF4 ameliorates β-conglycinin induced inflammation in intestinal epithelial cells of Lateolabrax maculatus." *Fish & Shellfish Immunology* 137 (2023): 108797.

Mao, Qing, Xueliang Sun, Jingfeng Sun, Feng Zhang, Aijun Lv, Xiucai Hu, and Yongjun Guo. "A candidate probiotic strain of Enterococcus faecium from the intestine of the crucian carp Carassius auratus." *AMB Express* 10, no. 1 (2020): 1–9.

Martínez Cruz, Patricia, Ana L. Ibáñez, Oscar A. Monroy Hermosillo, and Hugo C. Ramírez Saad. "Use of probiotics in aquaculture." *International Scholarly Research Notices* 2012 (2012) Volume 2012, Article ID 916845, 13 pages.

Masoomi Dezfooli, Seyedehsara, Noemi Gutierrez-Maddox, Andrea Alfaro, and Ali Seyfoddin. "Encapsulation for delivering bioactives in aquaculture." *Reviews in Aquaculture* 11, no. 3 (2019): 631–660.

Nargesi, Akbari Erfan, Bahram Falahatkar, and Mir Masoud Sajjadi. "Dietary supplementation of probiotics and influence on feed efficiency, growth parameters and reproductive performance in female rainbow trout (Oncorhynchus mykiss) broodstock." *Aquaculture Nutrition* 26, no. 1 (2020): 98–108.

Pandiyan, Priyadarshini, Deivasigamani Balaraman, Rajasekar Thirunavukkarasu, Edward Gnana Jothi George, Kumaran Subaramaniyan, Sakthivel Manikkam, and Balamurugan Sadayappan. "Probiotics in aquaculture." *Drug Invention Today* 5, no. 1 (2013): 55–59.

Pant, Bonika, Vibha Lohani, Ashutosh Mishra, M. D. Trakroo, and Hema Tewari. "Effect of probiotic supplementation on growth of carp fingerlings." *National Academy Science Letters* 42 (2019): 215–220.

Pereira, Wellison Amorim, Anna Carolina M. Piazentin, Rodrigo Cardoso de Oliveira, Carlos Miguel N. Mendonça, Yara Aiko Tabata, Maria Anita Mendes, Ricardo Ambrósio Fock et al. "Bacteriocinogenic probiotic bacteria isolated from an aquatic environment inhibit the growth of food and fish pathogens." *Scientific Reports* 12, no. 1 (2022): 5530.

Qi, Zizhong, Xiao-Hua Zhang, Nico Boon, and Peter Bossier. "Probiotics in aquaculture of China—Current state, problems and prospect." *Aquaculture* 290, no. 1–2 (2009): 15–21.

Rohani, Md Fazle, S. M. Majharul Islam, Md Kabir Hossain, Zannatul Ferdous, Muhammad A. B. Siddik, Mohammad Nuruzzaman, Uthpala Padeniya, Christopher Brown, and Md Shahjahan. "Probiotics, prebiotics and synbiotics improved the functionality of aquafeed: Upgrading growth, reproduction, immunity and disease resistance in fish." *Fish & Shellfish Immunology* 120 (2022): 569–589.

Schar, Daniel, Cheng Zhao, Yu Wang, D. G. Joakim Larsson, Marius Gilbert, and Thomas P. Van Boeckel. "Twenty-year trends in antimicrobial resistance from aquaculture and fisheries in Asia." *Nature Communications* 12, no. 1 (2021): 5384.

Soltani, Mehdi, Koushik Ghosh, Seyed Hossein Hoseinifar, Vikash Kumar, Alan J. Lymbery, Suvra Roy, and Einar Ringø. "Genus *Bacillus*, promising probiotics in aquaculture: Aquatic animal origin, bioactive components, bioremediation and efficacy in fish and shellfish." *Reviews in Fisheries Science & Aquaculture* 27, no. 3 (2019): 331–379.

Spherical Insights LLP. "Global probiotics in animal feed market size to grow USD 10.7 billion by 2032 | CAGR of 7.6%. GlobeNewswire News Room. (2023, May 22). https://www.globenewswire.com/news-release/2023/05/22/2673699/0/en/Global-Probiotics-in-Animal-Feed-Market-Size-To-Grow-USD-10-7-Billion-By-2032-CAGR-of-7-6.html

Tran, Ngoc Tuan, Wei Yang, Xuan Truong Nguyen, Ming Zhang, Hongyu Ma, Huaiping Zheng, Yueling Zhang, Kok-Gan Chan, and Shengkang Li. "Application of heat-killed probiotics in aquaculture." *Aquaculture* 548 (2022): 737700.

Tran, Ngoc Tuan, Zhongzhen Li, Hongyu Ma, Yueling Zhang, Huaiping Zheng, Yi Gong, and Shengkang Li. "Clostridium butyricum: A promising probiotic confers positive health benefits in aquatic animals." *Reviews in Aquaculture* 12, no. 4 (2020): 2573–2589.

Van Doan, Hien, Seyed Hossein Hoseinifar, Einar Ringø, Maria Angeles Esteban, Maryam Dadar, Mahmoud A. O. Dawood, and Caterina Faggio. "Host-associated probiotics: A key factor in sustainable aquaculture." *Reviews in Fisheries Science & Aquaculture* 28, no. 1 (2020): 16–42.

Verschuere, Laurent, Geert Rombaut, Patrick Sorgeloos, and Willy Verstraete. "Probiotic bacteria as biological control agents in aquaculture." *Microbiology and Molecular Biology Reviews* 64, no. 4 (2000): 655–671.

Wang, Anran, Chao Ran, Yanbo Wang, Zhen Zhang, Qianwen Ding, Yalin Yang, Rolf Erik Olsen, Einar Ringø, Jérôme Bindelle, and Zhigang Zhou. "Use of probiotics in aquaculture of China—A review of the past decade." *Fish & Shellfish Immunology* 86 (2019): 734–755.

Wu, Xuexiang, Tsegay Teame, Qiang Hao, Qianwen Ding, Hongliang Liu, Chao Ran, Yalin Yang et al. "Use of a paraprobiotic and postbiotic feed supplement (HWF™) improves the growth performance, composition and function of gut microbiota in hybrid sturgeon (Acipenser baerii x Acipenser schrenckii)." *Fish & Shellfish Immunology* 104 (2020): 36–45.

Yang, J., Li, Y., Wen, Z., Liu, W., Meng, L., and Huang, H. "Oscillospira - a candidate for the next-generation probiotics". *Gut Microbes* 13, no. 1. https://doi.org/10.1080/ 19490976.(2021). 1987783

Zaman, Md Farid Uz, and Sung Hwoan Cho. "Dietary inclusion effect of various sources of phyto-additives on growth, feed utilization, body composition, and plasma chemistry of olive flounder (Paralichthys olivaceus), and challenge test against Edwardsiella tarda compared to a commercial probiotic (super lacto®)." *Journal of the World Aquaculture Society* 54, no. 5 (2023): 1121–1136.

Zhang, Yingying, Tongwei Ji, Yinan Jiang, Chen Zheng, Hui Yang, and Qiuning Liu. "Long-term effects of three compound probiotics on water quality, growth performances, microbiota distributions and resistance to Aeromonas veronii in crucian carp Carassius auratus gibelio." *Fish & Shellfish Immunology* 120 (2022): 233–241.

Zhou, Sheng, Deli Song, Xiaofeng Zhou, Xinliang Mao, Xuefeng Zhou, Sunli Wang, Jingguang Wei et al. "Characterization of *Bacillus* subtilis from gastrointestinal tract of hybrid Hulong grouper (Epinephelus fuscoguttatus× E. lanceolatus) and its effects as probiotic additives." *Fish & Shellfish Immunology* 84 (2019): 1115–1124.

Zorriehzahra, Mohammad Jalil, Somayeh Torabi Delshad, Milad Adel, Ruchi Tiwari, K. Karthik, Kuldeep Dhama, and Carlo C. Lazado. "Probiotics as beneficial microbes in aquaculture: An update on their multiple modes of action: A review." *Veterinary Quarterly* 36, no. 4 (2016): 228–241.

8 Microbiome in Aquaponics *Emergent Roles and Potential Applications*

Rameshori Yumnam and Maibam Birla Singh

8.1 INTRODUCTION

Microorganisms are of great importance for healthy aquaponics. They recycle nutrients, degrade organic matter, enrich nutrients, and increase mineralization (Goddek et al., 2019; Joyce et al., 2019; Kasozi et al., 2021; Schmautz et al., 2022; Bartelme et al., 2018; Yep and Zheng, 2019). They are the drivers of a symbiotic ecosystem relationship in an aquaponic system. They play a critical role in the system's health and affect the overall performance of the system. Also, in certain cases, they may have a negative impact on the system, such as lowering the plant productivity. Further, they may cause infection in plants and fish, leading to failure of the system. Recent studies on aquaponics have focused on unraveling the complex microbial communities or the microbiota and, further, the microbiome (Kasozi et al., 2021; Schmautz et al., 2022; Bartelme et al., 2018). However, in the context of aquaponics, the definition of microbial communities or microbiota and microbiome is less clear. Microbial communities or microbiota have commonly been defined as the collection of microorganisms living together in a given region, whether it is a specific body, habitat, or ecosystem (Berg et al., 2020). Another definition of microbiome is simply "a community of commensal, symbiotic, and pathogenic microorganisms within a body space or other environment" (Lederberg and McCray, 2001; Lorgen-Ritchie et al., 2023). A microbiome comprises a community of microbes, including bacteria, viruses, fungi, microeukaryotic and metazoan parasites, and archaea (collectively known as the microbiota) as well as the downstream products and functionality of the microbiota (Lorgen-Ritchie et al., 2023). Thus, the microbiome is the collection of genomes and gene products of the microbiota residing within the host or environment (Berg et al., 2020). Further, "microbiome" defines a microbial community with distinct properties and functions and its interactions with its environment, resulting in the formation of specific ecological niches. In the case of aquatic ecosystems, microbial communities mainly include bacteria, fungi and protozoa. There is a complex and dynamic microbiota associated with fish and plants in an aquaponic system that can participate in host metabolism, development and immune response. The microbiota in an aquaponic system can reflect the ecological interactions and can play an important role in water quality regulation and system stability, and influence fish and plant growth and health. In recent years, researchers have made considerable efforts to understand the microbial communities critical for the smooth functioning of an aquaponic system. This has resulted in an increasing number of publications detailing the identification, characterization and description of the microbial communities. There exist comprehensive reviews on microorganism distribution, diversity, and monitoring in the context of aquaculture and recirculating aquaculture system (RAS) bio-filtration, excluding the plant hydroponic unit. However, there is an increasing need to shift from exploring and cataloguing the microbial communities towards understanding the factors affecting the emergence of microbial communities, and their structure and function. A shift from microorganisms to microbiomes is a new development in aquaponics.

DOI: 10.1201/9781003408543-8

The use of microorganisms in the aquaculture industry is not new (Martinez-Cordova et al., 2016; Bentzon-Tilia et al., 2016), and microorganisms are extensively used in aquacultural wastewater treatments and bioremediation (Kim et al., 2020; Yang et al., 2019). For example, in nature, biological nitrogen fixation, which essentially drives the nitrogen cycle improving soil nutrients, is due to heterotrophic bacteria, and this is a key process on which our agriculture depends (Nardi et al., 2020). Some aquacultural systems that utilize microbes include bioflocs (Avnimelech, 2015; Bagi et al., 2023; Li et al., 2023), FLOCponics (Pinho et al., 2022) and aquaponics (Joyce et al., 2019; Kasozi et al., 2021; Schmautz et al., 2022; Bartelme et al., 2018; Yep and Zheng, 2019). It may be noted that the success of these aquaculture systems depends on the growth, sustenance and interaction of microbial communities and their interaction with environmental conditions. The importance of microorganisms has been increasingly recognized for their role in the health and overall performance of aquatic organisms, providing an opportunity in the aquaculture industry for efficient utilization of resources for sustainability.

At present, there is increasing interest in aquaponics microbiome research due to the important role of bacterial communities. Aquaponics is an ecosystem involving a symbiotic interaction between fish, microbes and plants in a micro-environment. It is based on the concept of integrating farming methods with a view to reusing the waste water for plant growth and thereby, reducing the environmental impact. The working principle of aquaponics is based on the concept of RAS and the use of nutrient-rich aquacultural wastewater as a source of nitrogenous compounds for growing plants in a hydroponic culture unit, which in the process depurates the water that is returned to the aquaculture tanks (Goddek et al., 2019; Yep and Zheng, 2019; Singh and Yumnam, 2023). It simply combines the advantages of RAS and hydroponic systems and overcomes the individual standalone drawbacks. The system results in a symbiosis between fish, microorganisms, and plants and encourages sustainable use of water and nutrients, including their recycling. Within this synergistic interaction, the respective ecological weaknesses of aquaculture and hydroponics are converted into strengths. This substantially minimizes the need for input of mineral fertilizers and output of waste, unlike when they are run as separate systems. The role of microbes lies at the heart of aquaponic systems (Figure 8.1).

Recent research highlights the interrelation of microbiota in fish tanks, bio-filters and hydroponic units (Joyce et al., 2019; Kasozi et al., 2021; Schmautz et al., 2022; Bartelme et al., 2018). This has further emphasized the need to understand the complex role and interconnected nature of microbial communities at the microbiome level. However, the full potential of microbiomes in making aquaponics a sustainable waste-to-wealth technology and the creation of a bio-circular economy

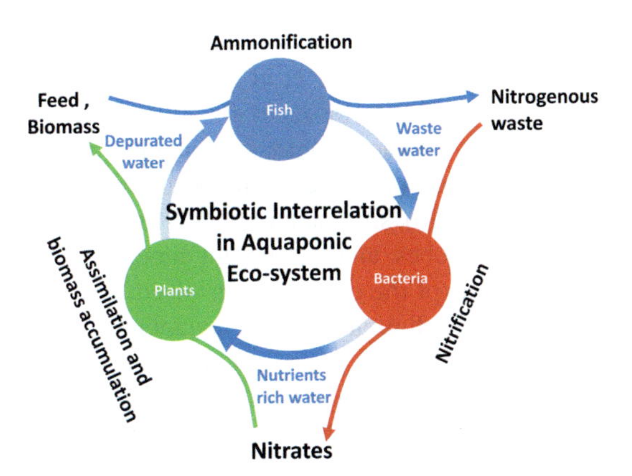

FIGURE 8.1 Schematic diagram showing material flow and the symbiotic interrelation between plants, fish and bacteria in an aquaponic ecosystem. (From Singh, M.B. and Rameshori, Y., *Fisheries and Aquaculture of the Temperate Himalayas,* 2023.)

is yet to materialize. This chapter brings together the current status of microbiome research pertaining to aquaponics and the existing knowledge, gaps, and challenges involved in understanding the symbiotic ecosystem relationship in fish, plants, and associated microbiomes. This is done with a view to providing a holistic approach integrating the multi-process of nitrogen fixation, the phosphorus cycle, sludge digestion and mineralization via microbiome assembly assisting crop production, which will have important implications for the aquaponic system's health and long-term sustainability. Particularly, we review the microbes identified, their classification and the diversity of the microbial communities, and the associated microbial role in nitrogen fixation, nutrient cycling, sludge digestion and mineralization. We also outline the emergent potential applications of the aquaponics microbiome in the biocontrol of pests and its uses as fish feed and for the improvement of fish health. We conclude with a discussion on possible applications of the microbiome in aquaponic systems for sustainability.

8.2 IMPORTANCE OF MICROBIOME IN AQUAPONICS

An aquaponic system consists of many subunits, and each compartment has distinct process conditions, which are caused by different configurations, dimensions, and flow rates (Schmautz et al., 2022; Bartelme et al., 2017; Eck et al., 2019b; Menezes-Blackburn et al., 2021). On account of the different environmental conditions in the subunits, this may result in the development of unique conditions suitable for localized microbial communities. As a result of the interconnection between the compartments and the flow of water, the localized communities may interact with or influence each other (Bartelme et al., 2019, 2018), which can result in changing process conditions over time. This process may enhance or reduce productivity, and the understanding of compartment-specific microbial communities becomes paramount in terms of water quality, fish health and growth, plant health, and food safety (Schmautz et al., 2022; Bartelme et al., 2018; Blancheton et al., 2013). The bacterial community in an aquaponic system is diverse, with different families and genera, each occurring more specifically in the different components of the system (Schmautz et al., 2017; Schmautz et al., 2022; Bartelme et al., 2017, 2018, 2019; Eck et al., 2019b; Menezes-Blackburn et al., 2021). Table 8.1 shows the list of various bacteria in the bio-filter and plant roots in a hydroponic system, the periphyton from a fish tank, fish feces and a sump tank. The knowledge of the microbial diversity and functional distribution of the microbes in aquaponics subsystems is the key to understanding microbial community dynamics and interactions and to enhance the performance of the system. Recent microbial research on aquaponics has focused on bacterial diversity in periphyton, plant roots, bio-filter and fish feces (Schreier et al., 2010; Eck et al., 2019b; Schmautz et al., 2017; Hu et al., 2015; Wongkiew et al., 2018). Further, microorganisms in aquaponics subunits are involved in multiple processes that include nitrification, organic matter decomposition, denitrification, phosphorus mineralization and iron cycling (Schreier et al., 2010; Eck et al., 2019b; Schmautz et al., 2017; Hu et al., 2015; Wongkiew et al., 2018; Kasozi et al., 2021). The crucial establishment of the bacterial ecosystem during the setup of an aquaponics system undermines the successful operation of aquaponics and can be done both in the presence and in the absence of fish (Brailo et al., 2019; De Long and Losordo, 2012; Sallenave, 2016), since aquaponics is an ecosystem with a symbiotic relationship between fish, plants and microbes, and the subunit component environment is engineered specifically to fulfill the requirements of the organisms involved, i.e., microorganisms, plants and fish. Due to the design, function and configuration of the aquaponics system, each of the individual units presents a different micro-environment. These conditions will influence the various microbial processes occurring in each compartment and thus, also affect the chemistry of water. Since a symbiotic balance between fish, microbes and plants is essential for the successful operation of an aquaponics system (including fish welfare and plant growth), it is paramount to know how different microbes in the subunit interact. As the microbes and microbial communities form an essential component of an aquaponics ecosystem, understanding their role in converting the fish waste in water to usable nitrates for plant growth is of the utmost importance.

TABLE 8.1

Summary of Compartments, Functions, Their Targeted Parameters, Niches, Biological Process and Predominant Contributing Taxa (Microorganism Phyla/Family/Order/Genera) in Aquaponic System

	Compartment										
Microorganism	**FT**	**BF**	**FF**	**ST**	**DS**	**HP**	**RT**	**SP**	**Host/Source**	**Microbial Process Involved**	**Ref.**
Abditibacteriota (FBP)		+						+	Aquaculture wastewater		
Acidibacter						+	+		Plant (root)	Iron cycling	Kasozi et al. (2020)
Acidobacteria	+	+				+	+	+	Plant (root)	Role in nitrogen cycle	Schmautz et al. (2017), Bartelme et al. (2019), Kasozi et al. (2021)
Acidovorax spp		+						+		Iron Cycling	Eck et al. (2019a)
Actinobacteria	+	+	+			+				Mineralisation, organic matter degradation	Schmautz et al. (2017)
Actinomycetospora							+		Plant (root)	Associative nitrogen fixation	Kasozi et al. (2020)
Aeromicrobium							+		Plant (root)	Nitrogen fixation	Kasozi et al. (2020)
Alcaligenaceae	+								Fish (faeces)	Nitrate reduction	Schmautz et al. (2022)
Aquaspirillum			+						Fish (faeces)		Schmautz et al. (2017)
Arcobacter		+							Aquaculture wastewater	Mineralisation, dinitrification	Eck et al. (2019a), Wongkiew et al. (2018)
Bacteroidetes	+		+			+	+	+	Fish (guts and feces)	Nitrification, denitrification	Schmautz et al. (2017), Eck et al. (2019a), Wongkiew et al. (2018)
Bacillus						+			Plants (Carrot, chrysanthemum, cucumber, lettuce, pepper tomato)		Lee and Lee (2015)
Blastococcus							+			Decomposition of organic matter	Kasozi et al. (2020)
Brocadia	+									Anaerobic ammonium oxidation (Anammox)	Schmautz et al. (2022)

(*Continued*)

TABLE 8.1 (CONTINUED)

Summary of Compartments, Functions, Their Targeted Parameters, Niches, Biological Process and Predominant Contributing Taxa (Microorganism Phyla/Family/Order/Genera) in Aquaponic System

Microorganism	Compartment								Host/Source	Microbial Process Involved	Ref.
	FT	BF	FF	ST	DS	HP	RT	SP			
Burkholderiales		+					+		Plants roots	Plants disease protection	Eck et al. (2019a), Wongkiew et al. (2018), Schmautz et al. (2017)
Candidatus saccharibacteria	+									Degradation of organic compounds under aerobic-, nitrate reducing- and anaerobic conditions.	Schmautz et al. (2017)
Chloracidobacterium	+									Photosynthetic bacteria	Schmautz et al. (2022)
Chlorobi	+							+	Fish		Wang et al. (2023)
Chloroflexi *Herpetosiphonaceae*		+					+	+	Aquaculture wastewater	Organic matter decomposition	Eck et al. (2019a)
Clavibacter		+						+	Fish (catfish and eel)	Phytopathogenic	Eck et al. (2019a)
Cetobacterium spp			+						Fish faeces	Fish health improvement	Schmautz et al. (2022)
Clostridium			+						Fish guts and feces	Fish health improvement	Schmautz et al. (2022)
Comamonadaceae						+	+		Plant root	Bioremediation	Schmautz et al. (2022)
Cyanobacteria	+					+		+	Fish and plant	Bacterial Photosynthesis	Schmautz et al. (2017), Schmautz et al. (2022)
Deinnococcus-Thermus,	+	+				+		+	Fish and plant	Bioremediation	Schmautz et al. (2017), Eck et al. (2019a)
Deinococcaceae Meiothermus	+	+				+		+	Fish	Bioremediation	Schmautz et al. (2017)
Devosia spp								+	Plant root	Nitrogen fixiation	Eck et al. (2019a)
Dokdonella		+					+		Aquaculture wastewater	Denitrification	Schmautz et al. (2017), Kasozi et al. (2020)
Enterobacter							+		Plant (cucumber)		Lee and Lee (2015)
Enterobacteriales			+						Fish faeces		Schmautz et al. (2017)
Firmicutes *Cetobacterium spp*	+	+	+			+		+	Fish guts (tilapia)	Nitrification	Schneider et al. (2007), Schmautz et al. (2017)

(Continued)

TABLE 8.1 (CONTINUED)
Summary of Compartments, Functions, Their Targeted Parameters, Niches, Biological Process and Predominant Contributing Taxa (Microorganism Phyla/Family/Order/Genera) in Aquaponic System

Microorganism	Compartment								Host/Source	Microbial Process Involved	Ref.
	FT	BF	FF	ST	DS	HP	RT	SP			
Flavobacterium							+		Plant root	Mineralisation, degrade complex organic molecules, denitrification	Eck et al. (2019a), Wongkiew et al. (2018)
Flavobacteriales							+		Plant root (lettuce)	Mineralisation, degrade complex organic molecules	Schmautz et al. (2017)
Flexivirga						+	+		Plant root	Degradation of organic matter	Kasozi et al. (2020)
Friedmanniella						+	+		Plant root	Production of antimicrobial substances	Kasozi et al. (2020)
Fusibacter		+								Sulphate reduction	Schreier et al. (2010), Somerville et al. (2014)
Fusobacteria		+	+			+		+			Schmautz et al. (2017)
Germmatimonadetes	+	+				+	+	+	Fish, Plant roots and Aquaculture wastewater		Schmautz et al. (2017), Kasozi et al. (2021)
Gracilibacteria (GN02)		+						+	Aquaculture waste water		
Gliocladium						+			Plant (cucumber, tomato)	Promoting Plant Health	Lee and Lee (2015)
Heliimonas							+		Plant root	Breakdown of complex organic compounds	Kasozi et al. (2020)
Herpetosiphonaceae						+	+		Plant root		Schmautz et al. (2022)
Hymenobacter							+		Plant root	Decomposition of plant residues	Kasozi et al. (2020)
Hyphomicrobium		+							Aquaculture waste water	Bioremediation	Schmautz et al. (2022)
Jatrophihabitans							+		Plant roots	Production of secondary metabolites	Kasozi et al. (2020)
Macellibacteroides			+						Fish faeces		Schmautz et al. (2022)
Marmoricola							+		Plant root	Denitrification	Kasozi et al. (2020)
Methanonobacteriacea		+							Aquaculture wastewater	Methanogenesis	Schmautz et al. (2022)
Methanoscarcinaceae		+							Aquaculture waste water	Methanogenesis	Schmautz et al. (2022)
Methanoregulaceae		+							Aquaculture wastewater	Methanogenesis	Schmautz et al. (2022)

(Continued)

TABLE 8.1 (CONTINUED)

Summary of Compartments, Functions, Their Targeted Parameters, Niches, Biological Process and Predominant Contributing Taxa (Microorganism Phyla/Family/Order/Genera) in Aquaponic System

	Compartment										
Microorganism	**FT**	**BF**	**FF**	**ST**	**DS**	**HP**	**RT**	**SP**	**Host/Source**	**Microbial Process Involved**	**Ref.**
Methylophilales						+			Plant root (lettuce)		Hu et al. (2015), Schmautz et al. (2017), Wongkiew et al. (2018)
Microbacteriaceae		+					+	+	Fish (catfish)		Eck et al. (2019a), Wongkiew et al. (2018)
Modestobacter							+		Plant root	Phosphate solubilization	Kasozi et al. (2020)
Nitrobacter		+							Aquaculture wastewater	Nitrite oxidation	Kasozi et al. (2020)
Nitrosomonadales		+					+		Aquaculture wastewater	Nitrification	Kasozi et al. (2020)
Nitrosomonas	+				+				Aquaculture wastewater	Nitrification	Ebeling et al. (2006),
Nitrosococcus		+							Aquaculture waste water	Nitrite oxidation	Rurangwa and
Nitrosospira	+	+				+	+	+	Aquaculture wastewater	Nitrification, COMAMMOX (COMplete AMMonia OXidation)	Verdegem (2015), Schreier et al. (2010), Schmautz et al. (2017),
Nitrosolobus		+				+			Aquaculture waste water	Nitrification	Daims et al. (2015),
Nitrosovibrio		+							Aquaculture wastewater	Nitrification	Bartelme et al. (2019)
Novosphingobium spp								+	Aquaculture waste water	Bioremediation	Eck et al. (2019a), Schmautz et al. (2022)
Nitrospina		+					+	+	Aquaculture wastewater	Nitrite oxidation	Ebeling et al. (2006), Schmautz et al. (2017), Wongkiew et al. (2018)
Nitrososphaeras							+		Plant root	Nitrification	Bartelme et al. (2019)
Nocardioides							+		Plant root	Production of antibiotics, lignocellulose decomposition	Kasozi et al. (2020)
Opitutus							+		Plant root	Denitrification, polysaccharide degradation	Kasozi et al. (2020)

(Continued)

TABLE 8.1 (CONTINUED)

Summary of Compartments, Functions, Their Targeted Parameters, Niches, Biological Process and Predominant Contributing Taxa (Microorganism Phyla/Family/Order/Genera) in Aquaponic System

Microorganism	Compartment								Host/Source	Microbial Process Involved	Ref.
	FT	BF	FF	ST	DS	HP	RT	SP			
Planctomycetes	+	+				+	+	+	Fish, plant, aquaculture wastewater	ANAMMOX (Anaerobic ammonium oxidation)	Hu et al. (2015), Schmautz et al. (2017), Wongkiew et al. (2018)
Plesiomonas spp.			+						Fish faeces		Schmautz et al. (2017)
Proteobacteria	+	+	+			+	+	+	Fish, plant, aquaculture wastewater	Nitrification, denitrification	Hu et al. (2015), Schmautz et al. (2017), Wongkiew et al. (2018)
Pontibacter							+		Plant root	Nitrogen fixation	Kasozi et al. (2020)
Porphyromonadaceae	+		+						Fish gut, faeces, Aquaculture wastewater	Fish health improvement	Larsen et al. (2014)
Pseudomonadota (*Dokdonella spp*) *Comamonadaceae*	+	+							Aquaculture waste water	Fish health improvement	Schmautz et al. (2017)
Pseudomonas						+			Plant (bean, carnation, chickpea, cucumber, lettuce, pepper, potato, radish, tomato	Mineralisation, solubilisation of P and K.	Hu et al. (2015), Schmautz et al. (2017), Wongkiew et al. (2018), Lee and Lee (2015), Eck et al. (2019a)
Pseudomonadales							+		Plant root (lettuce)		Schmautz et al. (2017)
Pseudonocardia							+		Plant root	Nitrogen fixation	Kasozi et al. (2020)
Ralstonia spp								+	Aquaculture waste water	Improvement of fish health	Eck et al. (2019a)
Rhizobiaceae								+	Rhizobiaceae	Nitrogen fixing	Eck et al. (2019a)
Rhizobiales	+	+							Aquaculture waste water	Nitrogen fixation	Schmautz et al. (2017), Bartelme et al. (2019)
Rhodobacteraceae *Rhodobacter*	+								Aquaculture wastewater	Bacterial Photosynthesis, Bioremediation	Schmautz et al. (2022)

(Continued)

TABLE 8.1 (CONTINUED)

Summary of Compartments, Functions, Their Targeted Parameters, Niches, Biological Process and Predominant Contributing Taxa (Microorganism Phyla/Family/Order/Genera) in Aquaponic System

	Compartment										
Microorganism	**FT**	**BF**	**FF**	**ST**	**DS**	**HP**	**RT**	**SP**	**Host/Source**	**Microbial Process Involved**	**Ref.**
Rhodocyclales		+						+	Fish(catfish)	Nutrient recycling (Phosphorus cycle)	Eck et al. (2019a), Wongkiew et al. (2018)
Rhodoluna							+		Aquaculture wastewater	Ability to produce secondary metabolites that have antibiotic properties	Kasozi et al. (2020)
Rudaea							+		Aquaculture waste water	Decomposition of plant residues	Kasozi et al. (2020)
Saprospiraceae			+					+	Fish (eel)	Breakdown of complex organic compound in the environment	Eck et al. (2019a)
Sphingobacterium		+		+				+	Aquaculture waste water	Mineralisation	Eck et al. (2019a), Wongkiew et al. (2018)
Sphingomonadaceae		+							Aquaculture wastewater	Nutrient Recycling (Nitrogen cycle)	Schmautz et al. (2022)
Sphingomonasspp								+	Aquaculture wastewater	Bioremediation	Eck et al. (2019a)
Streptomyces	+					+			Plant (cumber, tomato)		Lee and Lee (2015)
Synechococcus	+								Aquaculture water	Bacterial photosynthetic process	Schmautz et al. (2022)
Thermomonas spp	+				+				Wastewater	Denitrification	Schmautz et al. (2017)
Trichoderma						+	+		Plant (bean, cotton, cucumber, maize, rice)	Degradation of organic matter	Lee and Lee (2015)
Turicibacter			+						Intestinal tract fish		Schmautz et al. (2022)
Verrucomicrobia		+	+			+		+	Fish Gut	Biogeochemical cycle	Eck et al. (2019a), Wongkiew et al. (2018)
Xanthomonadales	+								Fish faeces		Schmautz et al. (2017)

FT: Fish tank (Periphytons), BF: Bio-filter, FF: Fish feces, ST: Radial flow settler tank, SD: Anaerobic sludge digester, HP: Hydroponic compartment, RT: Root zone and mineralized zone, SP: Sump.

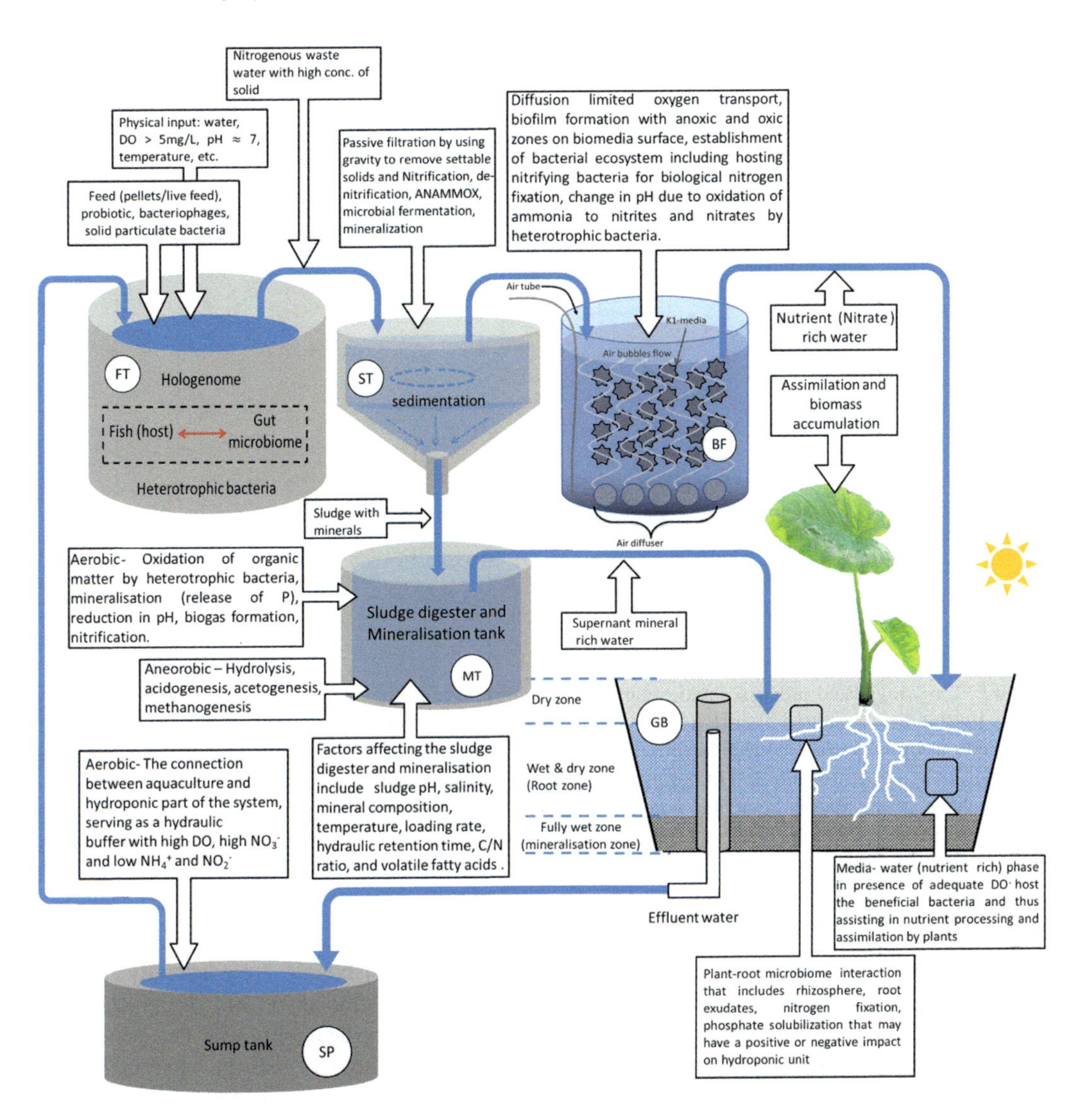

FIGURE 8.2 Schematic conceptual diagram of a model aquaponic system showing various processes in compartments that are responsible for sustaining the microbial communities associated with the inputs (feed, probiotics, bacteriophages, and solid particulate bacteria), fish (host)–gut interaction in fish tank (FT), bio-filter (BF), sump (SP) and grow bed (GB) in a hydroponic system in a symbiotic aquaponic ecosystem.

The microbial communities in an aquaponic ecosystem may be listed as (i) microbes in fish tanks, (ii) microbes in bio-filter, (iii) fish gut and feces microbiota, (iv) microbes in hydroponic sub-units and roots of plants, (v) microbes in sump tank, (vi) microbes in sludge digester and in mineralization tank, and (vii) feed-associated microbiota. In Figure 8.2, we show the various components and their associated processes pertaining to microbial communities in an aquaponic system. The role of the various compartments in connection with the microbial communities and diversity is discussed in the next section.

8.2.1 Role of Compartments in Aquaponics and Microbial Diversity

8.2.1.1 Fish Tank

The primary purpose of a fish tank is to hold the fish and to provide the essential conditions for fish survival. The conditions for fish survival include dissolved oxygen (DO) close to or at saturation,

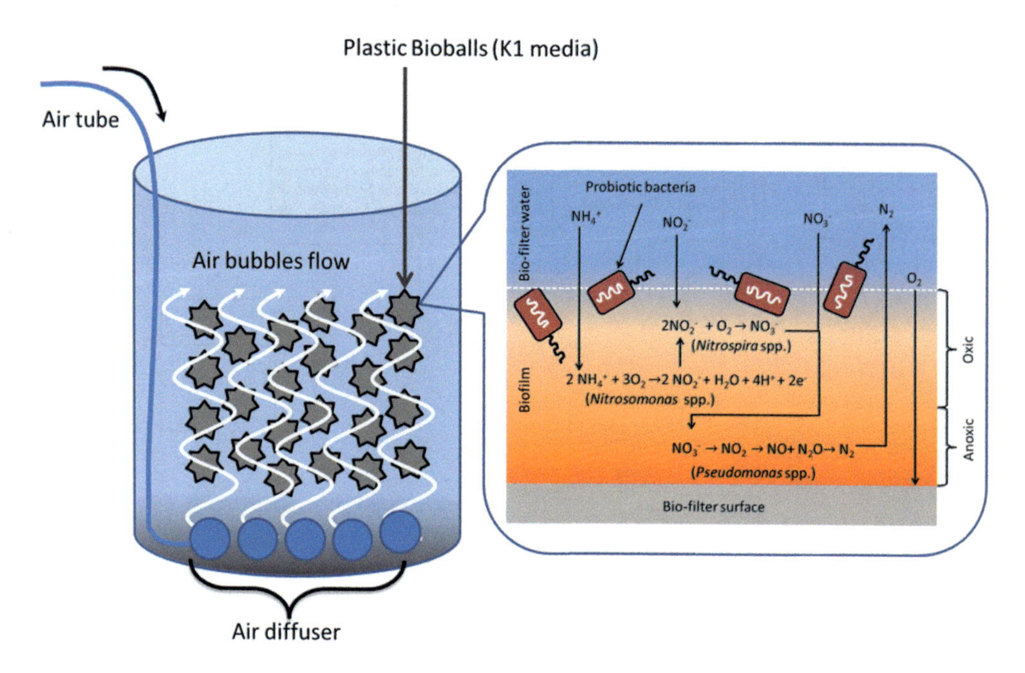

FIGURE 8.3 Suggested processes and organisms to incorporate in the design of a synthetic biofilm community in a biological aerated filter (BAF) for use in microbial reconditioning of rearing water. Denitrification is carried out in the anoxic, bottom layer by heterotrophs or autotrophs, whereas nitrification takes place in the upper, oxic part of the biofilm. Other processes and organisms could be included, e.g., anammox bacteria or archaea, in the bottom layer, and potentially, probiotic bacteria could be embedded in the upper layer, seeding the rearing water upon release from the biofilm. (Adapted from Bentzon-Tilia, M. et al., *Microbial Biotechnology*, 9, 5, 576–584, 2016.)

low total nitrogen, total organic carbon, NH4+, and NO2–. Since fish tanks are constantly aerated, the process of nitrification and aerobic respiration takes place (Schmautz et al., 2021). Also, the system design, mode of operation and hydraulic loading rate affect the quality of water, which further also affects the distribution of microbes (Endut et al., 2010). Table 8.1 lists the bacteria that are found in a fish tank and some microbial processes with which they are associated.

8.2.1.2 Bio-filter

The bio-filter is specifically designed to host microbes and is one of the most important and studied compartments of RAS and aquaponics. In general, a bio-filter is a compartment that includes a fixed medium for microbial attachments and growth, i.e., via biofilms or allows microbial growth to be suspended (Avnimelech, 2006; Gutierrez-Wing and Malone, 2006). The bio-filter integrates aerobic and anaerobic microbial processes for the elimination of nitrogenous waste products (like ammonia and total ammonia nitrogen (TAN)) excreted by fish and the carbon and nitrogen from uneaten feed and fecal matter (Avnimelech, 2006; Gutierrez-Wing and Malone, 2006; van Rijn et al., 2006). Figure 8.3 shows the major processes associated with an aerated bio-filter. The largest diversity of bacteria is seen in the bio-filter, showing that it is one of the most important components of an aquaponic system (Schreier et al., 2010; Schumatz et al., 2017; Eck et al., 2019b). Various bacteria that are found in the bio-filter and associated microbial processes are listed in Table 8.1.

8.2.1.3 Hydroponic Subunit and Plant Microbes

The hydroponic unit in an aquaponic system acts as a subunit to hold and sustain the growth of plants. The hydroponic units are considered to host their own microbial communities, showing density and diversity between systems design and type of substrate (Lee and Lee, 2015), nutrient sources and types of plants. There are three types of plant-rearing units used in aquaponics: (i) deep water culture (DWC), (ii) media bed (MB), and nutrient film techniques (NFT). These three hydroponic systems have their own advantages and disadvantages and can host microbial communities differently (Junge et al., 2020). The media bed is composed of a medium (gravel, perlite, sand, etc.) that is specifically designed to host a wide range of bacteria as well as provide a medium for root growth. The media bed provides a niche for a diverse population of micro- and macro-organisms with high bacterial diversity and spatial distribution of microbes (Kasozi et al., 2020). Distinct microbial communities and interaction with the host plant occur at different locations in the (i) root zone and (ii) media-water (nutrient-rich) phase in the wet and dry zones and (iii) mineralized zone in the fully wet zone due to physico-chemical variation in environmental parameters like pH, electrical conductivity (EC), sunlight, irrigation, temperature and root exudates. The flood and drain media bed can maintain high oxygen concentration, high NO_3^-, low NH_4^- and NO_2^- for efficient nitrification, better plant nutrient uptake, carbon input by rhizodeposits and good microbial respiration. About 32 phyla, with 17 phyla having relative abundance >0.5%, are reported (Kasozi et al., 2020). The bacteria with their specific functions are listed in Table 8.1. The microbial community in hydroponics mainly resides on inert substrates and in plant roots (Bartelme et al., 2018; Thomas et al., 2023). The microbial communities can promote the growth and health of plants. These microbial communities, also known as plant growth-promoting microorganisms (PGPM), have emerged as a great resource that can contribute to the precise management of nutrients, control of plant stress, and protection of plants from disease (Dhawi, 2023; Mourouzidou et al., 2023). The region around the root where PGPM reside is also called the rhizosphere, which can produce exudates and molecules that can act as a bacterial inoculum for enhancing nutrient availability and improving fish and plant health (Lee and Lee, 2015; Rivas-Garcia et al., 2020; Dhawi, 2023; Mourouzidou et al., 2023; Chiaranunt and White et al., 2023). However, their exact mechanism is still not yet known. It is important to understand the role of PGPM in the context of aquaponics.

8.2.1.4 Sludge Digester and Mineralization Tank

A very distinctive feature of aquaponics is the use of aquacultural wastewater and sludge for growing food and in the process minimizing waste. The sludge digester is a subunit where unused nutrients consisting of fish feces, uneaten feed and other solid waste rich in nitrogen and phosphorus in the aquaponic system are accumulated. The nutrients in the sludge are due to the deposition of organic matter and accumulation in the due course of aquaponic cycling. The nutrient cycling, recovery from the sludge and mineralization are done by the microorganisms present in the system (Eck et al., 2019b; Schmautz et al., 2017; Wongkiew et al., 2018; Bartelme et al., 2018). Mineralization is the breaking down of complex substances like proteins, carbohydrates and lipids into smaller molecules like amino acids, sugars, fatty acids or alcohol first and then into minerals in order to close the elemental cycles (van Lier et al., 2008). However, many of the nutrients in sludge are not readily available in the forms accessible by plants. At present, most aquaponic systems can only recycle liquid effluent, and a large quantity of nutrients is filtered out as sludge (Zhanga et al., 2021). The potential of nutrient recovery from fish sludge in aquaponics has been anticipated in various studies (Cerozi and Fitzsimmons, 2017; Goddek et al., 2019; Delaide et al., 2017; Gichana et al., 2018; Monsees et al., 2017; Mirzoyan et al., 2010; Delaide et al., 2018). The physical conditions in the sludge digester are an important factor in deciding the composition of microbes, which in turn affects the nutrient recovery process. Studies by Strauch et al. (2018) indicated that 7.1–9.9% of the fish feed input is left as sediment. Cerozi and Fitzsimmons (2017) estimated that 25–35% of the feed might remain in the water as suspended solids and daily sludge, which accounts for 5–20% of the total volume of a recirculating aquaculture system. According to Fu et al. (2018), the residual feed and fish feces, 17% of the dry matter is protein, 3% is carbohydrates, and the remaining 62% consists of other components. Cripps and Bergheim (2000) mentioned that the sludge solids contain 7–32% of the total nitrogen and 30–84% of the total phosphorus of the wastewater. Goddek et al. (2019) noticed that among macronutrients, 6% of the nitrogen, 18% of the phosphorus, 6% of the potassium, 16% of the calcium and 89% of the magnesium are contained in the sludge; while for micronutrients, 24% of the iron, 86% of the manganese, 47% of the zinc and 22% of the copper were measured (based on dry matter). The phosphorus is higher in sludge, as reported by Delaide et al. (2017). Most macronutrients and micronutrients that are essential for the plant can be obtained from aquaculture solid waste (Eck et al., 2019a). Thus, from an economic as well as an environmental perspective, the recovery of nutrients from fish sludge and their reintroduction in aquaponics can enhance the sustainability of an aquaponics system (Zhanga et al., 2021). The present methods of nutrient recovery via the mineralization of fish sludge include the use of aerobic and anaerobic digesters involving various microbial communities (Joyce et al., 2019; Delaide et al., 2019; Mirzoyan et al., 2010). In an aerobic digester, the fish sludge is constantly aerated using an air pump or blower to create favourable conditions for organic decomposition by heterotrophic bacteria along with ammonia-oxidizing bacteria, while nitrite-oxidizing bacteria are used to break down organic matter in solids and release minerals (Joyce et al., 2019; Delaide et al., 2019; Mirzoyan et al., 2010; Monsees et al., 2017; Khiari et al., 2019; Zhanga et al., 2021).The heterotrophic organisms decompose organic matter in aquaculture sludge, producing ammonium (NH_4^+) through a process called ammonification, which is followed by conversion of ammonium to nitrate (NO_3^-) by autotrophic bacteria through the process of nitrification. In this process, there is an increase in the respiration of heterotrophic bacteria, with an increase in CO_2, which further dissolves in water and forms carbonic acid, which dissociates and lowers the pH. Since NO_3^- is the preferred source of nitrogen for a wide variety of plant species, this process has a significant impact on the overall aquaponics system performance. Also the nitrification process can also contribute to a decrease in pH. The pH drop in the aerobic digestion process can promote the mineralization of bound minerals trapped in the sludge. Although aerobic mineralization can achieve high nitrogen nutrient recovery efficiency, nitrogen loss still occurs because the accumulation of nitrate may promote denitrification or anaerobic ammonium oxidation in a local anaerobic environment (Monsees et al., 2017; Zhanga et al., 2021).

Kiari et al. (2019) employed a microbial-assisted aerobic bioprocess, using endogenous heterotrophic and nitrifying bacteria for bioconversion of aquaculture solid waste into liquid fertilizer. Apart from the heterotrophic and nitrifying bacteria, the possible use of anoxygenic phototrophic bacteria (APB) has been suggested to recover nutrients from the environment (George et al., 2020) and aquaculture sludge in aquaponics (Zhanga et al., 2021). A review of the possible phototrophic bioconversion of aquaculture sludge from aquaponics and nutrient recovery is provided by Zhanga et al. (2021).

8.2.1.5 Aerobic Degradation

Another method of sludge processing is anaerobic digestion. In anaerobic digestion, there is a biological process of degradation of organic matter by microbes under anaerobic conditions, which is carried out by facultative and obligatory anaerobic bacteria (Appels et al., 2008; Delaide et al., 2019). The role of bacteria is crucial, and the digestion is done at one of three temperature regimes: thermophilic (45–65 C), mesophilic (25–45 C) and psychrophilic (10–25 C) (Marchaim, 1992), employing a variety of methanogenic archaea (Garcia et al., 2000). During the process of anaerobic digestion, the organic sludge undergoes considerable changes in its physical, chemical and biological properties (Appels et al., 2008; Delaide et al., 2019). Anaerobic digestion is divided into four main stages: hydrolysis, acidogenesis, acetogenesis, and methanogenesis (Figure 8.4). The by-products of anaerobic digestion are biogas composed of methane and carbon dioxide with small levels of hydrogen sulfide and ammonia (Appels et al., 2008). Apart from the microbial community involved in the anaerobic digester, the biogas quality is found to be affected by the concentration of organic matter (Choudhury et al., 2023), temperature, C/N ratio (Mir et al., 2016), pH, salinity, loading rate, hydraulic retention time and volatile fatty acid content (Delaide et al., 2019). Thus, understanding the anaerobic digestion mediated by novel microbial consortia in relation to the factors affecting

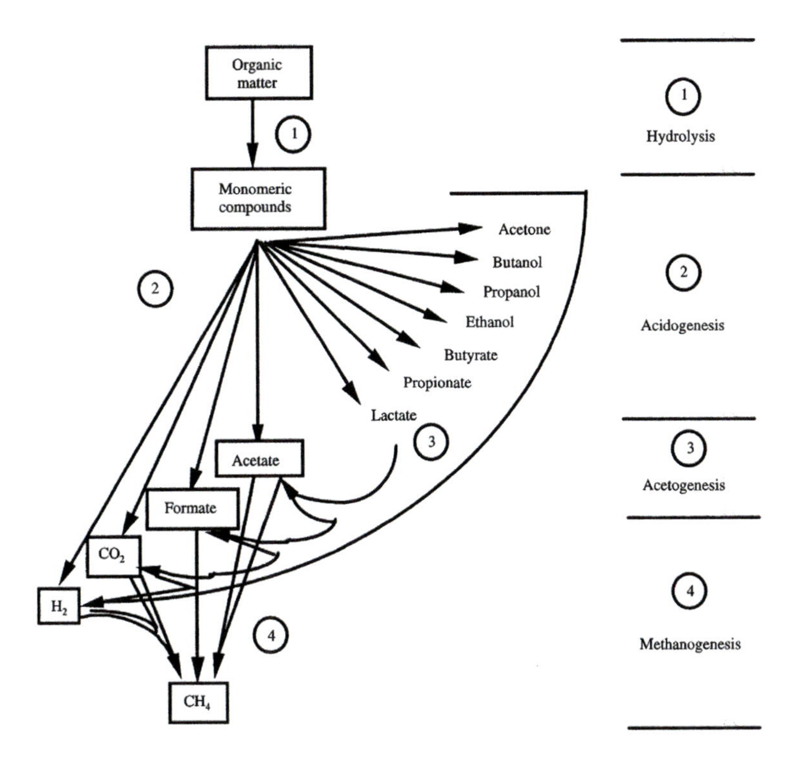

FIGURE 8.4 Schematic diagram showing anaerobic degradation of organic matter. (Based on Garcia, J-L. et al., *Anaerobe*, 6, 4, 205–226, 2000.)

biogas productivity is very important. This can help convert different wastes to green energy and make aquaponics more economically sustainable.

8.2.2 Factors That Influence Microbiome in Aquaponics

In aquaponics, the compartments are designed to host various microbial communities. The microbial diversity depends on nutrient availability and environmental conditions. The conditions in each compartment are different, and thus, the microbial interactions in each compartment must also be different. It is reported by Eck et al. (2019) that the different aquaponics system setups had a large influence on the bacterial communities, and the diversity of bacterial communities present in coupled and decoupled aquaponic systems is also different. There is limited research on microbes and microbiota interactions among the compartments in aquaponics. The establishment of a microbiome composition in an aquaponics system is a complex process that involves not only host factors, such as genotype and the host immune system, but also external environmental influences, such as diet (Perry et al., 2020; Robaina et al., 2019), physical parameters of culture systems (Goddek et al., 2019; Bartleme et al., 2019; Schmautz et al., 2022; Bartelme et al., 2018; Singh and Yumnam, 2023), microbes in the aquatic environment (Kasozi et al., 2021), and physico-chemical parameters of water, including temperature, pH and dissolved oxygen (Kasozi et al., 2021; Singh and Yumnam, 2023). Apart from these factors, the operation of aquaponics and design can play a role in the growth and sustenance of microbial communities.

The environmental factors are influenced by the micro-environment (Bartelme et al., 2019), system biography (Bartelme et al., 2019), and water parameters (Bartelme et al., 2019; Kasozi et al., 2021; Goddek et al., 2019; Schmautz et al., 2022; Singh and Yumnam, 2023). In all the subunits, water acts as a reservoir of microorganisms that constitute the microbiome in aquaponics.

The physico-chemical water parameters – pH, temperature, DO, salinity, chemical oxygen demand (COD), total nitrogen, total phosphorus, total carbon and inorganic nitrogen (C/N ratio) – are factors that affect the bacterial activity in an aquaponic system (Kasozi et al., 2021).

8.2.2.1 pH

In an aquaponic system, the pH is the main factors that controls microbial activity and affects nutrient availability. It can affect the health of both fish and plants (Wang et al., 2023). The pH of water affects the nitrification efficiency of bacteria. Nitrification is more efficient at pH 7.5 or higher and practically ceases at pH values less than 6.0 (Gerardi, 2006; Singh and Yumnam, 2023; Yep and Zheng, 2019). In wastewater, the biological oxidation of NH_4^+ and NO_2^-, and the activity of nitrifiers, decreases when the pH is below 6.4 or above 9.0 (Ruiz et al., 2003; Wongkiew et al., 2018). Also, the nitrogen utilization efficiency (NUE) in a media-based aquaponic system was found to reach a maximum of 50.9% when the pH was maintained at 6.4, and NUE dropped to 47.3% and 44.7% when the pH was increased to 7.4 and 8.0, respectively (Zou et al., 2016; Wongkiew et al., 2018). If the pH is below 6.0, this also affects the solubility of micronutrients such as calcium, phosphorus, potassium, magnesium, molybdenum, etc., which affects the bioavailability of nutrients for uptake by plants (Raviv et al., 2008; Schwarz, 1995). Lowering the pH can also promote the mineralization of bound minerals trapped in sludge (Delaide et al., 2019; Monsees et al., 2017). For example, if the pH is lower than 6, phosphorus can be recovered (Goddek et al., 2018). Usually, plants and fish have different pH requirements; therefore, it is recommended that pH = 7.0 should be maintained in an aquaponics system. However, the optimum nutrient availability pH range of 5 to 6 is by far the best suited for plants (Singh and Yumnam, 2023).

8.2.2.2 Temperature

The temperature of water is one of the important critical parameters in an aquaponics system. Water temperature affects the health of the fish, growth of plants, bacteria and thus, the microbiome in an aquaponics system. A proper temperature range is essential for the breakdown of sludge, uptake of nutrients by plants, and mineralization of wastes generated in aquaponics. It is essential to maintain a temperature range of 14–34 C for proper growth of bacteria (Gerardi, 2006), and the optimal growth temperature for most nitrifiers ranges between 25 and 30 C; ammonia-oxidizing bacteria (AOB) populations increase faster than nitrate-oxidizing bacteria (NOB) within this temperature range (Sallenave, 2016; Tyson et al., 2011). Also, high temperature can restrict the absorption of nutrients, such as calcium, by plants. Moreover, the temperature of water has an effect on DO as well as on the toxicity (ionization) of ammonia (Timmons et al., 2018) and, hence, the pH. The nitrification rate depends on the temperature, and a temperature change from 20 to 10 C showed an average decrease of 58% in the nitrification rate (Kasozi et al., 2021). Further, it is reported that *Nitrosomonas* and *Nitrospira* are dominant at 10 C and that *Nitrosospira* dominates at 5 C, suggesting that *Nitrosospira* might be more resistant to low-temperature stress conditions (Karkman et al., 2011; Kasozi et al., 2021). The role of temperature in sludge digestion and mineralization is also very important. Particularly in an anaerobic digester, sludge digestion is done at one of three temperature regimes: thermophilic (45–65 C), mesophilic (25–45 C) and psychrophilic (10–25 C) (Marchaim, 1992). During the process of anaerobic digestion, the organic sludge undergoes considerable changes in its physical, chemical and biological properties (Appels et al., 2008; Delaide et al., 2019), producing biogas composed of methane, carbon dioxide and small quantities of hydrogen sulfide and ammonia (Marchaim, 1992; Appels et al., 2008). A healthy microbiome in an aquaponics system can be achieved if the water temperature is kept in the range that is safe for fish, plants, and bacteria.

8.2.2.3 Dissolved Oxygen

Since aquaponics is a closed intensive system, a high level of DO is necessary for fish, plants and bacteria for maximum health and growth. A DO level of 5 mg L^{-1} or higher should be maintained in the fish-rearing tank and in the water surrounding plant roots. The process of nitrification requires DO, and it is also essential to maintain healthy populations of nitrifying bacteria, which convert toxic levels of ammonia and nitrite to relatively nontoxic nitrate ions. At a temperature of about 30 C, a DO level of >3.0 mg L^{-1} is required for maximum nitration. In aquaponic system, DO decreases in the bio-filter around the root zone of plants and in fish tanks due to the metabolic activity of aerobic microorganisms (nitrifiers and heterotrophs) and fish (Kasozi et al., 2021; Wongkiew et al., 2017; Hagopian and Riley, 1998). However, the effect of DO levels in aquaponics is not well studied, but studies using synthetic wastewater (225–450 mg L^{-1} of TAN) showed that the activity of AOB was reduced when the DO level was below 4 mg L^{-1}, while the activity of NOB decreased at DO below 2 mg L^{-1} (Kasozi et al., 2021; Wongkiew et al., 2017). A DO concentration above 1.7 mg L^{-1} was recommended in bio-filters to maintain the activity of nitrifiers (Ruiz et al., 2003; Kasozi et al., 2021; Wongkiew et al., 2017). However, the recommended DO in fish tanks and the inlet of grow beds is 5–6 mg L^{-1} to avoid stress on fish and plants (Somerville et al., 2014). Maintaining a high level of DO in the water is extremely important for intensive root respiration and optimal plant growth. The growth and development of root systems in deep water culture (DWC) are also affected by the level of DO. In DWC, plant roots may be very well developed if an oxygen concentration of 5 mg L^{-1} is maintained (Raviv et al., 2008; Schwarz, 1995). If DO is low and the temperature is high, the water absorption and nutrient uptake will decrease, resulting in plant root rot (Rakocy, 2007; Stouvenakers et al., 2019). This leads to reduced plant growth. Also, low DO levels correspond to high concentrations of CO_2, which in turn promote pathogen growth in plant roots (Losordo et al., 2000). If the aeration is inadequate, there may be anoxic zones and sludge build-up. In such cases, anaerobic bacteria such as sulfate-reducing bacteria may dominate and produce toxic metabolites (Somerville et al., 2014). This may further prevent the growth of nitrifying bacteria.

8.2.2.4 C/N Ratio

The C/N ratio in an aquaculture system is an important parameter that dictates how the system is affected by the structure and function of bacterial communities (Avnimelech, 1999; Martínez-Córdova et al., 2016; Khanjani et al., 2022). Manipulation of microbial communities and the associated microbial loop can be achieved by the C/N ratio. At high C/N ratio, there is a reduction in the abundance of nitrifiers and the nitrification efficiency because the growth rate of nitrifiers is lower than that of heterotrophs (Ebeling et al., 2006; Michaud et al., 2014; Wongkiew et al., 2017). At a C/N ratio of 6, there is a 90% decrease in NH_4^- removal (Ebeling et al., 2006). Further, in a system where heterotrophs and nitrifiers coexist, at C/N ratios of 0.5, 1.0 and 1.5, the rates of NO_3^- production/TAN consumption were reduced by 24, 56 and 73%, respectively (Ebeling et al., 2006). The heterotrophic bacteria assimilate NH_4^+ and NO_3^- in the presence of organic carbon for cell growth and turn into sludge, which reduces the nitrogen availability for plant uptake (Avnimelech, 1999; Martínez-Córdova et al., 2016; Khanjani et al., 2022). At a high C/N ratio, a high concentration of heterotrophs in bio-filters can drastically lower the DO and promote anoxic conditions, which causes denitrification and nitrogen loss via N_2O and N_2 (Wongkiew et al., 2017). Thus, the C/N ratio can be used to create conditions either to reduce ammonium concentration faster than the nitrification process or to transform ammonium and other nitrogenous organic waste compounds into a nutritional source with bacterial biomass for aquatic organisms, according to whether its value is high or low.

8.3 EMERGENT COMPLEX INTERACTIONS IN AQUAPONICS MICROBIOME

Figure 8.5 shows a conceptual diagram of the ecological domains and the symbiotic interactions of fish–microbes–plants with the environmental microbiota in the aquaponics microbiome. The interaction of fish–microbes–plants at the microbiome level increases not only the complexity but also the size of the ecological domain. The emergent ecological niches and the process of interactions

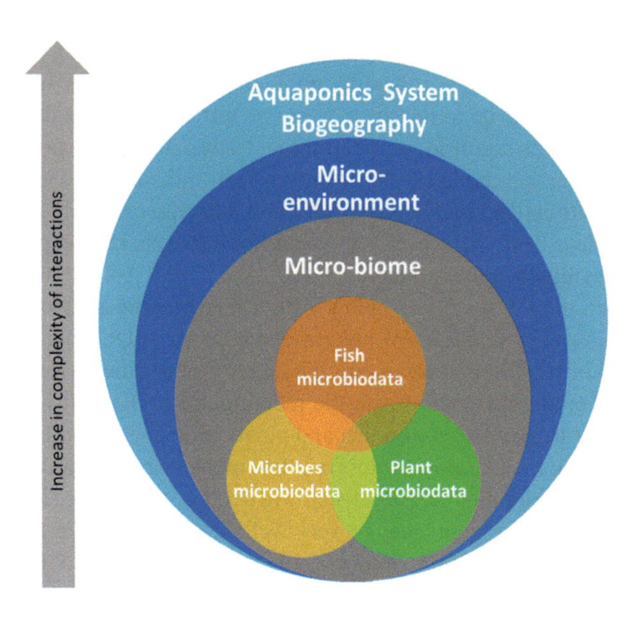

FIGURE 8.5 Microbial domains and evolving interrelations in emerging aquaponic microbiome.

of the fish, microbes, plants and feed with environmental microbiota are the cause of the emergent microbiome in aquaponic systems.

8.3.1 Fish-Associated Interactions

The host genetics may play a major role in the establishment of the microbial communities, associated with microbial recognition, immunity selection and determination of biochemical niche (Perry et al., 2020; Chen et al., 2022; Kim et al., 2021). The fish factors that can affect the microbiome include feed, gut microbiome and feeding protocol (Perry et al., 2020). The fish gut microbiome composition is a major contributor to the composition of the emergent aquaponics microbiome. However, the fish gut microbiome composition is largely influenced by the type of fish species, feed (diet) provided, health status and environmental factors (Chen et al., 2022; Kim et al., 2021). Further, the gut microbiome is also affected by the composition of the water microbiome (Chen et al., 2022; Kim et al., 2021). A recent study analyzed 227 individual fish representing 14 orders, 42 families, 79 genera, and 85 species and found that the fish gut microbiota was dominated by proteobacteria (51.7%), firmicutes (13.5%), and cyanobacteria (4.8 %) (Kim et al., 2021). It was found that the host habitat was the major factor for the microbiome in the gut of the fish. This study indicates the importance of understanding fish–feed–environment interactions.

8.3.1.1 Feed–Fish Interaction

The input feed is the source of both nutrients and solid waste in the aquaponic system. As feed is introduced directly into the water of the fish tank, the feed microbiota may have a significant effect on the composition of the environmental microbiota. Further, the composition of the environmental microbiota may be dependent on the host intestinal microbiota because of defecation in the water. The microorganisms in the water easily spread to the micro-units in the aquaponic system – bio-filter, fish tank surface, radial flow settler, hydroponic units (including roots) and sludge digester – including the fish intestine because of osmoregulation, during feeding or by active uptake. Thus, probing the relationship between the host microbiota from both the feed microbiota and the environmental microbiota is a relevant question. The composition of feed has been associated with the composition of the gut microbiome and the commensal relationship between the microbes and the host (fish) (Perry et al., 2020). Dietary changes can alter the fish gut microbiome. The diet→immunity→micobiome process demonstrates the impact of diet on fish immunity and thus, the microbial composition of the gut. This relationship shows how the growth of microbiome in fish is related to the feed supplied and is a very important aspect for understanding the feed–fish–microbiome interaction in an aquacultural system like aquaponics. There is increasing interest in understanding the influence of alternative plant-protein sources on the host–microbe interaction (Perry et al., 2020). Particularly, insect meal (chitin rich) is being explored as a protein source. An example is the inclusion of black soldier fly (*Hermetia illucens*) larvae (BSLFL) frass, which enhanced the production of channel catfish (*Ictalurus punctatus*) juveniles, stevia (*Stevia rebaudiana*) and lavender (*Lavandula angustifolia*) in an aquaponic system (Romano et al., 2023). Increased growth of catfish is linked to higher hepatic expression of genes for growth, feed intake and reduced intestinal inflammation (Romano et al., 2023). The potential uses of chitin-rich insect meal in manipulating the gut microbiome have been reported, and supplementation with black soldier fly larva in rainbow trout shows a positive effect through an increase in commensal bacteria such as *Pseudomonas* sp. and *Lactobacillus* sp., which improves health and immunity (Bruni et al., 2018). However, in tilapia hybrids (*Oreochromis niloticus* × *O. aureus*), a decrease in growth rates was found (Ringo et al., 2012). In this regard, the effect of chitin on growth and feed utilization, bacterial capacity to degrade chitin, modulation of the gut microbiota, and the effect on the innate immune system and disease resistance must be studied (Ringo et al., 2012). The influence of insect meal on the gut microbial community and its mediated functions are not clearly understood (Perry et al., 2020; Ringo et al., 2012). The understanding of microbially mediated feed–fish interaction is crucial to potentially enhance plant and fish production as well as to create a circular economy.

8.3.1.2 Fish Gut Microbes–Environment Interactions

The microbe–microbe interactions and host–microbe interactions in a host gut could affect the environmental microbiome (Akbar et al., 2022). The gut microbiome of fish is influenced by various processes (Figure 8.6) and is affected by both intrinsic and extrinsic factors. The intrinsic factors include host genetics, developmental stages, physiological conditions and starvation (Chen et al., 2022). The extrinsic factors may include diet and environmental factors such as water quality, bacteria, feed and feces (Figure 8.6(a), green colour). Further processes involving the fish gut environment, physiology and interaction with the surrounding environmental conditions, and fish–microbe interactions take place (Figure 8.6(a) brown colour). Also, the bacterial communities of the surrounding water environment influence the composition and structure of the fish gut microbiota, and the gut bacteria also influence the bacterial community of the surrounding water environment (Figure 8.6(b)).

In an aquatic ecosystem, the host microbiomes are determined by both environmental and host-associated factors, and the environmental parameters can also reciprocally impact each other as well as the host, particularly the microbial community within the water (Lorgen-Ritchie et al., 2023). But, the extent to which these factors contribute to the microbiome composition of fish is not clear (Perry et al., 2020; Kim et al., 2021; Chen et al., 2022). The microbial communities may be found in fish (host) mucosal tissues, including the gut, gill, and skin. The host microbiome composition has important implications for host health, but the functional and mechanistic pathways underlying these associations are, however, poorly understood in aquaculture (Lorgen-Ritchie et al., 2023; Chen et al., 2022). The gut microbiome is affected by the diet (fishmeal), starvation, immunity and the environment (Perry et al., 2020). Since the gut microbial community of aquatic animals can respond to changes in the aquatic environment by secreting specific metabolites, the influence of the aquaponics micro-environment needs proper study. Although metabolites of the gut microbiota can be produced by microbes and microbe–host and microbial interactions, the metabolism of the gut bacterial community is one of the fundamental modes of environmental factor–gut microbiome interactions that need to be accounted for in understanding the interactions in the complex aquaponics microbiome.

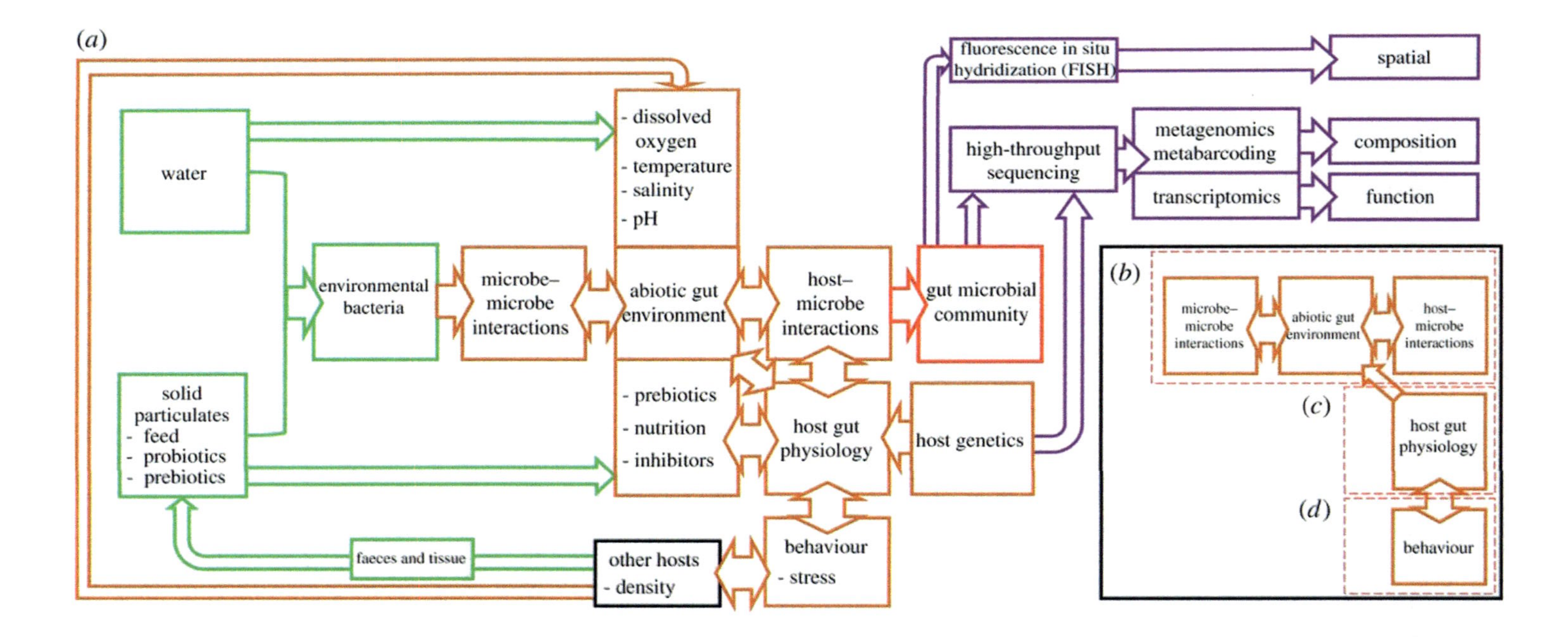

FIGURE 8.6 (a) Schematic view of the deterministic processes that influence gut microbial communities in fish. The community assemblage of bacteria in the gut starts with inputs from the environment (green), such as the bacteria within the water column or in solid particulates of biofilm, sediment, and feed. Once ingested, these bacteria are influenced by interacting deterministic processes (brown), such as the host's abiotic gut environment, interaction with the host's physiology through the gut lining and its secretions, as well as interactions between other microbiomes. The outcome (red) is final community assembly, which can be characterized using an array of cutting-edge molecular techniques (purple). A subset of the broader interactions is provided, with focus on (b) microbe–environment–host interactions, (c) host gut physiology and (d) behaviour. (Reprinted from Perry, W.B. et al., *Proceedings of the Royal Society B*, 287, 1926, 20200184, 2020.)

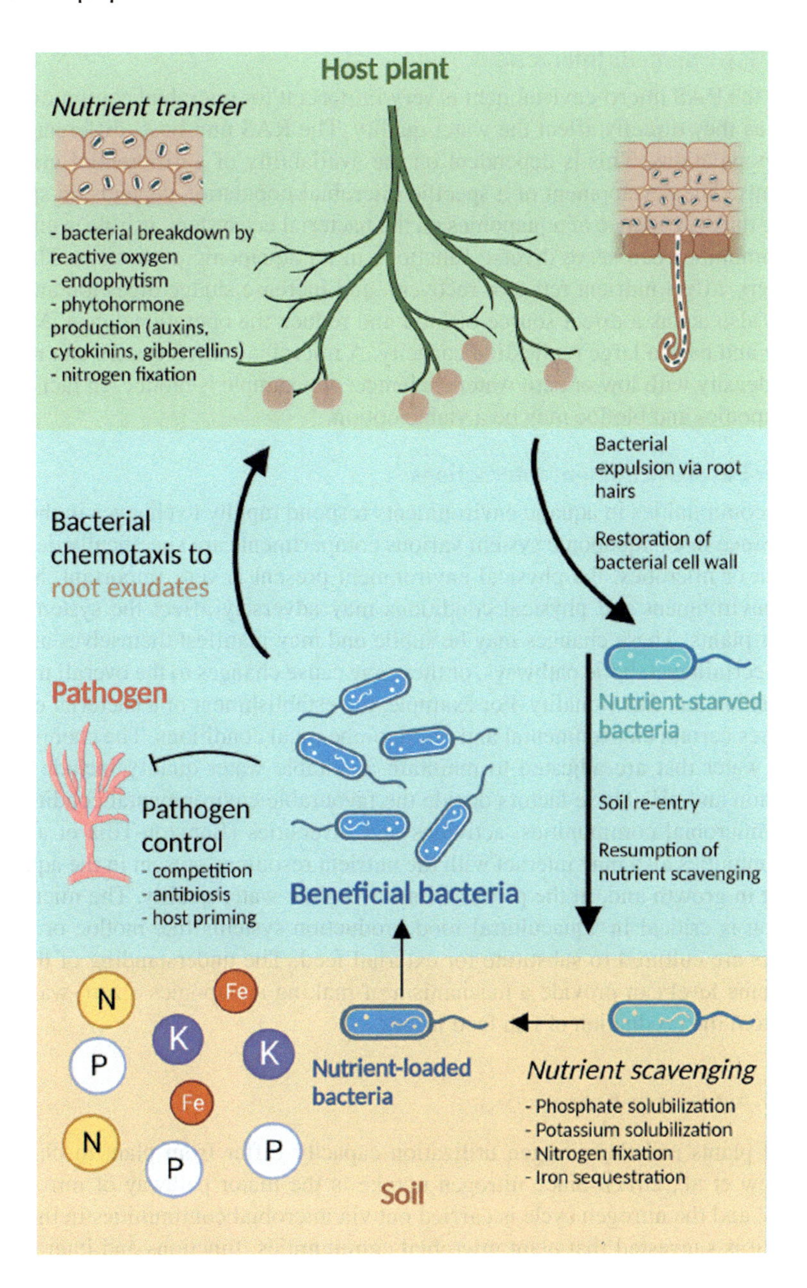

FIGURE 8.7 Schematic representation of the growth-promotional and defensive functions provided by beneficial bacteria, including participation in the rhizophagy cycle. The host plant (labelled in green) breaks down soil bacteria with reactive oxygen species (ROS), allowing endophytism and transfer of nutrients and phytohormones. Following this, nutrient-starved bacteria are expelled via root hairs, where they can restore their cell walls. In soil, bacteria resume nutrient scavenging, which includes phosphate and potassium solubilization, nitrogen fixation, and iron sequestration. Nutrient-loaded bacteria (labelled in blue) are subsequently attracted back to the host plant via root exudates, where they are degraded by ROS, and nutrient transfer can occur again. Throughout this cycle, beneficial bacteria may also participate in pathogen control through competition, antibiosis and priming of the host plant's resistance. (From Chiaranunt, P. and White, J.F., *Plants* 12, 2, 400, 2023.)

8.3.1.3 Fish–Environment Interactions

In aquaponics, the RAS micro-environment is very important for microbial communities, similarly to that of fish, as they directly affect the water quality. The RAS unit hosts different micro-niches for microbial populations. This is dependent on the availability of a differential gradient of oxygen and nutrients. The development of a specific microbial population depends on specific micro-niches. Despite the dependence of aquaponics on the bacterial ecosystem, its role is underestimated. Microbial communities can serve diverse functions in an aquaponic ecosystem. They can act as nutrient recyclers, affect nutrient reuse or recovery, and increase sludge digestion and mineralization. They can also act as a direct source of food and reduce the operational cost. Moreover, they are sustainable and have a large replication capacity. A microbial-based system can manage a very high stocking density with low or zero water exchange; an example is biofloc. In fact, an integrated system of aquaponics and biofloc may be a viable option.

8.3.1.4 Microbes–Environment Interactions

The microbial communities in aquatic environments respond rapidly to changes in their immediate environment. Since in an aquaponic system various compartments are specifically designed to host certain bacteria or microbes, the physical environment present is very important. Small changes in the micro-environment and physical conditions may adversely affect the system's health and may kill fish or plants. These changes may be subtle and may manifest themselves as activation or inactivation of certain metabolic pathways, or they may cause changes to the overall microbial community composition and functionality. For example, the establishment of a bacterial ecosystem in a bio-filter requires certain environmental and physico-chemical conditions. The chemical properties of the rearing water that are adjusted to maintain a suitable water quality include salinity, oxygen concentration and pH. These factors decide the favourable environmental conditions and thus, shape aquatic microbial communities, activities and diversities (Bentzon-Tilia et al., 2016). The microbial communities also may interact with the nutrient resources present in the aquatic environment, resulting in growth and, in the process, improving the water quality. The microbe–environment interaction is critical in aquacultural food production systems like biofloc or FLOCponics, where microbes are cultured to substitute for external feed. The understanding of this interaction at the microbiome level can provide a mechanism of making aquaponics a zero-waste generation system apart from the production of fish feed in situ.

8.3.2 Plant-Associated Interactions

The growth of plants and the nitrogen utilization capacity differ from plant to plant (Hu et al., 2015; Wongkiew et al., 2017). Since nitrogen uptake is the major pathway of nitrogen recycling into vegetables, and the nitrogen cycle is carried out via microbial communities in the surface area of plant roots, it is suggested that plant microbial communities, functions and interactions among microorganisms can have an impact on the microbial functions and nitrogen transformation in an aquaponics system (Lobanov et al., 2021). In particular, the region in and around the plant root, termed the rhizosphere, is a region where root microbiome, root exudates and nutrients interact, where electron transfer takes place and where most of the bacterial activity takes place (Vives-Peris et al., 2020; Senff et al., 2022). Plants with a well-developed root system with a large root area can provide ideal habitats for nitrifying bacteria (Hu et al., 2015). A number of factors can affect the rhizobiome composition, due to both the plant (genotype, life and stage) and the environment (water parameters and dissolved nutrients) (Lobanov et al., 2021). Since the microbial communities present in RAS and aquaponic systems are system condition dependent (Bartelme et al., 2019), a similar system-specific microbiome is also formed within the rhizosphere in the hydroponic compartment in aquaponics (Lobanov et al., 2021).

In a hydroponic system, microorganisms such as *Gliocladium* spp., *Trichoderma* spp., *Pseudomonas* spp. and *Bacillus* spp. (Lee and Lee, 2015) are naturally present in hydroponics, but the introduction of plant growth-promoting rhizobacteria (PGPR) has also been reported (Rivas-Garcia et al., 2020; Dhawi, 2023; Mourouzidou et al., 2023; Chiaranunt and White et al., 2023). Some of the PGPRs introduced include *Bacillus* sp., *Pseudomonas* sp., *Streptomyces griseoviridis, Pseudomonas chlororaphis, Bacillus cereus* (Lee and Lee, 2015), and *Bacillus amyloliquefaciens* (Lovanov et al., 2022), and *Bacillus licheniformis* has been reported to enhance yields (Lee and Lee, 2015). Further discussion on the interactions, mechanism and potential role of PGPM follows.

8.3.2.1 Microbes in Roots of Plants

The interactions between plants and microbes and their relationship with environmental factors can play a crucial role in plant adaptation to a toxic environment, promotion of growth, augmentation of phytoremediation or abiotic stress (Mehmood et al., 2021). It is found that rhizodeposits, or root exudates, into the rhizosphere influence a multitude of processes, both biotic and abiotic (Vives-Peris et al., 2020; Senff et al., 2022). Root exudates include various compounds that interact with fungi, bacteria, and other plants and create compositions of the root micro- and microbiome, including nitrogen-fixing bacteria (Vives-Peris et al., 2020) and beneficial mutualistic arbuscular mycorrhizal fungi (Vives-Peris et al., 2020; Bouwmeester et al., 2007). It is not very clear whether soluble plant exudate into an aqueous milieu diminishes their effect on the microbial communities, or how the microbial community in the aqueous environment contributes to a greater capacity for root colonization (Lobanov et al., 2021). Although the role of root exudates is not known clearly, it is anticipated that root exudates would play a role inside the hydroponic unit of an aquaponic system, as they shape the microbial communities of the rhizosphere (Senff et al., 2022; Lobanov et al., 2021). Particularly when different crop species are planted, exuded chemicals can influence biomass production and nutrient uptake negatively or positively. Further, Figure 8.8 shows the possible areas of influence of root exudates on the plant, animal and microbial compartments of an aquaponic system.

An aquaponic system depends on beneficial microorganisms, and the present research is mostly focusing on nitrifying bacteria or PGPR or plant growth-promoting bacteria (PGPB) or microbes (PGPM). Sanchez et al. (2019) identified PGPB including bacteria from the phyla Proteobacteria, Actinobacteria and Firmicutes, isolated from tilapia-rearing water from an RAS. Also, in a 16S rRNA meta-barcoding study of lettuce root PGPB including the phyla Proteobacteria, Actinobacteria and Firmicutes as well as the genera *Acidovorax, Sphingobium, Flavobacterium* and *Pseudomonas*, these genera have been found in the rhizosphere of lettuce grown in an aquaponics system integrated with *Oreochromis niloticus* and lettuce (Schmautz et al., 2022). These microorganisms in the medium interact with plant roots by promoting uptake of nutrients, stimulating plant growth and acting as antagonists against phytopathogens (Rivas-Garcia et al., 2020). For example, *Pseudomonas* sp. shows antagonistic and antifungal activity against *Fusarium graminearum* (Lee and Lee, 2015). Also, plant-associated bacteria may serve different roles with respect to a host plant: mutualistic, commensal or parasitic (Chiaranunt and White, 2023). Specifically, PGPMs may be used to enhance plant growth, environmental stress tolerance and phytoremediation (Chiaranunt and White, 2023; Dhawi, 2023). In controlled systems like hydroponics, aquaponics or vertical farming, PGPMs can increase the system's climatic resilience (Chiaranunt and White, 2023; Dhawi, 2023). While PGPMs can enrich plant growth, enhance resistance to disease and drought, produce beneficial molecules, and supply nutrients and trace metal to the plant's rhizosphere, they may compete with plants for certain nutrients like iron, thus requiring external supplementation, adding production cost and operational complexity (Bartelme et al., 2018; Kasozi et al., 2019). However, *P. fluorescens* Pf-5 is a PGPM known to increase the bioavailability of iron through siderophore production. Also, Radzki et al. (2013) reported that siderophore production by *Chryseobacterium* sp. was effective in supplying Fe to iron-starved tomato plants in hydroponics culture. In aquaponic systems, the production of siderophores occurs in various microbes of the genera *Pseudomonas, Bacillus, Enterobacter,*

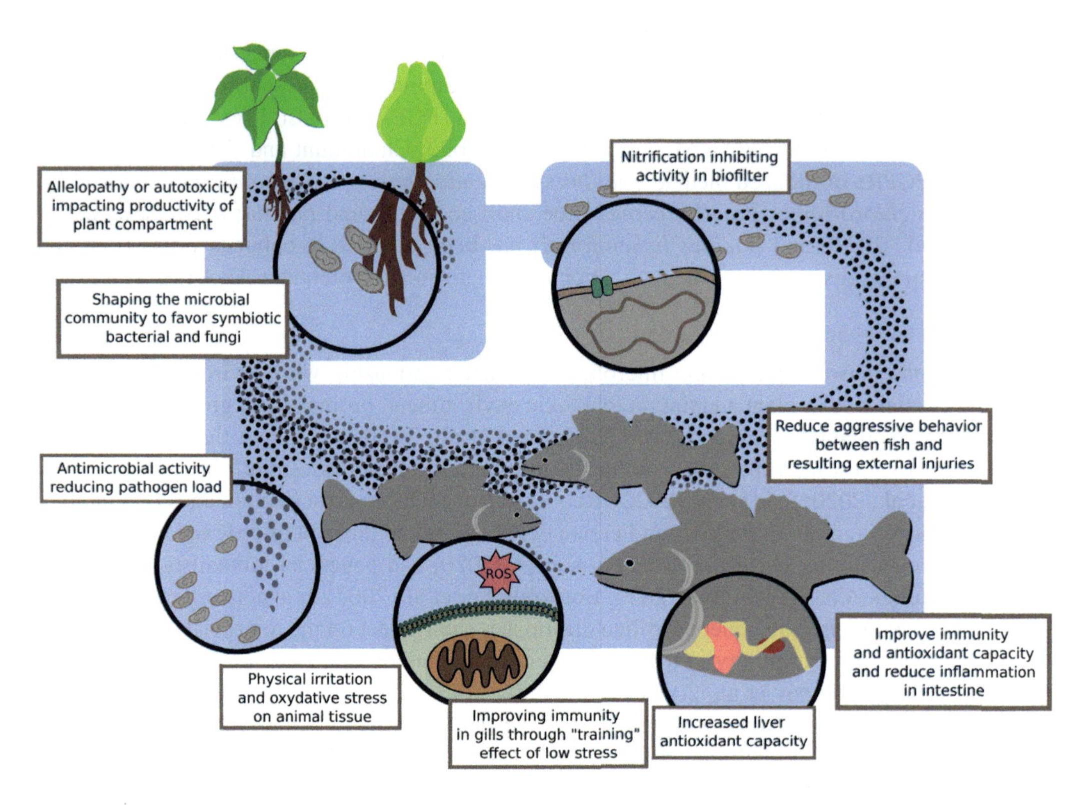

FIGURE 8.8 Possible areas of influence of root exudates on the plant, animal and microbial compartments of an aquaponic system. See text for details on the original studies. (Reprinted from Senff, P. et al., *Reviews in Aquaculture*, 2022.)

Streptomyces, *Gliocladium* and *Trichoderma* (Lee and Lee, 2015; da Silva Cerozi and Fitzsimmons, 2016; Schmautz et al., 2017). In recent years, there have been various metagenomic studies on the complex dynamics of rhizome development. However, the role of bacteria in roots and root exudates is still not clear (Senff et al., 2022).

8.3.2.2 Plant–Environment–Microbiome Interactions

Another important aspect of microbe–plant interactions in aquaponic systems is the occurrence of plant diseases. The aquaponic system is specifically designed for high plant density, high nutrient availability, and a warm, humid environment, which facilitates the growth of microorganisms, plants, and fish. It also forms a perfect environment for pathogen proliferation that can be harmful to fish as well as plants. Such cases have been reported in aquaculture and hydroponic systems. Disease occurrence in plants of aquaponic systems might be similar to those in hydroponics because of similar conditions, such as the continuous presence of water, constant temperature, and humid conditions (Stouvenakers et al., 2019). Also, the oomycetes constitute one of the most common pathogens on plants such as tomatoes and cucumbers (Stouvenakers et al., 2019). *Pythium* sp. and *Phytophthora* sp. are oomycetes well adapted to humid/aquatic conditions and are the most common plant root pathogens (Stouvenakers et al., 2019). These oomycetes have a motile structure, called a zoospore, which allows dissemination by moving independently in aqueous media (Walker and van West., 2007; Stouvenakers et al., 2019). *Phytophthora* sp. and *Pythium* sp. can quickly spread by zoospores in recirculating water, causing rotting of roots (Lee and Lee, 2015; Stouvenakers et al., 2019). *Pythium* sp. can cause 100% and 69% of mortality in spinach under a hydroponic system at 30 and 20 C, respectively (Bates and Stanghellini, 1984). It is reported that *Pythium aphanidermatum*, after three days of inoculation in a hydroponic cucumber, infected 100% of the plantation (Goldberg et al., 1992). The understanding of plant–environment–microbiome interaction is important for the biocontrol of phytopathogens in aquaponic systems (Rivas-Garcia et al., 2020). From the commercial point of view, a plant-centric approach towards aquaponics results in more profitability, and thus, the understanding of plant-microbe interaction will potentially increase the commercial feasibility of an aquaponic business (Love et al., 2015a, b, c). However, there is limited research on microbes, especially on plant-microbe interactions in aquaponic systems. PGPM research on soilless systems, hydroponics, and aquaponics must be carried out, as it has the potential to increase the understanding of rhizosphere microorganism associations (Bartelme et al., 2018).

8.3.3 Fish–Plant Interaction

The interaction of fish and plants may be through either gut microbiome–plant (rhizosphere) microbiome interaction or plant root exudates. It is anticipated that understanding the interaction/overlap between the fish gut microbiome and the plant (rhizosphere) microbiome could be important for manipulating it as a biological control agent to benefit both plants and fish (Hacquard et al., 2015). However, there is a lack of tools for research on how microorganisms, applied or introduced in aquaculture or aquaponic systems with benefits for fish, will also be beneficial for plant growth and health, and vice versa (Senff et al., 2022). Plants are known to improve fish growth and immunity when applied as feed additives or baths. Although no studies have directly investigated the effect of root exudates on fish, some observations from aquaponic experiments suggest positive effects (Senff et al., 2022). For example, intercropping of lettuce and red chicory, the latter known for its high polyphenol content and presence of antioxidants in the roots, resulted in higher growth of Pangasius (*Pangasianodon hypothalmus*) in an aquaponic system (Maucieri et al., 2017). Further, the culture of African catfish (*Clarias gariepinus*) with basil (*Ocimum basilicum*) or cucumber (*Cucumis sativus*) in coupled aquaponics led to reduced skin lesions on the fish, suggesting less aggressive behaviour (Senff et al., 2022; Baßmann et al., 2017; Baßmann et al., 2020).

8.4 MICROBES AND MICROBIOLOGICAL PROCESS IN AQUAPONICS

Recent studies have indicated an increasingly important role of microorganisms in aquaponics systems. The microorganisms are involved in various processes in aquaponics subunits, including the nitrogen cycle (nitrification, denitrification), sludge decomposition and mineralization, and macro- and micro-nutrient cycling. Often, molecular techniques have been used to identify the microbial communities present in different compartments and their interactions to understand their possible role in plant growth, fish health and overall improvement of the aquaponics ecosystem. As mentioned earlier, aquaponic systems have many components with diverse micro-environmental conditions. Understanding the compartments, functions, their targeted parameters, niches, biological process and predominant contributing taxa (microorganism phyla/family/order/genera) in an aquaponic system forms an important aspect of microbiome research. The microbial communities are essential components in aquaponic systems due to their important role in nutrient recycling, degradation of organic matter, improvement of water quality, and treatment and control of diseases. Most of the research on microbial communities in aquaponic systems has been confined to nitrifying bacteria (Wongkiew et al., 2017; Wongkiew et al., 2018) and plant growth promotion (Dhawi, 2023; Stegelmeier et al., 2022), but their mechanistic process remains elusive. Some of the possible emerging applications of the microbiome include (i) enhancing nutrient recovery, (ii) using microbial communities as biocontrol agents (BCAs) to improve plant and fish health, and (iii) use of microbial communities in feed generation, waste reduction, and the creation of a bio-circular economy and making aquaponics sustainable. For these emerging applications to materialize, we need to understand the associated microbial processes involved.

8.4.1 Microbial-Assisted Nutrient Recovery

The microbial community in aquaponics plays a paramount role in the recycling of nutrients. The process of nutrient recovery and bioremediation of aquacultural wastewater in aquaponics is a complex process and includes microbial processes such as the nitrogen fixation and nitrification process, phosphorus solubilization, potassium solubilization, and the use of probiotics. These microbial processes take place in compartments of aquaponics like the bio-filter, sludge digester and mineralization tanks. At present, not all microbial processes in aquaponics are completely understood.

8.4.1.1 Biological Nitrogen Fixation and Oxidation

Nitrogen is the most important macronutrient for plant growth. Nitrogen enters the aquaponic system through feed, and further, after ingestion by fish, is excreted as ammonia nitrogen (Wongkiew et al., 2017). The conversion of ammonia nitrogen to nitrites and nitrates by microorganisms, termed biological nitrogen fixation or nitrification, constitutes one of the most important microbiological processes involved in aquaponics. The nitrification process is a two-step process, during which the ammonia (NH_3) or ammonium (NH_4^+) excreted by the fish is transformed first into nitrite (NO_2^-) and then into nitrate (NO_3^-) by specific aerobic chemosynthetic autotrophic bacteria. Nitrifying bacteria (also known as autotrophic bacteria) are the most widely known and characterized bacteria in an aquaponics system. The main role of these bacteria is to transform ammonia excreted by the fish into nitrite and nitrate. High availability of DO is required for efficient nitrification, as it consumes oxygen. The first step of this transformation is carried out by ammonia-oxidizing bacteria (AOB) such as *Nitrosomonas*, *Nitrosococcus*, *Nitrosospira*, *Nitrosolobus* and *Nitrosovibrio*. The second step is conducted by nitrite-oxidizing bacteria such as *Nitrobacter*, *Nitrococcus*, *Nitrospira* and *Nitrospina* (Rurangwa and Verdegem, 2015; Timmons et al., 2018; Wongkiew et al., 2017). The process of nitrogen transformation in aquaponics has been reviewed excellently by Wongkiew et al. (2017). It is concluded that quantitative evaluation of nitrogen transformation is still not yet completely understood, and thus, it has not been possible to incorporate the NUE factor in designing a high nitrogen recovery, low nitrogen loss and low greenhouse gas emission aquaponic system. There is a requirement to benchmark the nutrient budget of nitrogen transformation for a wide variety of crops to make aquaponics economically feasible on a commercial scale. Since the nitrogen cycle in aquaponics depends on the microbial communities present in the system, supplementation with beneficial bacteria or probiotics is a potential approach towards enhancing the nitrate concentration. Kasozi et al. (2021a) have shown that the use of *Bacillus* in aquaponics enhances the nitrate concentration, consequently improving lettuce growth in a tilapia–lettuce aquaponic system.

8.4.1.2 Phosphorus Solubilization

Since phosphorus enters the aquaponic system through the fish feed, the amount present in the system is very much dependent on the type and quality of feed (Joyce et al., 2019). If the feed is plant-based, then phosphorus will mainly be found in the form of insoluble phytates (i.e., insoluble organic phosphate) (da Silva Cerozi and Fitzsimmons, 2017). Thus, to make the phosphorus available to plants, it must be converted to forms that plants can easily use, e.g., orthophosphates. However, soluble orthophosphates are prone to react with organic and inorganic compounds that are present in the environment, soil and aquatic environment and thus, may be unavailable to plants (Prabhu et al., 2019; Rawat et al., 2021). Studies by da Silva Cerozi and Fitzsimmons (2017), using the P mass balance analysis, identified and quantified several phosphorus pools with 89% recovery, demonstrating that aquaponic systems can maximize overall phosphorus utilization (71.7% of total P input), with fish and plants assimilating 42.3% and 29.4% of the phosphorus input in the feed, respectively. However, 13.1% of the phosphorus input occurred in a form unavailable to fish and plants. It is reported that a change in pH can result in the availability or enrichment of phosphorus in plants (Silva Cerozi and Fitzsimmons, 2017). Since plants can only assimilate P as the free orthophosphate ions H_2PO_4 and

HPO_4^{2-} (Prabhu et al., 2019), it depends on the forms in which phosphorus exists in solution, which depends on the pH. The rate of phosphate uptake decreases as the pH increases due to a reduction in the concentration of H2PO4–, while a decrease in pH can increase the activity of proton-coupled solute transporters and enhance anion uptake (Cerozi and Fitzsimmons, 2016; Schachtman et al., 1998). By lowering the pH from 8.0 to 4.0, the phosphate uptake increases by a factor of 3 in maize roots (Sentenac and Grignon, 1985). As pH increases above 7.0, the dissolved phosphorus can react with calcium, forming calcium phosphates, and making the free phosphate species form insoluble compounds that cause phosphate to become unavailable (Cerozi and Fitzsimmons, 2016).

Apart from the influence of pH, the availability of phosphorus may be increased by the use of phosphate-solubilizing microorganisms (PSO). PSO are microorganisms capable of solubilizing inorganic phosphorus from insoluble compounds, which play the role of bioremediation of phosphorus from aquaculture sludge. The microorganisms that solubilize phosphate include bacteria (*Bacillus*, *Pseudomonas*, *Agrobacterium*, *Azotobacter*), fungi (*Aspergillus* spp., *Penicillium* spp., *Trichoderma* sp., *Saccharomyces*, *Rhizoctonia* spp.) (Prabhu et al., 2019), algae (cyanobacteria) and actinomycetes (*Actinomyces*, *Micromonospora* and *Streptomyces*). There are various mechanisms by which phosphate is solubilized by microorganisms. These include the release of organic and inorganic acids, chelation, siderophore production, exopolysaccharide production, anions, protons, and hydroxyl ions (Prabhu et al., 2019). Further mechanisms are enzymatic and substrate degradation. For example, *Bacillus* spp. transform the complex form of phosphorus into bio-available forms such as dihydrogen phosphate ($H_2PO_4^-$) and phosphate (PO_4^{3-}), which can be readily absorbed by plant roots (Gouda et al., 2018; Kasozi et al., 2023). There is also a report of an increase in orthophosphate concentration by the addition of nitrifying bacteria (*Nitrobacter* and *Nitrosomonas*) (Ajijah et al., 2021). PSOs can play the role of bio-fertilizer formation by transforming the insoluble phosphorous content within organic materials to soluble forms, improving its bioavailability to plants. Recently, there has been growing interest in PSOs in aquaponics due to their potential to enhance plant growth (Kasozi et al., 2023).

8.4.1.3 Potassium Solubilization

The source of potassium (K) in aquaponics is mainly from fish feed (Delaide et al., 2017), but the quantity of potassium is generally low in commercial feed, as it is required in low amounts by the fish (Graber and Junge, 2009; Seawright et al., 1998; Somerville et al., 2014). It is important to have an optimum potassium concentration both in fish feed and for plants. So, maintaining and finding sources that can meet the requirements of plants is an important challenge. Instead of supplementing potassium externally, the use of K-solubilizing microbes to increase the K concentration is an important approach to increasing the availability of K concentration for uptake by plants. Although there is little information concerning potassium solubilization in aquaponic systems, there are several microorganisms known to possess the ability to solubilize potassium that may be useful to aquaponics. Examples of potassium-solubilizing microorganisms include saprophytic bacteria, some fungi, and actinomycetes, which are naturally present in the soil and in the rhizosphere (Etesami et al., 2017). Some of the bacteria (Etesami et al., 2017) and fungi (Sattar et al., 2019) are known for their K-solubilizing abilities. Some of the main potassium-solubilization mechanisms used by microorganisms are: "(i) lowering the pH; (ii) enhancing chelation of the cations bound to K; and (iii) acidolysis of the surrounding area of microorganism" (Meena et al., 2016). However, the detailed K-solubilization mechanisms are less understood than in the case of P-solubilization (Etesami et al., 2017), and this is also true in the case of aquaponics. Unlike P solubilization, where the main mechanism is the acidification of the environment via the production of organic acids, inorganic acids and protons (Etesami et al., 2017), little information is available as to the form of K captured in aquaponic sludge, so it is difficult to identify a specific solubilization process which could be of interest in aquaponics.

8.4.2 Microbial Communities as Biocontrol Agents (BCAs)

A major problem related to aquaponic systems is the potential dissemination of pests and pathogens due to repeated water recirculation and the controlled environment, making it suitable for pathogen growth (Goddek et al., 2015; Rivas-Garcia et al., 2020; Stouvenakers et al., 2019; Folorunso et al., 2021). Evidence of disease outbreaks and pathogen proliferation includes *Streptococcus iniae* causing 40% mortality in Barramundi fish (Bromage et al., 1999; Rivas-Garcia et al., 2020) and *Pythium aphanidermatum*, after three days of inoculation in a hydroponic cucumber, infecting 100% of the plantation (Goldberg et al., 1992). However, no pesticide or bio-pesticide has been developed especially for aquaponic systems to date (Rivas-Garcia et al., 2020; Folorunso et al., 2021). The conventional methods of protection against phytopathogens include disinfecting the water, involving the use of ultraviolet (UV) irradiation, media filtration, heat, sonication, and chemical methods such as chlorination and ozonation (Rivas-Garcia et al., 2020; Folorunso et al., 2021). However, the use of disinfecting methods could have negative effects on fish, plants, and beneficial microorganisms residing in the system. The use of microorganisms isolated from aquaponic systems as biocontrol agents (BCA) or antagonists to manage plants has great potential (Rivas-Garcia et al., 2020). It is important to explore the potential of indigenous microbial communities in aquaponics for disease management (Sirakov et al., 2016; Folorunso et al., 2021; Stouvenakers et al., 2022; Stouvenakers et al., 2023). There are recent studies that have explored the aquaponics microbial community for disease control (Schreier et al., 2010; Schmautz et al., 2017; Wongkiew et al., 2018; Bartelme et al., 2018; Eck et al., 2019; Sirakov et al., 2016; Folorunso et al., 2021; Stouvenakers et al., 2022; Stouvenakers et al., 2023). Among the prominent microorganisms reported in aquaponic systems are proteobacteria (42%), bacteroides (15%) and actinomycetes (13%) (Folorunso et al., 2021). Heterotrophic bacteria like *Pseudomonas* and *Bacillus* have been extensively tested as potential microbial inoculants against pathogens due to their broad-spectrum efficiency (Munguia-Faragozo et al., 2015; Folorunso et al., 2021). *Pseudomonas* spp., known to produce antimicrobial compounds, compete for nutrients and space and may also enhance induced resistance and/or parasitism (Flury et al., 2016; Hass and Defago, 2005). Recent studies have shown the use of original biocontrol agents, SHb30 (*Sphingobium xenophagum*), G2 (*Aspergillus flavus*) and Chito13 (*Mycolicibacterium fortuitum*), isolated from aquaponic microbiota for control of *Pythium aphanidermatum* on lettuce through the mining of high throughput sequencing data (Stouvenakers et al., 2022; Stouvenakers et al., 2023). Thus, the indigenous microbiome in aquaponics can be a source of BCA that can be used in soilless conditions to improve plant and fish health. Currently, there is limited study on taxonomic identification of micro-phyla and their possible BCA characteristics. A list of BCA for various plant pathogens in hydroponics is given by Folorunso et al. (2021). On the other hand, the potential antagonistic effects of bacteria against fish pathogens and their use as BCA of fish disease in aquaculture are also reported. For example, *Pseudomonas aeruginosa* strains exhibit antagonistic activity against the fish pathogen *Saprolegnia parasitica*, and *Lactobacillus plantarum* shows inhibitory effects against *Saprolegnia parasitica* (Sirakov et al., 2016). However, it may be noted that studies are lacking on the mechanism of interaction between plants, fish and the aquaponic microbial community as biocontrol agents against pathogens.

8.4.3 Microbial Communities in Feed Production

8.4.3.1 Microbiome as Fish Food

Apart from the diverse function of microorganisms in aquatic ecosystems as nutrient recyclers, bioremediation and biocontrol agents, they also serve as a direct source of food for higher trophic levels in aquaculture systems (Martinez-Cordova et al., 2017). The pathway

$$\text{Dissolved C + N} \rightarrow \text{C + N in microbes} \rightarrow \text{C + N in farmed organisms}$$

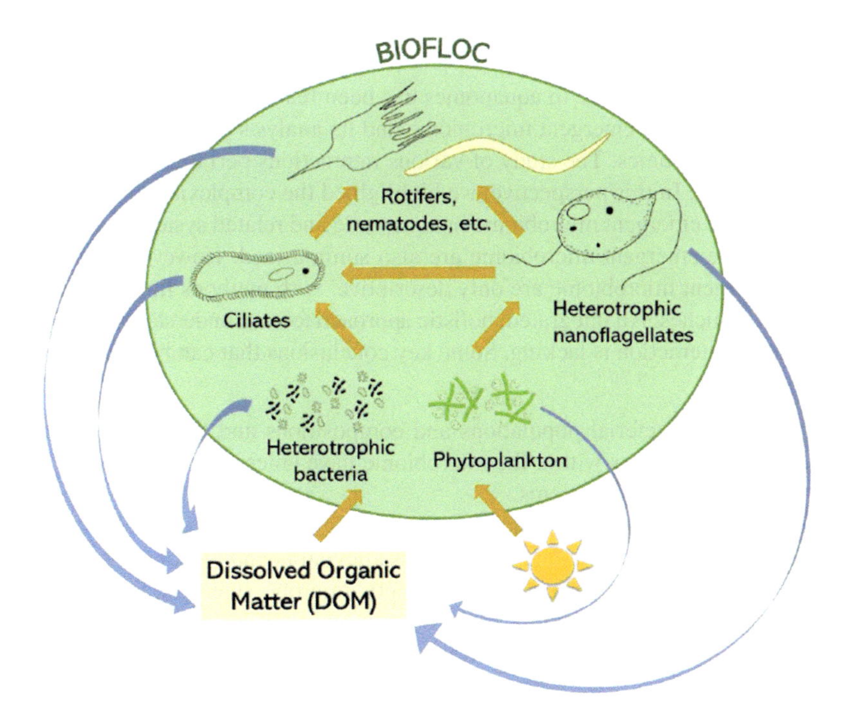

FIGURE 8.9 Schematic illustration of a simplified microbial loop and food web within a biofloc system in an outdoor setting, where sunlight is expected to promote microalgal growth (algae utilizing nitrogenic compounds, i.e., ammonia released from the organic matter). (From Bagi, A. et al., A desktop study on biofloc technology, 2023.)

is responsible for the conversion of waste nitrogen to fish food. This microbial loop can be a source of food and ammonia removal (Figure 8.9). Aquaculture systems like biofloc and FLOCponics rely on similar microbial loops, including the feed and/or fecal matter–microbe interaction (Pinho et al., 2022; Bagi et al., 2023). This microbial loop can also be promoted in aquaculture systems by modifying the cultural conditions based on manipulation of the C/N ratio (see Section 8.3). At a high C/N ratio, the microbial loop produces a high amount of biomass, composed of heterotrophic and autotrophic bacteria, archaea and organic matter, also known as biofloc. This biomass can be used by cultured fish and shrimp as a direct source of food (Martinez-Cordova et al., 2017; Martinez-Cordova et al., 2023; Minaz and Kubilay, 2021). There are two ways in which microbial biomass (biofloc) can be used as a direct food source for cultured organisms – in situ and ex-situ. In an in-situ system, growth is stimulated within the cultures by maintaining a high proportion of C/N through the addition of carbon sources or using feed with a low N content. In an ex-situ system, the production of the microbial biomass is done separately using the wastewater from the fish culture, containing inorganic nitrogen, by adding a carbon source. The biofloc is harvested and incorporated into the fish culture units as food. Another third alternative is the use of separate reactors in which the ex-situ bioflocs are harvested and incorporated into the feedstuff formulations (Martinez-Cordova et al., 2017; Martinez-Cordova et al., 2023). Integration of a fish food/floc production unit in aquaponics using the aquaponic microbiome is highly desirable, as it can reduce the input of feed protein (Martinez-Cordova et al., 2017, 2023; Bagi et al., 2023). The method will increase nutrient recovery, mitigate waste generation, and make aquaponics more sustainable. Also, it can help in the integration of the aquaponic system into an integrated multi-trophic aquaculture system (Martinez-Cordova et al., 2023).

8.5 CONCLUSIONS

The importance of the microbiome in aquaponics has been realized in recent years, and it is clear that the understanding of the emergent microbiome and its analysis are paramount in the development of sustainable aquaponics. The study of various interactions pertaining to aquaponic ecosystems is not yet complete. In this perspective, we highlighted the complexity, the present knowledge and the potential of the emergent microbiome in aquaponic and related systems. Some of the potential applications of the emergent microbiome are also summarized. However, most of the studies on the complex emergent microbiome are only descriptive, with methods limited to a few phyla or species or ecological niches. An integrated, holistic approach toward understanding the host–microbiome–environment interaction is lacking. Some key conclusions that can be made are:

1. The knowledge of bacterial populations and compositions and the understanding of the relationship and functions within the microbiome have emerged as an untapped resource for sustainable aquaponics systems.
2. There is a need for the exact identification of species and to understand their specific functions in making bio-fertilizers and as salt stress alleviators and biocontrol agents (BCA).
3. A clear understanding of specific plant variety–microbe interactions and the role of microbes as plant growth-promoting microorganisms is still lacking.
4. The role of environmental conditions in fish–plant–microbiota interactions needs further study.
5. The understanding of the role of microbial populations and gene expression in specific biological processes involved in aquaponics, like the nitrogen cycle, phosphorus cycle, etc., is still far from complete.
6. A meaningful correlation between various micro-units in micro-environments and the microbial community in aquaponics is still to be discovered.
7. The important role of root-released organic compounds – root exudates – needs further study in the context of mechanisms of stress response in plants, the physiology and behaviour of farmed fish, and their beneficial or toxic effects.
8. The potential adaptation of PGPM, probiotics, and other beneficial microorganisms needs to be explored in the future.

REFERENCES

Ajijah, Nur, Arina Yuthi Apriyana, Lies Sriwuryandari et al. "Beneficiary of nitrifying bacteria for enhancing lettuce (*Lactuca sativa*) and vetiver grass (*Chrysopogon zizanioides* L.) growths align with carp (*Cyprinus carpio*) cultivation in an aquaponic system." *Environmental Science and Pollution Research* 28 (2021): 880–889.

Akbar, Siddiq, Lei Gu, Yunfei Sun, Lu Zhang, Kai Lyu, Yuan Huang, and Zhou Yang. "Understanding host-microbiome-environment interactions: Insights from *Daphnia* as a model organism." *Science of the Total Environment* 808 (2022): 152093.

Appels, Lise, Jan Baeyens, Jan Degrève, and Raf Dewil. "Principles and potential of the anaerobic digestion of waste-activated sludge." *Progress in Energy and Combustion Science* 34, no. 6 (2008): 755–781.

Avnimelech, Yoram. "Carbon/nitrogen ratio as a control element in aquaculture systems." *Aquaculture* 176, no. 3–4 (1999): 227–235.

Avnimelech, Yoram. "Bio-filters: the need for an new comprehensive approach." *Aquacultural Engineering* 34, no. 3 (2006): 172–178.

Avnimelech, Yoram. *Biofloc technology, a practical guidebook,* 3rd Edition, pp. 258. World Aquaculture Soc, 2015.

Bagi, Andrea, Luís Henrique Poersch, and Elisa Ravagnan. "A desktop study on biofloc technology." (2023). NORCE Norwegian Research Centre, 9-2023 – NORCE Climate and Environment. Pp1-40, ISBN978-82-8408-300-1

Bartelme, Ryan P., Ben O. Oyserman, Jesse E. Blom, Osvaldo J. Sepulveda-Villet, and Ryan J. Newton. "Stripping away the soil: Plant growth promoting microbiology opportunities in aquaponics." *Frontiers in Microbiology* 9 (2018): 8.

Bartelme, Ryan P., Matthew C. Smith, Osvaldo J. Sepulveda-Villet, and Ryan J. Newton. "Component microenvironments and system biogeography structure microorganism distributions in recirculating aquaculture and aquaponic systems." *Msphere* 4, no. 4 (2019): 10–1128.

Bartelme, R. P., McLellan, S. L., and Newton, R. J. "Freshwater recirculating aquaculture system operations drive biofilter bacterial community shifts around a stable nitrifying consortium of ammonia-oxidizing archaea and comammox *Nitrospira*." *Frontiers in Microbiology* 8 (2017): 241488.

Baßmann, Björn, Harvey Harbach, Stephan Weißbach, and Harry W. Palm. "Effect of plant density in coupled aquaponics on the welfare status of African catfish, *Clarias gariepinus*." *Journal of the World Aquaculture Society* 51, no. 1 (2020): 183–199.

Baßmann, Björn, Matthias Brenner, and Harry W. Palm. "Stress and welfare of African catfish (*Clarias gariepinus* Burchell, 1822) in a coupled aquaponic system." *Water* 9, no. 7 (2017): 504.

Bates, M. L., and M. E. Stanghellini. "Root rot of hydroponically grown spinach caused by *Pythium aphanidermatum* and *P. dissotocum*." *Plant Disease* 68, no. 11 (1984): 989–991.

Bentzon-Tilia, Mikkel, Eva C. Sonnenschein, and Lone Gram. "Monitoring and managing microbes in aquaculture–Towards a sustainable industry." *Microbial Biotechnology* 9, no. 5 (2016): 576–584.

Berg, Gabriele, Daria Rybakova, Doreen Fischer, Tomislav Cernava, Marie-Christine Champomier Vergès, Trevor Charles, Xiaoyulong Chen et al. "Microbiome definition re-visited: Old concepts and new challenges." *Microbiome* 8, no. 1 (2020): 1–22.

Blancheton, J. P., K. J. K. Attramadal, L. Michaud, E. Roque d'Orbcastel, and O. Vadstein. "Insight into bacterial population in aquaculture systems and its implication." *Aquacultural Engineering* 53 (2013): 30–39.

Bouwmeester, Harro J., Christophe Roux, Juan Antonio Lopez-Raez, and Guillaume Becard. "Rhizosphere communication of plants, parasitic plants and AM fungi." *Trends in Plant Science* 12, no. 5 (2007): 224–230.

Brailo, Marina, Harold J. Schreier, Ryan McDonald, Jasna Maršić-Lučić, Ana Gavrilović, Marijana Pećarević, and Jurica Jug-Dujaković. "Bacterial community analysis of marine recirculating aquaculture system bioreactors for complete nitrogen removal established from a commercial inoculum." *Aquaculture* 503 (2019): 198–206.

Bromage, E. S., A. Thomas, and L. Owens. "Streptococcus iniae, a bacterial infection in barramundi Lates calcarifer." *Diseases of Aquatic Organisms* 36, no. 3 (1999): 177–181.

Bruni, Leonardo, Roberta Pastorelli, Carlo Viti, Laura Gasco, and Giuliana Parisi. "Characterisation of the intestinal microbial communities of rainbow trout (*Oncorhynchus mykiss*) fed with *Hermetia illucens* (black soldier fly) partially defatted larva meal as partial dietary protein source." *Aquaculture* 487 (2018): 56–63.

Cerozi, B. da S., and K. Fitzsimmons. "Phosphorus dynamics modeling and mass balance in an aquaponics system." *Agricultural Systems* 153 (2017): 94–100.

Chen, Cheng-Zhuang, Ping Li, Ling Liu, and Zhi-Hua Li. "Exploring the interactions between the gut microbiome and the shifting surrounding aquatic environment in fisheries and aquaculture: A review." *Environmental Research* 214 (2022): 114202.

Chiaranunt, Peerapol, and James F. White. "Plant beneficial bacteria and their potential applications in vertical farming systems." *Plants* 12, no. 2 (2023): 400.

Choudhury, Abhinav, Christine Lepine, and Christopher Good. "Methane and hydrogen sulfide production from the anaerobic digestion of fish sludge from recirculating aquaculture systems: Effect of varying initial solid concentrations." *Fermentation* 9, no. 2 (2023): 94.

Cripps, Simon J., and Asbjørn Bergheim. "Solids management and removal for intensive land-based aquaculture production systems." *Aquacultural Engineering* 22, no. 1–2 (2000): 33–56.

da Silva Cerozi, Brunno, and Kevin Fitzsimmons. "The effect of pH on phosphorus availability and speciation in an aquaponics nutrient solution." *Bioresource Technology* 219 (2016): 778–781.

Daims, H., Lebedeva, E. V., Pjevac, P., Han, P., Herbold, C., Albertsen, M., ... and Wagner, M. "Complete nitrification by Nitrospira bacteria." *Nature* 528, no. 7583 (2015): 504–509.

Delaide, Boris, Guillaume Delhaye, Michael Dermience, James Gott, Hélène Soyeurt, and M. Haissam Jijakli. "Plant and fish production performance, nutrient mass balances, energy and water use of the PAFF Box, a small-scale aquaponic system." *Aquacultural Engineering* 78 (2017): 130–139.

Delaide, Boris, Hendrik Monsees, Amit Gross, and Simon Goddek. "Aerobic and anaerobic treatments for aquaponic sludge reduction and mineralisation." In *Aquaponics Food Production Systems*, edited by Simon Goddek, Alyssa Joyce, Benz Kotzen, and Gavin M. Burnell, Springer Nature Switzerland AG, 2019: 247.

Delaide, Boris, Simon Goddek, Karel Keesman, and Haissam Jijakli. "A methodology to quantify aerobic and anaerobic sludge digestion performances for nutrient recycling in aquaponics." *Biotechnologie, Agronomie, Société Et Environnement* 22, no. 2 (2018).

DeLong, Dennis P., and Thomas Losordo. *How to Start a Biofilter.* Southern Regional Aquaculture Center, 2012.

Dhawi, Faten. "The role of Plant Growth-Promoting Microorganisms (PGPMs) and their feasibility in hydroponics and vertical farming." *Metabolites* 13, no. 2 (2023): 247.

Ebeling, James M., Michael B. Timmons, and J. J. Bisogni. "Engineering analysis of the stoichiometry of photoautotrophic, autotrophic, and heterotrophic removal of ammonia–nitrogen in aquaculture systems." *Aquaculture* 257, no. 1–4 (2006): 346–358.

Eck, Mathilde, Abdoul Razack Sare, Sébastien Massart, Zala Schmautz, Ranka Junge, Theo H. M. Smits, and M. Haïssam Jijakli. "Exploring bacterial communities in aquaponic systems." *Water* 11, no. 2 (2019a): 260.

Eck, Mathilde, Oliver Körner, and M. Haïssam Jijakli. "Nutrient cycling in aquaponics systems." In *Aquaponics Food Production Systems: Combined Aquaculture and Hydroponic Production Technologies For the Future*, edited by Simon Goddek, Alyssa Joyce, Benz Kotzen, Gavin M. Burnell, Springer Nature Switzerland AG, 2019b: 231–246.

Endut, Azizah, A. Jusoh, N. Ali, W. B. Wan Nik, and A. Hassan. "A study on the optimal hydraulic loading rate and plant ratios in recirculation aquaponic system." *Bioresource Technology* 101, no. 5 (2010): 1511–1517.

Etesami, Hassan, Somayeh Emami, and Hossein Ali Alikhani. "Potassium solubilizing bacteria (KSB): Mechanisms, promotion of plant growth, and future prospects: A review." *Journal of Soil Science and Plant Nutrition* 17, no. 4 (2017): 897–911.

Flury, Pascale, Nora Aellen, Beat Ruffner, Maria Péchy-Tarr, Shakira Fataar, Zane Metla, Ana Dominguez-Ferreras et al. "Insect pathogenicity in plant-beneficial pseudomonads: Phylogenetic distribution and comparative genomics." *The ISME Journal* 10, no. 10 (2016): 2527–2542.

Folorunso, Ewumi Azeez, Koushik Roy, Radek Gebauer, Andrea Bohatá, and Jan Mraz. "Integrated pest and disease management in aquaponics: A metadata-based review." *Reviews in Aquaculture* 13, no. 2 (2021): 971–995.

Fu, C., H. Liu, and Z. Yang. "Analysis of pollution and treatment status of residual bait feces in the development of domestic recirculating aquaculture." *Journal of Anhui Agricultural Sciences* 46 (2018): 76–79.

Garcia, Jean-Louis, Bharat K. C. Patel, and Bernard Ollivier. "Taxonomic, phylogenetic, and ecological diversity of methanogenic Archaea." *Anaerobe* 6, no. 4 (2000): 205–226.

George, Drishya M., Annette S. Vincent, and Hamish R. Mackey. "An overview of anoxygenic phototrophic bacteria and their applications in environmental biotechnology for sustainable resource recovery." *Biotechnology Reports* 28 (2020): e00563.

Gerardi, Michael H. *Wastewater Bacteria.* John Wiley & Sons, 2006.

Gichana, Z.M., Liti, D., Waidbacher, H. *et al.* "Waste management in recirculating aquaculture system through bacteria dissimilation and plant assimilation." *Aquacult Int* 26 (2018): 1541–1572.

Goddek, Simon, Boris Delaide, Utra Mankasingh, Kristin Vala Ragnarsdottir, Haissam Jijakli, and Ragnheidur Thorarinsdottir. "Challenges of sustainable and commercial aquaponics." *Sustainability* 7, no. 4 (2015): 4199–4224.

Goddek, Simon, Boris P. L. Delaide, Alyssa Joyce, Sven Wuertz, M. Haïssam Jijakli, Amit Gross, Ep H. Eding et al. "Nutrient mineralization and organic matter reduction performance of RAS-based sludge in sequential UASB-EGSB reactors." *Aquacultural Engineering* 83 (2018): 10–19.

Goddek, S., Joyce, A., Kotzen, B., and Burnell, G. M. *Aquaponics food production systems: combined aquaculture and hydroponic production technologies for the future.* pp. 619. Springer Nature, 2019.

Goldberg, N. P., M. E. Stanghellini, and S. L. Rasmussen. "Filtration as a method for controlling *Pythium* root rot of hydroponically grown cucumbers." *Plant Disease* 76, no. 8 (1992): 777–779.

Gouda, Sushanto, Rout George Kerry, Gitishree Das, Spiros Paramithiotis, Han-Seung Shin, and Jayanta Kumar Patra. "Revitalization of plant growth promoting rhizobacteria for sustainable development in agriculture." *Microbiological Research* 206 (2018): 131–140.

Graber, Andreas, and Ranka Junge. "Aquaponic systems: Nutrient recycling from fish wastewater by vegetable production." *Desalination* 246, no. 1–3 (2009): 147–156.

Gutierrez-Wing, Maria Teresa, and Ronald F. Malone. "Biological filters in aquaculture: Trends and research directions for freshwater and marine applications." *Aquacultural Engineering* 34, no. 3 (2006): 163–171.

Haas, Dieter, and Geneviève Défago. "Biological control of soil-borne pathogens by fluorescent pseudomonads." *Nature Reviews Microbiology* 3, no. 4 (2005): 307–319.

Hacquard, S., Garrido-Oter, R., González, A., Spaepen, S., Ackermann, G., Lebeis, S., ... and Schulze-Lefert, P. "Microbiota and host nutrition across plant and animal kingdoms." *Cell Host & Microbe* 17, no. 5 (2015): 603–616.

Hagopian, Daniel S., and John G. Riley. "A closer look at the bacteriology of nitrification." *Aquacultural Engineering* 18, no. 4 (1998): 223–244.

Hu, Zhen, Jae Woo Lee, Kartik Chandran, Sungpyo Kim, Ariane Coelho Brotto, and Samir Kumar Khanal. "Effect of plant species on nitrogen recovery in aquaponics." *Bioresource Technology* 188 (2015): 92–98.

Joyce, Alyssa, Mike Timmons, Simon Goddek, and Timea Pentz. "Bacterial relationships in aquaponics: New research directions." In *Aquaponics Food Production Systems: Combined Aquaculture and Hydroponic Production Technologies for the Future*, edited by Simon Goddek, Alyssa Joyce, Benz Kotzen, Gavin M. Burnell, Springer Nature Switzerland AG, (2019): 145–161.

Junge, Ranka, Nadine Antenen, Morris Villarroel, Tjaša Griessler Bulc, Andrej Ovca, and Sarah Milliken. "Aquaponics textbook for higher education." (2020). Zenodo. https://doi.org/10.5281/zenodo.3948179

Karkman, A., K. Mattila, M. Tamminen, and M. Virta. "Cold temperature decreases bacterial species richness in nitrogen-removing bioreactors treating inorganic mine waters." *Biotechnology and Bioengineering* 108, no. 12 (2011): 2876–2883.

Kasozi, Nasser, Benjamin Abraham, Horst Kaiser, and Brendan Wilhelmi. "The complex microbiome in aquaponics: Significance of the bacterial ecosystem." *Annals of Microbiology* 71, no. 1 (2021): 1–13.

Kasozi, Nasser, Gerald Degu Iwe, John Walakira, and Sandra Langi. "Integration of probiotics in aquaponic systems: An emerging alternative approach." *Aquaculture International* 32, no. 2 (2024): 2131–2150.

Kasozi, Nasser, Horst Kaiser, and Brendan Wilhelmi. "Metabarcoding analysis of bacterial communities associated with media grow bed zones in an aquaponic system." *International Journal of Microbiology* 2020 (2020): Article ID 8884070, 10 pages.

Kasozi, N., Tandlich, R., Fick, M., Kaiser, H., and Wilhelmi, B. "Iron supplementation and management in aquaponic systems: A review." *Aquaculture Reports* 15 (2019): 100221.

Khanjani, Mohammad Hossein, Alireza Mohammadi, and Maurício Gustavo Coelho Emerenciano. "Microorganisms in biofloc aquaculture system." *Aquaculture Reports* 26 (2022): 101300.

Khiari, Zied, Soba Kaluthota, and Nick Savidov. "Aerobic bioconversion of aquaculture solid waste into liquid fertilizer: Effects of bioprocess parameters on kinetics of nitrogen mineralization." *Aquaculture* 500 (2019): 492–499.

Kiari, B. K. K., M. Moussa, M. M. Inoussa, A. T. Abasse, S. Atta, and Y. Bakasso. "Effect of Di-Ammonium Phosphate on the agro-morphological parameters of the sorrel ecotypes (*Hibiscus sabdariffa* L.) in two agro-climatic zones of Niger." *International Journal of Biological and Chemical Sciences* 13, no. 3 (2019): 1596–1612.

Kim, Kyochan, Jun Wook Hur, Soohwan Kim, Joo-Young Jung, and Hyon-Sob Han. "Biological wastewater treatment: Comparison of heterotrophs (BFT) with autotrophs (ABFT) in aquaculture systems." *Bioresource Technology* 296 (2020): 122293.

Kim, Pil Soo, Na-Ri Shin, Jae-Bong Lee, Min-Soo Kim, Tae Woong Whon, Dong-Wook Hyun, Ji-Hyun Yun, Mi-Ja Jung, Joon Yong Kim, and Jin-Woo Bae. "Host habitat is the major determinant of the gut microbiome of fish." *Microbiome* 9, no. 1 (2021): 166.

Lederberg, Joshua, and Alexa T. McCray. "Ome sweetomics – A genealogical treasury of words." *The Scientist* 15, no. 7 (2001): 8–8.

Lee, Seungjun, and Jiyoung Lee. "Beneficial bacteria and fungi in hydroponic systems: Types and characteristics of hydroponic food production methods." *Scientia Horticulturae* 195 (2015): 206–215.

Li, Changwei, Xiaoyu Zhang, Yu Chen, Shiyu Zhang, Limin Dai, Wenjing Zhu, and Yuan Chen. "Optimized utilization of organic carbon in aquaculture biofloc systems: A review." *Fishes* 8, no. 9 (2023): 465.

Lobanov, Victor P., Doriane Combot, Pablo Pelissier, Laurent Labbé, and Alyssa Joyce. "Improving plant health through nutrient remineralization in aquaponic systems." *Frontiers in Plant Science* (2021): 1064.

Lobanov, V., Keesman, K. J., and Joyce, A. "Plants dictate root microbial composition in hydroponics and aquaponics." *Frontiers in Microbiology* 13, (2022): 848057.

Lorgen-Ritchie, Marlene, Tamsyn Uren Webster, Jamie McMurtrie, David Bass, Charles R. Tyler, Andrew Rowley, and Samuel A. M. Martin. "Microbiomes in the context of developing sustainable intensified aquaculture." *Frontiers in Microbiology* 14 (2023): 1200997. doi: 10.3389/fmicb.2023.1200997

Losordo, Thomas M., Michael P. Masser, and James Rakocy. *Recirculating Aquaculture Tank Production Systems*. Southern Regional Aquaculture Center, 2000.

Love, David C., Jillian P. Fry, Ximin Li, Elizabeth S. Hill, Laura Genello, Ken Semmens, and Richard E. Thompson. "Commercial aquaponics production and profitability: Findings from an international survey." *Aquaculture* 435 (2015a): 67–74.

Love, David C., Laura Genello, Ximin Li, Richard E. Thompson, and Jillian P. Fry. "Production and consumption of homegrown produce and fish by noncommercial aquaponics gardeners." *Journal of Agriculture, Food Systems, and Community Development* 6, no. 1 (2015b): 161–173.

Love, David C., Michael S. Uhl, and Laura Genello. "Energy and water use of a small-scale raft aquaponics system in Baltimore, Maryland, United States." *Aquacultural Engineering* 68 (2015c): 19–27.

Marchaim, Uri. *Biogas Processes for Sustainable Development.* No. 95–96. Food & Agriculture Org., 1992.

Martinez-Cordova, Luis R., Mauricio G. C. Emerenciano, Anselmo Miranda-Baeza, Sara M. Pinho, Estefanía Garibay-Valdez, and Marcel Martínez-Porchas. "Advancing toward a more integrated aquaculture with polyculture, aquaponics, biofloc technology, FLOCponics." *Aquaculture International* 31, no. 2 (2023): 1057–1076.

Martínez-Córdova, Luis Rafael, Marcel Martínez-Porchas, Maurício Gustavo Coelho Emerenciano, Anselmo Miranda-Baeza, and Teresa Gollas-Galván. "From microbes to fish the next revolution in food production." *Critical Reviews in Biotechnology* 37, no. 3 (2017): 287–295.

Maucieri, C., Nicoletto, C., Schmautz, Z., Sambo, P., Komives, T., Borin, M., and Jung, R. "Vegetable intercropping in a small-scale aquaponic system." *Agronomy* 7, no. 4 (2017): 63.

Meena, Vijay Singh, Indra Bahadur, Bihari Ram Maurya, Ashok Kumar, Rajesh Kumar Meena, Sunita Kumari Meena, and Jay Prakash Verma. "Potassium-solubilizing microorganism in evergreen agriculture: An overview." *Potassium Solubilizing Microorganisms For Sustainable Agriculture*, edited by Meena, V., Maurya, B., Verma, J., and Meena, R., Springer, New Delhi, 2016. https://doi.org/10.1007/978-81-322-2776-2

Mehmood, Tariq, Gajendra Kumar Gaurav, Liu Cheng, Jiří Jaromír Klemeš, Muhammad Usman, Awais Bokhari, and Jie Lu. "A review on plant-microbial interactions, functions, mechanisms and emerging trends in bioretention system to improve multi-contaminated stormwater treatment." *Journal of Environmental Management* 294 (2021): 113108.

Menezes-Blackburn, Daniel, Nahad Al-Mahrouqi, Buthaina Al-Siyabi, Adhari Al-Kalbani, Ralf Greiner, and Sergey Dobretsov. "Bacterial communities associated with the cycling of non-starch polysaccharides and phytate in aquaponics systems." *Diversity* 13, no. 12 (2021): 631.

Michaud, Luigi, Angelina Lo Giudice, Filippo Interdonato, Sebastien Triplet, Liu Ying, and Jean Paul Blancheton. "C/N ratio-induced structural shift of bacterial communities inside lab-scale aquaculture biofilters." *Aquacultural Engineering* 58 (2014): 77–87.

Minaz, Mert, and Aysegul Kubilay. "Operating parameters affecting biofloc technology: Carbon source, carbon/nitrogen ratio, feeding regime, stocking density, salinity, aeration, and microbial community manipulation." *Aquaculture International* 29, no. 3 (2021): 1121–1140.

Mir, Muzaffar Ahmad, Athar Hussain, and Chanchal Verma. "Design considerations and operational performance of anaerobic digester: A review." *Cogent Engineering* 3, no. 1 (2016): 1181696.

Mirzoyan, Natella, Yossi Tal, and Amit Gross. "Anaerobic digestion of sludge from intensive recirculating aquaculture systems." *Aquaculture* 306, no. 1–4 (2010): 1–6.

Monsees, Hendrik, Jonas Keitel, Maurice Paul, Werner Kloas, and Sven Wuertz. "Potential of aquacultural sludge treatment for aquaponics: Evaluation of nutrient mobilization under aerobic and anaerobic conditions." *Aquaculture Environment Interactions* 9 (2017): 9–18.

Mourouzidou, Snezhana, Georgios K. Ntinas, Aphrodite Tsaballa, and Nikolaos Monokrousos. "Introducing the power of plant growth promoting microorganisms in soilless systems: A promising alternative for sustainable agriculture." *Sustainability* 15, no. 7 (2023): 5959.

Munguia-Fragozo, P., Alatorre-Jacome, O., Rico-Garcia, E., Torres-Pacheco, I., Cruz-Hernandez, A., Ocampo-Velazquez, R. V., … and Guevara-Gonzalez, R. G. "Perspective for aquaponic sytems: "omic" technologies for microbial community analysis." *BioMed Research International* 2015 (2015).

Nardi, Pierfrancesco, Hendrikus J. Laanbroek, Graeme W. Nicol et al. "Biological nitrification inhibition in the rhizosphere: Determining interactions and impact on microbially mediated processes and potential applications." *Fems Microbiology Reviews* 44, no. 6 (2020): 874–908.

Perry, William Bernard, Elle Lindsay, Christopher James Payne, Christopher Brodie, and Raminta Kazlauskaite. "The role of the gut microbiome in sustainable teleost aquaculture." *Proceedings of the Royal Society B* 287, no. 1926 (2020): 20200184.

Pinho, Sara M., Luiz H. David, Fabiana Garcia, Maria Célia Portella, and Karel J. Keesman. "Sustainability assessment of FLOCponics compared to stand-alone hydroponic and biofloc systems using emergy synthesis." *Ecological Indicators* 141 (2022): 109092.

Prabhu, Neha, Sunita Borkar, and Sandeep Garg. "Phosphate solubilization by microorganisms: Overview, mechanisms, applications and advances." *Advances in Biological Science Research* (2019): 161–176.

Radzki, W., F. J. Gutierrez Mañero, E. Algar, J. A. Lucas García, A. García-Villaraco, and B. Ramos Solano. "Bacterial siderophores efficiently provide iron to iron-starved tomato plants in hydroponics culture." *Antonie Van Leeuwenhoek* 104 (2013): 321–330.

Rakocy, James. "Ten guidelines for aquaponic systems." *Aquaponics J* 46 (2007): 14–17.

Raviv, M., Lieth, J. H., and Bar-Tal, A. "Significance of soilless culture in agriculture." *Soilless Culture* (2008): 1–11.

Rawat, Pratibha, Sudeshna Das, Deepti Shankhdhar, and S. C. Shankhdhar. "Phosphate-solubilizing microorganisms: Mechanism and their role in phosphate solubilization and uptake." *Journal of Soil Science and Plant Nutrition* 21 (2021): 49–68.

Ringø, E., Z. Zhou, R. E. Olsen, and S. K. Song. "Use of chitin and krill in aquaculture–the effect on gut microbiota and the immune system: A review." *Aquaculture Nutrition* 18, no. 2 (2012): 117–131.

Rivas-García, Tomás, Ramsés Ramón González-Estrada, Roberto Gregorio Chiquito-Contreras, Juan José Reyes-Pérez, Uriel González-Salas, Luis Guillermo Hernández-Montiel, and Bernardo Murillo-Amador. "Biocontrol of phytopathogens under aquaponics systems." *Water* 12, no. 7 (2020): 2061.

Robaina, L., Pirhonen, J., Mente., Sánchez, J., and Goosen, N. "Fish diets in aquaponics." *Aquaponics Food Production Systems: Combined Aquaculture and Hydroponic Production Technologies for the Future* (2019): 333–352.

Romano, Nicholas, Carl Webster, Surjya Narayan Datta, Gde Sasmita Julyantoro Pande, Hayden Fischer, Amit Kumar Sinha, George Huskey, Steven D. Rawles, and Shaun Francis. "Black Soldier Fly (*Hermetia illucens*) Frass on Sweet-Potato (*Ipomea batatas*) Slip Production with Aquaponics." *Horticulturae* 9, no. 10 (2023): 1088.

Ruiz, G., D. Jeison, and R. Chamy. "Nitrification with high nitrite accumulation for the treatment of wastewater with high ammonia concentration." *Water Research* 37, no. 6 (2003): 1371–1377.

Rurangwa, Eugene, and Marc C. J. Verdegem. "Microorganisms in recirculating aquaculture systems and their management." *Reviews In Aquaculture* 7, no. 2 (2015): 117–130.

Sallenave, Rossana. *Important Water Quality Parameters In Aquaponics Systems*. College of Agricultural, Consumer and Environmental Sciences, 2016.

Sanchez, Francia A., Valerie R. Vivian-Rogers, and Hidetoshi Urakawa. "Tilapia recirculating aquaculture systems as a source of plant growth promoting bacteria." *Aquaculture Research* 50, no. 8 (2019): 2054–2065.

Sattar, Annum, Muhammad Naveed, Mohsin Ali, Zahir A. Zahir, Sajid M. Nadeem, M. Yaseen, Vijay Singh Meena et al. "Perspectives of potassium solubilizing microbes in sustainable food production system: A review." *Applied Soil Ecology* 133 (2019): 146–159.

Schachtman, Daniel P., Robert J. Reid, and Sarah M. Ayling. "Phosphorus uptake by plants: from soil to cell." *Plant Physiology* 116, no. 2 (1998): 447–453.

Schmautz, Zala, Andreas Graber, Sebastian Jaenicke, Alexander Goesmann, Ranka Junge, and Theo H. M. Smits. "Microbial diversity in different compartments of an aquaponics system." *Archives of Microbiology* 199 (2017): 613–620.

Schmautz, Zala, Carlos A. Espinal, Andrea M. Bohny, Fabio Rezzonico, Ranka Junge, Emmanuel Frossard, and Theo H. M. Smits. "Environmental parameters and microbial community profiles as an indication towards microbial activities and diversity in aquaponic system compartments." *Bmc Microbiology* 21, no. 1 (2021): 1–11.

Schmautz, Zala, Jean-Claude Walser, Carlos A. Espinal, Florentina Gartmann, Ben Scott, Joël F. Pothier, Emmanuel Frossard, Ranka Junge, and Theo H. M. Smits. "Microbial diversity across compartments in an aquaponic system and its connection to the nitrogen cycle." *Science of the Total Environment* 852 (2022): 158426.

Schneider, O., Chabrillon-Popelka, M., Smidt, H., Haenen, O., Sereti, V., Eding, E. H., and Verreth, J. A. "HRT and nutrients affect bacterial communities grown on recirculation aquaculture system effluents." *FEMS Microbiology Ecology* 60, no. 2 (2007): 207–219.

Schreier, Harold J., Natella Mirzoyan, and Keiko Saito. "Microbial diversity of biological filters in recirculating aquaculture systems." *Current Opinion In Biotechnology* 21, no. 3 (2010): 318–325.

Schwarz, Meier. "Some future aspects." In *Soilless Culture Management*, pp. 162–164. Berlin: Springer, 1995.

Seawright, Damon E., Robert R. Stickney, and Richard B. Walker. "Nutrient dynamics in integrated aquaculture–hydroponics systems." *Aquaculture* 160, no. 3–4 (1998): 215–237.

Senff, Paula, Björn Baßmann, Frederik Kaiser, Harvey Harbach, Christophe Robin, and Pascal Fontaine. "Root-released organic compounds in aquaponics and their potential effects on system performance." *Reviews in Aquaculture* 15, no. 5491 (2022): 1–7.

Sentenac, Hervé, and Claude Grignon. "Effect of pH on orthophosphate uptake by corn roots." *Plant Physiology* 77, no. 1 (1985): 136–141.

Singh, Maibam Birla, and Yumnam Rameshori. "En route to aquaponics in cold water: Identifying the gaps in principles and system design." In *Fisheries and Aquaculture of the Temperate Himalayas*. Springer, Singapore, (2023): 371–406.

Sirakov, Ivaylo, Matthias Lutz, Andreas Graber et al. "Potential for combined biocontrol activity against fungal fish and plant pathogens by bacterial isolates from a model aquaponic system." *Water* 8, no. 11 (2016): 518.

Somerville, Christopher, Moti Cohen, Edoardo Pantanella, Austin Stankus, and Alessandro Lovatelli. "Small-scale aquaponic food production: integrated fish and plant farming." *Fao Fisheries and Aquaculture Technical Paper* 589 (2014): I.

Stegelmeier, Ashley A., Danielle M. Rose, Benjamin R. Joris, and Bernard R. Glick. "The use of PGPB to promote plant hydroponic growth." *Plants* 11, no. 20 (2022): 2783.

Stouvenakers, Gilles, Peter Dapprich, Sebastien Massart, and M. Haïssam Jijakli. "Plant pathogens and control strategies in aquaponics." In *Aquaponics Food Production Systems*, edited by Simon Goddek, Alyssa Joyce, Benz Kotzen, and Gavin M. Burnell, Springer Nature Switzerland AG, 2019: 353–378.

Stouvenakers, Gilles, Sébastien Massart, and Haissam Jijakli. "Aquaponics as future urban food production systems: Phytopathological challenges and opportunities thanks to aquaponic microbiota characterization and original biocontrol agent isolation." *Bulletin Oilb/Srop* 165 (2023): 79–82.

Stouvenakers, Gilles, Sébastien Massart, and M. H. Jijakli. "First study case of microbial biocontrol agents isolated from aquaponics through the mining of high-throughput sequencing data to control *Pythium aphanidermatum* on lettuce." *Microbial Ecology* 86, no. 2 (2023): 1107–1119.

Stouvenakers, G., Massart, S., and Jijakli, M. H. "Application of aquaponic microorganisms alone or in consortium as original biocontrol method of lettuce root rots in soilless culture." In *XXXI International Horticultural Congress (IHC2022): International Symposium on Sustainable Control of Pests and Diseases 1378* (pp. 289–294), 2022, August.

Strauch, Sebastian Marcus, Lisa Carolina Wenzel, Adrian Bischoff et al. "Commercial African catfish (*Clarias gariepinus*) recirculating aquaculture systems: Assessment of element and energy pathways with special focus on the phosphorus cycle." *Sustainability* 10, no. 6 (2018): 1805.

Thomas, Phil, Oliver G. G. Knox, Jeff R. Powell, Brian Sindel, and Gal Winter. "The hydroponic Rockwool root microbiome: Under control or underutilised?." *Microorganisms* 11, no. 4 (2023): 835.

Timmons, M. B., T. Guerdat, and B. J. Vinci. *Recirculating aquaculture* 4th Ed." (p. 779). Ithaca, NY: Ithaca Publishing Company. (2018).

Tyson, Richard V., Danielle D. Treadwell, and Eric H. Simonne. "Opportunities and challenges to sustainability in aquaponic systems." *Horttechnology* 21, no. 1 (2011): 6–13.

Van Lier, Jules B., Nidal Mahmoud, and Grietje Zeeman. "Anaerobic wastewater treatment." In *Biological Wastewater Treatment: Principles, Modelling and Design*, edited by Mogens Henze, Mark C. M. van Loosdrecht, G. A. Ekama, and Damir Brdjanovic. IWA Publishing, (2008: 415–456.

Van Rijn, Jaap, Yossi Tal, and Harold J. Schreier. "Denitrification in recirculating systems: Theory and applications." *Aquacultural Engineering* 34, no. 3 (2006): 364–376.

Vives-Peris, Vicente, Carlos De Ollas, Aurelio Gómez-Cadenas, and Rosa María Pérez-Clemente. "Root exudates: from plant to rhizosphere and beyond." *Plant Cell Reports* 39 (2020): 3–17.

Walker, Claire A., and Pieter van West. "Zoospore development in the oomycetes." *Fungal Biology Reviews* 21, no. 1 (2007): 10–18.

Wang, Yi-Ju, Teng Yang, and Hye-Ji Kim. "pH dynamics in aquaponic systems: Implications for plant and fish crop productivity and yield." *Sustainability* 15, no. 9 (2023): 7137.

Wongkiew, Sumeth, Brian N. Popp, and Samir Kumar Khanal. "Nitrogen recovery and nitrous oxide (N2O) emissions from aquaponic systems: Influence of plant species and dissolved oxygen." *International Biodeterioration & Biodegradation* 134 (2018): 117–126.

Wongkiew, Sumeth, Zhen Hu, Kartik Chandran, Jae Woo Lee, and Samir Kumar Khanal. "Nitrogen transformations in aquaponic systems: A review." *Aquacultural Engineering* 76 (2017): 9–19.

Yang, Limin, Fufeng Chen, Yan Xiao, Huankai Li, Jun Li, Lijian Leng, Hui Liu et al. "Microbial community-assisted water quality control and nutrients recovery: Emerging technologies for the sustainable development of aquaponics." *Journal of Chemical Technology & Biotechnology* 94, no. 8 (2019): 2405–2411.

Yep, Brandon, and Youbin Zheng. "Aquaponic trends and challenges–A review." *Journal of Cleaner Production* 228 (2019): 1586–1599.

Zhanga, Hong, Yueshu Gaoa, Junyan Liua et al. "Recovery of nutrients from fish sludge as liquid fertilizer to enhance sustainability of aquaponics: A review." *Chemical Engineering* 83 (2021): 55–60.

Zou, Yina, Zhen Hu, Jian Zhang, Huijun Xie, Christophe Guimbaud, and Yingke Fang. "Effects of pH on nitrogen transformations in media-based aquaponics." *Bioresource Technology* 210 (2016): 81–87.

9 Microbial Community of Biofloc

M. Rajkumar

9.1 INTRODUCTION

Aquaculture systems mainly depend on the exploitation of autotrophic and heterotrophic microbial food webs. A heterotrophic food web consists of the decomposition of organic matter by microorganisms, leading to the formation of assimilable detritus and inorganic nutrients. The detritus and associated microbes are directly consumed by the cultured animals or by other small animals on which the cultured species feed (Moriarty, 1987; Moriarty, 1997). The heterotrophic food web consistently appears as a major contributor to the total production of the target animals (Schroeder, 1987). The fish assimilate only 15–30% of the nitrogen feed in a pond environment (Davenport et al., 2009), and the remaining quantity is lost to the system as ammonia and organic-N in the form of feces and feed residue. The organic-N in feces and uneaten feed undergoes decomposition, resulting in ammonia production. This shows that a high protein level in fish feed contributes to a high concentration of ammonia in the water column, which is detrimental to the cultured animals and therefore, needs to be minimized. Microbial proteins are generated in ponds with organic matter as manure or feed and are decomposed by microorganisms such as bacteria and protozoa under both aerobic and anaerobic conditions. The organic matter decomposition process under aerobic conditions is faster than in anaerobic conditions (Reddy and Patrick, 1975). Microorganisms play a major role concerning natural productivity, nutrient cycling, water quality, and nutrition of cultured animals (McIntosh et al., 2000).

9.2 BIOFLOC DEFINITION

Biofloc is the core of this technology, and it is defined as macro-aggregates composed of protozoa, diatoms, zooplankton, macroalgae, feces, exoskeletons, and remains of dead organisms, bacteria, and invertebrates (Decamp et al., 2007). It is the suspended growth in ponds, which consists of aggregates of living and dead particulate organic matter, phytoplankton, bacteria, and grazers of the bacteria (Hargreaves, 2006). It can provide nutrients such as "native protein" (Emerenciano et al., 2011), lipids (Wasielesky et al., 2006), amino acids (Ju et al., 2008), and fatty acids (Ekasari et al., 2010) in the form of diverse microorganisms. It is used as a natural feed for many filter feeders, like *Litopenaeus vannamei* and *Oreochromis mossambicus* or *niloticus*. Biofloc is also called microbial protein. Microbial protein has higher availability than feed protein in the fish culture system (Avnimelech, 2007).

9.3 MICROBIAL BIOFLOC COMPONENTS AND STRUCTURES

Microbial floc is made up of filamentous bacteria, protozoans, zooplankton, colloids, organic polymers, and dead cells, encased in a gelatinous matrix containing extracellular polymeric substances that encapsulate the microbial cells and play an important role in flocculation (Avnimelech, 2007). According to Ray et al. (2010), a biofloc particle contained microbial components such as high abundance chlorophytes, three types of diatoms, heterotrophic dinoflagellates, a nematode, a rotifer,

DOI: 10.1201/9781003408543-9

and others. The zeta potential and Van der Waals forces are primarily responsible for bacterial grouping within the floc (Sobeck and Higgins, 2002). Microbial flocs are composed of a heterogeneous mixture of microorganisms (floc formers and filamentous bacteria), particles, colloids, organic polymers, cations, and dead cells (Jorand et al., 1995). Typical flocs are irregular in shape, finely divided, easily compressible, highly porous (up to more than 99% porosity), and permeable to fluids (Chu and Lee et al., 2004). Individual bacterial cells are about 1 μm in size (Madigan and Martinko, 2006). This implies that these organisms are generally surrounded by a layer of liquid, which impedes the mass transfer of nutrients and waste (Logan and Hunt, 1988). According to Wilen et al. (2003), the floc contains 2–20% live microbial cells, 60–70% total organic matter, and 30–40% total inorganic matter. Extracellular polymeric substances are the primary components of the floc matrix. These structures help to bind the floc's constituent parts together by forming a matrix that encases the microbial cells and they can make up to 80% of the mass of activated sludge systems. Polysaccharides, proteins, humic substances, nucleic acids, and lipids are their main components (Zita and Hermansson, 1994). They are produced as slime or capsule layers under a variety of nutritional conditions, but especially when nutrients such as nitrogen are limited (Steiner et al., 1976).

9.4 MICROBIAL COMMUNITY IN BIOFLOC

Biofloc is rich in microbes, and several microbes have been isolated from it. Proteobacteria, Bacteroidetes, Cyanobacteria, Planctomycetes, Actinobacteria, *Verrucomicrobia*, *Firmicutes*, *Chlorobi*, *Gemmatimonadetes*, *Chloroflexi*, NKB19, OD1, Lentisphaerae, GN02, and *Acidobacteria* were identified in biofloc. Among these, Proteobacteria and Bacteroidetes are the dominant bacteria in aquaculture (Wei et al., 2016; Rajkumar et al., 2016). The majority of Proteobacteria found in both intensive and extensive prawn ponds were symbiotic bacteria in aquaculture (Sakami et al., 2008). *Proteobacteria* degrade organic matter (Miura et al., 2007), particularly in biofloc wastewater treatment. *Proteobacteria* are also important in bacterial composition (Daims et al., 1999). This demonstrates that when a biofloc is used in a culture system, it can effectively regulate the quality of the aquaculture water. *Actinobacteria* is a common probiotic that can produce beneficial substances in certain conditions (Das et al., 2008). According to Gerardi (2006), *Flavobacterium* can produce glue-like extracellular polymers that can bind cells together. Rajkumar et al. (2018) isolated *Saccharophagus*, *Sulfitobacter*, *Rugeria*, *Pseudomonas*, *Phaeobacter*, *Vibrio*, and *Proteobacterium*, and the total heterotrophic bacteria population in biofloc ranged from 6.0×10^2 to 3.87×10^8 cfu mL^{-1}. Yale Deng et al. (2019) found four core bacteria in biofloc water, namely, *Rhodopirellula*, *Paracoccus*, *Nocardioides*, and *Blastopirellula*, which influenced the gut microbiota of *Litopenaeus vannamei* grown in the biofloc system.

According to Panigrahi et al. (2018), the carbon–nitrogen ratio in biofloc waters may determine bacterial dominance. This study revealed that *Vibrio* accounted for 79% in both control and C: N5 treated water; in C: N10, *Thauera* (62%) was the most represented taxon; and in C: N15, Attheyaceae (56%) and Peridiniaceae (30%) were the most dominant phyla. The phyla *Psychrobacter* (26%), *Proteobacteria* (25%), and *Peridiniaceae* (20%) were found to be dominant in C: N20. Jiang et al. (2020) investigated how light and carbon resources in the biofloc affect microbial structure in the biofloc system. According to the study, the most abundant bacterial classes were *Flavobacteria*, *Gammaproteobacteria*, *Sphingobacteria*, and *Alphaproteobacteria*. The addition of carbon sources increased bacterial richness but not diversity. Denitrifying bacteria of the genus *Paracoccus* were identified in the systems with both light and carbon sources, but some harmful bacteria, such as the genus *Leucothrix*, were identified in systems with carbon sources but no light.

Abakari et al. (2021) reported that bioflocs are abundant and more diverse in bacterial communities, and different carbon sources can influence the structure of bacterial communities. The study reported *Chloroflexi*, Actinobacteria, and *Norank_f_Caldilineaceae* in bioflocs. *Proteobacteria* and *Bacteroidetes*, which are important for organic degradation, were commonly dominant bacterial phyla in activated sludge systems and biofloc technology (BFT) aquaculture systems (Zhao et al., 2012). According to Li et al. (2019), the three most dominant phyla – *Proteobacteria*, *Bacteroidetes*, and *Firmicutes* – were present in biofloc and were isolated using 16S rRNA genes. Protozoa were also found to be the most abundant phylum within BFT systems (Miao et al., 2017; Deng et al., 2018). Proteobacteria have previously been shown to aid in the removal of organic matter, particularly in activated sludge systems (Miura et al., 2007), which are similar to the biofloc system. *Actinobacteria* are found to be a common probiotic that produce beneficial substances for biofloc growth (Das et al., 2008).

The low concentration of NO_2-N in systems containing MPs (microplastics) suggested that the presence of MPs might inhibit ammonia oxidation while promoting nitrite oxidation by altering the structure and function of the microbial community. These findings suggested that aggregates of bioflocs and MPs could form in aquaculture water, inhibiting their settlement and changing the nitrogen transformation function by influencing the microbial community composition.

9.5 METHODS FOR CHARACTERIZATION OF MICROBIAL COMMUNITY IN BIOFLOC SYSTEM

The most abundant microbes in the floc particles are chlorophytes (green algae), diatoms, dinoflagellates, nematodes, rotifers, and cyanobacteria (Ray et al., 2010). Heterotrophic ammonia-assimilative and chemoautotrophic nitrifying bacteria are primarily responsible for maintaining water quality and lead to minimal exchange of water in intensive systems (Hargreaves, 2006). In natural aquatic ecosystems, microorganisms play important roles in nutrient cycling, such as methane production (Chan et al., 2005), sulfate reduction (Li et al., 1999), and ammonia oxidation (Hastings et al., 1998). Microbiological aspects, particularly bacterial growth patterns, characterization of biofloc, and possible manipulation of the microbial community, are necessary for the successful design and operation of BFT (Azim et al., 2008). The easiest and most common method to determine the presence and to differentiate the type of microorganisms in a sample is microscopy. Since the method is based on visual morphology, it is generally not possible to identify all of them. It can be used to gain a value for the proportion of filamentous and zoogloeal flocs within a water sample. The fluorescence in-situ hybridization (FISH) procedure is based on the binding of fluorescently labeled DNA probes with the ribosomal RNA of bacteria (Amann et al., 1995). The DNA probes can be designed to exclusively bind to the rRNA of a chosen type of microorganism and thus allow the detection of a certain species in a community. Since rRNA is present only in biologically active organisms, it only allows the detection of those that are performing a specific task (non-active cells are not detected). Real-time polymerase chain reaction (RT-PCR) is a molecular technique that allows simultaneous amplification and quantification of the extracted DNA from a sample (Heid et al., 1996). This technique is commonly used to determine the amount and type of microorganisms present in a sample. A quantitative array allows simultaneous quantification of phylogenetic and functional genes involved in the activity of interest, e.g., nitrification and denitrification processes (Geets et al., 2007). As such, the evolution of a complete system can be analyzed.

Denaturing gradient gel electrophoresis (DGGE) is a molecular approach that provides information about the genetic microbial diversity in samples such as water, sludge, air, etc. (Muyzer et al., 1993). This technique is based on the separation of extracted DNA and PCR-amplified genes (mostly 16S rRNA genes), which is unique for a group of microorganisms. The analysis of an environmental sample using DGGE results in a band pattern in which roughly every band represents a specific microorganism. The information that can be obtained from a DGGE band pattern is limited, except for comparative purposes. For instance, only bacteria that are present in more than 1% of the total community are detected. The technique is mostly used as a research tool to visualize shifts in the microbial population composition over time. Relatively new concepts offer the possibility to make use of the DGGE band patterns in an alternative way, comparing the bacterial diversity in biofluids. Moving window analysis is a technique based on DGGE to detect shifts in the microbial community over time (Wittebolle et al., 2005). Through Bionumerics software (Applied Maths, Sint-Martens-Latem, Belgium), DGGE patterns can be analyzed and compared, thereby quantifying the differences. Both techniques can be applied in BFT. Relations between the shifts in microbial populations and changes in performance may be established.

Studies employing DGGE to compare the diversity of 16S rRNA gene amplicons obtained directly from samples are very rare. The DGGE method can indicate the existence of common groups of microorganisms in natural aquatic systems, but it does not permit more precise identification of these microbes to the species or sub-species level. DGGE has proved to be an exceptional tool to study species diversity and bacterial community dynamics. But, a more comprehensive understanding of the physiology of these organisms and their complex biogeochemical processes will require their cultivation, isolation, and characterization (Bae et al., 2005). It is well known that more than 90% of the microorganisms existing in nature are resistant to selective enrichment cultures (Ward et al., 1990). To overcome the constraints of these culture-dependent methods, researchers currently focus on the use of molecular biological techniques, with their powerful capacity to allow the analysis of microorganisms in their natural habitats. In this context, analysis of the 16S rRNA gene has been the most widely used approach in the last decade. Of the 16S rRNA gene–based methods used for studying complex microbial populations, DGGE has received the most attention and has been successfully applied to several natural habitats. Profiling of microbial communities from the soil, marine environments, hydrothermal vents, the gastrointestinal tract of humans, and even the study of a microbial community resident in a medieval wall painting (Nicol et al., 2003) has already been done using the DGGE technique. More recently, DGGE has been introduced into food microbiology for the identification of microorganisms isolated in food and for the evaluation of microbial diversity during food fermentation. Ray et al. (2010) characterized microbial communities in minimal exchange and intensive aquaculture systems and studied the effects of suspended solids management. The microbial community structure in shrimp biofloc cultures was characterized using biomarkers and analysis of floc amino acid profiles (Ju et al., 2008; Ray et al., 2010). The predominant microbe analyzed with DGGE was characterized by *Bacillus* sp. in the bioflocs treatment group but by *Vibrio* sp. in the relative control group (Zhao et al., 2012).

REFERENCES

Abakari, G., G. Luo, L. Shao, Y. Abdullateef, and S. J. Cobbina. "Effects of biochar on microbial community in bioflocs and gut of *Oreochromis niloticus* reared in a biofloc system." *Aquaculture International* 29 (2021): 1295–1315.

Amann, Rudolf I., Wolfgang Ludwig, and Karl-Heinz Schleifer. "Phylogenetic identification and in situ detection of individual microbial cells without cultivation." *Microbiological Reviews* 59, no. 1 (1995): 143–169.

Avnimelech, Yoram. "Feeding with microbial flocs by tilapia in minimal discharge bio-flocs technology ponds." *Aquaculture* 264, no. 1–4 (2007): 140–147.

Azim, Mohammed Ekram, David C. Little, and J. E. Bron. "Microbial protein production in activated suspension tanks manipulating C: N ratio in feed and the implications for fish culture." *Bioresource Technology* 99, no. 9 (2008): 3590–3599.

Bae, Jin-Woo, Sung-Keun Rhee, Ja Ryeong Park, Byung-Chun Kim, and Yong-Ha Park. "Isolation of uncultivated anaerobic thermophiles from compost by supplementing cell extract of Geobacillus toebii in enrichment culture medium." *Extremophiles* 9 (2005): 477–485.

Chan, On Chim, Peter Claus, Peter Casper, Andreas Ulrich, Tillmann Lueders, and Ralf Conrad. "Vertical distribution of structure and function of the methanogenic archaeal community in Lake Dagow sediment." *Environmental Microbiology* 7, no. 8 (2005): 1139–1149.

Chu, C. P., and D. J. Lee. "Advective flow in a sludge floc." *Journal of Colloid and Interface Science* 277, no. 2 (2004): 387–395.

Daims, Holger, Andreas Brühl, Rudolf Amann, Karl-Heinz Schleifer, and Michael Wagner. "The domain-specific probe EUB338 is insufficient for the detection of all Bacteria: Development and evaluation of a more comprehensive probe set." *Systematic and Applied Microbiology* 22, no. 3 (1999): 434–444.

Das, Surajit, Louise R. Ward, and Chris Burke. "Prospects of using marine actinobacteria as probiotics in aquaculture." *Applied Microbiology and Biotechnology* 81 (2008): 419–429.

Davenport, John, Kenneth D. Black, Gavin Burnell, Tom Cross, Sarah Culloty, Suki Ekaratne, Bob Furness, Maire Mulcahy, and Helmut Thetmeyer. *Aquaculture: The Ecological Issues*. John Wiley & Sons, 2009.

Decamp, Olivier, Lytha Conquest, Jeff Cody, Ian Forster, and Albert G. J. Tacon. "Effect of shrimp stocking density on size-fractionated phytoplankton and ecological groups of ciliated protozoa within zero-water exchange shrimp culture systems." *Journal of the World Aquaculture Society* 38, no. 3 (2007): 395–406.

Deng, Min, Jieyu Chen, Jingwei Gou, Jie Hou, Dapeng Li, and Xugang He. "The effect of different carbon sources on water quality, microbial community and structure of biofloc systems." *Aquaculture* 482 (2018): 103–110.

Deng, Yale, Xiangyang Xu, Xuwang Yin, Huifeng Lu, Guangshuo Chen, Jianhai Yu, and Yunjie Ruan. "Effect of stock density on the microbial community in biofloc water and Pacific white shrimp (Litopenaeus vannamei) gut microbiota." *Applied Microbiology and Biotechnology* 103 (2019): 4241–4252.

Ekasari, Julie, Roselien Crab, and Willy Verstraete. "Primary nutritional content of bio-flocs cultured with different organic carbon sources and salinity." *HAYATI Journal of Biosciences* 17, no. 3 (2010): 125–130.

Emerenciano M., Ballester E.L., Cavalli R.O. & Wasielesky W. Biofloc technology application as a food source in a limited water exchange nursery system for pink shrimp *Farfantepenaeus brasiliensis* (Latreille, 1817). *Aquaculture Research* **43** (2011) 447–457.

Geets, Joke, Michaël De Cooman, Lieven Wittebolle, Kim Heylen, Bram Vanparys, Paul De Vos, Willy Verstraete, and Nico Boon. "Real-time PCR assay for the simultaneous quantification of nitrifying and denitrifying bacteria in activated sludge." *Applied Microbiology and Biotechnology* 75 (2007): 211–221.

Gerardi, Michael H. *Wastewater Bacteria*. John Wiley & Sons, 2006.

Hargreaves, John A. "Photosynthetic suspended-growth systems in aquaculture." *Aquacultural Engineering* 34, no. 3 (2006): 344–363.

Hastings, Richard C., Jon R. Saunders, Grahame H. Hall, Roger W. Pickup, and Alan J. McCarthy. "Application of molecular biological techniques to a seasonal study of ammonia oxidation in a eutrophic freshwater lake." *Applied and Environmental Microbiology* 64, no. 10 (1998): 3674–3682.

Heid, Christian A., Junko Stevens, Kenneth J. Livak, and P. Mickey Williams. "Real time quantitative PCR." *Genome Research* 6, no. 10 (1996): 986–994.

Jiang, Wenwen, Wenjing Ren, Li Li, Shuanglin Dong, and Xiangli Tian. "Light and carbon sources addition alter microbial community in biofloc-based Litopenaeus vannamei culture systems." *Aquaculture* 515 (2020): 734572.

Jorand, F., F. Zartarian, F. Thomas, J. C. Block, J. Y. Bottero, G. Villemin, V. Urbain, and J. Manem. "Chemical and structural (2D) linkage between bacteria within activated sludge flocs." *Water Research* 29, no. 7 (1995): 1639–1647.

Ju, Zhi Yong, Ian Forster, Lytha Conquest, Warren Dominy, Wenhao Cedric Kuo, and Floyd David Horgen. "Determination of microbial community structures of shrimp floc cultures by biomarkers and analysis of floc amino acid profiles." *Aquaculture Research* 39, no. 2 (2008): 118–133.

Li, Jian-hua, Kevin J. Purdy, Susumu Takii, and Hidetake Hayashi. "Seasonal changes in ribosomal RNA of sulfate-reducing bacteria and sulfate reducing activity in a freshwater lake sediment." *FEMS Microbiology Ecology* 28, no. 1 (1999): 31–39.

Li, C., Li, J., Liu, G., Deng, Y., Zhu, S., Ye, Z., Shao, Y. and Liu, D.. Performance and microbial community analysis of Combined Denitrification and Biofloc Technology (CDBFT) system treating nitrogen-rich aquaculture wastewater. *Bioresource technology, 288*, (2019) p.121582.

Logan, Bruce E., and James R. Hunt. "Bioflocculation as a microbial response to substrate limitations." *Biotechnology and Bioengineering* 31, no. 2 (1988): 91–101.

Madigan, Michael T., and John M. Martinko. *Brock mikrobiologie*. Vol. 11. München: Pearson Studium, 2006.

McIntosh, Dennis, T. M. Samocha, E. R. Jones, A. L. Lawrence, D. A. McKee, S. Horowitz, and A. Horowitz. "The effect of a commercial bacterial supplement on the high-density culturing of Litopenaeus vannamei with a low-protein diet in an outdoor tank system and no water exchange." *Aquacultural Engineering* 21, no. 3 (2000): 215–227.

Meng, Liu-Jiang, Xin Hu, Bin Wen, Yuan-Hao Liu, Guo-Zhi Luo, Jian-Zhong Gao, and Zai-Zhong Chen. "Microplastics inhibit biofloc formation and alter microbial community composition and nitrogen transformation function in aquaculture." *Science of the Total Environment* 866 (2023): 161362.

Miao, Shuyan, Longsheng Sun, Hongyi Bu, Jinyu Zhu, and Guohong Chen. "Effect of molasses addition at C: N ratio of 20: 1 on the water quality and growth performance of giant freshwater prawn (Macrobrachium rosenbergii)." *Aquaculture International* 25 (2017): 1409–1425.

Miura, Yuki, Mirian Noriko Hiraiwa, Tsukasa Ito, Takanori Itonaga, Yoshimasa Watanabe, and Satoshi Okabe. "Bacterial community structures in MBRs treating municipal wastewater: Relationship between community stability and reactor performance." *Water Research* 41, no. 3 (2007): 627–637.

Moriarty, D. J. W. *Detritus and Microbial Ecology in Aquaculture: Proceedings of the Conference on Detrital Systems for Aquaculture, 26–31 August 1985, Bellagio, Como, Italy.* no. 387. WorldFish, 1987.

Moriarty, David J. W. "The role of microorganisms in aquaculture ponds." *Aquaculture* 151, no. 1–4 (1997): 333–349.

Muyzer, Gerard, Ellen C. De Waal, and A. G. Uitterlinden. "Profiling of complex microbial populations by denaturing gradient gel electrophoresis analysis of polymerase chain reaction-amplified genes coding for 16S rRNA." *Applied and Environmental Microbiology* 59, no. 3 (1993): 695–700.

Nicol, Graeme W., L. Anne Glover, and James I. Prosser. "Spatial analysis of archaeal community structure in grassland soil." *Applied and Environmental Microbiology* 69, no. 12 (2003): 7420–7429.

Panigrahi, A., C. Saranya, M. Sundaram, S. R. Vinoth Kannan, Rasmi R. Das, R. Satish Kumar, P. Rajesh, and S. K. Otta. "Carbon: Nitrogen (C: N) ratio level variation influences microbial community of the system and growth as well as immunity of shrimp (Litopenaeus vannamei) in biofloc based culture system." *Fish & Shellfish Immunology* 81 (2018): 329–337.

Rajkumar, M., Pandey P. K, Radhakrishnapillai Aravind R., Vennila A., Vivekanand Bharti, and Purushothaman C. S. "Effect of different biofloc system on water quality, biofloc composition and growth performance in *Litopenaeus vannamei* (Boone, 1931)." *Aquaculture Research* 47, no. 11 (2016): 3432–3444.

Rajkumar M., Aravind R., Pandey P. K. Characterization of Microbial Populations in *Litopenaeus vannamei* (Boone, 1931) Based Biofloc System Using Denaturing Gradient Gel Electrophoresis Profiles of 16s rRNA. Asian Journal of Microbiology, *Biotechnology & Environmental Sciences* (2018) 20(4):1246–1252.

Ray, Andrew J., Beth L. Lewis, Craig L. Browdy, and John W. Leffler. "Suspended solids removal to improve shrimp (Litopenaeus vannamei) production and an evaluation of a plant-based feed in minimal-exchange, superintensive culture systems." *Aquaculture* 299, no. 1–4 (2010): 89–98.

Reddy, K. R., and W. H. Patrick Jr. "Effect of alternate aerobic and anaerobic conditions on redox potential, organic matter decomposition and nitrogen loss in a flooded soil." *Soil Biology and Biochemistry* 7, no. 2 (1975): 87–94.

Sakami, Tomoko, Yoshimi Fujioka, and Toru Shimoda. "Comparison of microbial community structures in intensive and extensive shrimp culture ponds and a mangrove area in Thailand." *Fisheries Science* 74 (2008): 889–898.

Schroeder G. L. Carbon pathways in aquatic detrital systems. In *Detritus and Microbial Ecology in Aquaculture.* ICLARM Conference Proceedings No. 14 (ed. by D. J. W. Moriarty & R. S. V. Pullin), (1987) pp. 217–236. ICLARM, Manila.

Sobeck, David C., and Matthew J. Higgins. "Examination of three theories for mechanisms of cation-induced bioflocculation." *Water Research* 36, no. 3 (2002): 527–538.

Steiner, A. E., D. A. McLaren, and C. F. Forster. "The nature of activated sludge flocs." *Water Research* 10, no. 1 (1976): 25–30.

Ward, David M., Roland Weller, and Mary M. Bateson. "16S rRNA sequences reveal numerous uncultured microorganisms in a natural community." *Nature* 345, no. 6270 (1990): 63–65.

Wasielesky Jr, Wilson, Heidi Atwood, Al Stokes, and Craig L. Browdy. "Effect of natural production in a zero exchange suspended microbial floc based super-intensive culture system for white shrimp Litopenaeus vannamei." *Aquaculture* 258, no. 1–4 (2006): 396–403.

Wei, YanFang, Shao-An Liao, and An-li Wang. "The effect of different carbon sources on the nutritional composition, microbial community and structure of bioflocs." *Aquaculture* 465 (2016): 88–93.

Wilén, Britt-Marie, Bo Jin, and Paul Lant. "The influence of key chemical constituents in activated sludge on surface and flocculating properties." *Water Research* 37, no. 9 (2003): 2127–2139.

Wittebolle, Lieven, Nico Boon, Bram Vanparys, Kim Heylen, Paul De Vos, and Willy Verstraete. "Failure of the ammonia oxidation process in two pharmaceutical wastewater treatment plants is linked to shifts in the bacterial communities." *Journal of Applied Microbiology* 99, no. 5 (2005): 997–1006.

Zhao, Pei, Jie Huang, Xiu-Hua Wang, Xiao-Ling Song, Cong-Hai Yang, Xu-Guang Zhang, and Guo-Cheng Wang. "The application of bioflocs technology in high-intensive, zero exchange farming systems of Marsupenaeus japonicus." *Aquaculture* 354 (2012): 97–106.

Zita, Anna, and Malte Hermansson. "Effects of ionic strength on bacterial adhesion and stability of flocs in a wastewater activated sludge system." *Applied and Environmental Microbiology* 60, no. 9 (1994): 3041–3048.

10 Extremophiles in Aquatic Environments and Their Ecological Significance

Sumanta Kumar Mallik, Richa Pathak,
Satya Narayan Sahoo, and Neetu Shahi

10.1 INTRODUCTION

Within the expansive mosaic of life on Earth, there is a remarkable category of organisms that challenge our conventional notions of where life can thrive. These remarkable creatures flourish in environments that would typically be deemed hostile or even lethal for most life forms. As their name implies, extremophiles are organisms capable of thriving in extreme conditions. These extremes may manifest in the form of exceptionally high or low temperatures, extreme acidity or alkalinity, extreme salinity, crushing pressure, or intense radiation. Extremophiles have been found in a diverse array of habitats, ranging from the scalding waters of deep-sea hydrothermal vents to the frozen expanses of Antarctica's ice sheets. What unites these organisms is their exceptional ability not only to endure but often to thrive in conditions that appear utterly inhospitable.

The term "extremophile" was initially coined by MacElroy in 1974. Since the mid-1970s, advancements in technology have enabled scientists to explore and investigate a growing number of extreme environments that were previously beyond their reach due to technical limitations. This progress has led to the discovery and characterization of numerous new organisms (Howland, 1998; Rainey and Aharon, 2006). Extremophiles are categorized and named based on the specific extreme conditions in which they are found.

Extremophiles can be found across all three major domains of life, i.e., bacteria, archaea, and eukarya. They play a crucial role in the study of evolutionary relationships and offer a unique window into some of the most ancient life forms on Earth (Abbamondi et al., 2019). Moreover, these microorganisms are a rich source of industrially significant enzymes, biomolecules, biomaterials, and metabolites (Margesin and Schinner, 1994; Salwan and Kasana, 2013). In the domain of archaea, we encounter organisms with distinctive adaptations to extreme conditions. This group includes halophiles, alkali and acidophiles, as well as hyperthermophiles like *Methanopyrus kandleri* and *Picrophilus torridus* (D'Amico et al., 2006; Gupta et al., 2011; Moyer and Morita, 2007; Reddy et al., 2004; Shivaji et al., 1992). Bacteria, on the other hand, exhibit an impressive array of extremophiles, including *Arthrobacter*, *Flavobacterium*, *Idiomarina*, *Micrococcus*, *Pseudomonas*, *Rheinheimera*, *Sphingobacterium*, and *Vibrio*, which have been identified in cold regions and high-altitude alkaline lakes.

Cyanobacteria showcase their ability to form microbial mats in environments as diverse as cold deserts, hot springs, acidic mines, and hypersaline lakes, allowing them to thrive even under water-stress conditions. Similarly, eukaryotic organisms, such as algae, fungi, and lichens, demonstrate remarkable adaptability to a wide range of extreme environments, although they are less equipped to handle extreme high-temperature conditions (Rampelotto, 2013). The history of extremophiles within the context of evolution is extensive, and archaea are among the earliest organisms discovered

DOI: 10.1201/9781003408543-10

in extreme environments, offering invaluable insights into the early stages of life's development on our planet.

Research on aquatic extremophiles has provided crucial insights into the fundamental principles of life, the potential for life beyond Earth, biotechnology applications, and the conservation of extreme aquatic ecosystems.

10.2 CLASSIFICATION OF EXTREMOPHILES

The taxonomy of microorganisms involves the categorization of these entities based on relatedness through a combination of techniques. Chemotaxonomic approaches analyze fatty acids, lipids, quinones, and proteins, while phenotypic methods rely on observable morphological characteristics. Phylogenetic methods utilize molecular markers such as 16S rDNA, DNA probes, DNA-DNA hybridization, and total G+C content to assign microorganisms to specific taxonomic groups. When the DNA-DNA hybridization values fall below 70% and the 16S rDNA similarity is less than 97% with a representative strain, the microorganism is assigned to a distinct group or species (Tindall et al., 2010). The 16S rDNA is widely accepted as a gold standard for classifying microorganisms into new or higher taxa (Nogi, 2011). Techniques like denaturing gradient gel electrophoresis, terminal restriction fragment length polymorphism (T-RFLP), and metagenomics are used to study extremophiles adapted to low-temperature environments (Junge et al., 2019).

Extremophiles, which thrive in extreme habitats, span all three domains of life and occupy diverse ecological niches worldwide. These microorganisms have developed specialized strategies to survive in environments characterized by extreme factors like temperature, pH, salt concentration, metal ions, and radiation. Extremophiles are categorized based on their tolerance to these harsh conditions, leading to various groups. For instance, psychrophiles adapted to cold environments, like *Colwellia*, *Deinococcus*, *Desulfuromonas*, and *Moritella*, thrive at low temperatures. Conversely, thermotolerant microorganisms such as *Bacillus thermophilus* and *Thermus aquaticus* can endure high temperatures. Halophiles, which flourish in saline environments, include diverse bacterial species (e.g., *Rhodospirillum* and *Halothiobacillus*) and archaea (e.g., *Halobacterium* and *Haloferax*). Acidophiles, capable of surviving highly acidic environments with a pH below 3, are found within bacterial classes like α-proteobacteria, β-proteobacteria, γ-proteobacteria, and archaea from phyla Euryarchaeota and Crenarchaeota (Hallberg and Johnson, 2001; Mirete et al., 2017). Alkaliphiles, adapted to extremely alkaline pH, are widespread in environments like soda lakes, Lonar Lake, Sambhar Lake, Mono Lake, Wadi Natrun Lakes, and Saline Qinghai Lake, and exhibit various physiological activities, including aerobic, anaerobic, methanogenic, chemolithotrophic, and phototrophic. High-pressure conditions also create extreme environments, altering physiological parameters and metabolic reactions, and piezophiles adapt by producing specific molecules like unsaturated fatty acids. Extremophiles are also found in environments with characteristics such as toxic waste, heavy metals, organic solvents, low nutrient availability, methane, sulfur, iron, and high-pressure conditions, making them promising candidates for industrial applications under harsh conditions.

10.3 TYPES OF EXTREMOPHILES

Extremophiles can be categorized into several distinct groups based on the type of extreme environment they inhabit.

10.3.1 Thermophiles

In 1966, Thomas Brock made a groundbreaking discovery, uncovering thriving microorganisms in the scalding hot springs of Yellowstone National Park. These "thermophiles" have a global geothermal presence and significant biotechnological relevance. Notably, thermophilic species like

Thermus aquaticus and *Thermococcus litoralis* play a crucial role by providing the DNA polymerase enzyme used in DNA fingerprinting through polymerase chain reactions (PCR) (Holland et al., 1991). Thermophiles are categorized by their optimal growth temperatures: "moderate thermophiles" (below 50 C), "extreme thermophiles" (above 50 C), and "hyperthermophiles" (above 80 C). These categories encompass various microorganisms, including photosynthetic bacteria, Firmicutes, and primarily archaea. Some extremophiles, such as *Geogemma* and *Pyrolobus fumarii*, exhibit remarkable resilience to extreme conditions, thriving at maximum growth temperatures of 121 C and 113 C, respectively. Notable thermophiles include *Pyrococcus furiosus*, which thrives at 100 C, and *Pyrobaculum aerophilum*, adapted to conditions near the boiling point (Hafenbradl et al., 1997; Sievert and Vetriani, 2012). Methanogenic archaea like *Methanobacterium thermoautotrophicus* and *Methanopyrus kandleri* demonstrate adaptability to varying growth temperatures, with the latter capable of thriving at 122 C under pressure (Zeikus and Wolee, 1972; Slesarev et al., 1994). Extreme thermophilic acidophiles like *Picrophilus* and *Acidianus* flourish at pH 0.7 and 60 C. Among thermophilic bacteria, prominent examples come from the *Deinococcus–Thermus* group, *Chloroflexus aurantiacus*, and *Thermoleophilum* species, which prefer temperatures ranging from 35 to 70 C and employ n-alkanes (C_{13} to C_{20}) as substrates (Zeikus, 1979; Allgood and Perry, 1986). Despite the challenges posed by high temperatures, including protein and nucleic acid denaturation, thermophiles possess exceptionally heat-stable proteins with distinct amino acid sequences that bolster rigidity and stability. Further adaptations include modifications in RNA purine content, enhanced ribosomal RNA secondary structure stability, reduced protein structural disorder, and genome size reduction, all of which facilitate survival among thermophilic bacteria.

10.3.2 Acidophiles

Highly acidic environments result from geological and acidophile metabolic processes. Acidophiles, thriving below pH 5, have specific pH requirements. To endure extreme acidity, they employ mechanisms to maintain near-neutral intracellular pH. Acidophiles produce extremozymes and alter transmembrane potential to survive. Intracellular pH typically ranges from 4.6 to 7, while *Picrophilus oshimae* and *P. torridus* are archaea species adapted to pH 0.8–4. Eukaryotic acidophiles, like *Dunaliella acidophila* and *Chlamydomonas acidophila*, perform photosynthesis at low pH. Acidophilic bacteria, including *Acidimicrobium*, *Sulfobacillus*, and *Leptospirillum*, dominate acid mine drainage (AMD) environments. Recent studies suggest that *Leptospirillum* spp. and *Ferroplasma* spp. are more significant than *Acidithiobacillus ferrooxidans* in AMD systems (Valdés et al., 2008).

10.3.3 Alkaliphiles

Alkaliphiles, capable of thriving between pH 7.5 and 14, are divided into "essential alkaliphiles" and "haloalkaliphiles." Some require pH 9.0 and higher for optimal growth, while facultative alkaliphiles grow within the pH range of 7.5 to 11. Haloalkaliphiles need both high pH ($\geq$9) and elevated NaCl ($\geq$33%). Thriving in extremely alkaline settings poses a challenge, as biomolecules like RNA degrade beyond pH 8. To adapt, alkaliphiles maintain cytoplasm closer to neutrality and employ various adaptations, including enhanced proton-capturing transporters and enzymes (Krulwich et al., 2011). *Arthrospira platensis* is an alkaliphilic cyanobacterium, used in industries for protein and pigment production (Berry et al., 2003).

10.3.4 Halophiles

High salinity reduces water activity and disturbs charge distribution around macromolecules like proteins and DNA, leading to denaturation. Organisms thriving in these conditions are known as "halophiles," categorized as "slightly" (0.2–0.85 M, ~1–5% NaCl), "moderately" (0.85–3.4 M,

5–20% NaCl), or "extremely halophilic" (3.4–5.1 M, 20–30% NaCl) based on their salt requirements. Halotolerant organisms can grow in high salinity or its absence. To cope with extreme salt, they employ strategies like accumulating inorganic solutes (e.g., potassium and chloride ions) and osmolytes (small organic molecules). *Halobacteria* are known for their high intracellular potassium concentrations. Additionally, some halophiles use acidic proteins to counter the denaturing effects of salt. While many halophiles belong to the archaea domain, bacterial and eukaryotic halophiles exist. They are found in various phylogenetic groups, with moderate halophiles being more common than extreme ones. Halophilic bacteria exist within *Proteobacteria*, *Cyanobacteria*, *Flavobacterium–Cytophaga*, *Spirochaetes*, and *Actinomycetes*. In the Firmicutes, there are both aerobic and anaerobic halophilic representatives, including *Salinibacter* (Madigan and Orent, 1999; Ordoñez et al., 2013). Lake Magadi and the Cave of the Crystals house unique halophilic environments, hosting a diverse population of bacterial and archaeal representatives. Microbes in these environments actively contribute to the formation of geological structures.

10.3.5 Psychrophiles

Psychrophiles are cold-adapted organisms, thriving in icy environments, like the Arctic, Antarctica, and deep oceans. They've evolved remarkable adaptations to survive and reproduce at temperatures between –20 C and +20 C, with their optimal growth around 15 C or lower. These adaptations help them surmount various challenges, such as reduced enzymatic activity, stiffening of cell membranes, altered nutrient transport, slower genetic processes (transcription, translation, and cell division), protein denaturation, misfolded proteins, and even internal ice formation. One vital adaptation is the production of glycoproteins and peptides with antifreeze properties. This trait, well documented in polar fish, insects, plants, and fungi, is less studied in prokaryotes. Some cold-adapted bacteria in places like Antarctica possess antifreeze proteins (AFPs) and antifreeze lipoproteins. These molecules lower the freezing point of solutions below their melting point (termed thermal hysteresis) and prevent the growth of large ice crystals at sub-zero temperatures. Another key adjustment involves maintaining flexible cell membranes (Yamashita et al., 2002; Gilbert et al., 2004). Psychrophiles often do this by increasing the proportion of unsaturated fatty acids in their lipid composition. This strategy keeps their membranes pliable and semi-liquid, unlike saturated or monounsaturated fatty acids, which stiffen in the cold (Bowman et al., 1998; Nichols and McMeekin, 2002).

In contrast, psychrotolerant organisms are less specialized and can grow within a broader temperature range, typically between 20 and 40 C. They are more widely distributed in various environments, including soil, water, and refrigerated products. Psychrotolerants belong to diverse groups within bacteria, archaea, fungi (especially yeasts), and microalgae. Common genera among psychrotolerant bacteria include *Hyphomonas*, *Psychrobacter*, *Pseudomonas*, *Halomonas*, *Vibrio*, *Coryneform*, *Arthrobacter*, and *Micrococcus* (Jagannadham et al., 2000; Vazquez et al., 2004; Zhang et al., 2007; Dib et al., 2013). Some psychrotolerant microorganisms exhibit remarkable adaptations, like *Moritella profunda*, which thrives in both cold and high-pressure conditions, with its best growth at 2 C and a maximum of 12 C (Xu et al., 2003). Psychrotolerant bacteria have even been found in unique environments, like Antarctica's subglacial Lake Vostok, where they've survived for hundreds of thousands of years (Mitskevich et al., 2001). These findings shed light on microorganisms' resilience in extreme cold and prompt questions about their potential to exist on celestial bodies or during space travel.

Psychrophiles are more abundant and diverse than archaea, though some archaea genera, like *Methanogenium* and *Methanococcus*, are common among cold-adapted archaea. Additionally, psychrophilic yeasts, especially Cryptococcus species, are frequently isolated from soil samples and are pivotal in Antarctic desert soils (Bartlett, 1999; D'Amico et al., 2006).

10.3.6 Barophiles

The upper limit of pressure for supporting life on Earth remains uncertain, as environments with pressures exceeding 1,100 bar remain unexplored, but macromolecules and cellular components appear to denature only at pressures of 4,000 to 5,000 bar. Recent space missions have demonstrated bacteria's ability to endure extended periods in a vacuum (Stojanović et al., 2008).

"Piezophiles" or "barophiles," organisms that thrive under high pressures like those in the deep sea, exhibit optimal growth at pressures exceeding 1 atm, often surpassing 380 atm. "Hyperpiezophiles" require high or extreme pressure for growth. "Obligate piezophiles" thrive at 70 to 80 MPa (equivalent to 700 to 800 atm), and for some, the optimal range is 100 to 200 MPa, with pressures lower than 50 MPa being detrimental (Nogi et al., 2004). High-pressure environments on Earth include deep lakes, seas, and subsurface locations, often accompanied by extreme temperatures, categorizing piezophiles as "psychrophilic," "mesophilic," or "thermophilic" based on optimal growth temperatures.

"Piezothermophilic" organisms include *Thermococcus barophilus, Palaeococcus ferrophilus, Marinitoga piezophila, M. kandleri*, and *Pyrococcus yayanosii*, which exhibits optimal reproduction rates at 52 MPa and 98 C (Kurr et al., 1991; Marteinsson et al., 1999; Takai et al., 2000; Alain et al., 2002). Elevated pressure poses survival challenges, compressing lipids in cellular membranes, restraining membrane fluidity, and limiting chemical reactions and protein functions (Murakami et al., 2011). The majority of piezophiles belong to genera like *Shewanella, Colwellia, Photobacterium, Moritella*, and *Psychromonas. Colwellia* sp. MT41 from the Mariana Trench thrives optimally at 690 bar, tolerating pressures up to 1,035 bar, and grows at a temperature of 4 C (Nogi et al., 2004).

10.3.7 Radioresistant

Radiation threats encompass UV radiation (UVR) and ionizing radiation. UVR includes UV-A (320–400 nm), UV-B (280–320 nm), and UV-C (100–280 nm) and poses biological risks, with UV-C being most harmful to DNA (Coohill, 1996). Ionizing radiation causes damage via ionization and free radicals (Dartnell, 2011a). Microbes in radioactive environments, termed "radiophiles," exhibit resistance to both ionizing radiation and UV radiation. They've been found in areas with elevated UVR and are affected by ozone layer depletion and nuclear accidents (Persson et al., 1987; Salbu and Krekling, 1998). Microorganisms, with vulnerable haploid genomes and no protective cell walls, use DNA repair mechanisms like photoreactivation, base excision repair, nucleotide excision repair, recombinational repair, and double-strand break repair to resist radiation (Rosenstein, 1984).

Some UV-resistant bacteria, like *Acinetobacter* strains, use efficient DNA repair mechanisms and metabolic products to withstand radiation. *Deinococcus radiodurans* is a well-studied radiation-resistant microbe, utilizing stress-response systems, chromosomal redundancy, and efficient repair pathways (Blasius et al., 2008; Hua et al., 2008). It also has a unique cell envelope with multiple layers (Sleytr and Messner, 1988; Battista, 1997; Eltsov and Dubochet, 2005).

The classification of extremophiles as stated by Horikoshi and Grant (1998) is shown in Table 10.1.

10.4 EXTREMOPHILES IN THE MARINE ENVIRONMENT

Earth's vast oceans, covering over 70% of its surface, house an incredibly diverse array of life, representing more than 95% of the entire biosphere. Within marine ecosystems, which can be divided into coastal and open ocean regions teeming with diverse microorganisms, lies the primary reservoir of global biodiversity. These environments offer immense potential for discovering new natural products, notably enzymes, although only a small portion (1–10%) of prokaryotic species have been identified (Bull et al., 2000).

Extreme conditions in these marine settings, encompassing factors like pressure, temperature fluctuations, pH variations, salinity changes, radiation exposure, and the presence of various chemicals and metals, including toxic ones, pose significant challenges to life. Nevertheless, extremophilic microorganisms have demonstrated remarkable adaptability, due to their highly flexible metabolisms, enabling them to flourish in these harsh circumstances (Ambily and Bharathi, 2011).

TABLE 10.1
Classification of Extremophiles

Environmental Parameter	Type	Definition	Paramount Microbe
Temperature	Hyperthermophile	Growth >80 C	*Pyrolobus fumarii*, 113 C
	Thermophile	Organisms thriving in high-temperature environments, typically above 45 C (113 F)	*Synechococcus lividis*
	Mesophile	15–60 C	
	Psychrophile	Organisms adapted to cold conditions, with optimal growth temperatures below 15 C (59 F)	*Psychrobacter immobilis*
Radiation	Radiophiles	Resistant to ionizing radiation	*Deinococcus radiodurans*
Pressure	Barophile (Piezophile)	Adapted to high-pressure environments, like those found in the deep ocean or beneath the Earth's crust	*Pyrococcus yayanosii*
Gravity	Hypergravity	>1 g	
	Hypogravity	<1 g	
Vacuum		Tolerates vacuum	
Desiccation	Xerophiles	Organisms capable of surviving in dry and arid conditions, with low water availability	
Salinity	Halophile	These extremophiles flourish in high-salinity environments, such as salt flats, salt lakes, and salt mines	*Halobacterium* spp.
pH	Alkalophile	Thrive in strongly alkaline conditions, typically with a pH above 9	*Arthrospira* sp.
	Acidophile	Thrive in highly acidic conditions, often with pH levels below 3	*Acidithiobacillus ferrooxidans*
Oxygen tension	Anaerobe	Cannot tolerate O_2	*Clostridium difficile*
	Microaerophile	Tolerates O_2 deprivation	
	Aerobe	Requires O_2	
Osmotic pressure	Osmophile	Thrive in environments with high osmotic pressure, such as sugary solutions	
Chemicals	Metallophile	Withstand and even thrive in environments with high levels of heavy metals, such as mines and metal-contaminated soils	*Amycolatopsis tucumanensis*
	Toxitolerant	Tolerate toxic and xenobiotic chemicals like benzene	

Source: adapted from Horikoshi and Grant (1998).

Consequently, in recent decades, researchers have shown keen interest in extremophiles as sources of bioactive substances, including enzymes and biocides, with applications spanning agriculture, chemicals, food production, textiles, pharmaceuticals, bioenergy, and cosmetics.

The industrial enzyme market, valued at nearly US$ 4.8 billion in 2013, is anticipated to reach US$ 7.1 billion by 2018, with a compound annual growth rate (CAGR) of 8% over this five-year period, as per BCC Research (Dewan, 2014). Microbial enzymes offer a multitude of advantages, such as environment-friendly processes and easy cultivation in bioreactors with precise control over growth conditions, including pH, temperature, aeration, and medium composition, ensuring reproducibility. Conversely, enzymes derived from plants and animals face limitations related to soil variations, light exposure, seed uniformity, pathogen control, and other factors, leading to less consistent processes (Díaz-Tena et al., 2013; Vermelho and Noronha, 2013).

Competition for resources and space within the marine environment serves as a driving force for evolution, giving rise to multiple enzyme systems adapted to diverse conditions. Many marine extremophiles excel in thriving in demanding circumstances like hypersaline habitats, high-pressure zones, and extreme temperatures, making them valuable sources of enzymes with distinct properties (Chandrasekaran and Kumar, 1997). The rich array of marine habitats has led to the discovery of new hydrolases with unique specificities and the capacity to function under extreme conditions, catering to various industrial needs. Metagenomic studies have unveiled that prokaryotic extremophiles in marine environments harbor novel genes, providing a source of new bioproducts, including enzymes and active metabolites (Russo et al., 2010; Trincone, 2010).

Hydrolases from extremophiles offer several advantages over chemical biocatalysts, as they enable clean, eco-friendly, highly specific, and gentle reactions, including the ability to function in the presence of organic solvents—an essential feature for producing single-isomer chiral drugs. These enzymes find diverse applications, ranging from the production of pharmaceutical intermediates to the synthesis of nitrogenized compounds. Furthermore, peptidases from marine extremophiles serve in medical and industrial domains, with their properties proving invaluable for advancements in pharmaceuticals.

10.4.1 Amylases from Marine Extremophiles

Amylases from marine extremophiles are enzymes produced by microorganisms found in extreme aquatic environments such as deep-sea hydrothermal vents, high-pressure ocean depths, and salt-saturated brine pools. These enzymes, specifically amylases, play a crucial role in breaking down starch and related compounds into simpler sugars. Marine extremophiles have adapted to extreme conditions, including high pressures, low temperatures, and high salinity, making their amylases unique and valuable for industrial applications, such as biofuel production and bioremediation, where traditional enzymes may not function effectively (Dalmaso et al., 2015).

10.4.2 Hydrolytic Enzymes Targeting Cell Walls in Marine Extremophiles

Hydrolases that target cell walls, originating from marine extremophiles, represent a group of enzymes that possess remarkable abilities to break down complex cell wall structures in extreme aquatic environments. These enzymes play a crucial role in the degradation of cellulosic and chitinous materials, which are abundant in marine ecosystems. What sets these cell wall–degrading hydrolases apart is their adaptation to extreme conditions, including high pressures, low temperatures, and high salinity levels. Their unique characteristics make them valuable tools for various biotechnological applications, including biofuel production, wastewater treatment, and the development of environmentally friendly processes (Dalmaso et al., 2015).

10.4.3 Proteases in Marine Extremophiles

Marine extremophilic proteases refer to enzymes found in microorganisms inhabiting extreme marine environments. These proteases have adapted to function effectively in challenging conditions, such as high or low temperatures, extreme pH levels, high salinity, and high pressure. They play a crucial role in breaking down proteins, which is essential for the survival and growth of extremophiles in their extreme habitats. The study of these proteases not only expands our understanding of extremophiles and their remarkable adaptations but also holds significant potential for various biotechnological applications. These applications may include their use in industrial processes, bioremediation, and pharmaceuticals due to their unique properties and stability in extreme conditions (Dalmaso et al., 2015).

10.4.4 Enzymes with Lipolytic Activity from Extremophiles in Marine Environments

"Enzymes with lipolytic activity from extremophiles in marine environments" refers to a group of enzymes that exhibit lipolytic or lipid-degrading properties and are sourced from extremophilic microorganisms found in marine environments. Lipases and esterases are specific types of enzymes capable of breaking down lipids (fats and oils) into smaller components. Extremophiles are microorganisms that thrive in extreme conditions, such as high or low temperatures, high salinity, or high-pressure environments. These enzymes play a significant role in various industrial applications, particularly those involving the breakdown of lipids, like the production of biodiesel, removal of oil spills, and wastewater treatment. Understanding these enzymes and their sources can lead to advancements in biotechnology and environmental remediation (Dalmaso et al., 2015).

10.5 EXTREMOPHILES IN FRESHWATER

Freshwater extremophiles represent a captivating group of microorganisms that have remarkably adapted to thrive in challenging conditions within these aquatic ecosystems. Among these extraordinary freshwater extremophiles, acidophiles, capable of thriving in highly acidic waters with pH levels as low as 1–2, are frequently found in acidic mine drainage (AMD) environments (Baker and Banfield, 2003). Thermophiles are heat-loving microorganisms that can endure scorching temperatures exceeding 40 C, typically inhabiting hot springs, owing their survival to their robust heat-resistant enzymes and proteins (Brock, 1967). Halophiles, usually associated with hypersaline surroundings, have demonstrated the ability to adapt to salinity fluctuations within specific freshwater habitats such as saline lakes or estuaries (Oren, 2002). Oligotrophic extremophiles have evolved to flourish in low-nutrient settings, such as certain deep freshwater lakes, where they efficiently utilize trace nutrients to sustain life (Lauro et al., 2009). In anaerobic sediments of lakes and wetlands, methanogenic archaea, often referred to as methanogens, thrive by producing methane, contributing to the complex biogeochemical cycles of these environments (Conrad, 1999). Remarkably, even the most frigid of environments are not devoid of life, as psychrophiles flourish in icy settings, including polar lakes and subglacial waters, demonstrating an astounding adaptation to extremely low temperatures (Cavicchioli et al., 2019). These freshwater extremophiles greatly expand our comprehension of life's resilience and hold significant promise for various biotechnological and astrobiological applications due to their unique and versatile survival strategies. Some examples of extremophiles found in freshwater environments are:

a) *Acidiuliprofundum boonei*: This extremophilic bacterium thrives in freshwater lakes with extremely low pH levels, such as Lake Tuz in Turkey.
b) *Ferroplasma acidarmanus*: These acidophilic extremophiles inhabit acidic environments, including AMD sites. They are known for their ability to thrive in highly acidic conditions, contributing to the acidification of their surroundings.
c) *Alicyclobacillus acidocaldarius*: This bacterium is an acidothermophile found in acidic hot springs and geothermal environments. It can endure both high temperatures and low pH levels, making it a remarkable extremophile in freshwater hot springs.

10.6 EXTREMOPHILES IN BRACKISHWATER

Extremophiles in brackishwater environments, such as the halotolerant bacterium *Vibrio parahaemolyticus* (Letchumanan et al., 2014) and the halophilic archaeon *Haloquadratum walsbyi* (Dyall-Smith et al., 2011), have successfully adapted to the demanding conditions of fluctuating salinity between freshwater and seawater. These remarkable microorganisms have evolved unique strategies to thrive in this transitional habitat, rendering them invaluable subjects for extensive studies on extremophiles. Their ability to withstand variable salt concentrations not only contributes to our understanding of microbial diversity and adaptation in complex ecosystems but also opens the door to numerous biotechnological and environmental applications. These applications encompass bioremediation, pharmaceuticals, and biofuel production, among others (DasSarma and DasSarma, 2015; Sorokin et al., 2008). Investigating extremophiles in brackishwater not only reveals their ecological significance but also showcases their potential to address real-world challenges and drive advances in biotechnology, positioning them as pivotal players in this unique ecological niche. A few examples of extremophiles found in brackishwater environments are:

a) *Halobacterium salinarum*: This halophilic archaeon is often found in environments with high salinity, such as salt flats, salt mines, and brine pools. *H. salinarum* has adapted to thrive in extremely salty conditions and is known for its distinctive reddish pigmentation due to bacteriorhodopsin, a light-driven proton pump.
b) *Vibrio parahaemolyticus*: This halotolerant bacterium is commonly found in brackish coastal waters and estuaries. It has developed mechanisms to cope with changing salinity levels and is often associated with seafood-related illnesses.
c) *Haloquadratum walsbyi*: Another halophilic archaeon, *H. walsbyi*, is typically found in salt flats and salterns. It forms square-shaped cells.
d) *Salicornia europaea*: This is a halophytic plant, often known as glasswort or samphire, that grows in brackish water habitats and salt marshes. It has specialized mechanisms for salt tolerance and is used as a food source in some regions.

10.7 ADAPTATIONS OF AQUATIC EXTREMOPHILES

Aquatic extremophiles, which are microorganisms specially equipped to thrive in the most demanding aquatic environments, boast a diverse array of extraordinary adaptations. These unique adaptations span a wide spectrum of strategies that aid in preserving critical cellular functions, enabling reproduction, and safeguarding against harm in environments defined by their extremity. These adaptations not only carve out a niche for extremophiles in some of the planet's harshest settings but also furnish valuable revelations regarding the molecular and physiological mechanisms that form the foundation of their survival. Furthermore, these insights hold significance across various scientific domains, from unraveling the origins of life to applications in biotechnology and astrobiology.

Some of the key adaptations of aquatic extremophiles include:

a) **Enzyme stability:** Many extremophiles have evolved enzymes and proteins that exhibit exceptional stability at extreme temperatures, pH levels, or pressure conditions. These specialized biomolecules enable extremophiles to conduct vital metabolic reactions even in the harshest of environments. For example, thermophiles have enzymes that function optimally at high temperatures, allowing them to thrive in hot springs and hydrothermal vents.
b) **Membrane flexibility:** Extremophiles often possess unique lipid membranes that remain fluid and functional in extreme temperature ranges. These adaptable cell membranes are crucial for maintaining cellular integrity, especially in thermophiles and psychrophiles living in extremely hot and cold environments, respectively (Horneck et al., 2012).

c) **Osmoregulation:** Halophiles, a category of extremophiles, have developed intricate mechanisms for managing osmotic stress induced by high salt concentrations in their surroundings. These microorganisms employ specialized transporters and osmoprotectants to balance water and ion concentrations within their cells, ensuring osmotic equilibrium (Oren, 2008).
d) **DNA repair mechanisms:** Extremophiles exposed to high levels of radiation, such as ionizing radiation, have evolved highly efficient DNA repair systems. These mechanisms can effectively mend the extensive damage caused by ionizing radiation, ensuring the genetic stability of the extremophile's populations (Dartnell, 2011b).
e) **Metabolic flexibility:** Many extremophiles exhibit metabolic versatility, enabling them to switch between different metabolic pathways based on the availability of resources. This adaptability allows them to thrive in environments with fluctuating or limited nutrient sources.
f) **Antioxidant defenses:** To counteract the increased levels of oxidative stress in some extreme environments, extremophiles often have robust antioxidant defense systems. These systems help protect their cellular components from damage caused by reactive oxygen species.
g) **Symbiotic relationships:** Some extremophiles engage in symbiotic relationships with other microorganisms, which can provide them with advantages, such as access to essential nutrients or protection from environmental stresses.

10.8 RESEARCH WORKS CONDUCTED ON EXTREMOPHILES

Research works on extremophiles from the perspective of historical events and recent advances are presented in Tables 10.2 and 10.3.

10.9 APPLICATIONS

Extremophiles, microorganisms capable of thriving in extreme environments, have significant applications in various fields, from scientific research to industrial and environmental applications. Some notable applications of extremophiles are stated here:

TABLE 10.2
Research Works on Extremophiles: Historical Events

S. No.	Time Period	Research Milestones and Discoveries
1.	1970s	Discovery of Hydrothermal Vent Communities: The late 1970s saw the discovery of hydrothermal vent ecosystems, revealing diverse extremophiles inhabiting extreme environments.
2.	1980s	Thermophilic Enzymes: In the 1980s, thermophilic bacteria were isolated from Yellowstone National Park's hot springs, leading to the discovery of heat-stable enzymes like Taq polymerase, revolutionizing molecular biology.
3.	1990s	Microbial Communities in Saline Lakes: Research in saline lakes, like the Great Salt Lake, unveiled the diversity of halophiles and their ecological roles.
4.	2000s	Genomic Studies: Genomics in the 2000s allowed in-depth exploration of extremophile genomes, shedding light on their genetic adaptations. Barophilic Adaptations: Studies on barophilic extremophiles in deep ocean trenches, such as *Moritella profunda* and *Shewanella piezotolerans*, revealed their mechanisms of pressure resistance.

TABLE 10.3
Highlights of Recent Advances in Aquatic Extremophiles

S. No.	Year	Research Work	Highlights
1.	2023	"Exploring Microbial Diversity in Antarctic Sub-glacial Lakes"	• Discovery of novel extremophiles in subglacial lakes. • Insights into microbial adaptations to extreme cold and high-pressure conditions. • Psychrophiles such as *Psychrobacter* spp., *Polaribacter* spp.
2.	2022	"Metagenomic Profiling of Hydrothermal Vent Microbiomes"	• Metagenomic analysis revealing the diversity of hydrothermal vent extremophiles. • Identification of unique metabolic pathways and extremophile interactions. • Hydrothermal vent extremophiles like *Thermococcus* spp., *Methanothermococcus* spp.
3.	2021	"Bioprospecting Halophiles for Sustainable Bioplastic Production"	• Application of halophilic extremophiles in bioplastics. • Environmentally friendly bioplastic synthesis using extremophile enzymes. • Halophilic extremophiles, including *Haloferax* spp. and *Halomonas* spp.
4.	2020	"Genomic Insights into Psychrophile Cold-Adaptation"	• Genome analysis of psychrophiles providing insights into cold-adaptation mechanisms. • Potential biotechnological applications for cold-resistant enzymes. • Psychrophilic bacteria like *Psychromonas* spp., *Colwellia* spp., and *Shewanella* spp.
5.	2019	"Microbial Communities in Acidic Lakes: Biodiversity and Functions"	• Metagenomic study of extremophile microbiomes in highly acidic lakes. • Understanding the roles of acidophiles in ecosystem processes. • Acidophilic extremophiles, such as *Acidithiobacillus* spp. and *Leptospirillum* spp.
6.	2018	"Piezophiles in the Deep Sea: Exploring Pressure Adaptations"	• Research on extremophiles in deep ocean trenches. • Insights into barophilic adaptations and their potential applications. • Piezophilic extremophiles, like *Piezophiles* spp. and *Moritella* spp.
7.	2017	Extremophiles and Astrobiology: Implications for Space Exploration"	• Extremophiles as analogs for extraterrestrial life in extreme aquatic environments. • Contributions to astrobiology and future space missions.
8.	2016	"Bioremediation with Halophilic Extremophiles in Polluted Waters"	• Utilizing halophilic extremophiles in bioremediation processes. • Cleansing polluted aquatic environments with extremophile enzymes. • Halophilic extremophiles used for bioremediation, including *Halomonas* spp., *Halobacterium* spp., and *Salinibacter* spp.
9.	2015	"Metagenomic Study of Microbial Diversity in Hydrothermal Vents"	• Exploration of microbial diversity in hydrothermal vent ecosystems. • Identification of extremophiles and their unique genetic adaptations. • Thermophiles, such as *Thermococcus* spp., *Methanothermococcus* spp., and *Pyrococcus* spp.
10.	2014	"Thermophiles in Extreme Geothermal Environments"	• Study of thermophilic extremophiles thriving in geothermal hot springs. • Applications of heat-resistant enzymes from extremophiles in biotechnology. • Thermophilic extremophiles like *Thermus* spp., *Aquifex* spp., and *Geothermobacter* spp.

a) Biotechnology: Extremophiles are a treasure trove of enzymes and biomolecules that have revolutionized biotechnology. For instance, thermophiles, which thrive in high-temperature environments, produce heat-stable enzymes like DNA polymerases, such as *Taq polymerase*. These enzymes are essential for techniques like PCR and DNA sequencing, playing a pivotal role in genetic research and diagnostics. Thermophilic enzymes are also used in the production of biofuels, such as cellulosic ethanol.
b) Bioremediation: Extremophiles have been harnessed for bioremediation, the process of using microorganisms to clean up contaminated environments. Acidophiles, which thrive in highly acidic conditions, are employed to remediate AMD and detoxify metals and metalloids. Halophiles are used to treat saline wastewater in regions like desalination plants.
c) Pharmaceuticals: Extremophiles are a source of bioactive compounds with pharmaceutical potential. These microorganisms produce antibiotics, anti-cancer agents, and other medically important molecules. For instance, thermophiles are known for their ability to produce antimicrobial peptides, while other extremophiles provide enzymes crucial in pharmaceutical production.
d) Astrobiology: Astrobiology heavily relies on studying extremophiles, which are organisms thriving in extreme conditions on Earth. They act as important models for investigating life beyond our planet, providing vital clues in the quest for extraterrestrial life. Understanding extremophiles helps scientists improve space exploration tools. Additionally, studying these creatures helps us speculate about the possibility of life on other celestial bodies like Mars, Europa, and Enceladus, expanding our grasp of potential life in the universe.
e) Biofuels: Certain extremophiles, such as thermophiles and acidophiles, play a key role in the production of biofuels. They assist in breaking down lignocellulosic biomass at high temperatures, which enhances the efficiency of biofuel production processes.
f) Food Preservation: Extremophiles are employed in the food industry to improve food safety and preservation. Acidophilic bacteria are used in the fermentation of certain foods like kimchi and sauerkraut, while halophilic bacteria are used in the fermentation of foods like pickles and olives. These microorganisms help create an environment that is hostile to pathogenic bacteria, preserving the food.
g) Mining and Resource Extraction: Extremophiles are employed in biomining or bioleaching, where they are used to extract valuable minerals from ores. Acidophiles, for example, can thrive in acidic mine waters and help dissolve minerals, making it easier to extract valuable metals like copper and gold.
h) Bioplastics: Extremophiles play a role in the development of bioplastics, which are biodegradable alternatives to conventional plastics. Microorganisms like thermophiles are used in the production of biodegradable plastics through fermentation processes.
i) Agriculture: Some extremophiles, such as halophiles, are being investigated for their potential to improve crop yields in saline soils, which are often unproductive for traditional agriculture. These microorganisms can help enhance soil fertility and make it suitable for farming.

The enzymatic activities of extremophiles are listed in Table 10.4.

Thus, extremophiles have a broad range of applications that span various fields. Their ability to thrive in extreme conditions has opened up new possibilities for innovation and sustainable solutions in multiple fields.

TABLE 10.4
Applications of Extremophiles in Biotechnology, Medicine, and Industry

Source	Biomolecule	Process
Thermophiles	DNA polymerase	Polymerase chain reaction (PCR) (diagnostic)
Thermophiles Psychrophiles	proteases	food processing (baking, brewing) cheese making and dairy production
Thermophiles	α-amylase	paper bleaching
AlkaliphilesPsychrophiles	Proteases, cellulases, amylases, lipases	detergents - polymer breakdown
Alkaliphiles	antibiotics	treatment of infections
Psychrophiles	unsaturated fatty acids	food supplement
Psychrophiles	dehydrogenases	biosensors
Halophiles	compatible solutes glycerol	pharmaceuticals
Halophiles	carotene	food additive
Psychrophiles		bioremediation of oil spills
Radiation-resistant		bioremediation of radioactive waste

Source: Irwin, J.A. and Baird, A.W., *Irish Veterinary Journal,* 57, 1–7, 2004.

10.10 ADAPTATION TO HIGHLY CHALLENGING ENVIRONMENTS

Extremophiles employ two primary strategies to thrive in harsh and extreme environments. The first approach involves protecting their cytoplasm from the adverse effects of environmental stress, and this protection is achieved through the unique properties of their cell membrane. In this context, the specific adaptations are primarily required for the cytoplasmic membrane, as well as for the periplasmic and excreted proteins of the entire cell, enabling them to withstand the challenges posed by the extreme environmental conditions.

Halophilic or halotolerant organisms adapt to a wide range of salt concentrations (2 to 5 M NaCl) through two methods: accumulating salts inside the cell, or excluding salts from the cytoplasm and producing organic solutes like ectoine and glycine-betaine to maintain osmotic balance (Oren, 2000). Acidophiles and alkalophiles maintain a near-neutral intracellular pH with low proton permeability in their membranes. Some psychrophiles have high levels of polyunsaturated fatty acids in their membranes, while UV radiation–resistant radiophiles employ UV-absorbing pigments to protect their DNA from radiation damage (Ventosa et al., 1998). Hyperthermophiles growing at temperatures over 100 C, typically archaea, have stable ether bond–based lipids and monolayer membranes with covalent bonds, providing extra stability in high-temperature conditions (Rainey and Oren, 2006).

Additionally, microbially produced extracellular polysaccharides (EPS) protect cells from various stresses (Wharton, 2007; Nichols et al., 2005). EPS forms a hydrated matrix that affects the diffusion of compounds in and out of cells, acting as a protective layer against toxic substances, extreme acidity, UV radiation, or predation (Bitton and Freihofer, 1977; Aguilera et al., 2008; Wang et al., 2007; Caron, 1987). The specific role of EPS depends on the microorganism's environment. Changes in pH and salinity minimally affect EPS viscosity and stability. EPS can also reduce water loss during desiccation and aid water uptake during rehydration (Ritsema and Smeekens, 2003). High-polyhydroxyl EPS in low-temperature, high-salinity environments serve as a cryoprotectant, lowering the freezing point of water. For instance, the EPS of Antarctic fungal strain *Phoma herbarum* acts as a cryoprotective agent in the harsh Antarctic conditions with low liquid water availability and extremely low temperatures (Selbmann et al., 2002).

10.10.1 Utilization of Extremophiles for the Production of Exopolysaccharides (EPS)

EPS synthesized by extremophiles possess exceptional physico-chemical and rheological characteristics, rendering them valuable biomaterials applicable to a wide range of industry. These EPS serve various purposes, including roles in textiles, detergents, adhesives, microbial-enhanced oil recovery, wastewater treatment, dredging, brewing, pharmaceuticals, cosmetology, and food additives (Kumar et al., 2004). Notably, they function as emulsifiers, bioflocculants, and absorbers of inorganic ions essential for microbial metabolism.

Among the primary EPS producers are bacteria, with notable examples including *Xanthomonas*, *Leuconostoc*, *Pseudomonas*, and *Alcaligenes*, responsible for generating EPS varieties like xanthan, dextran, gellan, and curdlan. Lately, there is an increasing focus on EPS produced by lactic acid bacteria, recognized for their safety in food applications. Fungal polysaccharides such as pullulan from *Aureobasidium pullulans* and scleroglucan from *Sclerotium glucanicum* have also garnered attention and are produced on a technical scale.

Despite the considerable potential of microbial EPS, challenges persist, notably their higher production costs compared with synthetic polymers. To address this, researchers are actively refining biopolymer production processes by optimizing upstream-to-downstream engineering strategies, encompassing metabolic and cellular enhancements, efficient fermentation techniques, recovery processes, and post-production modifications (Lee et al., 2005).

While numerous microbial polysaccharides have been characterized, only a select few match the robustness of commercially employed EPS like xanthan gum, acknowledged for its stability under diverse conditions (Lee et al., 2005). Researchers are now directing their efforts toward extremophiles capable of producing unique EPS varieties adapted to extreme conditions. These extremophiles, comprising bacteria, archaea, and even eukaryotic microorganisms, hold particular intrigue due to their capacity to thrive in environments characterized by desiccation, temperature fluctuations, extreme pressures, salinity, acidity, heavy metal presence, and radiation. The EPS produced by these microorganisms demonstrate promise for diverse industrial applications, given their remarkable adaptation to harsh conditions.

10.11 CONSERVATION AND ENVIRONMENTAL IMPACT

Extremophiles are a valuable asset, contributing to both conservation efforts and environmental impact. They not only preserve unique ecosystems on Earth but also provide solutions to pollution, environment-friendly technologies, and sustainable energy sources. The study and application of extremophiles are pivotal in ensuring the preservation of our planet and the development of a cleaner and more sustainable future.

Extremophilic microorganisms have ecological and biotechnological significance, showcasing extraordinary adaptations to extreme conditions. Thermophiles, for example, boast proteins and cell membranes capable of withstanding high temperatures, while psychrophiles and barophiles maintain membrane and cell wall stability in the face of low temperatures and high pressure. Halophiles thrive in high-salt environments by accumulating inorganic ions and compatible solutes, and acidophiles and alkaliphiles regulate pH through specific ion expulsion mechanisms. In these challenging surroundings, extremophiles reveal their exceptional ability to maintain membrane stability and fluidity, and safeguard their genetic material, underscoring the presence of unique genes that enable them to thrive under such extreme conditions. These extremophiles play a pivotal role in producing stable biomolecules that can endure extreme temperatures, pH variations, high pressure, and exposure to pollutants. Their diverse applications encompass providing stable enzymes for extreme temperature and pressure conditions, contributing to biodegradation and bioremediation in extreme environments, serving as sources of biofuel and bioenergy, and supplying specialized pigments for solar cells designed to function under extreme conditions like polar caps. The utilization of extremophiles is positioned to be instrumental in achieving sustainability and advancing the bio-based economy (Arora and Panosyan, 2019). Extremozymes, derived from extremophiles, are already pivotal biomolecules in various industries, and as industries expand, climate conditions evolve, and pollutants diversify, their value in the bio-based market is set to rise significantly. Realizing the full potential of extremophiles for industrial purposes necessitates innovative techniques to overcome challenges associated with uncultured extremophiles and the large-scale production of their metabolites and enzymes. An illustrative instance of industrial extremozyme application is the use of thermophilic DNA polymerases, such as *Taq polymerase* from *Thermus aquaticus*, in PCR for diagnostics. Extremophiles offer a rich source of hydrolases, including amylases, cellulases, esterases, lipases, peptidases, and xylanases, finding applications in detergents, petroleum, pulp and paper, food processing, beverages, and bioremediation. Beyond their industrial advantages, research on extremophiles enhances our comprehension of extreme ecological systems, facilitating insight into microbial composition and biogeochemical cycles. This knowledge can be harnessed to address global challenges, including the restoration of polluted ecosystems and the improvement of yields in degraded habitats. Countless valuable species and metabolites remain undiscovered, ensuring that research on extremophiles will continue to play a central role in ensuring environmental sustainability (Arora and Panosyan, 2019).

The conservation and environmental impact of extremophiles are substantial. These organisms not only contribute to the preservation of unique ecosystems on Earth but also provide effective solutions to pollution, promote environmentally friendly technologies, and offer sources of sustainable energy. Their study and application are pivotal for both safeguarding our planet and advancing a cleaner and more sustainable future. In terms of conservation, extremophiles are integral to the preservation of extreme ecosystems, such as hydrothermal vents, acidic lakes, and saline environments. Their study is crucial for the understanding and conservation of these unique habitats, ensuring the maintenance of biodiversity and the protection of sensitive environments. Extremophiles also contribute to our understanding of microbial biodiversity, particularly in extreme environments. This knowledge is invaluable for conservation efforts aimed at preserving the genetic diversity of life on Earth and maintaining ecosystem health by identifying potential ecological imbalances.

On the environmental front, extremophiles show promise in bioremediation, a process that employs microorganisms to eliminate or neutralize pollutants in the environment. Their enzymes can break down contaminants, even in extreme conditions, contributing to cleaner ecosystems. Furthermore, extremophiles are a valuable resource for developing green technologies, as they provide stable enzymes and biocatalysts capable of functioning under extreme conditions, reducing the environmental impact of various industries. They also offer alternatives to fossil fuels, particularly in the production of biofuels and bioenergy, which can reduce greenhouse gas emissions and contribute to a cleaner environment. Extremophiles play a role in biogeochemical cycles within extreme ecosystems, helping scientists monitor the movement of elements in the environment and understand the effects of climate change and anthropogenic activities. Additionally, these microorganisms can aid in the restoration of polluted ecosystems by breaking down pollutants and contaminants. Their unique metabolic capabilities make them effective agents in environmental restoration efforts. Finally, extremophiles have implications for astrobiology, with their survival in extreme conditions on Earth suggesting the potential for life in analogous environments on other planets or moons, a concept with significant ramifications for space exploration and the search for extraterrestrial life (Arora and Panosyan, 2019).

10.12 CONCLUSION

Extremophiles are a remarkable category of organisms that challenge our conventional notions of where life can thrive. They flourish in environments that are typically considered hostile or even lethal for most life forms, demonstrating exceptional adaptability. They are found in various extreme conditions, including high and low temperatures, extreme acidity or alkalinity, salinity, pressure, and radiation. These microorganisms span all three domains of life, including bacteria, archaea, and eukarya. They play a crucial role in understanding evolutionary relationships and provide valuable resources for industrial applications, such as enzymes and biomaterials. Extremophiles are categorized based on their tolerance to harsh conditions, leading to various groups, such as thermophiles, acidophiles, alkaliphiles, halophiles, psychrophiles, barophiles, and radioresistant organisms. Research on extremophiles has led to significant discoveries in the field of aquatic microbiology, shedding light on the adaptability of life in extreme conditions and its potential implications for astrobiology.

REFERENCES

Abbamondi, Gennaro Roberto, Margarita Kambourova, Annarita Poli, Ilaria Finore, and Barbara Nicolaus. "Quorum sensing in extremophiles." In *Quorum Sensing*, edited by Tommonaro, G., 97–123. Academic Press: Cambridge, MA, USA, 2019.

Aguilera, Angeles, Virginia Souza-Egipsy, Patxi San Martín-Úriz, and Ricardo Amils. "Extracellular matrix assembly in extreme acidic eukaryotic biofilms and their possible implications in heavy metal adsorption." *Aquatic Toxicology* 88, no. 4 (2008): 257–266.

Alain, Karine, Viggó Thór Marteinsson, Margarita L. Miroshnichenko et al. "Marinitoga piezophila sp. nov., a rod-shaped, thermo-piezophilic bacterium isolated under high hydrostatic pressure from a deep-sea hydrothermal vent." *International Journal of Systematic and Evolutionary Microbiology* 52, no. 4 (2002): 1331–1339.

Allgood, G. S., and J. J. Perry. "Characterization of a manganese-containing catalase from the obligate thermophile Thermoleophilum album." *Journal of Bacteriology* 168, no. 2 (1986): 563–567.

Ambily Nath, I. V., and P. A. Loka Bharathi. "Diversity in transcripts and translational pattern of stress proteins in marine extremophiles." *Extremophiles* 15 (2011): 129–153.

Arora, Naveen Kumar, and Hovik Panosyan. "Extremophiles: Applications and roles in environmental sustainability." *Environmental Sustainability* 2 (2019): 217–218.

Baker, Brett J., and Jillian F. Banfield. "Microbial communities in acid mine drainage." *FEMS Microbiology Ecology* 44, no. 2 (2003): 139–152.

Bartlett, Douglas H. "Microbial adaptations to the psychrosphere/piezosphere." *Journal of Molecular Microbiology and Biotechnology* 1, no. 1 (1999): 93–100.

Battista, John R. "Against all odds: The survival strategies of Deinococcus radiodurans." *Annual Review of Microbiology* 51, no. 1 (1997): 203–224.

Berry, S., Y. V. Bolychevtseva, M. Rögner, and N. V. Karapetyan. "Photosynthetic and respiratory electron transport in the alkaliphilic cyanobacterium Arthrospira (Spirulina) platensis." *Photosynthesis Research* 78 (2003): 67–76.

Bitton, Gabriel, and Vic Freihofer. "Influence of extracellular polysaccharides on the toxicity of copper and cadmium toward Klebsiella aerogenes." *Microbial Ecology* 4 (1977): 119–125.

Blasius, Melanie, Ulrich Hübscher, and Suzanne Sommer. "Deinococcus radiodurans: What belongs to the survival kit?." *Critical Reviews in Biochemistry and Molecular Biology* 43, no. 3 (2008): 221–238.

Bowman, John P., Sharee A. McCammon, Tom Lewis, Jennifer H. Skerratt, Janelle L. Brown, David S. Nichols, and Tom A. McMeekin. "Psychroflexus torquis gen. nov., sp. nov. a psychrophilic species from Antarctic sea ice, and reclassification of Flavobacterium gondwanense (Dobson et al. 1993) as Psychroflexus gondwanense gen. nov., comb. nov." *Microbiology* 144, no. 6 (1998): 1601–1609.

Brock, Thomas D. "Life at high temperatures: Evolutionary, ecological, and biochemical significance of organisms living in hot springs is discussed." *Science* 158, no. 3804 (1967): 1012–1019.

Bull, Alan T., Alan C. Ward, and Michael Goodfellow. "Search and discovery strategies for biotechnology: The paradigm shift." *Microbiology and Molecular Biology Reviews* 64, no. 3 (2000): 573–606.

Caron, David A. "Grazing of attached bacteria by heterotrophic microflagellates." *Microbial Ecology* 13 (1987): 203–218.

Cavicchioli, Ricardo, William J. Ripple, Kenneth N. Timmis, Farooq Azam, Lars R. Bakken, Matthew Baylis, Michael J. Behrenfeld et al. "Scientists' warning to humanity: Microorganisms and climate change." *Nature Reviews Microbiology* 17, no. 9 (2019): 569–586.

Chandrasekaran, M., and S. Rajeev Kumar. "Marine microbial enzymes." *Biotechnology* 9 (1997): 47–79.

Conrad, Rolf. "Contribution of hydrogen to methane production and control of hydrogen concentrations in methanogenic soils and sediments." *FEMS Microbiology Ecology* 28, no. 3 (1999): 193–202.

Coohill, Thomas P. "Stratospheric ozone loss, ulraviolet effects and action spectroscopy." *Advances in Space Research* 18, no. 12 (1996): 27–33.

Dalmaso, Gabriel Zamith Leal, Davis Ferreira, and Alane Beatriz Vermelho. "Marine extremophiles: A source of hydrolases for biotechnological applications." *Marine Drugs* 13, no. 4 (2015): 1925–1965.

D'Amico, Salvino, Tony Collins, Jean-Claude Marx, Georges Feller, Charles Gerday, and Charles Gerday. "Psychrophilic microorganisms: Challenges for life." *EMBO Reports* 7, no. 4 (2006): 385–389.

Dartnell, Lewis R. "Ionizing radiation and life." *Astrobiology* 11, no. 6 (2011a): 551–582.

Dartnell, Lewis. "Biological constraints on habitability." *Astronomy & Geophysics* 52, no. 1 (2011b): 1–25.

DasSarma, Shiladitya, and Priya DasSarma. "Halophiles and their enzymes: Negativity put to good use." *Current Opinion in Microbiology* 25 (2015): 120–126.

Dewan, S. S. "Global markets for enzymes in industrial applications." BCC Research, Wellesley (2014).

Díaz-Tena, E., A. Rodríguez-Ezquerro, L. N. López de Lacalle Marcaide, L. Gurtubay Bustinduy, and A. Elías Sáenz. "Use of extremophiles microorganisms for metal removal." *Procedia Engineering* 63 (2013): 67–74.

Dib, Julián Rafael, Wolfgang Liebl, Martin Wagenknecht, María Eugenia Farías, and Friedhelm Meinhardt. "Extrachromosomal genetic elements in Micrococcus." *Applied Microbiology and Biotechnology* 97 (2013): 63–75.

Dyall-Smith, Mike L., Friedhelm Pfeiffer, Kathrin Klee et al. "Haloquadratum walsbyi: Limited diversity in a global pond." *PLoS One* 6, no. 6 (2011): e20968.

Eltsov, Mikhail, and Jacques Dubochet. "Fine structure of the Deinococcus radiodurans nucleoid revealed by cryoelectron microscopy of vitreous sections." *Journal of Bacteriology* 187, no. 23 (2005): 8047–8054.

Gilbert, Jack A., Philip J. Hill, Christine E. R. Dodd, and Johanna Laybourn-Parry. "Demonstration of antifreeze protein activity in Antarctic lake bacteria." *Microbiology* 150, no. 1 (2004): 171–180.

Gupta, Hemant Kumar, Rinkoo Devi Gupta, Ajit Singh, Nar Singh Chauhan, and Rakesh Sharma. "Genome sequence of Rheinheimera sp. strain A13L, isolated from Pangong Lake, India." *J Bacteriol* 193, (2011).https://doi.org/10.1128/jb.05636-11.

Hallberg, Kevin B., and D. Barrie Johnson. "Biodiversity of acidophilic prokaryotes." *Advances in Applied Microbiology, Academic Press* 49, (2001): 37–84.

Holland, Pamela M., Richard D. Abramson, Robert Watson, and David H. Gelfand. "Detection of specific polymerase chain reaction product by utilizing the 5′----3'exonuclease activity of Thermus aquaticus DNA polymerase." *Proceedings of the National Academy of Sciences* 88, no. 16 (1991): 7276–7280.

Horikoshi, K. & W. D. Grant. 1998. *Extremophiles-microbial life in extreme environments.* Wiley, New York. Howland, J. Edited by K Horikoshi and WD Grant.

Horneck, Gerda, Ralf Moeller, Jean Cadet, Thierry Douki, Rocco L. Mancinelli, Wayne L. Nicholson, Corinna Panitz et al. "Resistance of bacterial endospores to outer space for planetary protection purposes—Experiment PROTECT of the EXPOSE-E mission." *Astrobiology* 12, no. 5 (2012): 445–456.

Howland, J. "Extremophiles—Microbial life in extreme environments." Edited by K. Horikoshi and W. D. Grant, 322. New York: Wiley-Liss, 1998. $119.95 ISBN 0-471-02618-2." 331–331.

Hua, Xiaoting, Lifen Huang, Bing Tian, and Yuejin Hua. "Involvement of recQ in the ultraviolet damage repair pathway in Deinococcus radiodurans." *Mutation Research/Fundamental and Molecular Mechanisms of Mutagenesis* 641, no. 1–2 (2008): 48–53.

Irwin, Jane A., and Alan W. Baird. "Extremophiles and their application to veterinary medicine." *Irish Veterinary Journal* 57 (2004): 1–7.

Jagannadham, Medicharla V., Madhab K. Chattopadhyay, Chilukuri Subbalakshmi et al. "Carotenoids of an Antarctic psychrotolerant bacterium, Sphingobacterium antarcticus, and a mesophilic bacterium, Sphingobacterium multivorum." *Archives of Microbiology* 173 (2000): 418–424.

Junge, Karen, Karen Cameron, and Brook Nunn. "Diversity of psychrophilic Bacteria in sea and glacier ice environments—Insights through genomics, metagenomics, and proteomics approaches." In *Microbial Diversity in the Genomic Era*, 197–216. Academic Press, 2019.

Krulwich, Terry A., George Sachs, and Etana Padan. "Molecular aspects of bacterial pH sensing and homeostasis." *Nature Reviews Microbiology* 9, no. 5 (2011): 330–343.

Kumar, C. Ganesh, Han-Seung Joo, Jang-Won Choi, Yoon-Moo Koo, and Chung-Soon Chang. "Purification and characterization of an extracellular polysaccharide from haloalkalophilic Bacillus sp. I-450." *Enzyme and Microbial Technology* 34, no. 7 (2004): 673–681.

Kurr, Margit, Robert Huber, Helmut König et al. "Methanopyrus kandleri, gen. and sp. nov. represents a novel group of hyperthermophilic methanogens, growing at 110 C." *Archives of Microbiology* 156 (1991): 239–247.

Lauro, Federico M., Diane McDougald, Torsten Thomas, Timothy J. Williams, Suhelen Egan, Scott Rice, Matthew Z. DeMaere et al. "The genomic basis of trophic strategy in marine bacteria." *Proceedings of the National Academy of Sciences* 106, no. 37 (2009): 15527–15533.

Lee, Sang Yup, Si Jae Park, Jong Pil Park, Young Lee, and Seung Hwan Lee. "Economic aspects of biopolymer production." In *Biopolymers Online: Biology• Chemistry• Biotechnology• Applications.* Alexander Steinbüchel (Ed.), (2005). vol. 2. Wiley-VCH, Weinheim, Germany.

Letchumanan, Vengadesh, Kok-Gan Chan, and Learn-Han Lee. "Vibrio parahaemolyticus: A review on the pathogenesis, prevalence, and advance molecular identification techniques." *Frontiers in Microbiology* 5 (2014): 705.

Madigan, Michael T., and Aharon Orent. "Thermophilic and halophilic extremophiles." *Current Opinion in Microbiology* 2, no. 3 (1999): 265–269.

Margesin, R., and F. Schinner. "Properties of cold-adapted microorganisms and their potential role in biotechnology." *Journal of Biotechnology* 33, no. 1 (1994): 1–14.

Marteinsson, Viggó Thór, Jean-Louis Birrien, Anna-Louise Reysenbach et al. "Thermococcus barophilus sp. nov., a new barophilic and hyperthermophilic archaeon isolated under high hydrostatic pressure from a deep-sea hydrothermal vent." *International Journal of Systematic and Evolutionary Microbiology* 49, no. 2 (1999): 351–359.

Mirete, Salvador, Verónica Morgante, and José Eduardo González-Pastor. "Acidophiles: Diversity and mechanisms of adaptation to acidic environments." In *Adaption of Microbial Life to Environmental Extremes*, Stan-Lotter, H., Fendrihan, S. (eds), Springer, Cham. https://doi.org/10.1007/978-3-319-48327-6_9.

Mitskevich, I. N., M. N. Poglazova, S. S. Abyzov, N. I. Barkov, N. E. Bobin, and M. V. Ivanov. "Microorganisms found in the basal horizons of the Antarctic glacier above Lake Vostok." In *Doklady Biological Sciences*, vol. 381, 582–585. Kluwer Academic Publishers-Plenum Publishers, 2001.

Moyer, Craig L., and Richard Y. Morita. "Psychrophiles and psychrotrophs." *Encyclopedia of Life Sciences* 1, no. 6 (2007). (New York, NY: John Wiley & Sons): 1-6. doi:10.1002/9780470015902.a0000402.pub2.

Murakami, Chiho, Eiji Ohmae, Shin-ichi Tate, Kunihiko Gekko, Kaoru Nakasone, and Chiaki Kato. "Comparative study on dihydrofolate reductases from Shewanella species living in deep-sea and ambient atmospheric-pressure environments." *Extremophiles* 15 (2011): 165–175.

Nichols, C. A. Mancuso, Jean Guezennec, and J. P. Bowman. "Bacterial exopolysaccharides from extreme marine environments with special consideration of the southern ocean, sea ice, and deep-sea hydrothermal vents: A review." *Marine Biotechnology* 7 (2005): 253–271.

Nichols, David S., and Tom A. McMeekin. "Biomarker techniques to screen for bacteria that produce polyunsaturated fatty acids." *Journal of Microbiological Methods* 48, no. 2–3 (2002): 161–170.

Nogi, Y., Hosoya, S., Kato, C., & Horikoshi, K. (2004). Colwellia piezophila sp. nov., a novel piezophilic species from deep-sea sediments of the Japan Trench. *International Journal of Systematic and Evolutionary Microbiology, 54*(5), 1627–1631.

Nogi, Yuichi. "Taxonomy of psychrophiles." In *Extremophiles Handbook*, edited by Koki Horikoshi, 777–792, 2011. Springer, Japan.

Ordoñez, Omar Federico, Esteban Omar Lanzarotti, Daniel German Kurth, Gorriti MF, Revale S, Cortez N, Vazquez MP, Farías ME, Turjanski AG. "Draft genome sequence of the polyextremophilic *Exiguobacterium* sp. Strain S17, Isolated from Hyperarsenic Lakes in the Argentinian Puna." *Genome Announc* (2013). 1:10.1128/genomea.00480-13.

Oren, A. "Life at high salt concentrations." In *The Prokaryotes a Handbook on the Biology of Bacteria Ecophysiology Isolation Identification Applications*, edited by M. Dworkin, S. Falkow, E. Rosenberg, K. H. Schleifer, and E. Stackebrandt, 421–440, 2000.

Oren, Aharon. "Microbial life at high salt concentrations: Phylogenetic and metabolic diversity." *Saline Systems* 4 (2008): 1–13.

Oren, Aharon. "Solar salterns." *Halophilic Microorganisms and Their Environments*, edited by Joseph Seckbach, (2002): 441–469. Kluwer Academic Publishers New York, Boston, Dordrecht, London, Moscow.

Persson, Christer, Henning Rodhe, and Lars-Erik De Geer. "The Chernobyl accident: A meteorological analysis of how radionuclides reached and were deposited in Sweden." *Ambio* 16 (1987): 20–31.

Rainey, Fred A., and Aharon Oren. "1 Extremophile microorganisms and the methods to handle them." *Methods in Microbiology* 35 (2006): 1–25.

Rampelotto, Pabulo Henrique. "Extremophiles and extreme environments." *Life* 3, no. 3 (2013): 482–485.

Reddy, Gundlapalli S. N., Genki I. Matsumoto, Peter Schumann, Erko Stackebrandt, and Sisinthy Shivaji. "Psychrophilic pseudomonads from Antarctica: Pseudomonas antarctica sp. nov., Pseudomonas meridiana sp. nov. and Pseudomonas proteolytica sp. nov." *International Journal of Systematic and Evolutionary Microbiology* 54, no. 3 (2004): 713–719.

Ritsema, Tita, and Sjef Smeekens. "Fructans: Beneficial for plants and humans." *Current Opinion in Plant Biology* 6, no. 3 (2003): 223–230.
Rosenstein, Barry S. "Inhibition of semiconservative DNA synthesis in ICR 2A frog cells by pyrimidine dimers and nondimer photoproducts induced by ultraviolet radiation." *Radiation Research* 100, no. 2 (1984): 378–386.
Russo, Roberta, Daniela Giordano, Alessia Riccio, Guido Di Prisco, and Cinzia Verde. "Cold-adapted bacteria and the globin case study in the Antarctic bacterium Pseudoalteromonas haloplanktis TAC125." *Marine Genomics* 3, no. 3–4 (2010): 125–131.
Salbu, Brit, and Trygve Krekling. "Characterisation of radioactive particles in the environment." *Analyst* 123, no. 5 (1998): 843–850.
Salwan, Richa, and Ramesh Chand Kasana. "Purification and characterization of an extracellular low temperature-active and alkaline stable peptidase from psychrotrophic Acinetobacter sp. MN 12 MTCC (10786)." *Indian Journal of Microbiology* 53 (2013): 63–69.
Selbmann, Laura, Silvano Onofri, Massimiliano Fenice, Federico Federici, and Maurizio Petruccioli. "Production and structural characterization of the exopolysaccharide of the Antarctic fungus Phoma herbarum CCFEE 5080." *Research in Microbiology* 153, no. 9 (2002): 585–592.
Shivaji, Sisinthy, M. K. Ray, N. Shyamala Rao, L. Saisree, M. V. Jagannadham, G. Seshu Kumar, G. S. N. Reddy, and Pushpa M. Bhargava. "Sphingobacterium antarcticus sp. nov., a psychrotrophic bacterium from the soils of Schirmacher Oasis, Antarctica." *International Journal of Systematic and Evolutionary Microbiology* 42, no. 1 (1992): 102–106.
Sievert, Stefan M., and Costantino Vetriani. "Chemoautotrophy at deep-sea vents: Past, present, and future." *Oceanography* 25, no. 1 (2012): 218–233.
Slesarev, Alexei I., James A. Lake, Karl O. Stetter, Martin Gellert, and Sergei A. Kozyavkin. "Purification and characterization of DNA topoisomerase V. An enzyme from the hyperthermophilic prokaryote Methanopyrus kandleri that resembles eukaryotic topoisomerase I." *Journal of Biological Chemistry* 269, no. 5 (1994): 3295–3303.
Sleytr, U. B., and P. Messner. "Crystalline surface layers in procaryotes." *Journal of Bacteriology* 170, no. 7 (1988): 2891–2897.
Sorokin, D. Yu, T. P. Tourova, Marc Mußmann, and G. Muyzer. "Dethiobacter alkaliphilus gen. nov. sp. nov., and Desulfurivibrio alkaliphilus gen. nov. sp. nov.: Two novel representatives of reductive sulfur cycle from soda lakes." *Extremophiles* 12 (2008): 431–439.
Stojanović, Dejan B., Oliver Fojkar, Aleksandra V. Drobac-Čik, Kristina O. Čajko, Tamara I. Dulić, and Zorica B. Svirčev. "Extremophiles: Link between earth and astrobiology." *Zbornik Matice srpske za prirodne nauke* 114 (2008): 5–16.
Takai, Ken, Akihiko Sugai, Toshihiro Itoh, and Koki Horikoshi. "Palaeococcus ferrophilus gen. nov., sp. nov., a barophilic, hyperthermophilic archaeon from a deep-sea hydrothermal vent chimney." *International Journal of Systematic and Evolutionary Microbiology* 50, no. 2 (2000): 489–500.
Tindall, Brian J., Ramón Rosselló-Móra, H.-J. Busse, Wolfgang Ludwig, and Peter Kämpfer. "Notes on the characterization of prokaryote strains for taxonomic purposes." *International Journal of Systematic and Evolutionary Microbiology* 60, no. 1 (2010): 249–266.
Trincone, Antonio. "Potential biocatalysts originating from sea environments." *Journal of Molecular Catalysis B: Enzymatic* 66, no. 3–4 (2010): 241–256.
Valdés, Jorge, Inti Pedroso, Raquel Quatrini, Robert J. Dodson, Herve Tettelin, Robert Blake, Jonathan A. Eisen, and David S. Holmes. "Acidithiobacillus ferrooxidans metabolism: From genome sequence to industrial applications." *BMC Genomics* 9 (2008): 1–24.
Vazquez, Susana C., Silvia H. Coria, and Walter P. Mac Cormack. "Extracellular proteases from eight psychrotolerant Antarctic strains." *Microbiological Research* 159, no. 2 (2004): 157–166.
Ventosa, Antonio, Joaquín J. Nieto, and Aharon Oren. "Biology of moderately halophilic aerobic bacteria." *Microbiology and Molecular Biology Reviews* 62, no. 2 (1998): 504–544.
Vermelho, A. B., E. F. Noronha, ., Filho, E.X.F., Ferrara, M.A., and Bon, E.P.S. "Diversity and biotechnological applications of prokaryotic enzymes." In *The Prokaryotes*, edited by Rosenberg, E., DeLong, E.F., Lory, S., Stackebrandt, E., Thompson, F. (2013), Springer, Berlin, Heidelberg. https://doi.org/10.1007/978-3-642-31331-8_112.
Wang, Hongyuan, Xiaolu Jiang, Haijin Mu, Xiaoting Liang, and Huashi Guan. "Structure and protective effect of exopolysaccharide from P. agglomerans strain KFS-9 against UV radiation." *Microbiological Research* 162, no. 2 (2007): 124–129.
Wharton, David A. *Life at the Limits: Organisms in Extreme Environments*. Cambridge University Press, 2007.

Xu, Ying, Yuichi Nogi, Chiaki Kato et al. "Moritella profunda sp. nov. and Moritella abyssi sp. nov., two psychropiezophilic organisms isolated from deep Atlantic sediments." *International Journal of Systematic and Evolutionary Microbiology* 53, no. 2 (2003): 533–538.

Yamashita, Yasuhiro, Norifumi Nakamura, Kazuhiro Omiya, Jiro Nishikawa, Hidehisa Kawahara, and Hitoshi Obata. "Identification of an antifreeze lipoprotein from Moraxella sp. of Antarctic origin." *Bioscience, Biotechnology, and Biochemistry* 66, no. 2 (2002): 239–247.

Zeikus, J. G., and R. S. Wolee. "Methanobacterium thermoautotrophicus sp. n., an anaerobic, autotrophic, extreme thermophile." *Journal of Bacteriology* 109, no. 2 (1972): 707–713.

Zeikus, J. Gregory. "Thermophilic bacteria: Ecology, physiology and technology." *Enzyme and Microbial Technology* 1, no. 4 (1979): 243–252.

Zhang, Gaosen, Xiaojun Ma, Fujun Niu, Maoxing Dong, Huyuan Feng, Lizhe An, and Guodong Cheng. "Diversity and distribution of alkaliphilic psychrotolerant bacteria in the Qinghai–Tibet Plateau permafrost region." *Extremophiles* 11 (2007): 415–424.

11 Algicidal Microbes in the Aquatic Environment

Kapil S. Sukhdhane and Pramod Kumar Pandey

11.1 INTRODUCTION

The focus on terrestrial plants in relation to the human experience often leads to overlooking the critical role of aquatic organisms in global biogeochemical cycles. While terrestrial plants receive significant attention, it is crucial to recognize the pivotal role of single-celled microalgae, known as phytoplankton, in aquatic ecosystems. These microscopic organisms are the primary photosynthetic organisms in aquatic environments, efficiently converting solar energy into biomass. They serve as the fundamental building blocks of the aquatic food web, supporting diverse forms of life. Moreover, phytoplankton play a vital role in biogeochemical cycles by acting as substantial sinks for inorganic nutrients and carbon dioxide (CO_2), contributing to the regulation of nutrient levels and climate balance in aquatic ecosystems. Extensive research has shown that these tiny organisms are responsible for approximately 50% of global primary production, effectively harnessing sunlight to convert it into organic matter (Field et al., 1998; Behrenfeld et al., 2006). Additionally, phytoplankton play a critical role in the assimilation of nitrogen, accounting for about 70% of the global level (Raven et al., 1993).

Phytoplankton inhabit a wide range of environments, from freshwater to marine, encompassing coastal zones as well as open waters. These organisms serve as a reservoir for inorganic nutrients, which become available through processes like predation, cell mortality, and subsequent breakdown. Planktonic species can be of macro and micro scale and are primarily influenced by a combination of biotic interactions and the fluctuations of temperature, resources, and nutrients at spatial scale (Dann et al., 2016). Biotic interactions, such as predation, competition, and symbiosis, play a significant role in shaping the assemblage of planktonic species. Research has demonstrated that in aquatic ecosystems, the interactions between phytoplankton and bacteria can have both positive or negative interactions, which are contingent upon the nutrient concentrations present in their environment (Azam et al., 1983; Wang and Priscu, 1994; Kamjunke et al., 1997). Interactions between phytoplankton and bacteria form essential ecological relationships within aquatic ecosystems, as they significantly contribute to the structure and population density of microbial communities, which limits the overall functioning of aquatic ecosystems (Seymour et al., 2017).

11.2 ALGICIDAL INTERACTIONS

Among these bacteria, algicidal bacteria are the group that can greatly inhibit the growth of algae or lyse algal cells (Meyer et al., 2017; Coyne et al., 2022). Algicidal bacteria can control algal blooms, which is attracting significant interest among researchers in different fields, such as biogeochemistry, algal physiology, ecology, the aquatic environment, and even aquaculture. Algicidal bacteria possess a wide array of potential in various environmental applications due to their diverse characteristics, including biodegradability, low toxicity, and biocompatibility (Fiechter, 1992). Algicidal bacteria have been isolated and characterized from various lakes and seas and also from land. Several studies have found that algicidal bacteria may only be effective against certain species of algae (Doucette et al., 1999).

DOI: 10.1201/9781003408543-11

Defining algicidal bacteria in the literature proves challenging due to their taxonomic diversity, leading to specific inhibitory actions against particular prokaryotic cyanobacteria or eukaryotic algae. The other algicidal bacteria that have been isolated are Bacteroidetes and Proteobacteria (Meyer et al., 2017). Most of the isolated algicidal bacteria effective against cyanobacteria are widely found in Actinobacteria, Proteobacteria, and Firmicutes (Coyne et al., 2022). Recent study has shown that algicidal bacteria vary in their degree of activity according to their type of species as well as target phytoplankton. Meyer et al. (2017) stated that the relationship between algicidal bacteria and their target planktons may involve cell-to-cell dependence, or it may be independent, which provides a wide range of algicides, such as alkaloids, amino acids derivatives, peptides and other metabolites.

Algicidal interactions are categorized into two different modes depending on the host, i.e., direct and indirect mode (Hu et al., 2019). The direct mode of algicidal interactions requires live bacterial cells to effectively antagonize the alga through direct contact (Xue et al., 2022; Zhang et al., 2019). Direct modes of algicidal interaction depend on different factors, mainly varying temperature, pH, and the concentration of algicidal bacteria. In the indirect mode of action, bacterial metabolites mediate the interactions between the bacteria and algal cells. The bacterial inoculum density and physiological status (growth phase) of the host algae are considered key factors for determining the algicidal activity of the bacteria (Doucette et al., 1999; Manage et al., 2000; Sigee et al., 1996). A study by McDowell et al. (2014) reported that the release and accumulation of strong oxidants in algal cells led to the production of reactive oxygen species (ROS) in freshwater. The current knowledge on algicidal bacteria is more biased towards culturable bacteria due to limited study. The evaluation of algicidal bacteria as target strains against different cyanobacteria or algae can be done in different ways.

11.3 ISOLATION OF ALGICIDAL BACTERIAL STRAINS

The isolation and screening of algicidal bacteria depend on the different aquatic environments, i.e., freshwater, brackishwater, and marine water. Algicidal bacteria can also be sourced from soil or estuarine sediments, as highlighted in studies by Luo et al. (2013) and Zhang et al. (2018, 2021). However, the ecological roles of compounds secreted by bacteria in sediment ecosystems are more likely linked to the abundance of dissolved and particulate organic matter present in these environments, rather than being primarily focused on inducing lysis in planktonic algal populations. Water samples from aquatic ecosystems are mainly collected from the surface water at ~0.5 m depth in sterilized bottles and are transported to the laboratory for bacterial isolation. In the case of long duration of transportation, samples should be shipped with ice packs to the laboratory. Isolation methods include bacterial enrichment followed by isolation. Different bacterial media with different incubation times are used based on the targeted bacterial strains and the water medium. A photosynthetic bacteria medium is commonly used as culture medium for the identification of different bacterial strains (Xue et al., 2022). Other agars, such as R2A agar (Reasoner and Geldreich, 1985), ½ TY agar (Beringer, 1974), and M9 agar (Miller, 1972), are used for isolation of the bacterial colonies from different freshwater lakes for algicidal study. The incubation period for the isolated bacteria varies from 24h to 1 week at room temperature (Ren et al., 2023). A serial dilution method also can be used for the enumeration of bacterial strains for algicidal study. Representative bacterial colonies, based on their colour, size, and morphology, are streaked several times for purification. Simultaneously, scientists have acknowledged the remarkable versatility of bacterial communities in generating hydrolytic ectoenzymes, as observed in the work of Martinez et al. (1996). Following this discovery, numerous algicidal bacteria have been documented, and novel isolates continue to be identified.

The screening of algicidal bacteria is carried out by mixing a pure culture of bacteria with an experimental algal mixture. In the initial screening, the inoculum ratio of bacterial cells with algal cells can be maintained at 100:1, and different inoculum ratios can then be tried based on the efficiency of bacterial strains (Ren et al., 2023). The algal-lysing activities are then evaluated based on the number of algae counted by light microscopy. Algal-lysing activities are calculated based on the following formula:

$$\text{Algal-lysing activities}(\%) = 1 - \frac{Tt}{Ct} \times 100$$

where T and C are concentrations of the algae in the treatment and algae groups, and t is the incubation time. Pure isolates with high algal-lysing activities can be selected for study of algicidal activities. Minocha et al. (2009) determined the algicidal rates using the chlorophyll a method for cyanobacteria, and the total chlorophyll content of algae was determined using the following formula:

$$\text{Algicidal Rates}(\%) = 1 - \frac{\text{Chlorophyll}_{+\text{bacteria}}}{\text{Chlorophyll}_{\text{control}}} \times 100$$

where $\text{Chlorophyll}_{+\text{bacteria}}$ is the chlorophyll content of the culture medium and $\text{Chlorophyll}_{\text{control}}$ is the chlorophyll content of the alga/cyanobacterium used for the study.

The evaluation of algicidal bacteria against target strains of cyanobacteria or algae can be done in different ways. The growth of the targeted cyanobacteria is monitored by measuring the fluorescence density of the algal culture at different wavelengths and also by microscopic observation. The other method of measuring algicidal bacteria is by calculation of the volume/volume ratio of bacterial cells in the seawater (Azam et al., 1992). Another method of investigation involves the characterization of the chemical composition of the exudates and bioassay-guided fractionation (Kim et al., 2015). The cell concentration of algicidal bacteria is assessed through both macroscopic observations and the sedgwick rafter method, which involves quantifying algal cell densities. A dose–response experiment can be performed to determine the activity of bacteria. The bacterial strain is considered to be algicidal when the desired cyanobacterial culture is killed. The following formula is used to calculate the algicidal effects:

$$\text{Algicidal activity}(\%) = \frac{(\text{Number of initial cells} - \text{Number of surviving cells})}{\text{Number of initial cells}} \times 100$$

The growth of targeted cyanobacteria is also monitored by measuring the fluorescence density of algal culture at different wavelengths, and also by microscopic observation or by following an electron microscopic method. Algicidal metabolites have also been studied against the selective bacterial strains by a few researchers. Wang et al. (2022) studied algicidal power sawdust as a selected carrier against algicidal bacteria. The effectiveness of sawdust was evaluated by adding 50 ml of *Noctiluca scintillans*, and it was found to be a novel metabolite to control harmful algal blooms.

11.4 IDENTIFICATION OF ALGICIDAL BACTERIA

Algicidal bacteria are identified by extraction and polymerase chain reaction (PCR) amplification of genomic 16s rDNA. For the amplification of 16s rDNA, 27F and 1492R primers are used. Near-complete 16S rDNA sequences are aligned using different software, such as CLUSTAL W software (Thompson et al., 1994), or by using different DNA sequencing systems for PCR products. The output of 16S rDNA sequencing is used for phylogenetic analysis by comparing it with the other bacterial sequences available in the gene bank database. By applying software like MEGA 5.0, a phylogenetic tree is constructed for neighbourhood analysis (Tamura and Nei, 1993). MEGA version 2.0 is used further for bootstrap analysis of 1000 replicates (Kumar et al., 2001).

11.5 CONTROL OF HARMFUL ALGAL BLOOMS BY ALGICIDAL BACTERIA

Globally pervasive, harmful algal blooms (HABs) pose a significant challenge, exerting a multitude of adverse impacts on aquatic ecosystems. HABs harm aquatic ecosystems by severely depleting the oxygen level in natural water bodies. Their duration can span from days to several months. The rapid and unchecked proliferation of algae is detrimental to aquatic ecosystems by emitting secondary metabolites and toxins. HABs are responsible for a decrease in water clarity (Capuzzo et al., 2015), which may lead to a decrease in dissolved oxygen levels (Mu et al., 2017). HABs are also responsible for the death of numerous aquatic organisms, which leads to biodiversity loss (Nasri et al., 2008) and also limits the aesthetic value of aquatic water bodies (Mitra and Flynn, 2006). The primary source of HABs is mainly excess nutrients in the ecosystem (i.e., N, P, K), which originate from point or nonpoint sources, resulting in excess growth of algae in aquatic ecosystems. Excessive algal growth stands as a contributing factor, triggering the discharge of secondary metabolites and toxins. These substances possess the potential to inflict harm to the environment and various organisms within it. The physical structure of algal cells or biomass accumulation contributes to harmful algal blooms (HABs), disrupting co-occurring organisms and food web dynamics..

Several measures, including physical, chemical, and biological methods, attracted attention after the diverse impacts of HABs were noticed in lentic ecosystems (Mohamed et al., 2014). Physical methods are the preferred form for control of algal blooms in many places. Physical methods include adsorption (Marzbali et al., 2017), the use of UV radiation (Alam et al., 2001), membrane-based filtration techniques (Zhao et al., 2017) and ultrasound techniques (Park et al., 2017), which have all been found to be efficient techniques to remove HABs. Chemical methods include the use of different herbicides (Nagai et al., 2016), photosensitizers (Pohl et al., 2015), and metals (Magdaleno et al., 2014). Physical and chemical method of remediation have several limitations. They may lead to toxicity in aquatic ecosystems, and the expenditure involved in the control operation is huge.

Biological control methods include the use of different aquatic plants, their derivatives, and algicidal microorganisms to control HABs. These approaches have garnered significant interest and are being regarded as a potent strategy due to their ability to control algae and their environmentally conscious nature (Backer et al., 2015; Harke et al., 2016). Among these techniques, the algicidal approach stands out as a widely adopted method for disrupting or diminishing the proliferation of harmful algae. The primary participants in this approach are microbial strains known for their algicidal properties, comprising bacteria, fungi, and actinomycetes (Mohamed et al., 2014). Algicidal bacteria kill cyanobacteria by two methods, direct and indirect. Direct algicidal activity is mainly based on the contact between the algae and HABs. In 1970, Shilo discovered the algicidal properties of myxobacteria, the first bacteria recognized for these properties. These myxobacteria exhibited the ability to eliminate both single-celled and filamentous Cladophora algae, relying on direct physical interaction between the myxobacteria and the Cladophora algae (Shilo, 1970). A yellow-pigmented strain named *Pseudoalteromonas* Y demonstrated the remarkable ability to disintegrate HABs belonging to the *Chattonella*, *Gymnodinium*, and *Heterosigma* genera (Lovejoy et al., 1998). A recently identified bacterium that produces chitinase (designated strain LY03) has demonstrated a notable algicidal influence on *Thalassiosira pseudonana*, primarily through a direct interaction mechanism. Strain LY03 exhibits a form of chemotaxis towards algal cells by attaching bacteria to the algal surface through their flagella. Subsequently, these bacteria secrete chitinase, an enzyme that breaks down the algal cell walls, ultimately resulting in the disintegration and demise of the algae (Li et al., 2016).

The substances released by bacteria known to possess algicidal properties predominantly encompass an array of compounds such as peptides, proteins, alkaloids, amino acids, antibiotics, pigments, and fatty acids, among other chemical constituents. Numerous investigations have been carried out to isolate diverse algicidal active compounds that are secreted by bacteria. Jeong et al. (2003) isolated a novel polypeptide, bacillamide, produced by a marine bacterium, *Bacillus* sp. SY-1, which has a specific algicidal effect on the harmful dinoflagellate *Cochlodinium polykrikoides*.

Similarly, Hare et al. (2005) and Pokrzywinski et al. (2012, 2017a, 2017b) indicated that the bacterium Shewanella sp. IRI-160 and its water-soluble algicide IRI-160AA were able to control the growth of dinoflagellates with no negative impact on cell densities of other phytoplankton species. Lee et al. (2000) found that a serine protease extracted from *Pseudomonas* A28 has algicidal activity against *Skeletonema costatum* NIES-324.

Utilizing algicidal bacteria for the control of HABs offers multiple benefits, including cost reduction and improved utilization of algicidal bacterial resources. Additionally, this approach has the potential to mitigate the risk of secondary pollution caused by the introduction of exogenous bacteria, as highlighted in a study by Sun et al. (2016). Continued exploration into the isolation and characterization of the active compounds present in bacterial filtrates holds the potential to enhance our comprehension of HAB dynamics and facilitate effective management.

11.6 POTENTIAL USES OF ALGAL–BACTERIAL INTERACTIONS

11.6.1 Greenwater Technology

The greenwater culture technique, widely utilized in aquaculture (Faulk and Holt, 2005; Hargreaves, 2006), involves cultivating animals in water that contains abundant levels of microalgae. In practice, greenwater culture does not use a single species, and the bacterial density can reach levels five to sevenfold higher than that of conventional clear water (Salvensen et al., 1999). Secretion of slime by fishes serves as a bioaugmentor by enhancing the production of plankton and promoting the development of greenwater rich in microbes. The microbial community within greenwater generally comprises specific bacterial species, exhibiting reduced diversity in comparison to clear water (Nicolas et al., 2004). Single-celled microalgae are favoured for production in greenwater culture due to their advantageous attributes. These microalgae possess natural dispersal capabilities, allowing them to spread effectively throughout the water column. Moreover, they remain buoyant for extended periods, contributing to their suitability for cultivation. Importantly, their presence in the cultured water does not lead to fouling or undesirable accumulation, further enhancing their desirability for greenwater culture. Chlorella, among a few other microalgae, stands out as one of the top choices for producing substantial quantities of greenwater. Greenwater proves to be an efficient approach applicable to various culture methodologies, including extensive, intensive, and mesocosm systems employed for larval fish rearing worldwide. Studies have shown that specific beneficial bacteria exhibit improved growth when co-cultured with microalgae compared with when grown alone in monoculture (Gomez-Gil et al., 2002).

Chithambaran et al. (2017) studied shrimp health status and bacterial load from a *Penaeus vannamei* farm cultured in greenwater and noticed improved shrimp health status as well as a significant reduction in the population of harmful bacteria. Until now, there has been minimal emphasis on the bacteria that coexist with microalgae in greenwater systems, and the process of bacterial colonization in these systems has largely been determined by random occurrences. Krishnani et al. (2019) isolated and characterized eight strains of *Bacillus* from milkfish surface mucus cultured in a greenwater system. The antagonistic effects were also evaluated against *Vibrio harveyi* by the agar well diffusion method and co-culture method of the study. These studies found that an antagonistic bacterial formulation, producing biomolecules with antibacterial activity, could be efficiently used against pathogenic bacteria in aquaculture and related aquatic environments through greenwater culture systems.

11.7 PRODUCTION OF MICROALGAE TO BE USED AS FEED

Microalgae form an important part of the diet for many fishes from their fry to adult stages (Hodar et al., 2020). Microalgae exhibit remarkable photosynthetic efficiency and rapid growth rates, which contribute to the generation of substantial yields of carbohydrates, proteins, and lipids (Abomohra et al., 2016; Chen et al., 2018; Su et al., 2017). Microalgae surpass all other terrestrial plants or animals in terms of net biomass productivity (Rizwan et al., 2018). Microalgae cultivation offers the advantage of not requiring fertile land; they can be cultivated in areas that are not suitable for traditional agriculture, such as non-arable land or non-potable water sources, including seawater and wastewater (Li et al., 2019). Microalgae serve as valuable dietary sources of both macro- and micronutrients. The biomass of these organisms holds the potential to serve as natural components and supplements within animal diets, addressing the growing need for protein and energy while also offering an alternative to synthetic additives in feed formulations. The nutritional content of different microalgae used in aquaculture varies from species to species.

Microalgae possess a balanced amino acid profile, negating the necessity for expensive amino acid supplements within the diet. Aquaculture employs a variety of microalgal species, including *Anabaena cylindrica*, *Botryococcus braunii*, *Chlamydomonas rheinhardii*, *Chlorella vulgaris*, *Dunaliella salina*, *Nannochloropsis granulata*, *Phaeodactylum tricornutum*, *Scenedesmus obliquus*, *Spirulina platensis*, *Spirogyra* sp., *Dunaliella* sp., and *Isochrysis*, among others. Incorporating small quantities of microalgal biomass into animal feed can have various physiological benefits. These include enhancing antiviral and antibacterial effects, enhancing immune response, boosting disease resistance, improving gut function, and stimulating the colonization of probiotics.

Bacteria play a vital role in bolstering the productivity and resilience of microalgal cultures by generating essential components such as vitamins, plant growth hormones, and CO_2. They also contribute to the recycling of inorganic nutrients and establish favourable environmental conditions, including the reduction of oxygen tension, which collectively enhance the overall performance and stability of microalgal cultures (Mouget et al., 1995; Croft et al., 2005; Azam and Malfatti, 2007; de-Bashan et al., 2008). However, thorough research is needed in order to select the bacterial strains with the best effect on microalgal activity and to establish to what extent these well-selected strains will increase productivity.

11.8 BIOFLOC TECHNOLOGY

Biofloc technology is a cost-effective, environment-friendly sustainable aquaculture technique with minimum or zero water exchange. In this technique, fish and shrimp are grown in an intensive way, with capacity for 300g biomass per square metre (Avnimelech, 2003). In biofloc culture systems, artificial feeding is reduced, which improves water quality as well as producing microbial protein for different aquatic species (Dinda et al., 2020; Gao et al., 2019). Maintaining the C/N ratio through the addition of carbon sources leads to the production of high-quality single-cell microbial protein (Crab et al., 2012). Adding a carbon source and maintaining the C/N ratio leads to the development of dense microorganisms, which function as a bioreactor controlling water quality (Avnimelech et al., 1989), and it also provides protein food sources to fish and shrimp for production. The uptake rate of inorganic nitrogen compounds by heterotrophic bacteria is higher than that of denitrifying bacteria; therefore, the growth rate and production of microbial biomass per unit substrate in heterotrophic bacteria are 10 times higher (Hargreaves, 2006). In a study conducted by Emerenciano et al. (2017), it was observed that within a biofloc system, certain groups of microorganisms, like bacteria and microalgae, exhibit the ability to engage in biological interactions. Consequently, there is a need to determine the extent to which bacteria and microalgae can generate bioflocs possessing the targeted nutritional characteristics. These microorganisms hold potential as a viable inoculum for initiating aquaculture systems utilizing biofloc technology.

11.9 CONCLUSION

Algicidal bacteria and their derivatives in aquatic environment are highly diverse. Isolation, characterization, and effective use of algicidal bacteria to control HABs as well as to control the growth of algae can be an effective tool of management. Gaining a deeper comprehension of the interactions between algae and bacteria within intensive fish culture systems could potentially unlock novel avenues of opportunity for advancing algal research in the coming years.

REFERENCES

Abomohra, Abd El-Fatah, Wenbiao Jin, Renjie Tu, Song-Fang Han, Mohammed Eid, and Hamed Eladel. "Microalgal biomass production as a sustainable feedstock for biodiesel: Current status and perspectives." *Renewable and Sustainable Energy Reviews* 64 (2016): 596–606.

Alam, M., and Ratner, D. "Cutaneous squamous-cell carcinoma." *New England Journal of Medicine* 344, no. 13 (2001): 975–983.

Avnimelech, Y. "Control of microbial activity in aquaculture systems: Active suspension ponds." *World Aquaculture-Baton Rouge* 34, no. 4 (2003): 19–21.

Avnimelech, Y., S. Mokady, and G. L. Schroeder. "Circulated ponds as efficient bioreactors for single cell protein production." *Israeli Journal of Aquaculture* 41, no. 2 (1989): 58–66.

Azam, Farooq, and Francesca Malfatti. "Microbial structuring of marine ecosystems." *Nature Reviews Microbiology* 5, no. 10 (2007): 782–791.

Azam, Farooq, David C. Smith, and Angelo F. Carlucci. "Bacterial transformation and transport of organic matter in the Southern California Bight." *Progress in Oceanography* 30, no. 1–4 (1992): 151–166.

Azam, Farooq, Tom Fenchel, John G. Field, John S. Gray, Lutz-Arend Meyer-Reil, and F. Thingstad. "The ecological role of water-column microbes in the sea." *Marine Ecology Progress Series* 10, no. 3 (1983): 257–263.

Backer, Lorraine C., Deana Manassaram-Baptiste, Rebecca LePrell, and Birgit Bolton. "Cyanobacteria and algae blooms: Review of health and environmental data from the harmful algal bloom-related illness surveillance system (HABISS) 2007–2011." *Toxins* 7, no. 4 (2015): 1048–1064.

Behrenfeld, Michael J., Robert T. O'Malley, David A. Siegel, Charles R. McClain, Jorge L. Sarmiento, Gene C. Feldman, Allen J. Milligan, Paul G. Falkowski, Ricardo M. Letelier, and Emmanuel S. Boss. "Climate-driven trends in contemporary ocean productivity." *Nature* 444, no. 7120 (2006): 752–755.

Beringer, John E. "R factor transfer in Rhizobium leguminosarum." *Microbiology* 84, no. 1 (1974): 188–198.

Capuzzo, Elisa, David Stephens, Tiago Silva, Jon Barry, and Rodney M. Forster. "Decrease in water clarity of the southern and central North Sea during the 20th century." *Global Change Biology* 21, no. 6 (2015): 2206–2214.

Chen, Hui, Jie Wang, Yanli Zheng, Jiao Zhan, Chenliu He, and Qiang Wang. "Algal biofuel production coupled bioremediation of biomass power plant wastes based on Chlorella sp. C2 cultivation." *Applied Energy* 211 (2018): 296–305.

Chithambaran, Sambhu, Mamdouh Harbi, Mohammad Broom, Khalid Khobrani, Osama Ahmad, Hazem Fattani, Abdulmohsen Sofyani, and Nasser Ayaril. "Green water technology for the production of Pacific white shrimp Penaeus vannamei (Boone, 1931)." *Indian Journal of Fisheries* 64, no. 3 (2017): 43–49.

Coyne, Kathryn J., Yanfei Wang, and Gretchen Johnson. "Algicidal bacteria: A review of current knowledge and applications to control harmful algal blooms." *Frontiers in Microbiology* 13 (2022): 871177.

Crab, R., Defoirdt, T., Bossier, P., and Verstraete, W. "Biofloc technology in aquaculture: beneficial effects and future challenges." *Aquaculture* 356, (2012): 351–356.

Croft, Martin T., Andrew D. Lawrence, Evelyne Raux-Deery, Martin J. Warren, and Alison G. Smith. "Algae acquire vitamin B12 through a symbiotic relationship with bacteria." *Nature* 438, no. 7064 (2005): 90–93.

Dann, Lisa M., James S. Paterson, Kelly Newton, Rod Oliver, and James G. Mitchell. "Distributions of virus-like particles and prokaryotes within microenvironments." *PloS One* 11, no. 1 (2016): e0146984.

De-Bashan, Luz E., Hani Antoun, and Yoav Bashan. "Involvement of indole-3-acetic acid produced by the growth-promoting bacterium Azospirillum spp. in promoting growth of chlorella vulgaris 1." *Journal of Phycology* 44, no. 4 (2008): 938–947.

Dinda, Riya, Amit Mandal, and S. K. Das. "Neem (Azadirachta indica A. Juss) supplemented biofloc medium as alternative feed in common carp (Cyprinus carpio var. communis Linnaeus) culture." *Journal of Applied Aquaculture* 32, no. 4 (2020): 361–379.

Doucette, Gregory J., Elizabeth R. McGovern, and John A. Babinchak. "Algicidal bacteria active against *Gymnodinium Breve* (dinophyceae). I. Bacterial isolation and characterization of killing activity1, 3." *Journal of Phycology* 35, no. 6 (1999): 1447–1454.

Emerenciano, Maurício Gustavo Coelho, Luis Rafael Martínez-Córdova, Marcel Martínez-Porchas, and Anselmo Miranda-Baeza. "Biofloc technology (BFT): A tool for water quality management in aquaculture." *Water Quality* 5 (2017): 92–109.

Faulk, Cynthia K., and G. Joan Holt. "Advances in rearing cobia Rachycentron canadum larvae in recirculating aquaculture systems: Live prey enrichment and greenwater culture." *Aquaculture* 249, no. 1–4 (2005): 231–243.

Fiechter, Armin. "Biosurfactants: Moving towards industrial application." *Trends in Biotechnology* 10 (1992): 208–217.

Field, Christopher B., Michael J. Behrenfeld, James T. Randerson, and Paul Falkowski. "Primary production of the biosphere: Integrating terrestrial and oceanic components." *Science* 281, no. 5374 (1998): 237–240.

Gao, Fangzhou, Shaoan Liao, Shanshan Liu, Hong Bai, Anli Wang, and Jianmin Ye. "The combination use of Candida tropicalis HH8 and Pseudomonas stutzeri LZX301 on nitrogen removal, biofloc formation and microbial communities in aquaculture." *Aquaculture* 500 (2019): 50–56.

Gomez-Gil, B., A. Roque, and G. Velasco-Blanco. "Culture of Vibrio alginolyticus C7b, a potential probiotic bacterium, with the microalga Chaetoceros muelleri." *Aquaculture* 211, no. 1–4 (2002): 43–48.

Hare, Clinton E., Elif Demir, Kathryn J. Coyne, S. Craig Cary, David L. Kirchman, and David A. Hutchins. "A bacterium that inhibits the growth of Pfiesteria piscicida and other dinoflagellates." *Harmful Algae* 4, no. 2 (2005): 221–234.

Hargreaves, John A. "Photosynthetic suspended-growth systems in aquaculture." *Aquacultural Engineering* 34, no. 3 (2006): 344–363.

Harke, Matthew J., Morgan M. Steffen, Christopher J. Gobler, Timothy G. Otten, Steven W. Wilhelm, Susanna A. Wood, and Hans W. Paerl. "A review of the global ecology, genomics, and biogeography of the toxic cyanobacterium, Microcystis spp." *Harmful Algae* 54 (2016): 4–20.

Hodar, A. R., R. J. Vasava, D. R. Mahavadiya, and N. H. Joshi. "Fish meal and fish oil replacement for aqua feed formulation by using alternative sources: A review." *Journal of Experimental Zoology India* Vol. 23, No. 1, pp. 13–21(2020).

Hu, Xiao-Juan, Yu Xu, Hao-Chang Su, Wu-Jie Xu, Li-Hua Wang, Yun-Na Xu, Zhuo-Jia Li, Yu-Cheng Cao, and Guo-Liang Wen. "Algicidal bacterium CZBC1 inhibits the growth of Oscillatoria chlorina, Oscillatoria tenuis, and Oscillatoria planctonica." *AMB Express* 9, no. 1 (2019): 1–13.

Jeong, Seong-Yun, Keishi Ishida, Yusai Ito, Shigeru Okada, and Masahiro Murakami. "Bacillamide, a novel algicide from the marine bacterium, *Bacillus* sp. SY-1, against the harmful dinoflagellate, Cochlodinium polykrikoides." *Tetrahedron Letters* 44, no. 43 (2003): 8005–8007.

Kamjunke, Norbert, W. Böing, and Hanno Voigt. "Bacterial and primary production under hypertrophic conditions." *Aquatic Microbial Ecology* 13, no. 1 (1997): 29–35.

Kim, Yun Sook, Hong-Joo Son, and Seong-Yun Jeong. "Isolation of an algicide from a marine bacterium and its effects against the toxic dinoflagellate Alexandrium catenella and other harmful algal bloom species." *Journal of Microbiology* 53 (2015): 511–517.

Krishnani, Kishore K., V. Kathiravan, M. Kailasam, A. Nagavel, and A. G. Ponniah. "Isolation and characterization of antagonistic bacteria against Vibrio harveyi from milkfish Chanos chanos." *Indian Journal of Fisheries* 66, no. 1 (2019): 124–130.

Kumar, Sudhir, Koichiro Tamura, Ingrid B. Jakobsen, and Masatoshi Nei. "MEGA2: Molecular evolutionary genetics analysis software." *Bioinformatics* 17, no. 12 (2001): 1244–1245.

Lee, Sun-og, Junichi Kato, Noboru Takiguchi, Akio Kuroda, Tsukasa Ikeda, Atsushi Mitsutani, and Hisao Ohtake. "Involvement of an extracellular protease in algicidal activity of the marine bacterium Pseudoalteromonas sp. strain A28." *Applied and Environmental Microbiology* 66, no. 10 (2000): 4334–4339.

Li, Kun, Qiang Liu, Fan Fang, Ruihuan Luo, Qian Lu, Wenguang Zhou, Shuhao Huo et al. "Microalgae-based wastewater treatment for nutrients recovery: A review." *Bioresource Technology* 291 (2019): 121934.

Li, Yi, Xueqian Lei, Hong Zhu, Huajun Zhang, Chengwei Guan, Zhangran Chen, Wei Zheng, Lijun Fu, and Tianling Zheng. "Chitinase producing bacteria with direct algicidal activity on marine diatoms." *Scientific Reports* 6, no. 1 (2016): 21984.

Lovejoy, Connie, John P. Bowman, and Gustaaf M. Hallegraeff. "Algicidal effects of a novel marine Pseudoalteromonas isolate (class Proteobacteria, gamma subdivision) on harmful algal bloom species of the genera Chattonella, Gymnodinium, and Heterosigma." *Applied and Environmental Microbiology* 64, no. 8 (1998): 2806–2813.

Luo, Jianfei, Yuan Wang, Shuishui Tang, Jianwen Liang, Weitie Lin, and Lixin Luo. "Isolation and identification of algicidal compound from Streptomyces and algicidal mechanism to Microcystis aeruginosa." *PloS One* 8, no. 10 (2013): e76444.

Magdaleno, Anahí, Carlos Guillermo Vélez, María Teresa Wenzel, and Guillermo Tell. "Effects of cadmium, copper and zinc on growth of four isolated algae from a highly polluted Argentina river." *Bulletin of Environmental Contamination and Toxicology* 92 (2014): 202–207.

Manage, Pathmalal M., Zenichiro Kawabata, and Shin-ichi Nakano. "Algicidal effect of the bacterium Alcaligenes denitrificans on Microcystis spp." *Aquatic Microbial Ecology* 22, no. 2 (2000): 111–117.

Martinez, Josefina, David C. Smith, Grieg F. Steward, and Farooq Azam. "Variability in ectohydrolytic enzyme activities of pelagic marine bacteria and its significance for substrate processing in the sea." *Aquatic Microbial Ecology* 10, no. 3 (1996): 223–230.

Marzbali, Mojtaba Hedayati, Ali Alinejad Mir, Maryam Pazoki, Reza Pourjamshidian, and Mahla Tabeshnia. "Removal of direct yellow 12 from aqueous solution by adsorption onto spirulina algae as a high-efficiency adsorbent." *Journal of Environmental Chemical Engineering* 5, no. 2 (2017): 1946–1956.

McDowell, Ruth E., Charles D. Amsler, Dale A. Dickinson, James B. McClintock, and Bill J. Baker. "Reactive oxygen species and the A ntarctic macroalgal wound response." *Journal of Phycology* 50, no. 1 (2014): 71–80.

Meyer, Nils, Arite Bigalke, Anett Kaulfuß, and Georg Pohnert. "Strategies and ecological roles of algicidal bacteria." *FEMS Microbiology Reviews* 41, no. 6 (2017): 880–899.

Miller, J. H. "Experiments in molecular genetics." ColdSpring Harbor Laboratory, NY (1972): 328–330.

Minocha, Rakesh, Gabriela Martinez, Benjamin Lyons, and Stephanie Long. "Development of a standardized methodology for quantifying total chlorophyll and carotenoids from foliage of hardwood and conifer tree species." *Canadian Journal of Forest Research* 39, no. 4 (2009): 849–861.

Mitra, Aditee, and Kevin J. Flynn. "Promotion of harmful algal blooms by zooplankton predatory activity." *Biology Letters* 2, no. 2 (2006): 194–197.

Mohamed, Zakaria A., Mohamed Hashem, and Saad A. Alamri. "Growth inhibition of the cyanobacterium Microcystis aeruginosa and degradation of its microcystin toxins by the fungus Trichoderma citrinoviride." *Toxicon* 86 (2014): 51–58.

Mouget, Jean-Luc, Azzeddine Dakhama, Marc C. Lavoie, and Joël de la Noüe. "Algal growth enhancement by bacteria: Is consumption of photosynthetic oxygen involved?." *FEMS Microbiology Ecology* 18, no. 1 (1995): 35–43.

Mu, Dongyan, Roger Ruan, Min Addy, Sarah Mack, Paul Chen, and Yong Zhou. "Life cycle assessment and nutrient analysis of various processing pathways in algal biofuel production." *Bioresource Technology* 230 (2017): 33–42.

Nagai, Takashi, Kiyoshi Taya, and Ikuko Yoda. "Comparative toxicity of 20 herbicides to 5 periphytic algae and the relationship with mode of action." *Environmental Toxicology and Chemistry* 35, no. 2 (2016): 368–375.

Nasri, Hichem, Soumaya El Herry, and Noureddine Bouaïcha. "First reported case of turtle deaths during a toxic Microcystis spp. bloom in Lake Oubeira, Algeria." *Ecotoxicology and Environmental Safety* 71, no. 2 (2008): 535–544.

Nicolas, J.-L., S. Corre, and J.-C. Cochard. "Bacterial population association with phytoplankton cultured in a bivalve hatchery." *Microbial Ecology* 48 (2004): 400–413.

Park, Jungsu, Jared Church, Younggyu Son, Keug-Tae Kim, and Woo Hyoung Lee. "Recent advances in ultrasonic treatment: Challenges and field applications for controlling harmful algal blooms (HABs)." *Ultrasonics Sonochemistry* 38 (2017): 326–334.

Pohl, J., I. Saltsman, A. Mahammed, Z. Gross, and Beate Röder. "Inhibition of green algae growth by corrole-based photosensitizers." *Journal of Applied Microbiology* 118, no. 2 (2015): 305–312.

Pokrzywinski, Kaytee L., Allen R. Place, Mark E. Warner, and Kathryn J. Coyne. "Investigation of the algicidal exudate produced by Shewanella sp. IRI-160 and its effect on dinoflagellates." *Harmful Algae* 19 (2012): 23–29.

Pokrzywinski, Kaytee L., Charles L. Tilney, Shannon Modla, Jeffery L. Caplan, Jean Ross, Mark E. Warner, and Kathryn J. Coyne. "Effects of the bacterial algicide IRI-160AA on cellular morphology of harmful dinoflagellates." *Harmful Algae* 62 (2017a): 127–135.

Pokrzywinski, Kaytee L., Charles L. Tilney, Mark E. Warner, and Kathryn J. Coyne. "Cell cycle arrest and biochemical changes accompanying cell death in harmful dinoflagellates following exposure to bacterial algicide IRI-160AA." *Scientific Reports* 7, no. 1 (2017b): 45102.

Raven, John A., Andrew M. Johnston, and David H. Turpin. "Influence of changes in CO2 concentration and temperature on marine phytoplankton 13C/12C ratios: An analysis of possible mechanisms." *Global and Planetary Change* 8, no. 1–2 (1993): 1–12.

Reasoner, Donald J., and E. E. Geldreich. "A new medium for the enumeration and subculture of bacteria from potable water." *Applied and Environmental Microbiology* 49, no. 1 (1985): 1–7.

Ren, Sanguo, Yuanpei Jin, Jianan Ma, Ningning Zheng, Jie Zhang, Xingyu Peng, and Bo Xie. "Isolation and characterization of algicidal bacteria from freshwater aquatic environments in China." *Frontiers in Microbiology* 14 (2023): 1156291.

Rizwan, Muhammad, Ghulam Mujtaba, Sheraz Ahmed Memon, Kisay Lee, and Naim Rashid. "Exploring the potential of microalgae for new biotechnology applications and beyond: A review." *Renewable and Sustainable Energy Reviews* 92 (2018): 394–404.

Salvesen, Ingrid, Jorunn Skjermo, and Olav Vadstein. "Growth of turbot (Scophthalmus maximus L.) during first feeding in relation to the proportion of r/K-strategists in the bacterial community of the rearing water." *Aquaculture* 175, no. 3–4 (1999): 337–350.

Seymour, Justin R., Shady A. Amin, Jean-Baptiste Raina, and Roman Stocker. "Zooming in on the phycosphere: The ecological interface for phytoplankton–bacteria relationships." *Nature Microbiology* 2, no. 7 (2017): 1–12.

Shilo, Miriam. "Lysis of blue-green algae by myxobacter." *Journal of Bacteriology* 104, no. 1 (1970): 453–461.

Sigee, D. C., R. Glenn, M. J. Andrews, E. G. Bellinger, R. D. Butler, H. A. S. Epton, and R. D. Hendry. "Biological control of cyanobacteria: Principles and possibilities." In *The Ecological Bases for Lake and Reservoir Management: Proceedings of the Ecological Bases for Management of Lakes and Reservoirs Symposium*, held 19–22 March 1996, Leicester, pp. 161–172. Springer Netherlands, 1999.

Su, Yujie, Kaihui Song, Peidong Zhang, Yuqing Su, Jing Cheng, and Xiao Chen. "Progress of microalgae biofuel's commercialization." *Renewable and Sustainable Energy Reviews* 74 (2017): 402–411.

Sun, Daquan, Yu Lan, Elvis Genbo Xu, Jun Meng, and Wenfu Chen. "Biochar as a novel niche for culturing microbial communities in composting." *Waste Management* 54 (2016): 93–100.

Tamura, Koichiro, and Masatoshi Nei. "Estimation of the number of nucleotide substitutions in the control region of mitochondrial DNA in humans and chimpanzees." *Molecular Biology and Evolution* 10, no. 3 (1993): 512–526.

Thompson, Julie D., Desmond G. Higgins, and Toby J. Gibson. "CLUSTAL w: Improving the sensitivity of progressive multiple sequence alignment through sequence weighting, position-specific gap penalties and weight matrix choice." *Nucleic Acids Research* 22, no. 22 (1994): 4673–4680.

Wang, Junyue, Xueyao Yin, Mingyang Xu, Yifan Chen, Nanjing Ji, Haifeng Gu, Yuefeng Cai, and Xin Shen. "Isolation and characterization of a high-efficiency algicidal bacterium Pseudoalteromonas sp. LD-B6 against the harmful dinoflagellate Noctiluca scintillans." *Frontiers in Microbiology* 13 (2022): 1091561.

Wang, Lizhu, and John C. Priscu. "Influence of phytoplankton on the response of bacterioplankton growth to nutrient enrichment." *Freshwater Biology* 31, no. 2 (1994): 183–190.

Xue, Gang, Xiaonuan Wang, Chenlan Xu, Binxue Song, and Hong Chen. "Removal of harmful algae by Shigella sp. H3 and Alcaligenes sp. H5: Algicidal pathways and characteristics." *Environmental Technology* 43, no. 27 (2022): 4341–4353.

Zhang, Chengcheng, Isaac Yaw Massey, Yan Liu, Feiyu Huang, Ruihuan Gao, Ming Ding, Lin Xiang et al. "Identification and characterization of a novel indigenous algicidal bacterium Chryseobacterium species against Microcystis aeruginosa." *Journal of Toxicology and Environmental Health, Part A* 82, no. 15 (2019): 845–853.

Zhang, Hengfeng, Yuanyuan Wang, Juan Huang, Qianlong Fan, Jingjing Wei, Fang Wang, Zijing Jia, Wensheng Xiang, and Wenyan Liang. "Inhibition of Microcystis aeruginosa using Brevundimonas sp. AA06 immobilized in polyvinyl alcohol-sodium alginate beads." *Desalination Water Treat* 111 (2018): 192–200.

Zhang, Yulei, Dong Chen, Jian Cai, Ning Zhang, Feng Li, Changling Li, and Xianghu Huang. "Complete genome sequence analysis of Brevibacillus laterosporus Bl-zj reflects its potential algicidal response." *Current Microbiology* 78 (2021): 1409–1417.

Zhao, Fangchao, Huaqiang Chu, Zhenjiang Yu, Shuhong Jiang, Xinhua Zhao, Xuefei Zhou, and Yalei Zhang. "The filtration and fouling performance of membranes with different pore sizes in algae harvesting." *Science of the Total Environment* 587 (2017): 87–93.

12 Microbial Indicators of Aquatic Pollution

Kundan Kumar, Saurav Kumar, Satya Prakash Shukla, and Rajive Kumar Bhramchari

12.1 INTRODUCTION

Two-thirds of the earth's surface is covered by water, which supports various life forms. Anthropogenic activities, such as urban and agricultural runoff and industrial discharge, affect water quality and associated ecosystems. Moreover, unrestrained human interference adversely affects wellness, damages local biota, and disturbs the ecological balance, causing pollution problems in air, water, and soil ecosystems. Specifically, aquatic pollution is one of the most serious environmental issues, having far-reaching consequences for both humans and other living organisms. Aquatic pollution from industry, agricultural activities, and other anthropogenic activities degrades the aquatic system and directly impacts our health, economic activities, and the natural environment. Thus, it is essential to lower pollution to a minimum attainable level. Monitoring methods serve a crucial role in reducing pollution levels. Multiple monitoring methods to check on environmental quality are known, but biological methods are the most precise, reliable, cost-effective, and significantly relevant to a given environment. Biological methods to indicate environmental health include several species of plants, insects, fish, and microbes. Among these, microbes are most desirable indicators because they facilitate early-stage detection, are cost-effective, and are easily transportable.

Microbial indicators of pollution are organisms or specific microbial genes used to detect the presence and/or activity of contamination in water. Generally, from the perspective of microbial pollution, the main concern associated with water is enteric disease, caused by microbes such as viruses, bacteria, and parasites leading to intestinal diseases. However, detection of pathogens that may contaminate water is challenging. Hence, water safety is determined by detecting indicators of fecal contamination. Indicator bacteria are types of bacteria that are used as valuable tools to detect and estimate the level of fecal contamination of water (Holcomb and Stewart, 2020). They are frequently detected in animal or human feces. If a high number are detected during an investigation, this indicates water pollution. Importantly, these microbial markers are not often human pathogens themselves. The key characteristic features of indicator microorganisms are as follows: (1) they should be present in the intestinal tract of warm-blooded animals, (2) they should represent the pathogenic enteric flora of warm-blooded animals, (3) they should be resilient and have a long life span, (4) they should not grow in natural water, (5) they should be able to reproduce and survive in all types of water, (6) they should have the ability to survive for a reasonable period under the conditions in which they are being monitored, and (7) they should be detectable using standard laboratory method. Total coliforms and fecal coliforms fulfill these essential requirements for indicator microorganisms. Hence, total coliforms or fecal coliforms are commonly employed as indicator microorganisms. *Escherichia coli*, *Enterococcus*, *Pseudomonas*, and *Vibrio* are the other bacteria used to detect feces or other contaminants from sewage, agricultural runoff, and other sources in natural waters. Even at low pollution rates, these microorganisms are predicted to adapt and provide important indicators of ecological changes.

DOI: 10.1201/9781003408543-12

Microbial indicators serve as a complementary means of assessing water quality, and their application, necessitating organism identification based on diversity indices. Microbial indicators are broadly applied in the field to monitor fecal pollution, sewage pollution, the efficiency of sewage water treatment, industrial or agricultural runoff, and the overall health of the aquatic ecosystem. Coliform bacteria, *Escherichia coli*, and *Streptococcus* sp. are common indicators of fecal pollution. Additionally, coliform bacteria are a common indicator of food quality and water sanitation. Sewage pollution is monitored using heterotrophic bacteria, coliforms, *Streptococcus* sp., and *Pseudomonas aeruginosa.* In addition, *Chromatium* sp. serve as bioindicators for crude oil contamination in the environment, whereas *Bacillus subtilis* serves as an indicator of toxicity and pentachlorophenol occurrence (Ayude et al., 2009; Jiang et al., 2016). *Thiobacillus* sp. indicates the presence of mercury (Hg) in the marine environment. Indicator microorganisms provide an early warning of pollution and information on the origins and extent of the contamination. Moreover, they are used to assess water treatment and its efficiency. With the development of molecular techniques, microbial pollution indicators are gaining importance in environmental monitoring and evaluation, and their use is projected to increase multi-fold in the future (Figure 12.1).

12.2 BIOINDICATOR MICROORGANISMS IN THE AQUATIC ENVIRONMENT

Bioindicator microorganisms are microorganisms employed to monitor the health of an aquatic ecosystem. These organisms indicate the presence of organic matter, toxins, heavy metals, and other pollutants. Bioindicator microorganisms include bacteria, algae, protozoa, viruses, and certain aquatic invertebrates. Bacterial species like *E coli*, *Vibrio cholerae*, *Salmonella typhi*, *Klebsiella*, and *Shigella* sp. are commonly used as indicator microbes. Algae are another form of bioindicator organism that can detect the nutrient load in an aquatic ecosystem. Likewise, fungi, viruses, and protozoa serve as crucial indicator organisms (Chahal et al., 2016; Tandukar et al., 2018).

12.2.1 Bacteria

Fecal indicator bacteria like *E. coli* and enterococci have been used for decades to detect fecal pollution in the aquatic environment and the associated health risks. Identification of diseases such as cholera, caused by *V. cholerae*, and typhoid, caused by *Salmonella* sp., extends the definition of microbiological pollution and linked health risks. Thus, assessing the presence of bacteria linked to health hazards in water defines the pollution. Fecal and non-fecal forms are the two major groups of bacteria used to measure environmental health and its risks. Fecal bacteria are a group of microbes that are commonly found in the intestines and feces of warm-blooded animals, such as humans. These bacteria serve as indicators of fecal contamination in a variety of environmental samples, especially water sources. The presence of fecal bacteria indicates aquatic pollution. Coliform bacteria, enterococci, and clostridia are some of the most prevalent fecal bioindicators. The basic fecal indicator should meet the following requirements: fecal bacteria should be detected only in feces-polluted environments; should be in the same host as pathogens; should correlate with pathogen abundance within and outside the host; and should reproduce inside the host. Non-fecal coliforms are bacteria with coliform characteristics but not present in fecal matter and are used to detect heavy metals, pesticides, toxins, and volatile organic compounds. Some of the prominent non-fecal bioindicators are *Enterobacter*, *Klebsiella*, and *Citrobacter.* Hence, assessing total coliforms provides an improved understanding of water quality throughout the drinking water distribution system, where coliform bacteria act as operational indicators. Further, they serve as a measure of sewage treatment efficiency. Together, the most prominent bacteria that transmit diseases in water include *E coli*, *V. cholerae*, *S. typhi*, *Campylobacter jejuni*, *Shigella* sp., and *Yersinia enterocolitica. E coli* is a dependable indicator of enteric pathogens, since it gives the most reliable indication of fecal contamination in drinking water (WHO, 2004) (Figure 12.2).

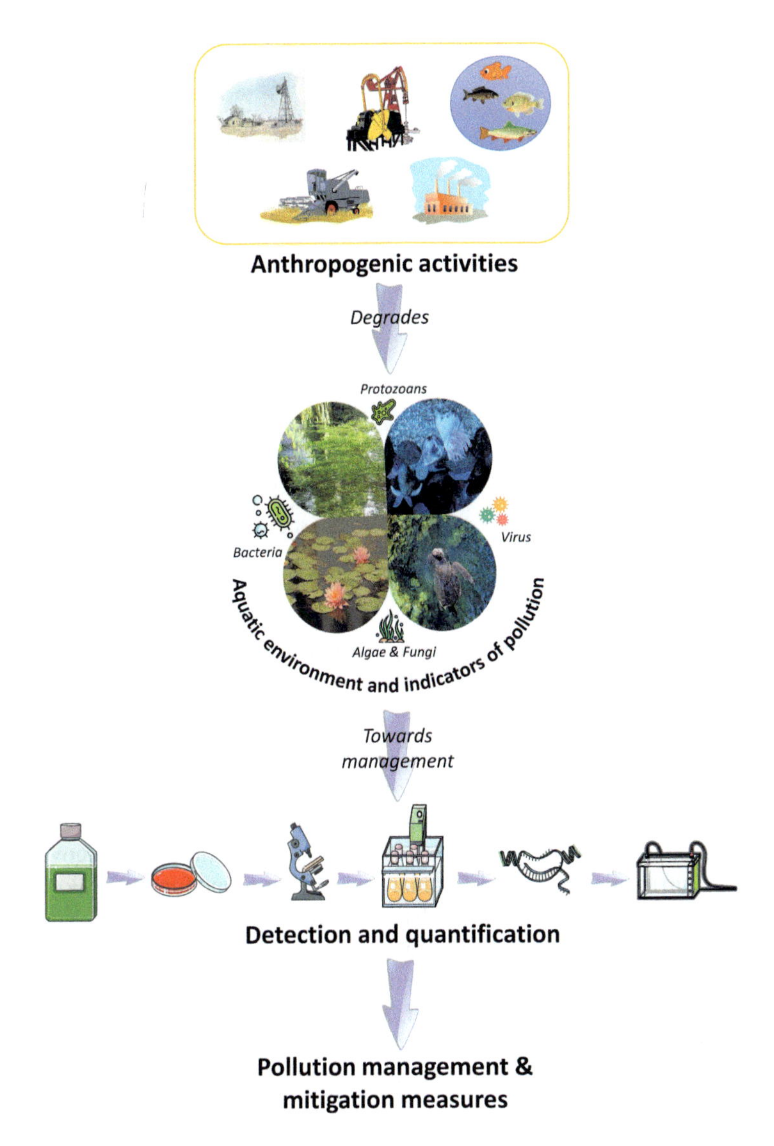

FIGURE 12.1 Schematic depiction of the role of microbial indicators of pollution in aquatic management. Aquatic pollution from industry, agricultural activities, and other anthropogenic activities degrades the aquatic system and directly impacts our health, economic activities, and the natural environment. Microbial indicators such as bacteria, viruses, algae, fungi, and protozoa provide a method for evaluating the quality of the environment, as they reflect the presence and abundance of pollutants and toxic compounds. These microbial indicators are collected from polluted waters and analysed with conventional and molecular methods to detect their presence and possible threat. This information is helpful in selecting the strategy and refining approaches for the effective removal/management of contaminants.

12.2.2 Fungi

Fungal communities are abundant in aquatic habitats and have a significant impact on their functioning, including their contribution to the mineralization and cycling of organic matter (Panzer et al., 2015). In addition, fungi are involved in the bio-transformation of xenobiotics and heavy metals, indicating a close relationship between the fungal community and anthropogenic activities in an aquatic ecosystem. Hence, the fungal community could serve as an indicator organism. Moulds

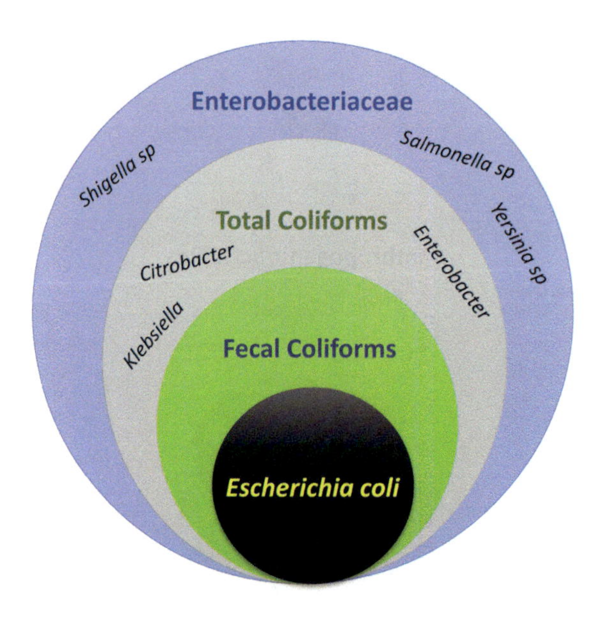

FIGURE 12.2 Schematic diagram showing the correlation among coliform bacteria. The presence of total coliforms suggests environmental contamination, whereas *E. coli* and fecal coliforms indicates fecal contamination.

such as *Trichoderma* sp., *Aspergillus versicolor*, *Exophiala* sp., *Stachybotrys* sp., *A. fumigatus*, *Phialophora* sp., *Fusarium* sp., *Ulocladium* sp., *Penicillium* sp., *A. niger*, *Candida albicans*, and specific yeasts are potential biological indicators for aquatic pollution. Moreover, Ascomycetes, Basidiomycetes, Chytridiomycetes, and Glomeromycetes account for the majority of freshwater fungi. The presence of these organisms in drinking water suggests the presence of organic pollutants, heavy metals, and other microorganisms. Aquatic pollution could be efficiently monitored using a range of fungal indicator species based on their resistance to specific aquatic conditions. A survey indicated that typical distribution patterns of hyphomycete species is related to the altitude and temperature. For example, the moss *Hylocomium splendens* is used as a natural fungal indicator for toxic elements in Tundra ecosystems (Hasselbach et al., 2005).

12.2.3 Algae

Algae are yet another type of microorganisms that can be effectively utilized as indicators of water pollution. They are sensitive to environmental changes and are found in fresh, brackish, and saltwater environments. The influx of pollutants like heavy metals and fertilizers significantly enhances the algal growth, which can measure the pollution level in the specific water body. Algal species commonly used to identify aquatic pollution include *Euglena* sp., *Chlorella* sp., *Chlamydomonas* sp., and *Scenedesmus* sp. However, polluted aquatic environments support only a few algal species. For instance, the diversity of phytoplankton species steadily declines as reservoirs and lakes become more eutrophicated, leading to the eventual dominance of cyanobacteria. The sensitivity of algal species could detect specific chemicals and toxins, as well as pH fluctuations, caused by other pollutants. Blue-green algae typically indicate a change in pH, while *Cladophora* sp., *Lemanea* sp., *Enteromorpha* sp., and *Nitella* sp. indicate the presence of toxic substances in an aquatic ecosystem. Likewise, *Euglena clastica*, *Phacus tortus*, and *Trachelon anas* indicate a degraded marine ecosystem (Zaghloul et al., 2020). *E. gracilis* is the most commonly used indicator species due to its ease of handling, tolerance of acidity, sensitivity to pollutants, and rapid response to stress factors induced by toxins and organic pollutants. In addition, cyanobacteria have been recognized as a prominent indicator of eutrophication in freshwater. Furthermore, the algal pollution index, where algal species present in an aquatic environment are summed up with an indicative score, is frequently used to measure the degree of pollution. An algal pollution index score of >20 indicates high organic pollution, 15–19 indicates organic pollution, and scores of <15 indicate low organic pollution and lack of nutrient enrichment (Palmer, 1977). Algal monitoring in aquatic environments provides additional information about the overall health of the aquatic system and facilitates in decision-making regarding pollution control.

12.2.4 Viruses

The presence of the viruses in an aquatic ecosystem indicates fecal pollution that poses a potential risk to the exposed population, as sick individuals excrete these viruses (Abad et al., 1997). Thus, viruses are considered to be significant indicators of water pollution. Additionally, fecal indicator bacteria are considered to be less resistant than enteric viruses. Hence, water quality monitoring techniques based exclusively on bacteria are inadequate. Enteric viruses are ubiquitous and are associated with waterborne disease outbreaks. People in direct or indirect contact with virus-infected waters are at high risk of getting viral infections. The Picornaviridae, Caliciviridae, Reoviridae, and Adenoviridae families are common viruses responsible for gastroenteritis. The viruses from the Caliciviridae family cause a substantial majority of gastroenteritis infections worldwide (Katayama and Vinjé, 2017). Viruses like rotaviruses, enteroviruses, sapoviruses, astroviruses, aichi virus, and hepatitis E virus are highly associated with wastewater contamination. Also, these viruses demonstrate seasonal fluctuations. For instance, enteroviruses show peaks during summer, while noroviruses and sapoviruses show peaks during winter. These seasonal variations could act as an indicator of aquatic pollution during a specific period (Rachmadi et al., 2016).

Recently, bacteriophages are gaining attention worldwide as an alternative indicator of fecal and viral contamination in water. These bacteriophages are prevalent and generally more resilient in the environment and offer greater and more accurate information about viral pathogens. They are typically divided into four main categories: somatic coliphages, F-specific coliphages, enterophages, and bacteriophages. In most aquatic environments, somatic coliphages are the most abundant heterogeneous group of indicator bacteriophages that can infect *E. coli* and other coliform bacteria (Jofre et al., 2016). Myoviridae, Siphoviridae, Podoviridae, and Microviridae are the most common families of somatic coliphages. F-specific bacteriophages potentially infect *E. coli* and other coliforms via sexual pili, encoded in the F-plasmid, which moves to enteric bacteria through conjugation (Davis et al., 1961). Bacteriophages infect Bacteroides sp., but their concentration in fecal-contaminated water samples appears to be lower than in coliphages. These bacteriophages infect bacteria via receptors in the cell wall. Enterophage is a newly discovered phage that infects *Enterococcus faecalis* and could be utilized as an additional indicator of fecal contamination (Bonilla et al., 2010). Viruses collectively constitute an influential group that can be employed to monitor aquatic pollution and associated health risks. Their ability to indicate the presence of pollutants and pathogens makes them ideal indicators for aquatic pollution.

12.2.5 Protozoans

Protozoans are unicellular organisms capable of surviving in various water conditions, making them promising bioindicator organisms for assessing aquatic pollution. Specifically, protozoans have demonstrated their effectiveness in determining the existence of pollution during the biological treatment of wastewater (Medeiros et al., 2019; Pereao et al., 2021; Schmitz et al., 2018). The protozoans *Cryptosporidium*, *Entamoeba histolytica*, *Naegleria fowleri*, and *Giardia lamblia* are widely used bioindicators. These protozoa can transmit diseases when present in substantial numbers.

12.3 BENEFITS OF MICROBIAL INDICATORS

Microbial indicators are used for assessing the safety of water. These microbiological indicators are simple to handle and relatively inexpensive to evaluate, making them a cost-effective option for monitoring water quality. They are capable of indicating the presence of specific contaminants that cannot be detected by chemical analysis. Further, microbial indicators are sensitive to environmental changes, demonstrating their ability as indicators of ever-changing aquatic ecosystems. Moreover, these indicators provide key information on the threats posed by contaminants to human health by giving early warning of potential hazards. More importantly, microbial indicators indicate the potential remediation strategy that must be adopted and help in assessing the efficacy of remediation activities (Xu et al., 2014).

12.4 INDICATOR MICROORGANISMS IN VARIOUS AQUATIC RESOURCES

Indicator microbes are microorganisms that reflect water quality in various water bodies. Indicator microbes, such as bacteria, protozoa, and viruses, are used to monitor water quality. They are used to determine potential sources of contamination and risk of waterborne diseases. Different water bodies have different characteristics, which can result in variations in the types and concentrations of indicator microorganisms present. Multiple microorganisms are used, based on the water body, to measure the quality or pollution, signifying the health of the respective aquatic systems and providing insight into the overall water quality. Microorganisms that are used as indicators in different water bodies are listed in Table 12.1.

Characteristically, most of the total coliforms are gram-negative, non-spore-forming, oxidase/indole-negative, rod-shaped facultative anaerobic bacteria. However, some species, like *K. pneumoniae*, may represent a non-fecal origin and may not be specific indicators of fecal pollution. Although all the microorganisms mentioned above can act as pollution indicators in specific water bodies, total coliforms, fecal coliforms, *E. coli*, and enterococci are recommended indicators in most water bodies (Ortega et al., 2009; Rodrigues and Cunha, 2017).

12.5 HEALTH RISKS ASSOCIATED WITH INDICATOR MICROORGANISMS

Water is the essence of life and supports diverse life forms. Industrial waste, agricultural runoff, aquaculture wastes, and other anthropogenic activities introduce a significant amount of nutrients,

TABLE 12.1
Indicator Microorganisms in Different Water Bodies

Water Type	Indicator Microorganism(s)
Drinking water	Total coliforms, *Clostridium perfringens*, somatic coliphages
Natural water body (Freshwater)	Fecal coliforms, Bacteroides HF 183, *Bifidobacterium*, *Pseudomonas*, *Salmonella*
Natural water body (Saltwater)	All coliforms and enterococci
Aquaculture ponds	All coliforms, streptococci, enterococci, coliphages, SRC (Sulfur-Reducing Clostridia)
Water for agriculture	Total coliforms
Wastewater effluent	Fecal coliforms
Disinfected water	MS2 coliphages, *Salmonella*

Source: Partially adopted from Tchobanoglous, G. et al., *Wastewater Engineering: Treatment and Reuse*, Metcalf & Eddy Inc.; McGraw-Hill, Inc., New York, 2003; Saxena, G. et al., *Reviews of Environmental Contamination and Toxicology*, 240, 31–69, 2017.

TABLE 12.2
Health Risks Associated with Indicator Microorganisms

Group	Pathogen	Diseases
Bacteria	*Escherichia coli*	Gastroenteritis
	Vibrio cholerae	Cholera
	Salmonella typhi	Typhoid
	Campylobacter jejuni	Gastroenteritis
	Shigella sp.	Bacillary dysentery
	Yersinia enterocolitica	Gastroenteritis
	Legionella pneumophila	Respiratory illness
Fungi	*Trichoderma* sp.	Sinusitis, pneumonia
	Aspergillus versicolor	Diarrhoea, stomach upset
	Penicillium sp.	Keratitis, urinary tract infection
	Phialophora sp.	Mycetoma, mycotic keratitis
Algae	*Chlorella* sp.	Diarrhoea, respiratory illness
Virus	*Adenovirus*	Eye infection, respiratory illness
	Rotavirus	Diarrhoea
	Astrovirus	Gastroenteritis
	Hepatitis	Hepatitis
	Caliciviruses	Gastroenteritis
	Coxsackievirus	Meningitis
	Echovirus	Fever, meningitis
Protozoans	*Giardia lamblia*	Nausea, anorexia
	Dientamoeba	Diarrhoea, abdominal pain
	Plasmodium sp.	Malaria
	Cystosporidium sp.	Cystosporidosis

Source: Partially adopted from Arnone, D. and Perdek Walling, J., *Journal of Water and Health*, 5, 1, 149–162, 2007; Gerba, C.P. and Smith, J.E., *Journal of Environmental Quality*, 34, 1, 42–48, 2005.

organic matter, pollutants, toxins, and heavy metals into the water. The presence of these compounds provides a favourable environment for the growth of microbes. The growth of pathogens poses greater health risks. People in direct or indirect contact with pathogens are at high risk of getting infections. Table 12.2 shows the various diseases caused by indicator microorganisms.

12.6 WATER QUALITY STANDARDS AND GUIDELINES

Water quality standards and guidelines on bioindicators of aquatic pollution are essential for protecting the health of the aquatic life and environment and avoiding associated health risks to humans. Water quality should be consistent with its intended use, and the suitability of water is evaluated based on permissible levels of different water quality parameters specified by guidelines and standards. These recommendations and standards are established on the basis of knowledge that contaminants can affect wildlife, humans, and aquatic ecosystems. Bioindicators of aquatic pollution serve as an early warning system for identifying pollution in aquatic ecosystems. By observing the presence of specific species and the health of their populations, it is possible to detect changes in water quality. Standards and guidelines enable the assessment and interpretation of aquatic pollution bioindicators. The results of bioindicators could be difficult to understand, and their accuracy

TABLE 12.3
Water Quality Standards as per Water Use

Indicator Microorganism	Water Use			
	Drinking Water	**Agriculture**	**Recreation**	**Livestock**
E. coli	<1/mL	<1000/100 mL	<100/mL	200/100 mL
Enterococci	<3/mL	250/100 mL	<3/mL	50/100 mL
P. aeruginosa	<4/mL	NA	<10/mL	NA
Fecal coliforms	<2/mL	<1000/100 mL	<200/100 mL	200/100 mL
Shigella sp.	Absent	NA	NA	NA
Vibrio sp.	Absent	NA	NA	NA
Salmonella sp.	Absent	NA	NA	NA
Fungi	Not listed as hazard species			
Algae	Not listed as hazard species			
Viruses	Absent	NA	NA	NA
Protozoans	Absent	NA	NA	NA

Collected from multiple sources. NA, Precise information not available.

could be disputed, without standards and guidelines. Fungi and algae are exempted from water quality regulations. The World Health Organization (WHO) and most world authorities classify bacteria, viruses, protozoa, and helminths as pathogens, as they can transmit disease through the water. Table 12.3 shows comprehensive information on water quality standards for microbiological indicators (Holcomb and Stewart, 2020; Saxena et al., 2017).

12.7 SAMPLING AND ANALYSIS OF MICROBIAL INDICATORS OF AQUATIC POLLUTION

The water quality and microbiological parameters must be evaluated per the guidelines suggested by Pollution Control Board (PCB) or American Public Health Association (APHA). The physical water quality parameters, such as temperature, turbidity, and colour, must be evaluated on-site. Subsequently, it is essential to collect samples to analyse total alkalinity, salinity, pH, dissolved oxygen, and conductivity. These physico-chemical parameters reveal the aquatic environment in which microbiological indicators thrive. The total coliform bacteria and *E. coli* contamination load are determined and enumerated as per the standard procedures. Fecal coliforms should be tested at a higher incubation temperature of 44.5 C, using m-Endo agar, whereas coliforms are examined at a lower incubation temperature of 35 C using endo agar. Gram staining should be performed for all the incubated cultures. Then, samples should be analysed for lactate fermentation and gas production. After these basic microbiological tests, samples must be analysed using specific assays.

12.8 DETECTION AND QUANTIFICATION OF INDICATOR MICROORGANISMS

Numerous pathogenic bacteria can be detected in the environment; nevertheless, it is essential to evaluate their viability to determine the likelihood that they pose a serious concern to public health. The conventional techniques to assess bacterial viability are determined by the potential of cells to form visible colonies on solid media. However, culture-dependent methods for detecting and quantifying waterborne pathogens are costly and time-consuming. Furthermore, sub lethally damaged organisms may cause viable organisms to be grossly underestimated under specific conditions. Despite drawbacks, the significance of conventional tracking methods is comparable to that of advanced tracking methods like molecular and immunological methods.

12.8.1 Conventional Methods

12.8.1.1 Standard Plate Count (SPC) and Heterotrophic Plate Count (HPC)

SPC entails the growth of collected microorganisms in a nutrient medium for a predetermined period. This is then followed by the estimation of the number of bacterial colonies on the plate. Similarly, HPC is performed to evaluate the presence of heterotrophic bacteria, which are capable of breaking down organic material, to assess the organic pollutants in water. However, some pathogens exist in a viable but nonculturable (VBNC) state, so culture-dependent methods can generate false-negative results.

12.8.1.2 Membrane Filtration

Membrane filtration involves passing a water sample through a filter that collects microorganisms. After filtration, the filter is transferred to a culture medium, incubated, and assessed.

12.8.1.3 Most Probable Number (MPN)

The MPN process requires inoculating collected samples in multiple nutrient broths over a predetermined period. After the incubation time, the number of positive tubes is evaluated to estimate the sample's bacterial content.

12.9 MOLECULAR METHODS

12.9.1 Polymerase Chain Reaction (PCR)

The PCR method is a powerful tool for detecting microbial pollution indicators. In this method, a specific gene or set of genes is amplified in a given sample to detect the target microbial species present in the sample. PCR is a rapid, sensitive, and specific technique that can be used to detect a broad range of microbiological indicators, including bacteria, viruses, and protozoa. Similarly, quantitative PCR (qPCR) is an advanced version of conventional PCR for detecting microbial contamination indicators. It relies on amplifying target DNA sequences using specific primers and fluorescent probes. qPCR enables the rapid and sensitive identification and quantification of microbial indicators and is employed to evaluate the relative abundance of various microbial species in a sample.

12.9.2 Gene Sequencing

DNA and RNA sequencing techniques are employed to sequence the transcriptome of a sample. This precisely assists in identifying the presence and abundance of different microbial species in a given sample. Recently, next-generation sequencing (NGS) has gained acceptance as a method for detecting microbial pollution indicators, since it offers a more accurate and comprehensive view of microbial communities in a specific environment. NGS involves metagenomics, metatranscriptomics, and metaproteomics to detect indicator microbes. Metagenomics identifies the total gene content of a microbial community, whereas metatranscriptomics and metaproteomics assess the expression of various genes and proteins, respectively. Collectively, NGS can provide insight into the interactions between microbial populations and the environment.

12.10 IMMUNOLOGICAL METHODS

Immunological techniques use antibodies to detect the presence of specific microbial species. ELISA (enzyme-linked immunosorbent assay) is the most widely used immunological method to detect indicator microbes (Nnachi et al., 2022). It is a highly sensitive and accurate method where specific antibodies bind to the target organism, and binding is detected by a colorimetric signal. Similarly, immunofluorescence techniques detect the indicator microorganisms by the binding of fluorescent antibodies to the target organisms, which can subsequently be observed using a fluorescence microscope. Recently, flow cytometry has acquired prominence in detecting indicator microorganisms, since it detects microorganisms by measuring their size, shape, or other properties. These insights provide an edge over other immunological methods (Cheswick et al., 2019; Vucinic et al., 2022).

12.11 GENOTYPICAL METHODS

Microbial source tracking (MST) is a method used to track the sources of bacterial contamination in a water body. Genetic markers like bacteria-specific enzymes and DNA markers are employed to identify and trace the origin of bacteria. Next, microarray-based techniques use a set of probes to detect the presence and abundance of microbial genes in a given sample. Further, fluorescence in situ hybridization (FISH) is another technique whereby microbial cells and their locations in a sample are visualized. This method determines the presence and abundance of specific microbial species in polluted waters. Indicator microorganisms found at low concentrations can be detected with mass spectrometry. A mass spectrometer can identify ionized molecules by measuring their mass-to-charge ratio. By assessing the ratio of the various molecules, it is possible to identify the microbial species in the sample.

12.12 LIMITATIONS AND FUTURE PERSPECTIVES IN DETECTING INDICATOR MICROORGANISMS

It is essential to understand the significance of detection to select the appropriate detection method, as the specificity of detection methods can vary. Additionally, there could be cross-reactivity during detection, which could lead to inconsistent observations. Further, these advanced techniques are believed to be accurate. However, they are expensive, as they need specific equipment and reagents for detecting indicator microorganisms. Still, accurate and cost-effective techniques based on microarrays and biosensors are being developed. Their affordability, multiplexing capability, and rapid detection make them valuable instruments for assessing the presence of bioindicators in the near future.

12.13 ROLE OF MICROBIAL INDICATORS OF POLLUTION IN BIOREMEDIATION

Microbial indicators of pollution are important tools in bioremediation, as they provide a means to assess the quality of the environment and the success of a bioremediation effort. Specifically, microbial indicators provide information on the presence or abundance of the contaminants or toxic compounds and their activities. This information can be utilized for selecting the optimal bioremediation strategy, refining approaches for effectively removing contaminants, and assessing the efficacy of bioremediation processes.

12.14 LIMITATIONS OF MICROBIAL INDICATORS OF AQUATIC POLLUTION

Microbial indicators are a useful tool for measuring the quality of aquatic resources, but they have certain limitations when it comes to identifying actual contaminants in aquatic systems. Microbiological indicators do not always generate reliable signals of pollution, as environmental variables like temperature, light, and nutrient availability can influence their growth, which may mask or alter their ability to indicate pollution. Further, when contaminants are present in low concentrations, microbial indicators may not be able to detect pollution or may not accurately reflect the presence of pollutants. Also, they do not establish the precise source of contamination and are unreliable for detecting long-term pollution. Notably, the absence of microbial indicators does not necessarily indicate a pollution-free aquatic ecosystem.

12.15 CONCLUSION

Indicators of microbiological pollution will be used extensively in environmental monitoring for years to come. They are efficient, cost-effective, and can give rapid and reliable findings, enabling management to make prompt and effective decisions. The development of advanced technologies like metagenomics and metatranscriptomics gives more accurate information on microbial communities and their activities which provides new insights into the effects of pollution on the environment. Additionally, new sensors and detection systems are being developed to improve the accuracy and sensitivity of microbial indicators for monitoring water quality. These advancements will enable more effective and efficient monitoring and management of pollution, ultimately leading to better environmental protection.

REFERENCES

Abad, F. X., R. M. Pintó, C. Villena, R. Gajardo, and A. Bosch. "Astrovirus survival in drinking water." *Applied and Environmental Microbiology* 63, no. 8 (1997): 3119–3122.

Arnone, Russell D., and Joyce Perdek Walling. "Waterborne pathogens in urban watersheds." *Journal of Water and Health* 5, no. 1 (2007): 149–162.

Ayude, María Alejandra, Elena Okada, J. F. González, Patricia Monica Haure, and Silvia Elena Murialdo. "Bacillus subtilis as a bioindicator for estimating pentachlorophenol toxicity and concentration." *Journal of Industrial Microbiology and Biotechnology* 36, no. 5 (2009): 765–768.

Bonilla, N., T. Santiago, P. Marcos, M. Urdaneta, J. Santo Domingo, and G. A. Toranzos. "Enterophages, a group of phages infecting Enterococcus faecalis, and their potential as alternate indicators of human faecal contamination." *Water Science and Technology* 61, no. 2 (2010): 293–300.

Chahal, Charndeep, Ben Van Den Akker, Fiona Young, Christopher Franco, J. Blackbeard, and Paul Monis. "Pathogen and particle associations in wastewater: Significance and implications for treatment and disinfection processes." *Advances in Applied Microbiology* 97 (2016): 63–119.

Cheswick, Ryan, Elise Cartmell, Susan Lee, Andrew Upton, Paul Weir, Graeme Moore, Andreas Nocker, Bruce Jefferson, and Peter Jarvis. "Comparing flow cytometry with culture-based methods for microbial monitoring and as a diagnostic tool for assessing drinking water treatment processes." *Environment International* 130 (2019): 104893.

Davis, James E., James H. Strauss, and Robert L. Sinsheimer. "Bacteriophage MS2: Another RNA phage." *Science* 134, no. 348 (1961): 1427.

Gerba, Charles P., and James E. Smith. "Sources of pathogenic microorganisms and their fate during land application of wastes." *Journal of Environmental Quality* 34, no. 1 (2005): 42–48.

Hasselbach, L., J. M. Ver Hoef, Jesse Ford, P. Neitlich, E. Crecelius, S. Berryman, B. Wolk, and T. Bohle. "Spatial patterns of cadmium and lead deposition on and adjacent to National Park Service lands in the vicinity of Red Dog Mine, Alaska." *Science of the Total Environment* 348, no. 1–3 (2005): 211–230.

Holcomb, David A., and Jill R. Stewart. "Microbial indicators of fecal pollution: Recent progress and challenges in assessing water quality." *Current Environmental Health Reports* 7 (2020): 311–324.

Jiang, Jian, Ling Gao, Xiaomei Bie, Zhaoxin Lu, Hongxia Liu, Chong Zhang, Fengxia Lu, and Haizhen Zhao. "Identification of novel surfactin derivatives from NRPS modification of Bacillus subtilis and its antifungal activity against Fusarium moniliforme." *BMC Microbiology* 16 (2016): 1–14.

Jofre, Juan, Francisco Lucena, Anicet R. Blanch, and Maite Muniesa. "Coliphages as model organisms in the characterization and management of water resources." *Water* 8, no. 5 (2016): 199.

Katayama, H., and J. Vinjé. "Norovirus and other Calicivirus." Global Water Pathogens Project (2017).

Medeiros, Raphael Corrêa, Luiz Antonio Daniel, Gabriela Laila de Oliveira, and Maria Teresa Hoffmann. "Performance of a small-scale wastewater treatment plant for removal of pathogenic protozoa (oo) cysts and indicator microorganisms." *Environmental Technology* 40, no. 26 (2019): 3492–3501.

Nnachi, Raphael Chukwuka, Ning Sui, Bowen Ke, Zhenhua Luo, Nikhil Bhalla, Daping He, and Zhugen Yang. "Biosensors for rapid detection of bacterial pathogens in water, food and environment." *Environment International* 166 (2022): 107357.

Ortega, Cristina, Helena M. Solo-Gabriele, Amir Abdelzaher, Mary Wright, Yang Deng, and Lillian M. Stark. "Correlations between microbial indicators, pathogens, and environmental factors in a subtropical estuary." *Marine Pollution Bulletin* 58, no. 9 (2009): 1374–1381.

Palmer, C. Mervin. "Algae and water pollution." The National Technical Information Service, Springfield, VA as PB-287 128, Price codes: A 07 in paper copy, A 01 in microfiche. Report (1977).

Panzer, Katrin, Pelin Yilmaz, Michael Weiß, Lothar Reich, Michael Richter, Jutta Wiese, Rolf Schmaljohann et al. "Identification of habitat-specific biomes of aquatic fungal communities using a comprehensive nearly full-length 18S rRNA dataset enriched with contextual data." *PLoS One* 10, no. 7 (2015): e0134377.

Pereao, Omoniyi, Michael Ovbare Akharame, and Beatrice Opeolu. "Effects of municipal wastewater treatment plant effluent quality on aquatic ecosystem organisms." *Journal of Environmental Science and Health, Part A* 56, no. 14 (2021): 1480–1489.

Rachmadi, Andri T., Jason R. Torrey, and Masaaki Kitajima. "Human polyomavirus: Advantages and limitations as a human-specific viral marker in aquatic environments." *Water Research* 105 (2016): 456–469.

Rodrigues, Carla, and Maria Ângela Cunha. "Assessment of the microbiological quality of recreational waters: Indicators and methods." *Euro-Mediterranean Journal for Environmental Integration* 2 (2017): 1–18.

Saxena, Gaurav, Ram Chandra, and Ram Naresh Bharagava. "Environmental pollution, toxicity profile and treatment approaches for tannery wastewater and its chemical pollutants." *Reviews of Environmental Contamination and Toxicology* 240 (2017): 31–69.

Schmitz, Bradley W., Hitoha Moriyama, Eiji Haramoto, Masaaki Kitajima, Samendra Sherchan, Charles P. Gerba, and Ian L. Pepper. "Reduction of Cryptosporidium, Giardia, and fecal indicators by Bardenpho wastewater treatment." *Environmental Science & Technology* 52, no. 12 (2018): 7015–7023.

Tandukar, Sarmila, Jeevan B. Sherchand, Dinesh Bhandari, Samendra P. Sherchan, Bikash Malla, Rajani Ghaju Shrestha, and Eiji Haramoto. "Presence of human enteric viruses, protozoa, and indicators of pathogens in the Bagmati River, Nepal." *Pathogens* 7, no. 2 (2018): 38.

Tchobanoglous, George, Franklin L. Burton, and H. David Stensel. *Wastewater Engineering: Treatment and Reuse.* Metcalf & Eddy Inc.; McGraw-Hill, Inc., New York, 2003. https://doi.org/10.0070418780.

Vucinic, Luka, David O'Connell, Rui Teixeira, Catherine Coxon, and Laurence Gill. "Flow cytometry and fecal indicator bacteria analyses for fingerprinting microbial pollution in karst aquifer systems." *Water Resources Research* 58, no. 5 (2022): e2021.

World Health Organization. *Guidelines for Drinking-Water Quality.* Vol. 1. World Health Organization, 2004.

Xu, Tingting, Nicole Perry, Archana Chuahan, Gary Sayler, and Steven Ripp. "Microbial indicators for monitoring pollution and bioremediation." In *Microbial Biodegradation and Bioremediation*, pp. 115–136. Elsevier, 2014.

Zaghloul, Alaa, Mohamed Saber, Samir Gadow, and Fikry Awad. "Biological indicators for pollution detection in terrestrial and aquatic ecosystems." *Bulletin of the National Research Centre* 44, no. 1 (2020): 1–11.

13 Impact of Emerging Pollutants on Freshwater Microbes

Saurav Kumar, Pritam Sarkar, Tapas Paul, and Kundan Kumar

13.1 INTRODUCTION

In recent times, aquatic pollution has become a global concern for mankind with increasing population and urbanization. Pollutant loads are increasing due to the expansion of industries and other human activities, which threatens ecosystem sustainability. Emerging pollutants (EPs) are not currently monitored under national and international regulations, and their life cycle and their potential ecotoxicity are not properly known. EPs include diverse groups of daily-use products such as personal care products, pharmaceuticals, hydrocarbons (aliphatic and aromatic), surfactants, and parts of automobiles. As per the NORMAN database, around 20 categories of EPs exist, which contain at least 700 pollutants (Gavrilescu et al., 2015). EPs are transported via air, water, and terrestrial pathways and reach the waterbodies, where they negatively affect both aquatic flora and fauna at levels of nanograms to micrograms per litre. Moreover, they are persistent, chemically stable, and reactive in nature, depending on their surface properties, and therefore resistant to biodegradation (TerLaak et al., 2006).

As EPs spread across the water column and sediment, aquatic microbes become the first organisms to be exposed. A broad array of aquatic microbes, including bacteria, fungi, viruses, and microalgae, play a vital role in the ecological process. Microalgae are the primary producers of the food chain, while bacteria and fungi work as the essential decomposers of the aquatic ecosystem. Due to their sensitivity to pollution, they are also excellent indicators of the environment. EPs disrupt the microecological balance of the environment. EPs can contribute significantly to the development of antimicrobial resistance (AMR), alteration in the enzymatic balance, inhibition of growth rate, and reduction in pigment production in the case of microalgae. They also create favourable conditions for opportunistic pathogens to invade aquatic organisms. This results in the disbalance of the ecosystem and may also cause economic loss in the case of aquaculture. The present chapter provides a holistic overview of different types of EPs and their impact on microbes in freshwater.

13.2 EMERGING POLLUTANTS IN THE FRESHWATER ENVIRONMENT

Atmospheric deposition is one of the important by which EPs reach the aquatic environment. EPs originate from various sources and enter wastewater treatment plants (WWTPs) before ultimately being released into aquatic systems through discharge channels (Figure 13.1). Physico-chemical properties such as water solubility, vapour pressure, and polarity of the EPs determine their fate in the receiving water. Most of the EPs are hydrophobic and lipophilic in nature, which make them bioaccumulative and persistent inside living organisms and the environment. There are some groups of EPs, such as steroids, hormones, and pharmaceuticals, that can be biodegraded by microbes, but the rate of degradation is very slow. Several pharmaceuticals and veterinary drugs enter streams and rivers after they are consumed and excreted in the form of non-metabolized parent compounds or metabolites through WWTPs (Pérez and Barceló, 2007). Surface water can be contaminated through runoff from digested sludge-treated fields and groundwater by leaching. The overuse and

DOI: 10.1201/9781003408543-13

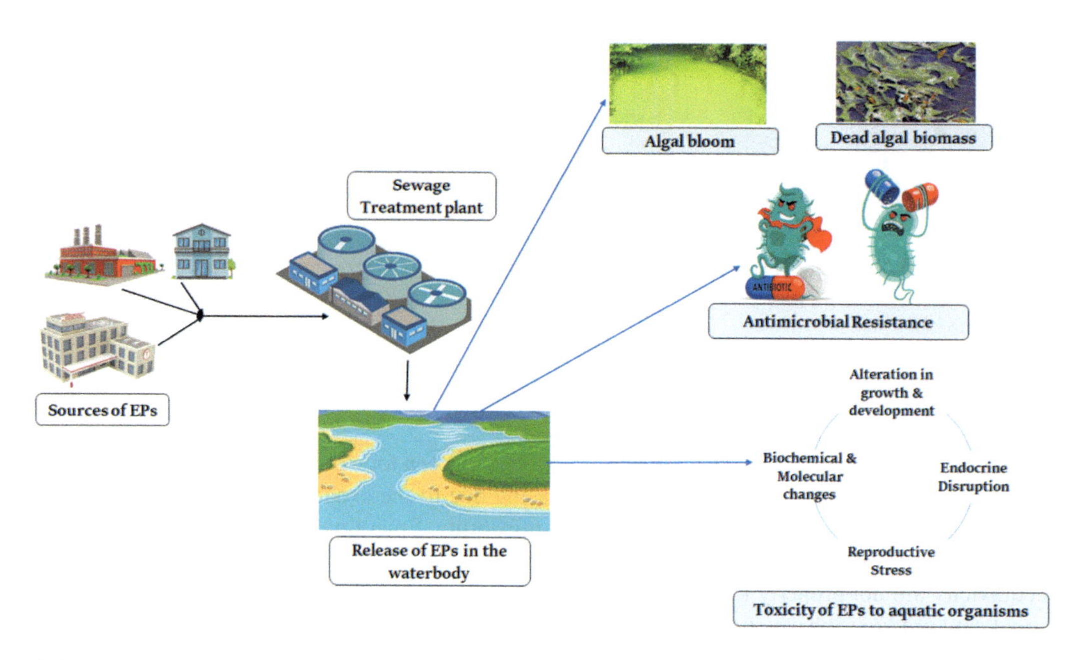

FIGURE 13.1 Fate of EPs in aquatic environments and their toxicity to aquatic organisms.

illegal use of banned drugs not only increase the chance of pollution but have given rise to AMR development among microbes.

Personal care products (PCPs), such as toothpaste, shower gel, face creams, shampoo, ointments, suncream, deodorant, eyeliner, etc., contain various antimicrobial agents, i.e., triclosan, trichloro carbanilide, chlorhexidine, triclocarban, benzalkonium chloride, and chlorhexidine (Halla et al., 2018). The residues of these active ingredients are washed out into the environment after domestic and industrial use. In WWTPs, they may get transformed and converted into their metabolites and may become more toxic to organisms. For instance, triclosan transforms into methyl triclosan after the methylation process in the WWTP and becomes more toxic to the organisms.

Chlorination processes have been widely used to disinfect water for human and veterinary use. After use, it was observed that chlorinated by-products (CBPs) could remain in the treated water for a long time. Disinfection by-products (DBPs) can also remain in the water and adversely affect aquatic microbes. Among DBPs, brominated products such as bromo nitromethane are highly toxic and act as potential carcinogens. There are also haloaldehydes, haloamides, and n-nitroso dimethylamine (NDMA), commonly found DBPs in drinking water and surface water, which do not contain bromine but are considered probable carcinogens (Richardson, 2005). These products show genotoxic and cytotoxic effects on aquatic organisms and microbiota. During the chlorination process, when chlorine reacts with ammonia, it forms chloramines, and these products give rise to the production of iodo acids, which are potentially genotoxic in nature (Gong et al., 2005).

Nanoparticles (NPs) (1–100 nm dimensions), used in sunscreen products, laundry, and paints as anti-fouling agents, adsorb other NPs or chemicals in WWTP effluents and form agglomerates, showing differential fate and toxicity in the receiving water (Zhu et al., 2012;). NPs released into aquatic ecosystems also undergo various transformations due to physical, chemical, and biological processes, based on type of NPs, abiotic factors, and natural organic matter (NOM) present in water (Biswas and Sarkar, 2019). Aggregation and agglomeration are the two major physical processes governing the transformation of NPs. All these factors result in change of persistence, mobility, and surface chemistry of NPs, causing deterioration of water quality.

Natural microbiota in rivers, marine environments, wetlands, and even soil play a critical role in ecosystem restoration. Microorganisms can be influenced by EPs even at trace levels, altering their diversity and functioning. Based on the persistent, bioaccumulative, and toxic (PBT) criteria, the ecological risks of EPs are determined. Numerous pollutants, when screened through the PBT approach, showed very high ecotoxicological risk to aquatic organisms (Paul et al., 2019, 2020; Bera et al., 2020; Droma et al., 2021). Various PCPs and pharmaceutical products can cause significant changes in the normal physiological and biochemical processes of living organisms even at low environmental concentrations (Kumar et al., 2021) (Table 13.1).

13.3 EFFECT ON BACTERIA

EPs have a considerable effect on the microbial population of the aquatic environment. Proia et al. (2013) observed that shifting of biofilm towards polluted areas resulted in reduction of photosynthetic efficiency and phosphatase activity, whereas the contribution of autotrophic biomass and peptidase activity increased. Several studies have suggested that EPs can replace dominant bacteria species in an ecosystem. Du et al. (2022) observed that the natural bacterial community (*Proteobacteria*, *Bacteroidetes*, and *Firmicutes*) of estuarine sediment was substituted by other groups of bacteria (*Haliangium*, *Altererythrobacter*, *Gaiella*, and *Erythrobacter*) with the increase in level of EPs. In a microcosm study, triclosan (TCS) exposure caused replacement of native bacterial genera by phenol-tolerant bacterial genera (*Flavobacterium*, *Anaoysphingobium*, *Comamonas*, and *Alicycliphilus*) (Tan et al., 2021). Interestingly, DeLorenzo et al. (2008) found that the marine bacterium *Vibrio fischeri* is more sensitive to triclosan than the host grass shrimp. *V. fischeri* was also found to be more sensitive to benzotriazole compounds. These compounds are used as anticorrosive agents in metals and can have mutagenic effects on bacteria (Pillard et al., 2001).

The European Commission classifies chemicals based on their median effective concentration (EC_{50}) value, such as less than 1 mg L^{-1} very toxic, 1 to 10 mg L^{-1} toxic, and 10 to 100 mg L^{-1} categorized as harmful (European Commission, 1996). The existing literature indicates that pharmaceuticals, namely antibiotics, may act on a specific group of bacteria or can affect the entire bacterial community of a particular site based on the microbial species, the load of antibiotics, and the environment. Sometimes, a narrow or broad spectrum of antibiotics affects the specific bacterial species, which as a result, affects the relative abundance of bacteria and disrupts the entire interrelationship among the microbial organisms of the site (Figure 13.2). The changes also decrease the enzymatic balance of the system, alter the natural biogeochemical processes, and finally, decrease microbial diversity (Grenni et al., 2018). The activity of the ecologically important sediment enzymes phosphatase, dehydrogenase, fluorescein diacetate enzyme, and urease decreased when the concentration of antibiotics increased in the sediment (Molaei et al., 2017; Han et al., 2020). These enzymes are important in ecosystem functioning because of their catalytic feature on nutrient cycling (Bandick and Dick, 1999). However, EPs such as sulfadiazine, oxytetracycline, vancomycin, sulfadiazine, and sulfadimethoxine can effectively disrupt the biogeochemical processes, reducing the nitrification and denitrification process of the nitrogen cycle (Grenni et al., 2018). Further, sulphonamides can give rise to the release of toxic methane gas in sediments (Conkle and White, 2012). Among other pharmaceutical products, antidepressants, antibacterials, and antipsychotic drugs are more toxic to aquatic microbes, including bacteria, fungi, and algae (La Farre et al., 2008). Exposure of paracetamol to bacteria in the absence of glutathione results in metabolization into hepatotoxic metabolites, primarily N-acetyl-p-benzoquinone imine products, which are more hazardous than the parent molecule (Isidori et al., 2005).

13.3.1 EPs and Antimicrobial Resistance

Due to the unregulated use of antimicrobial agents, AMR has become a global concern, with increasing health issues in animal and human populations. As a consequence of AMR, thousands

TABLE 13.1
Effect of Different Types of EPs on Aquatic Microorganisms

Microorganisms	Species	EPs		Effect	References
Bacteria	*Vibrio fischeri*	Pharmaceuticals	Aspirin	50% growth inhibited at 446 µg L^{-1} concentration after 15 min. exposure	Brun et al. (2006)
	Pseudomonas putida	Nanoparticles	Fullerene C60	Alteration in lipid composition	Fang et al. (2007)
	Bacillus subtilis			Growth inhibited at 0.75 mg L^{-1}	Fang et al. (2007)
	Bacterial consortium	Biocides	Chloroxylenol	Phase of the bacterial membrane altered and at higher concentration, lipid chain interdigitation observed	Poger and Mark (2008)
	Aeromonas hydrophila and *Edwardsiella tarda*		Triclosan	Reversible phenotypic antimicrobial resistance was observed	Karmakar et al. (2019)
	Aeromonas versicolor			Triclosan induced antimicrobial resistance through enzymatic breakdown	Taştan and Dönmez (2015)
Protists	*Spirostomum ambiguum*	Pharmaceuticals	Ibuprofen	In chronic exposure for 7 days, 25% mortality was observed at a concentration of 32 µg L^{-1}	Brun et al. (2006)
Fungi	*Rhizopus arrhizus*	Dye	Yellow RL	Bioadsorption of dye was observed inside the fungi, and the adsorption was concentration dependent	Aksu and Balibek (2010)
Algae	*Daphnia magna*	Pharmaceuticals	Acetaminophen	48h LC_{50} value was recorded >32 µg L^{-1}	Brun et al. (2006)
		Nanomaterials	Fullerene C_{60}	Heart rate was increased when exposed to 260 µg L^{-1}	Lovern et al. (2007)
			Carbon nanotubes	Carbon nanotubes were observed to be ingested by the zooplankton, and subsequent toxicological effects were observed	Roberts et al. (2007)
	Pseudokirchneriella subcapitata		ZnO	Acute effect with a 96 h EC_{50} value of 60 µg L^{-1}	Usenko et al. (2008)
	Chlorella vulgaris	Dye	Rhodamine B	96 h IC_{50} value was calculated as 31.29 mg L^{-1}, and in addition, pigment production was reduced	Sudarshan et al. (2022)
	Spirulina platensis	Pesticides	Chlorpyrifos	EC_{50} values were found to be 33.65 mg L^{-1}, and antioxidant enzyme activity was elevated	Bhuvaneswari et al. (2018)
	Haematococcus pluvialis	Nanomaterials	ZnO	Nanomaterial accumulated inside the microalgal cell	Djearamane et al. (2019)

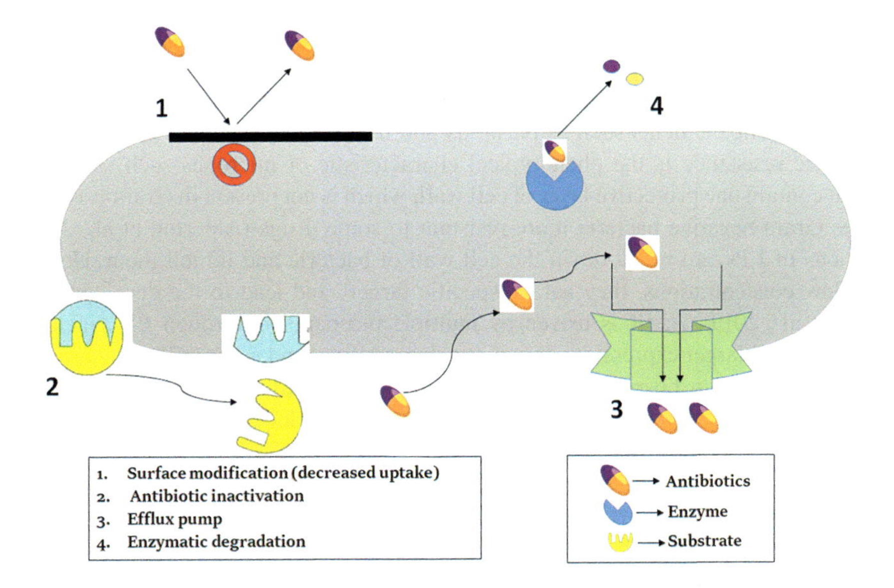

FIGURE 13.2 Different mechanisms of antimicrobial resistance due to chemicals and antibiotics.

of humans and animals die each year, and the number is increasing. Currently, the annual mortality due to antidrug resistance is over 700,000, which is projected to increase to 10 million by the end of 2050 (Pires et al., 2017). Resistance develops naturally either through mutation or through the transfer of genes by genetic exchange (Davies and Davies, 2010; Blakely, 2015). Excessive use of antibiotics is one of the main causes that can stimulate resistance. Recently, it was recognized that the environment had a considerable role in stimulating AMR, especially the EPs present in the environment. Untreated or semi-treated wastewater from urban sewage drains, hospitals, industries, agricultural fields, and aquaculture units is the main source of EPs in the aquatic system. These EPs facilitate the dissemination of antibacterial resistance (ABR) in the water system. The aquaculture industry is one of the leading sectors that use a huge quantity of drugs and pesticides in the culture system to make the species safe from any disease, but unfortunately, it is increasing the future risk by developing resistant genes. Oxytetracycline, florfenicol, sulphonamides, erythromycin, and sarafloxacin are widely used in culturing fish (Serrano, 2005), and abuse of these drugs may result in the rapid spread of AMRs. Among the various antimicrobial resistant genes (ARGs) in the aquatic ecosystem, *tet* genes are the most prominent. About 20 types of *tet* genes have been identified in aquatic ecosystems throughout the world (Zhang et al., 2009). Resistant genes spread among the bacterial flora in the aquatic ecosystem mainly through the horizontal gene transfer process. Transfer of mobile elements, including plasmids, transposons, and integrons, among bacterial communities can spread the ARGs. The other two processes are the transformation of naked DNA and transduction by bacteriophages (Zhang et al., 2009).

The use and discharge of EPs in the aquatic environment are largely unregulated due to improper sewage treatment, resulting in a considerable amount of organic pollutants and a diverse group of microbiota in effluent water. In the waterbodies, these microbiotas may mix with the indigenous bacterial flora and can spread AMR genes among them (Lupo et al., 2012). This phenomenon is triggered by the presence of EPs, already present in the same environment. EPs such as microplastics act as a platform for the biofilm-forming bacteria in the aquatic ecosystem. Moreover, they work as a substrate for the bacterial group and promote their growth, which in turn, can accelerate the horizontal gene transfer process. In this process, microplastic-associated biofilms can acquire ARGs from remote environments and may contribute to pathogenicity transfer and antibiotic resistance in the environment (Caruso, 2019). Therefore, the presence of EPs in areas with high anthropogenic input promotes the rapid transmission of resistance genes between bacterial populations, as opposed to the natural transfer of ARGs (Manaia, 2017).

AMR is either intrinsic or acquired. Resistance arises by genetic mutation or by the interchanging of mobile genetic elements such as plasmids and transposons (Klein et al., 2018; Heinemann, 1999). Intrinsic resistance is the phenotypical characteristic of microbes such as Gram-negative bacteria that contain one protective layer of cell wall, which is not present in Gram-positive bacteria. This makes Gram-negative bacteria more resistant to some drugs (Alderton et al., 2021). Higher concentrations of EPs can break down the cell wall of bacteria and inhibit them. However, when present in low concentrations, they act on specific targets and lead to the development of AMR (Karmakar et al., 2019). AMR is driven by multiple mechanisms through EPs, including target mutation, increased target expression, enzymatic degradation, and active efflux.

The wastewaters from hospitals and households are contaminated with biocides such as antiseptics, disinfectants, and preservatives. Common biocides include benzalkonium chloride, cetylpyridinium chloride, chlorhexidine, and triclosan. *Campylobacter*, *Escherichia coli*, *Salmonella typhimurium*, and *Edwardsiella tarda* are some of the bacterial strains that have been reported to form resistance to biocides (Mavri and Možina, 2013; Curiao et al., 2016; Karmakar et al., 2019). In the COVID-19 era, the use of medical and healthcare products has increased 10 to 20 times compared with the pre-COVID era (Bandala et al., 2021). This has led to the generation of huge quantities of biomedical waste. There have been reports that medical effluent bacteria contain ARGs, such as vancomycin (*van*A) and methicillin (*mec*A) (Heuer et al., 2002; Rowe et al., 2017). The primary drivers of the occurrence of AMR are misusing and overusing antibiotics, and it is recognized that EPs also play a crucial role in promoting AMR.

13.4 EFFECT ON FUNGI

Though fungi are predominant members of natural habitats, information regarding the effects of EPs on fungi is still scanty. Fungi, being important decomposers of the environment, break down organic matter and recycle many elements. Fungi form a close symbiotic relationship with other microbes in the ecosystem and take part in regulating many sensitive ecosystem services. Due to its sensitivity to extragenous elements and xenobionts, fungi experience disruptions in their symbiotic relationships with other microbes, which are essential for numerous ecological services. Textile dyes, along with several other industrial wastes released directly into the aquatic environment, react with the fungal species (Nadaroglu et al., 2015). Bragulat et al. (1991) found that commercially important dyes have a significant impact on a fungal colony. They reported a dose-dependent effect of dye, whereby the colony diameter reduced with increasing dye concentration. Maximum reduction was observed in the *Rhizopus stolonifer* strain; colony diameter was reduced by 98.6% by the dye water blue. Rose Bengal dye also has an inhibitory effect on some yeast and mould species (Jarvis, 1973). Further, excessive use of pesticides can alter the natural microbial composition of the system. Fungal species richness in the waterbody was reduced by 35 to 44% due to the introduction of fungicides, as well as sporulation rates significantly decreasing (Flores et al., 2014). Even though pesticides and fungicides have been designed to be as specific as possible, they have effects on aquatic organisms that are not their intended targets. Flores et al. (2014) reported that aquatic hyphomycetes were affected by the fungicide imazalil, which was not designed to target this organism. The fungicide also affected the production of ergosterol, an important biosynthesis process of fungi. Fungicides in the presence of prochloraz produced 12 times greater effects, which indicated that fungicides can have synergistic effects with other pesticides (Nørgaard and Cedergreen, 2010).

13.5 EFFECT ON MICROALGAE

Microalgae are the primary producers of the aquatic ecosystem. They are a diverse group of organisms forming the base of the aquatic food chain. In the presence of sunlight, microalgae convert inorganic elements into organic matter and provide it to higher organisms. In addition, they regulate biogeochemical cycles and carbon sequestration. Several microalgae contain a higher percentage of protein and have wide industrial value, as they contain many nutraceutical properties. On the other hand, they are also very sensitive to any changes in the ecosystem. The introduction of pollutants to aquatic bodies affects microalgae first, since they are present at the base of the food chain. Recently, Rempel et al. (2021) evaluated the toxic effect of five EPs on *Spirulina platensis*, *Chlorella homosphaera*, and *Scenedesmus obliquus* microalgae and found a concentration-dependent reduction in microalgal growth rate. Some metals are essential for every organism, but the non-essential metals are very toxic to organisms when present in the environment. Industries release various heavy metals into the aquatic environment, and through the microalgae, they reach higher trophic levels of the aquatic food chain. Metals such as Hg, Cr, Pb, Cu, etc. can accumulate inside the microalgal cell and can replace the co-factors of various vitamins and proteins. Exposure of microalgae to metals results in oxidative stress, which is evidenced by an increase in the activity of antioxidant enzymes, such as catalase (CAT), peroxidase (POD), and superoxide dismutase (SOD) (Xie et al., 2015). In addition, metals can transform into more toxic forms after entering the aquatic body. For example, mercury can undergo a methylation process and produce methyl mercury, which is many times more toxic and bioaccumulative than the parent form (Graham et al., 2017).

The production of nanoparticles (NPs) has increased significantly to meet the global demand, and they are being widely used in several fields, such as medical, textile, material production, automobile, and sports. NPs are discharged into water bodies after use, and due to their smaller dimensions and surface properties, they negatively affect the primary producers of the ecosystem. They can easily attach to the surface of the microalgae and prevent access of light and nutrients from the surrounding environment. NPs enter the algal cell and accumulate inside. The toxic effects of NPs greatly depend on their surface coating. Positively charged NPs are more toxic than negatively charged and neutral ones (Sendra et al., 2017). The interaction of NPs with the algal cell was verified by various authors through Fourier transmission infrared spectroscopy (FTIR) (Sadiq et al., 2011). Further, scanning electron microscopy (SEM) and transmission electron microscopy (TEM) analysis provide clear visuals of how these tiny particles enter the cells and damage various organelles. For example, Djearamane et al. (2019) evaluated the toxic effects of ZnO nanometals on the microalga *Haematococcus pluvialis*, and SEM images confirmed the presence of nanometals inside the cell. In the aquatic ecosystem, NPs can also work as a vector for many other hydrophobic pollutants. A synergistic effect was observed when NPs were exposed to lead. Growth and pigment production were reduced in marine microalgae (Hu et al., 2018).

Pesticides such as herbicides, insecticides, and fungicides are target specific. Insecticides are intended to kill insects, but unfortunately, some zooplankton perform similar kinds of physiological and metabolic activities as insects, so these non-target organisms are also inhibited by the insecticides. For instance, the insecticides chlorpyrifos and cypermethrin damaged the thylakoid and mitochondrial structure of *Selenastrum capricornutum*. As a stress response, lipids accumulated inside the cells, and starch granules multiplied in size and number (Fernandez et al., 2021). Herbicides block the electron transport mechanisms of photosystem II, and the photosynthetic efficiency of the microalgae is reduced (Lu et al., 2021).

The production of graphene and its oxides has increased in the last decade because of their widespread application in nanotechnology. Recently, it has been reported that graphene and graphene-based products are toxic to aquatic microbes. The 96 h EC_{50} value was found to be 21.2 and 8.8 mg L^{-1} against *Scenedesmus obliquus* and *Heterosigma akashiwo*, respectively. In addition, oxidative damage to the cells was induced, as the production of reactive oxygen species was elevated, which as a result, inhibited the protein synthesis process (Hu et al., 2016; Wang et al., 2022).

13.6 CONCLUSION

There has been unprecedented growth in the use of EPs, such as steroids, flame retardants, endocrine disruptors, hormones, food additives, and microplastics, over the past few decades, and the trend is expected to continue into the future. It is proven that EPs can affect the health of ecosystems by altering the diversity and functioning of natural microorganisms. They are bioaccumulative and persistent. However, information regarding their fate and biotransformation and its effect on the microbial community and function is scanty, posing a serious challenge. The scientific community needs to work on establishing frameworks for regulating the use and discharge of EPs and minimizing their negative effects. Though the existing technologies are partially capable of providing the desired results, new technologies need to be developed to remediate these molecules from effluent water discharges from various systems. This will reduce the entry of EPs into aquatic ecosystems.

REFERENCES

Aksu, Z., and E. Balibek. "Effect of salinity on metal-complex dye biosorption by Rhizopus arrhizus." *Journal of Environmental Management* 91, no. 7 (2010): 1546–1555.

Alderton, Izzie, Barry R. Palmer, Jack A. Heinemann, Isabelle Pattis, Louise Weaver, Maria J. Gutiérrez-Ginés, Jacqui Horswell, and Louis A. Tremblay. "The role of emerging organic contaminants in the development of antimicrobial resistance." *Emerging Contaminants* 7 (2021): 160–171.

Bandala, Erick R., Brittany R. Kruger, Ivana Cesarino, Alcides L. Leao, Buddhi Wijesiri, and Ashantha Goonetilleke. "Impacts of COVID-19 pandemic on the wastewater pathway into surface water: A review." *Science of the Total Environment* 774 (2021): 145586.

Bandick, Anna K., and Richard P. Dick. "Field management effects on soil enzyme activities." *Soil Biology and Biochemistry* 31, no. 11 (1999): 1471–1479.

Bera, Kuntal Krishna, Saurav Kumar, Tapas Paul, Kurcheti Pani Prasad, S. P. Shukla, and Kundan Kumar. "Triclosan induces immunosuppression and reduces survivability of striped catfish Pangasianodon hypophthalmus during the challenge to a fish pathogenic bacterium Edwardsiella tarda." *Environmental Research* 186 (2020): 109575.

Bhuvaneswari, G. Rathi, C. S. Purushothaman, P. K. Pandey, Subodh Gupta, H. Sanath Kumar, and S. P. Shukla. "Toxicological effects of chlorpyrifos on growth, chlorophyll a synthesis and enzyme activity of a cyanobacterium Spirulina (Arthrospira) platensis." *International Journal of Current Microbiology and Applied Sciences* 7 (2018): 2980–2990.

Biswas, J. K., and D. Sarkar. "Nanopollution in the aquatic environment and ecotoxicity: no nano issue!" *Current Pollution Reports* 5 (2019): 4–7.

Blakely, Garry W. "Mechanisms of horizontal gene transfer and DNA recombination." In *Molecular Medical Microbiology*, pp. 291–302, eds Yi-Wei Tang, Dongyou Liu, Paul Spearman, Musa Y. Hindiyeh, Andrew Sails and Jing-Ren Zhang, Academic Press, 2015.

Bragulat, M. R., M. L. Abarca, M. T. Bruguera, and F. J. Cabaes. "Dyes as fungal inhibitors: Effect on colony diameter." *Applied and Environmental Microbiology* 57, no. 9 (1991): 2777–2780.

Brun, Guy L., Marc Bernier, René Losier, Ken Doe, Paula Jackman, and Hing-Biu Lee. "Pharmaceutically active compounds in Atlantic Canadian sewage treatment plant effluents and receiving waters, and potential for environmental effects as measured by acute and chronic aquatic toxicity." *Environmental Toxicology and Chemistry: An International Journal* 25, no. 8 (2006): 2163–2176.

Caruso, Gabriella. "Microplastics as vectors of contaminants." *Marine Pollution Bulletin* 146 (2019): 921–924.

Conkle, Jeremy L., and John R. White. "An initial screening of antibiotic effects on microbial respiration in wetland soils." *Journal of Environmental Science and Health, Part A* 47, no. 10 (2012): 1381–1390.

Curiao, Tânia, Emmanuela Marchi, Denis Grandgirard, Ricardo León-Sampedro, Carlo Viti, Stephen L. Leib, Fernando Baquero, Marco R. Oggioni, José Luis Martinez, and Teresa M. Coque. "Multiple adaptive routes of Salmonella enterica Typhimurium to biocide and antibiotic exposure." *BMC Genomics* 17 (2016): 1–16.

Davies, Julian, and Dorothy Davies. "Origins and evolution of antibiotic resistance." *Microbiology and Molecular Biology Reviews* 74, no. 3 (2010): 417–433.

De Lorenzo, M. E., J. M. Keller, C. D. Arthur, M. C. Finnegan, H. E. Harper, V. L. Winder, and D. L. Zdankiewicz. "Toxicity of the antimicrobial compound triclosan and formation of the metabolite methyl-triclosan in estuarine systems." *Environmental Toxicology: An International Journal* 23, no. 2 (2008): 224–232.

Djearamane, Sinouvassane, Yang Mooi Lim, Ling Shing Wong, and PohFoong Lee. "Cellular accumulation and cytotoxic effects of zinc oxide nanoparticles in microalga Haematococcus pluvialis." *Peerj* 7 (2019): e7582.

Droma, Dawa, Saurav Kumar, Tapas Paul, Prasenjit Pal, Neelam Saharan, Kundan Kumar, and Nalini Poojary. "Biomarkers for assessing chronic toxicity of carbamazepine, an anticonvulsants drug on *Pangasianodon hypophthalmus* (Sauvage, 1878)." *Environmental Toxicology and Pharmacology* 87 (2021): 103691.

Du, Ming, Minggang Zheng, Aifeng Liu, Ling Wang, Xin Pan, Jun Liu, and Xiangbin Ran. "Effects of emerging contaminants and heavy metals on variation in bacterial communities in estuarine sediments." *Science of the Total Environment* 832 (2022): 155118.

European Commission. "Technical guidance document in support of commission directive 93/67/EEC on risk assessment for new notified substances and commission regulation (EC) No 1488/94 on risk assessment for existing substances." Part II. Environ. Risk Assessment, Luxemb (1996).

Fang, Jiasong, Delina Y. Lyon, Mark R. Wiesner, Jinping Dong, and Pedro J. J. Alvarez. "Effect of a fullerene water suspension on bacterial phospholipids and membrane phase behavior." *Environmental Science & Technology* 41, no. 7 (2007): 2636–2642.

Fernandez, Carolina, Viviana Asselborn, and Elisa R. Parodi. "Toxic effects of chlorpyrifos, cypermethrin and glyphosate on the non-target organism Selenastrum capricornutum (Chlorophyta)." *Anais Da Academia Brasileira De Ciências* 93 (2021): 20200233.

Flores, L., Z. Banjac, M. Farré, A. Larrañaga, E. Mas-Martí, I. Muñoz, D. Barceló, and A. Elosegi. "Effects of a fungicide (imazalil) and an insecticide (diazinon) on stream fungi and invertebrates associated with litter breakdown." *Science of the Total Environment* 476 (2014): 532–541.

Gavrilescu, Maria, Kateřina Demnerová, Jens Aamand, Spiros Agathos, and Fabio Fava. "Emerging pollutants in the environment: Present and future challenges in biomonitoring, ecological risks and bioremediation." *New Biotechnology* 32, no. 1 (2015): 147–156.

Gong, Huijuan, Zhen You, Qiming Xian, Xing Shen, Huixian Zou, Fei Huan, and Xu Xu. "Study on the structure and mutagenicity of a new disinfection byproduct in chlorinated drinking water." *Environmental Science & Technology* 39, no. 19 (2005): 7499–7508.

Graham, Andrew M., Keaton T. Cameron-Burr, Hayley A. Hajic, Connie Lee, Deborah Msekela, and Cynthia C. Gilmour. "Sulfurization of dissolved organic matter increases Hg–sulfide–dissolved organic matter bioavailability to a Hg-methylating bacterium." *Environmental Science & Technology* 51, no. 16 (2017): 9080–9088.

Grenni, Paola, Valeria Ancona, and Anna Barra Caracciolo. "Ecological effects of antibiotics on natural ecosystems: A review." *Microchemical Journal* 136 (2018): 25–39.

Halla, Noureddine, Isabel P. Fernandes, Sandrina A. Heleno, Patrícia Costa, Zahia Boucherit-Otmani, Kebir Boucherit, Alírio E. Rodrigues, Isabel C. F. R. Ferreira, and Maria Filomena Barreiro. "Cosmetics preservation: A review on present strategies." *Molecules* 23, no. 7 (2018): 1571.

Han, Lingxi, Yalei Liu, Kuan Fang, Xiaolian Zhang, Tong Liu, Fenglong Wang, and Xiuguo Wang. "Dissipation of chlorothalonil in the presence of chlortetracycline and ciprofloxacin and their combined effects on soil enzyme activity." *Environmental Science and Pollution Research* 27 (2020): 13662–13669.

Heinemann, Jack A. "How antibiotics cause antibiotic resistance." *Drug Discovery Today* 4, no. 2 (1999): 72–79.

Heuer, H., E. Krögerrecklenfort, E. M. H. Wellington, S. Egan, J. D. Van Elsas, L. Van Overbeek, J.-M. Collard et al. "Gentamicin resistance genes in environmental bacteria: Prevalence and transfer." *Fems Microbiology Ecology* 42, no. 2 (2002): 289–302.

Hu, Changwei, Naitao Hu, Xiuling Li, and Yongjun Zhao. "Graphene oxide alleviates the ecotoxicity of copper on the freshwater microalga Scenedesmus obliquus." *Ecotoxicology and Environmental Safety* 132 (2016): 360–365.

Hu, Ji, Zhechao Zhang, Cai Zhang, Shuxia Liu, Haifeng Zhang, Dong Li, Jun Zhao et al. "Al2O3 nanoparticle impact on the toxic effect of Pb on the marine microalga Isochrysis galbana." *Ecotoxicology and Environmental Safety* 161 (2018): 92–98.

Isidori, Marina, Margherita Lavorgna, Angela Nardelli, Alfredo Parrella, Lucio Previtera, and Maria Rubino. "Ecotoxicity of naproxen and its phototransformation products." *Science of the Total Environment* 348, no. 1–3 (2005): 93–101.

Jarvis, B. "Comparison of an improved rose bengal-chlortetracycline agar with other media for the selective isolation and enumeration of moulds and yeasts in foods." *Journal of Applied Microbiology* 36, no. 4 (1973): 723–727.

Karmakar, Sutanu, T. J. Abraham, Saurav Kumar, Sanath Kumar, S. P. Shukla, Utsa Roy, and Kundan Kumar. "Triclosan exposure induces varying extent of reversible antimicrobial resistance in Aeromonas hydrophila and Edwardsiella tarda." *Ecotoxicology and Environmental Safety* 180 (2019): 309–316.

Klein, Eili Y., Thomas P. Van Boeckel, Elena M. Martinez, Suraj Pant, Sumanth Gandra, Simon A. Levin, Herman Goossens, and Ramanan Laxminarayan. "Global increase and geographic convergence in antibiotic consumption between 2000 and 2015." *Proceedings of the National Academy of Sciences* 115, no. 15 (2018): E3463–E3470.

Kumar, Saurav, Tapas Paul, S. P. Shukla, Kundan Kumar, Sutanu Karmakar, and Kuntal Krishna Bera. "Biomarkers-based assessment of triclosan toxicity in aquatic environment: A mechanistic review." *Environmental Pollution* 286 (2021): 117569.

La Farre, Marinel, Sandra Pérez, Lina Kantiani, and Damià Barceló. "Fate and toxicity of emerging pollutants, their metabolites and transformation products in the aquatic environment." *TrAC Trends in Analytical Chemistry* 27, no. 11 (2008): 991–1007.

Lovern, Sarah B., J. Rudi Strickler, and Rebecca Klaper. "Behavioral and physiological changes in Daphnia magna when exposed to nanoparticle suspensions (titanium dioxide, nano-C60, and C60HxC70Hx)." *Environmental Science & Technology* 41, no. 12 (2007): 4465–4470.

Lu, Tao, Qi Zhang, Zhenyan Zhang, Baolan Hu, Jianmeng Chen, Jun Chen, and Haifeng Qian. "Pollutant toxicology with respect to microalgae and cyanobacteria." *Journal of Environmental Sciences* 99 (2021): 175–186.

Lupo, Agnese, Sébastien Coyne, and Thomas Ulrich Berendonk. "Origin and evolution of antibiotic resistance: The common mechanisms of emergence and spread in water bodies." *Frontiers in Microbiology* 3 (2012): 18.

Manaia, Célia M. "Assessing the risk of antibiotic resistance transmission from the environment to humans: Non-direct proportionality between abundance and risk." *Trends in Microbiology* 25, no. 3 (2017): 173–181.

Mavri, Ana, and Sonja Smole Možina. "Development of antimicrobial resistance in Campylobacter jejuni and Campylobacter coli adapted to biocides." *International Journal of Food Microbiology* 160, no. 3 (2013): 304–312.

Molaei, Ali, Amir Lakzian, Rahul Datta, Gholamhosain Haghnia, Alireza Astaraei, MirHassan Rasouli-Sadaghiani, and Maria T. Ceccherini. "Impact of chlortetracycline and sulfapyridine antibiotics on soil enzyme activities." *International Agrophysics* 31, no. 4 (2017): 499–505.

Nadaroglu, Hayrunnisa, Ekrem Kalkan, Neslihan Celebi, Esen Tasgin, and A. Dakovic. "Removal of Reactive Black 5 from wastewater using natural clinoptilolite modified with apolaccase." *Clay Minerals* 50, no. 1 (2015): 65–76.

Nørgaard, Katrine Banke, and Nina Cedergreen. "Pesticide cocktails can interact synergistically on aquatic crustaceans." *Environmental Science and Pollution Research* 17 (2010): 957–967.

Paul, Tapas, S. P. Shukla, Kundan Kumar, Nalini Poojary, and Saurav Kumar. "Effect of temperature on triclosan toxicity in *Pangasianodon hypophthalmus* (Sauvage, 1878): Hematology, biochemistry and genotoxicity evaluation." *Science of the Total Environment* 668 (2019): 104–114.

Paul, Tapas, Saurav Kumar, S. P. Shukla, Prasenjit Pal, Kundan Kumar, Nalini Poojary, Abhilipsa Biswal, and Archana Mishra. "A multi-biomarker approach using integrated biomarker response to assess the effect of pH on triclosan toxicity in Pangasianodon hypophthalmus (Sauvage, 1878)." *Environmental Pollution* 260 (2020): 114001.

Pérez, Sandra, and Damià Barceló. "Application of advanced MS techniques to analysis and identification of human and microbial metabolites of pharmaceuticals in the aquatic environment." *Trac Trends in Analytical Chemistry* 26, no. 6 (2007): 494–514.

Pillard, David A., Jeffrey S. Cornell, Doree L. DuFresne, and Mark T. Hernandez. "Toxicity of benzotriazole and benzotriazole derivatives to three aquatic species." *Water Research* 35, no. 2 (2001): 557–560.

Pires, D., M.E.A. de Kraker, E. Tartari, M. Abbas, and D. Pittet. "'Fight antibiotic resistance—It's in your hands': Call from the World Health Organization for 5th May 2017." *Clinical Infectious Diseases* 64, no. 12 (2017): 1780–1783.

Poger, David, and Alan E. Mark. "Effect of triclosan and chloroxylenol on bacterial membranes." *The Journal of Physical Chemistry B* 123, no. 25 (2019): 5291–5301.

Proia, Lorenzo, Victoria Osorio, S. Soley, Marianne Köck-Schulmeyer, Sandra Pérez, Damià Barceló, A. M. Romaní, and Sergi Sabater. "Effects of pesticides and pharmaceuticals on biofilms in a highly impacted river." *Environmental Pollution* 178 (2013): 220–228.

Rempel, Alan, Gabrielle Nadal Biolchi, Ana Carolina Farezin Antunes, Julia Pedó Gutkoski, Helen Treichel, and Luciane Maria Colla. "Cultivation of microalgae in media added of emergent pollutants and effect on growth, chemical composition, and use of biomass to enzymatic hydrolysis." *Bioenergy Research* 14 (2021): 265–277.

Richardson, S. D. "New disinfection by-product issues: Emerging DBPs and alternative routes of exposure." *Global NEST Journal* 7, no. 1 (2005): 43–60.

Roberts, Aaron P., Andrew S. Mount, Brandon Seda, Justin Souther, Rui Qiao, Sijie Lin, Pu Chun Ke, Apparao M. Rao, and Stephen J. Klaine. "In vivo biomodification of lipid-coated carbon nanotubes by Daphnia magna." *Environmental Science & Technology* 41, no. 8 (2007): 3025–3029.

Rowe, Will P. M., Craig Baker-Austin, David W. Verner-Jeffreys, Jim J. Ryan, Christianne Micallef, Duncan J. Maskell, and Gareth P. Pearce. "Overexpression of antibiotic resistance genes in hospital effluents over time." *Journal of Antimicrobial Chemotherapy* 72, no. 6 (2017): 1617–1623.

Sadiq, I. Mohammed, Sunandan Pakrashi, N. Chandrasekaran, and Amitava Mukherjee. "Studies on toxicity of aluminum oxide (Al2O3) nanoparticles to microalgae species: Scenedesmus sp. and Chlorella sp." *Journal of Nanoparticle Research* 13 (2011): 3287–3299.

Sendra, Marta, P. M. Yeste, Ignacio Moreno-Garrido, José Manuel Gatica, and Julián Blasco. "CeO_2 NPs, toxic or protective to phytoplankton? Charge of nanoparticles and cell wall as factors which cause changes in cell complexity." *Science of the Total Environment* 590 (2017): 304–315.

Serrano, Pilar Hernández. *Responsible Use of Antibiotics in Aquaculture.* Vol. 469. Food & Agriculture Org., 2005.

Sudarshan, Shanmugam, Vidya Shree Bharti, Sekar Harikrishnan, Satya Prakash Shukla, and Govindarajan RathiBhuvaneswari. "Eco-toxicological effect of a commercial dye Rhodamine B on freshwater microalgae Chlorella vulgaris." *Archives of Microbiology* 204, no. 10 (2022): 658.

Tan, Qiyang, Jinmei Chen, Yifan Chu, Wei Liu, Lingli Yang, Lin Ma, Yi Zhang, Dongru Qiu, Zhenbin Wu, and Feng He. "Triclosan weakens the nitrification process of activated sludge and increases the risk of the spread of antibiotic resistance genes." *Journal of Hazardous Materials* 416 (2021): 126085.

Taştan, Burcu Ertit, and Gönül Dönmez. "Biodegradation of pesticide triclosan by A. versicolor in simulated wastewater and semi-synthetic media." *Pesticide Biochemistry and Physiology* 118 (2015): 33–37.

TerLaak, T.L, W.A. Gebbink, and J. Tolls. "The effect of pH and ionic strength on the sorption of sulfachloropyridazine, tylosin, and oxytetracycline to soil." *Environmental Toxicology and Chemistry: An International Journal* 25, no. 4 (2006): 904–911.

Usenko, Crystal Y., Stacey L. Harper, and Robert L. Tanguay. "Fullerene C60 exposure elicits an oxidative stress response in embryonic zebrafish." *Toxicology and Applied Pharmacology* 229, no. 1 (2008): 44–55.

Wang, Jiayin, Xiaolin Zhu, Liju Tan, Ting Zhao, Ziqi Ni, Na Zhang, and Jiangtao Wang. "Single and combined nanotoxicity of ZnO nanoparticles and graphene quantum dots against the microalga Heterosigma akashiwo." *Environmental Science: Nano* 9, no. 8 (2022): 3094–3109.

Xie, Jun, Xiaocui Bai, Michel Lavoie, Haiping Lu, Xiaoji Fan, Xiangliang Pan, Zhengwei Fu, and Haifeng Qian. "Analysis of the proteome of the marine diatom Phaeodactylum tricornutum exposed to aluminum providing insights into aluminum toxicity mechanisms." *Environmental Science & Technology* 49, no. 18 (2015): 11182–11190.

Zhang, Xu-Xiang, Tong Zhang, and Herbert H. P. Fang. "Antibiotic resistance genes in water environment." *Applied Microbiology and Biotechnology* 82 (2009): 397–414.

Zhu, Z.J, H. Wang, B. Yan, H. Zheng, Y. Jiang, O.R.Miranda, V.M. Rotello, B. Xing, and R.W. Vachet. "Effect of surface charge on the uptake and distribution of gold nanoparticles in four plant species." *Environmental Science & Technology* 46, no. 22 (2012): 12391–12398.

14 Eutrophication in Freshwater and Its Microbial Implications

Debajit Sarma and Deepak Kumar

14.1 INTRODUCTION

The process by which a body of water, or certain areas within it, gradually becomes richer with minerals and nutrients, especially nitrogen and phosphorus, is known as eutrophication (Schindler et al., 2012. It has also been described as an "increase in phytoplankton productivity, caused by nutrients" (Chapin et al., 2002). Neither freshwater nor saltwater systems are immune to eutrophication. Almost always, high phosphorus is the cause in freshwater environments (Schindler, 2008). On the contrary, in coastal waters, nitrogen, or nitrogen and phosphorus together, is more frequently the major contributing nutrient (Elser et al., 2007). Water bodies are classified as mesotrophic (intermediate nutrient levels) or oligotrophic (low nutrient levels) depending on the amount of nutrients present. Early-stage eutrophication lakes are frequently characterised by low nutrient levels, restricted algal and plant productivity, and other physical characteristics. These oligotrophic lakes are nutrient-poor, with very clear water, and sustain high dissolved oxygen concentrations throughout the summer (Simpson, 1991; Moore & Thornton, 1988). High nutrient levels in eutrophic lakes encourage the growth of plants and algae in these extremely productive waterbodies (Simpson, 1991; Moore & Thornton, 1988). The range between oligotrophic and eutrophic lakes includes mesotrophic lakes. Due to the intermediate algal and plant development and intermediate water clarity caused by the intermediate nutrient availability in these lake, the ecological dynamics exhibit a delicate balance between productivity and transparency, fostering a diverse array of aquatic life forms and supporting complex food webs (Simpson, 1991; Moore & Thornton, 1988). Progressive eutrophication is also known as a dystrophic or hypertrophic situation (Robert, 1975). Hypereutrophic lakes are a subcategory of extremely eutrophic lakes that have "pea-soup quality". The lake basin may eventually become so heavily overgrown with vegetation and sediment that it transforms into a marsh, bog, or other wetland habitat (Addy & Green, 1996). In Sweden, Naumann (1919) classified the waters as oligotrophic, mesotrophic, or eutrophic based on the presence of mineral nutrients similar to those found in freshwaters, or brackish and marine waters. Whereas diatoms and cyanobacteria were considered the typical phytoplankters of eutrophic lowland lakes (Ptacnik, 2008; Anderson et al., 2012), desmids were regarded as the distinctive phytoplankters of oligotrophic lakes in hilly areas (Tomec et al., 2002). The amount of organic carbon produced by photosynthesis over the course of an annual cycle, or primary production, is the best way to define algal growth, which essentially defines the trophic status of lakes (Lindeman, 1942). By measuring carbon uptake in g C m^{-2} yr^{-1}, Rodhe (1969) identified different trophic levels: oligotrophic 7–25, eutrophic (natural), 75–250, and eutrophic (polluted) 350–700.

Eutrophication is a very slow naturally occurring process in which nutrients, particularly phosphorus compounds and organic waste, build up in water bodies (Addy & Green, 1996; Schindler, 2006). These nutrients come from the mineral-degradation and mineral-solution processes in rocks as well as from active nutrient-scavenging by lichens, mosses, and fungi on rocks (Sawyer, 1966). Algal blooms are a typical eutrophication phenomenon that might be seen. Algal blooms can be merely obtrusive to individuals wishing to use the water body, or they can develop into dangerous

 DOI: 10.1201/9781003408543-14

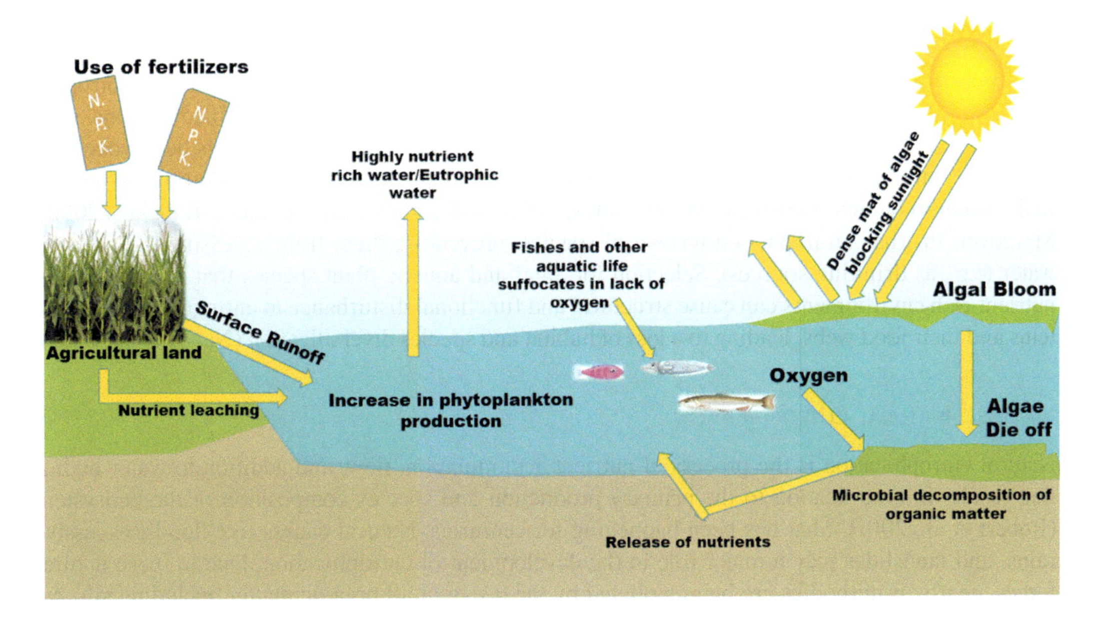

FIGURE 14.1 Process of eutrophication.

algal blooms that seriously impact the ecology of water bodies (Gilbert and Burford, 2017). As the algae are broken down by microbes in this process, the water body may become oxygen-depleted. Large concentrations of microscopic organisms and algae known as water blooms frequently form on the surface, blocking light and oxygen absorption, which are essential for underwater life (Schindler, 2008). Eutrophic waters are frequently murky and may not support as many large animals, such as fish and other aquatic organisms (Figure 14.1).

The idea of eutrophication was initially put forth by Lindeman in 1942, and a study came to the conclusion that a lake's eutrophic stage is its final stage of natural progression (Lindeman, 1942). Total phosphorus (TP) and nitrate nitrogen (NO_3-N) levels above 0.02 mg L^{-1} and 0.3 mg L^{-1}, respectively, indicate eutrophication in a water body (Namsaraev et al., 2018). It has been said that eutrophication of water bodies is similar to "ecological cancer", destroying their ecology and contributing to their ecological imbalance by homogenising the community structure (Ramachandra and Solanki, 2007). Freshwater eutrophication is a major environmental issue faced worldwide as a result of intensive human activity (Wang et al., 2019). Recent studies reveal that 54% of Asian lakes, 53% of European lakes, 48% of North American lakes, 41% of South American lakes, and 28% of African lakes are eutrophic (Zhang et al., 2020). Population explosion, economic growth, and new energy development plans have all increased the pressure on the world's water supplies, posing new environmental problems for long-term sustainability (Chislock et al., 2013). Each year, it is estimated that eutrophication costs the United States alone $2.2 billion (Dodds et al., 2009).

14.2 TYPES OF EUTROPHICATION

14.2.1 Cultural Eutrophication

Anthropogenic or "cultural eutrophication" is frequently a much faster process in which pollutants such as untreated or just partially treated sewage, industrial wastewater, and fertiliser from farming operations contribute nutrients to a waterbody. The primary source of eutrophication of surface waters is nutrient pollution, a type of water pollution in which excess nutrients, typically nitrogen or phosphorus, encourage the growth of algal and aquatic plants (Brian, 1983). One of the main factors contributing to the degradation of aquatic ecosystems is cultural eutrophication, which has resulted in devastating effects on freshwater resources, including fisheries. Hypoxia, or extremely low oxygen concentrations in bottom waters, is a common phenomenon in culturally eutrophic aquatic environments (Glibert and Burford, 2017). This is especially true of stratified systems, such as lakes, in the summer, when molecular oxygen concentrations may drop to levels below one milligram per litre, a threshold for a number of biological and chemical activities. Aerobic organisms in the waterbody, such as fish and invertebrates, suffer in this anoxic environment (Rabalais, 2002). Moreover, this has an impact on terrestrial species, preventing them from accessing the polluted water (e.g., as drinking sources). Selection for algal and aquatic plant species that can flourish in nutrient-rich environments can cause structural and functional disturbance to entire aquatic ecosystems and their food webs, leading to a loss of habitat and species diversification (Nancy et al., 2002).

14.2.2 Natural Eutrophication

Natural eutrophication is the process of nutrient accumulation, flow, and addition to water bodies that results in modification to the primary production and species composition of the ecosystem (Robert et al., 2001). That has been happening for centuries. Natural causes like floods, excessive rains, and landslides play a major role in the development of eutrophication. Due to these natural forces, nearby waterbodies are being polluted by the decay of all organic matter, including silt. An influx of organic material helps water bodies become more nutrient-rich (Walker, 2006; Whiteside, 1983).

14.3 SOURCES OF EUTROPHICATION

14.3.1 Point Source Pollution

Contamination caused by substances that enter a waterbody from a single, recognisable source, such as fixed places or infrastructure (Figure 14.2), is known as point source pollution. This includes discharges from industrial facilities, fish farms, or sewage treatment plants (van Leeuwen et al.,

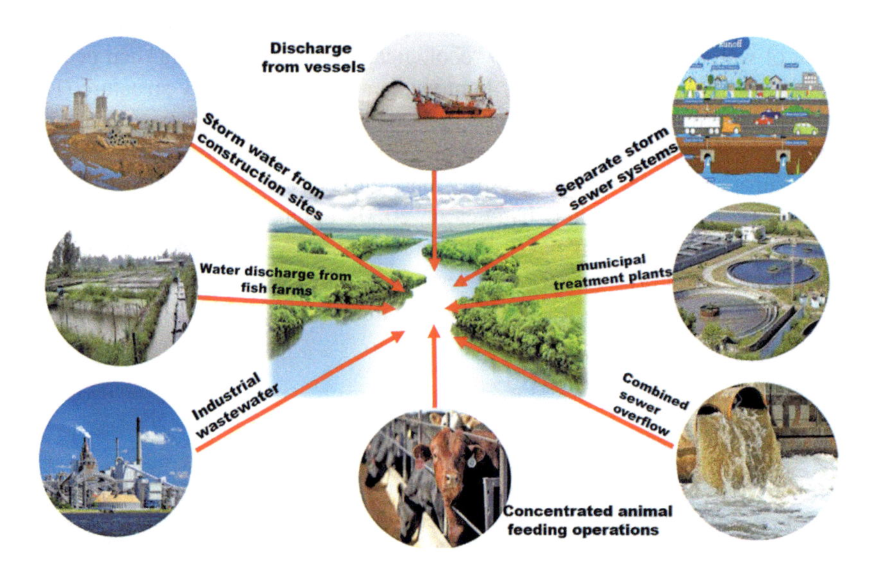

FIGURE 14.2 Point source pollution.

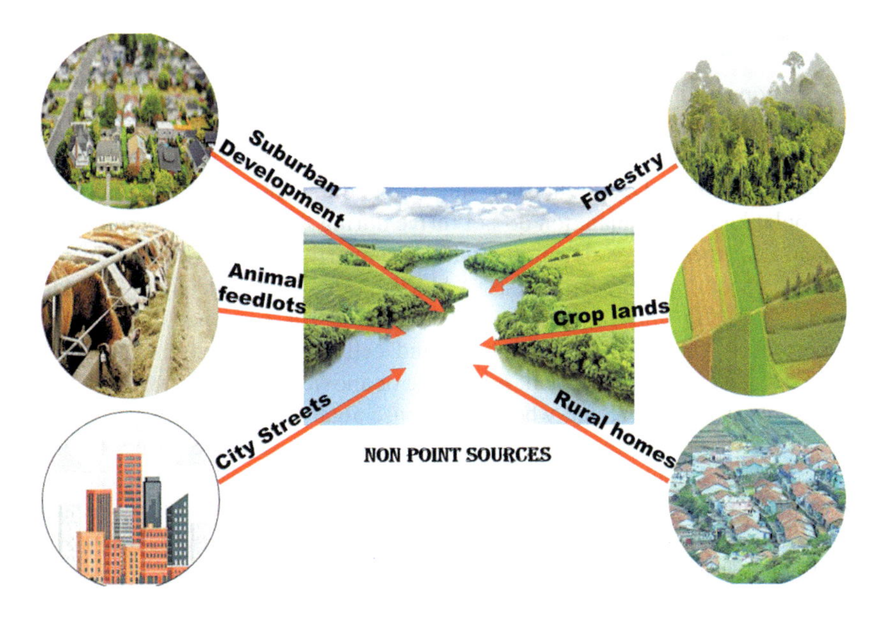

FIGURE 14.3 Non-point source pollution.

2007). Point source pollution is described by the U.S. Environmental Protection Agency (EPA) as "any single identifiable source of pollution from which pollutants are emitted, such as a pipe, ditch, ship, or manufacturing smokestack". Point sources include, among other things, factories and sewage treatment facilities.

14.3.2 Non-Point Source Pollution

Non-point source pollution is extensive pollution caused by human activities, including those without a single defined site of discharge or entry into receiving watersheds, such as excess nitrogen compounds from atmospheric deposition and agricultural regions that have been fertilised (Figure 14.3). According to the EPA, pollution from non-point sources is frequently referred to as "diffuse" contamination. It relates to inputs and impacts that span a large area and are difficult to link to a single source. In contrast to discrete point source discharges, they are frequently connected to specific land uses.

14.4 FACTORS ENCOURAGING EUTROPHICATION

Aquatic ecosystems are negatively impacted by the majority of human activities, which poses a huge environmental threat (EEA, 2018). Human activities are the main source of eutrophication because of our reliance on nitrate and phosphate fertilisers. The build-up of phosphate and nitrate nutrients is influenced by agricultural practices and the use of fertilisers in the fields. Controlling and managing water quality is therefore crucial for the wellbeing of ecosystems and human populations (UNESCO, 2009). Concentrated animal feeding operations (CAFOs) are the principal source of the nitrogen and phosphate nutrients that cause eutrophication. CAFOs typically release large quantities of nutrients into rivers, streams, lakes, and oceans, where they build up in high concentrations and plague the waterbodies with recurrent cyanobacterial and algal blooms. Algal blooms can grow excessively as a result of surplus nutrients being washed into water systems by natural events like floods and the regular flow of rivers and streams (Young, 2019).

14.5 WATER BLOOMS

A large aquatic population of microscopic photosynthetic organisms, known as water blooms, is created when there is an availability of nutritional salts in surface water and enough light for photosynthesis, when a species' blooming threshold is exceeded and there are more than 1000 cells per millilitre (Kim et al., 1993). Due to the short duration and low intensity of sunlight, and predation by grazing zooplankton, phytoplankton do not thrive during the winter. Upwelling replenishes the nutrient content near the surface during the winter, supporting rapid growth as increased insolation in the spring encourages photosynthesis. A bloom of this kind typically peaks in the Northern Hemisphere in April. By midsummer, the surface waters lack nutrients, and the phytoplankton population decreases. However, photosynthesis and cell division decline as winter approaches because of the decrease in solar radiation. A hypoxic or anoxic "dead zone" without enough oxygen to support most organisms is created when these dense algal blooms eventually die due to microbial breakdown, which drastically depletes dissolved oxygen. Several freshwater lakes, such as the Laurentian Great Lakes (including the central basin of Lake Erie; Arend et al., 2011), experience dead zones in the summer. Some algal blooms are more dangerous than others because they produce harmful toxins, e.g., microcystin and anatoxin-a (Chorus and Bartram, 1999). Harmful algal blooms (HABs) have been associated with three things throughout the past century: (1) declining water quality (Francis, 1878), (2) decimating economically significant fisheries (Burkholder et al., 1992), and (3) hazards to the public's health (Morris, 1999). Zhou et al. (2022) showed a coupling relationship between sulfur, iron, and phosphorus cycles in lake ecosystems (Figure 14.4). With an appropriate organic input from the decay processes of cyanobacteria blooms, the fast-rising sulfate concentration boosted the sulfate reduction to release a significant amount of $\sum S^{2-}$. The initial sulfate concentrations during the cyanobacterial breakdown exhibited a favourable association with the iron reduction. S^{2-} eventually trapped the Fe^{2+} that was produced during the iron reduction process, and the iron and phosphorus mixture was lowered as a result, stimulating the release of endogenous phosphorus. The authors discovered that cyanobacterial blooms are non-negligibly promoted by the significantly rising SO_2^- content in eutrophic lakes. Also, they discovered that cyanobacteria

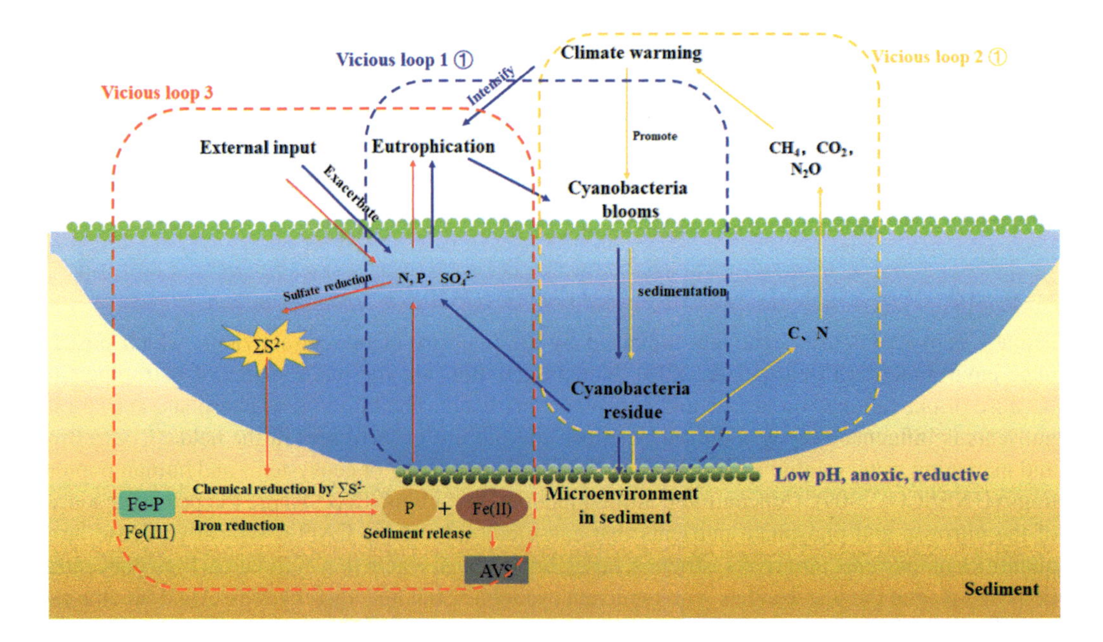

FIGURE 14.4 Extreme abiotic and biotic circumstances under scenarios of global warming favoured the emergence of cyanobacterial blooms. (From Zhou, C. et al., *Biogeosciences*, 19, 17, 4351–4360, 2022.)

generated a lot of organic material to encourage microbial development as they decayed and decomposed, which in turn aided in the anaerobic reduction of sulfur and iron (Holmer et al., 2001).

14.6 EFFECTS OF EUTROPHICATION

Freshwater ecosystems were the first to be threatened by eutrophication (Vollenweider, 1976), and then coastal and marine ecosystems in the early 1970s (Smith et al., 2006). Several marine environments around the world experienced acute eutrophication, including those in the Chesapeake Bay in the United States, the Black Sea, the Chinese coastal waters, and the Inland Sea of Japan (Anderson et al., 2002). Eutrophication can create substantial social-economic losses in addition to altering the biological and ecological aspects of the aquatic ecosystem (Kumagai, 2008). The sections that follow discuss how eutrophication harms both biotic and abiotic ecosystems.

14.6.1 Prevalence of Undesirable Substances or Ecological Effects

The turbidity and colour of the water are dependent on an abundance of particulate materials, including phytoplankton, zooplankton, bacteria, fungi, and debris, which are brought on by eutrophication. An unfavourable overpopulation of phytoplankton and their subsequent demise create a greenish slime layer on the water's surface that prevents light from penetrating (Khan and Ansari, 2005; Ansari et al., 2010). Aquatic plant death and decay result in an unpleasant odour and murky water (Beeton, 2002). In a polytrophic lake, Mutek Lake in Poland, the seasonal variation in zooplankton biomass and population correlated with O_2 availability (Widuto, 1988). Eutrophication was a direct cause of the seasonal variations in water quality along Karawang's northern coast in West Jawa. During the rainy season, the content of chlorophyll-a was raised by an excessive intake of organic waste containing significant amounts of dissolved inorganic nitrogen and phosphate from agriculture (Sachoemar and Yanagi, 1999). In drinking water treatment facilities, eutrophication causes an increase in inorganic chemicals such as ammonia, nitrites, hydrogen sulfide, etc., resulting in the development of hazardous substances like nitrosamines, which may be mutagenic. The most obvious result of cultural eutrophication is the development of dense blooms of toxic, foul-smelling phytoplankton that lower water clarity and degrade water quality. Algal blooms reduce light penetration, which slows plant growth and causes plant die-offs in littoral zones. They also make predators less successful because they need light to pursue and trap prey (Lehtiniemi et al., 2005).

14.6.2 Effects on Aquatic Organisms

Algal blooms are caused by the rapid growth of phytoplankton and other photosynthetic plants in aquatic habitats. As a result, the algal blooms reduce the amount of dissolved oxygen needed for other animal and plant species in the water to breathe. As the algae and plant life decay, oxygen levels are reduced. Shrimp, fish, and other aquatic biota species suffocate to death when the dissolved oxygen concentration falls below the hypoxic threshold. *Gymnodinium aureolum*, a dinoflagellate, repeatedly killed fish in aquaculture systems in Tunisian lagoons (Romdhane et al., 1998). According to Flemer et al. (1983), bottom-water hypoxia in Chesapeake Bay led to decreased submerged aquatic vegetation and fishery harvests as well as a continuous drop in the area's abundant native oysters, *Crassostrea virginica* (Kirby and Miller, 2005). Reduced oxygen levels and rising eutrophication in the North Sea's Skagerrak (Sweden) and Kattegat (Denmark) over the past 15 to 20 years (Andersson and Rydberg, 1993) reduced macroalgal growth, increased biomass, and altered the species diversity in benthic ecosystems (Anon, 1993). Certain diatom species secrete aldehydes that prevent various invertebrate species, including copepods and urchins, from growing (Sellner et al., 1996). Cyanobacterial blooms caused a significant decrease in copepod egg production in the Baltic Sea (Miralto et al., 1999). In the Bay of Brest, the standing crop of the toxic dinoflagellate *Gymnodinium cf. nagasakiense* significantly decreased (Cloern, 2001) the shifts in the types of phytoplankton, from smaller to larger dinoflagellates and diatoms, together with an increase in the concentration of dissolved inorganic nitrogen and phosphorus compounds (Furnas et al., 2005). Some tropical locations, including Singapore (Gin et al., 2000), Japan (Tada et al., 2003), Curacao (Van Duyl et al., 2002), New Caledonia (Jacquet et al., 2006), Hawaii (Cox et al., 2006), and Moorea have the most blatantly reported cases of this (Delesalle et al., 1993). In the course of the day, eutrophication's high rates of photosynthesis can drastically increase pH and deplete the amount of dissolved inorganic carbon in the water. In turn, increased pH can "blind" species that depend on detecting dissolved chemical stimuli for survival by reducing their chemosensory capabilities (Turner and Chislock, 2010). In extreme situations, the lack of oxygen promotes the growth of microorganisms that can poison birds and marine mammals. Less light reaches the lower levels of the ocean as a result of phytoplankton growth. In addition to reducing biodiversity, this can result in aquatic dead zones and the extinction of aquatic species. Significant alterations in the structure of aquatic communities are also linked to eutrophication. Small-bodied zooplankton typically predominate in plankton communities during cyanobacterial blooms, and previous observational studies have linked this trend to anti-herbivore characteristics of cyanobacteria (such as toxicity, shape, and poor food quality) (Porter, 1977). Due to their superior competitive abilities under conditions of high nutrient concentrations, low nitrogen-to-phosphorus ratios, low light levels, reduced mixing, and high temperatures, toxic cyanobacteria, such as Anabaena, *Cylindrospermopsis*, *Microcystis*, and *Oscillatoria* (Planktothrix), tend to dominate nutrient-rich freshwater systems (Downing et al., 2001; Paerl and Huisman, 2009; Paerl and Paul, 2012).

14.6.3 Effects on Water Quality

High-toxin algal blooms encourage the growth of more dangerous bacteria if anaerobic conditions prevail. Water bodies with low water quality and few viable uses are the outcome of these effects (Straskraba and Tundisi, 1999). When used as drinking water, high organic content gives the water unpleasant tastes or odours that chlorination can only just barely cover up. These pollutants combine to generate intricate chemical compounds that hinder regular purification procedures as well as accelerate corrosion and reduce flow rates by depositing on the inside walls of water purifier inlet tubes. The result is a severe decline in the quality of the water and a reduction in the supply of clean drinking water. Algal blooms and photosynthetic bacteria can develop densely and clog water systems, reducing the amount of piped water that is available. After an algal bloom, the decomposing debris also taints the desired water qualities and may encourage the growth of bacteria that cause illness. Rapid upwelling of a water body is caused by uncontrolled eutrophication (Ansari and Khan, 2009).

14.6.4 Effects on Human Health

Since the start of the second half of the 20th century, cyanobacteria have bloomed as the most prevalent harmful algae in many freshwater basins (Chorus and Bartram, 1999). These organisms can cause a wide range of environmental problems, including the formation of dense hyperscum mats (Zohary and Robarts, 1989), the production of impactful hepatotoxins and neurotoxins, and the death of livestock and other animals (Codd et al., 1997) and/or sometimes humans (Chorus and Bartram, 1999). Red tide is caused by dinoflagellates, which emit strong toxins into the water even at extremely low concentrations. The accelerated plant growth in the water creates anaerobic conditions that also cause the hazardous chemicals to multiply. The dangerous dinoflagellate *Alexandrium fundyense* affects other food sources like lobsters, fish, and marine mammals in the north-eastern United States and causes paralytic shellfish poisoning (PSP) (Anderson et al., 2008). Excessive nitrogen levels in drinking water have been linked to the tendency to reduce blood flow in new-borns, a condition known as "blue baby syndrome". Since Francis' (1878) initial observation of dead livestock associated with a cyanobacterial bloom, poisonings of domestic animals, wildlife, and even humans by blooms of toxic cyanobacteria have been observed around the world. Human sickness is caused by several HABs linked to eutrophication, including those in Narragansett Bay (Li and Smayda, 2000), Florida Bay (Glibert et al., 2004), the Texas coast (Buskey et al., 2001), and San Francisco Bay (Lehman et al., 2005).

14.6.5 Invasion of New Species

The species composition of an ecosystem changes as a result of the eutrophication process. For instance, a rise in nitrogen may encourage new, competitive species to invade and outcompete native ones. By making abundant a nutrient that is often scarce, eutrophication may lead to competitive release. During eutrophication, fish assemblages also shift, which may be a result of modifications to the plankton food web. In a similar manner, fish may have significant top-down influences on the plankton. High densities of planktivorous and omnivorous fish, including species that are voracious sight-feeding predators of zooplankton and filter-feeding omnivores like gizzard shad that eat benthos, phytoplankton, and zooplankton, are particularly supported by shallow eutrophic lakes (Jeppesen et al., 2007; Crisman and Beaver, 1990).

14.6.6 Effect on Climate

Greenhouse gas emissions (GHGE), which have risen significantly as a result of various anthropogenic activities, are major contributors to climate change. Freshwater ecosystems are crucial to these emissions (Tremblay, 2009). Due to shifting hydrology, wetlands, ponds, and special water bodies will be at risk. Out of all ecosystem types, freshwater aquatic environments appear to have the highest percentage of species threatened with extinction by climate change. Surface water quality and groundwater availability will be impacted by global warming, which will also put further stress on already overloaded water systems. A food web's stability, structure, interactions between predators and prey, and nutrient cycling are all significantly impacted by the coupling of habitats. The pelagic nutrient cycles, for instance, can be significantly aided by nutrient excretion by benthic macro- and meiofauna. Benthic resources also support carnivore populations, which have significant predatory consequences for plankton communities. These habitat relationships may be significantly changed by anthropogenic disturbances like eutrophication and climate change (Schindler and Scheuerell, 2002) (Figure 14.5).

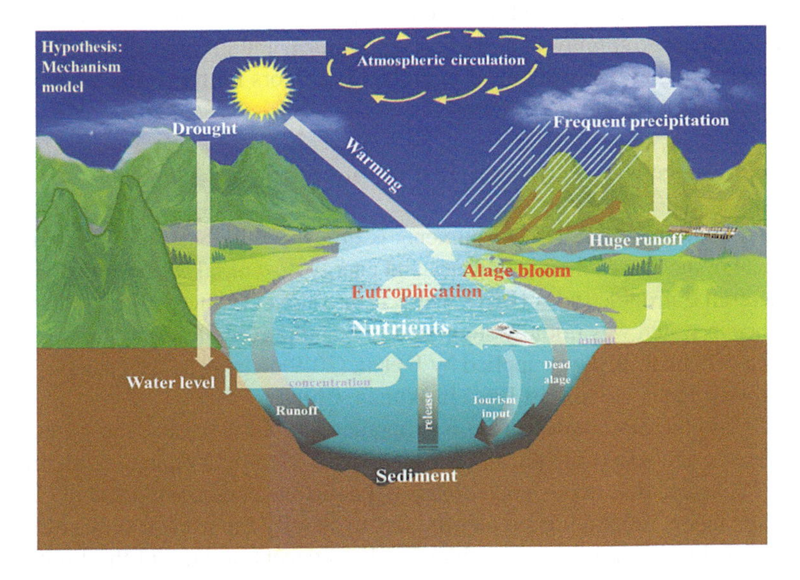

FIGURE 14.5 Diagrammatical representation of effect of eutrophication on climate. (From Lu, X. et al., *Journal of Environmental Sciences*, 75, 359–369, 2019.)

14.7 MICROBIAL IMPLICATIONS OF EUTROPHICATION

Microorganisms have a significant impact on the natural cycling of organic carbon, nitrogen, and phosphorus, as well as the features of water quality and the health of the aquatic environment. The makeup of the microbial population may significantly change as a result of eutrophication, which could have an impact on how the microbial loop functions and in turn, how the entire aquatic food web functions (Kiersztyn et al., 2019). Microorganisms, an essential component of the aquatic environment, are crucial for the movement and transformation of nutrients, organic matter, and other elements along the water–sediment interface. When the external environment in the aquatic environment changes, the species and diversity of the microbial community will also change. Microorganisms can affect the nutritional status in the water environment. The composition and functioning of the microbial biofilms that inhabit the rocks and sediments also provide evidence of the altered trophic status (Horn et al., 2010). The proportion of autotrophic to heterotrophic microbial activities declines significantly with increasing eutrophication (Reil et al., 2000). Visible indicators of eutrophication include drifting filamentous macroalgae, mats of sulfate-oxidising bacteria, and anaerobic phototrophic bacteria.

Cyanobacterial algal blooms, contaminated drinking water supplies, diminished recreational options, and hypoxia are some of the known effects of cultural eutrophication. In aquatic biota, microbes are often very sensitive to and profoundly affected by environmental disturbances. The vast majority of productivity and nutrient cycling in aquatic systems is carried out by microbes, which also make up the majority of the aquatic biomass. They adapt quickly to other physical, chemical, and biotic environmental changes as well as modest amounts of pollution. They provide sensitive, significant, and measurable indicators of ecological change from the detection and impact aspects.

The development and application of a new generation of environmental stress and ecological change indicators are being led by aquatic microbial ecology. The "arsenal" of molecular and chemotaxonomic identification, quantification, and characterisation approaches (such as species-specific rate measurements, biomass-specific rates of production, and nutrient conversions) holds great promise. The use of these methods at the ecosystem scale, where climate and human perturbations frequently coincide and interact, is emerging. Complementary usage of these techniques in dynamic estuarine and coastal ecosystems offers considerable potential in ensuring their effectiveness as qualitative and quantitative indicators suited for a wide variety of ecological applications. It has been extensively reported that the most prevalent phyla in high-quality freshwater are Actinobacteria (such as *Streptomyces* and *Micromonospora*) and Proteobacteria (Staley et al., 2013; Ji et al., 2018). Some other implications of microbial indicators for the assessment of water quality and ecology of aquatic systems are as follows:

a) Ecological change may be characterised in various aquatic ecosystems using microbial indicators.
b) The water quality of aquatic environments is influenced by bacteria, which also actively participate in the biogeochemical cycles and energy flow of lakes (Newton et al., 2011).
c) The Bacterial Eutrophic Index (BEI), which is the ratio of Cyanobacteria to Actinobacteria abundance in water, was initially suggested to quantitatively describe the water quality of a freshwater habitat. BEI was applied by Ji et al. (2019) to analyse the bacterial community and eutrophic index of the East Lake in Wuhan, China.
d) Carlson's Trophic State Index (TSI) is a popular tool for assessing the general health or trophic status of a lake (Carlson, 1977; Wen et al., 2019). Algal biomass with three components—TP, Secchi disc depth, and chlorophyll-a—is the main component of Carlson's trophic status index. Employing simpler metrics to classify the trophic status of freshwaters, such as Secchi disc transparency, chlorophyll-a concentration, and the limiting nutrient

concentration (often P or N) provides valuable insights into the overall nutrient levels, productivity, and ecological health of these aquatic systems (OECD 1982).

14.8 PREVENTION OF EUTROPHICATION

14.8.1 Reducing the Use of Fertilisers

Increased farm productivity and increased fertiliser use are required due to the growing population. It is anticipated that the usage of fertilisers will rise by 40% between 2002 and 2030 (FAO, 2000). The use of nitrate and phosphate fertilisers is the main cause of eutrophication. Composting can be used as a remedy to try to alleviate the problem. Composting is the process of turning organic materials, such as wasted food and decomposing plant matter, into compost manure. The cycle of eutrophication is prevented by compost fertiliser because all the needed elements are broken down and synthesised by the plants. Nutrient limitation is the name given to this strategy for reducing eutrophication.

14.8.2 Reducing Discharge of Waste into Waterbodies

According to estimates, sewage contributes 12% of the nitrogen added to rivers in the United States, 25% in Western Europe, 33% in China, and 68% in the Republic of Korea (MA, 2005). Urban trash is handled in sewage treatment facilities that operate under the premise of organic matter being oxidised by bacteria. All of the waste's principal components are oxidised in this fashion. The effluent from the treatment plant contains a high quantity of these elements, since they are now soluble. These effluents are phosphorus and nitrogen point sources. The more significant sources of industrial nutrient contamination include pulp and paper mills, food and meat processing facilities, agro industries, and direct sewage discharge. The primary urban sources of nutrient overload are industrial wastes and home sewage, which, combined, account for 50% of the total quantity of phosphorus discharged into lakes from human settlements (Smith et al., 2006). Phosphorus-rich wastewater effluents from about 15% of the U.S. population contribute to lake eutrophication (Hammer, 1986). The nutrient content in water systems can be decreased, which can then control eutrophication, if industries and municipalities can limit their waste discharge and pollution to a lower level. Large manufacturing companies and municipalities should stop discharging waste into water systems and reduce pollution in order to minimise the amount of pollutants and nutrients that end up in the waters. It is possible to modify sewage treatment facilities to remove nutrients biologically, which would result in substantially lower nitrogen and phosphorus discharges to receiving aquatic systems. Regulations governing the treatment and release of sewage have resulted in significant nutrient reductions for adjacent ecosystems (Figure 14.6).

14.8.3 Nutrient Bioextraction or Bioharvesting

The technique of raising and collecting shellfish and seaweed for the purpose of extracting nitrogen and other nutrients from natural water bodies is known as nutrient bioextraction or bioharvesting. Similarly to other nutrient trading scenarios, it has been proposed that nitrogen removal by oyster reefs could result in net advantages for sources facing nitrogen emission limitations. Specifically, oysters effectively save the sources from the compliance costs they would otherwise incur by keeping nitrogen levels in estuaries below thresholds that would result in the enforcement of emission controls. Several studies have demonstrated that oysters and mussels have the power to significantly alter estuary nitrogen levels. According to reports, duckweeds are promising macrophytes for the treatment of sewage. They have been used to treat prawn farm effluent and have been successful in removing nutrients and significant levels of ammonia (Ruenglertpanyakul et al., 2004). It has been discovered that a number

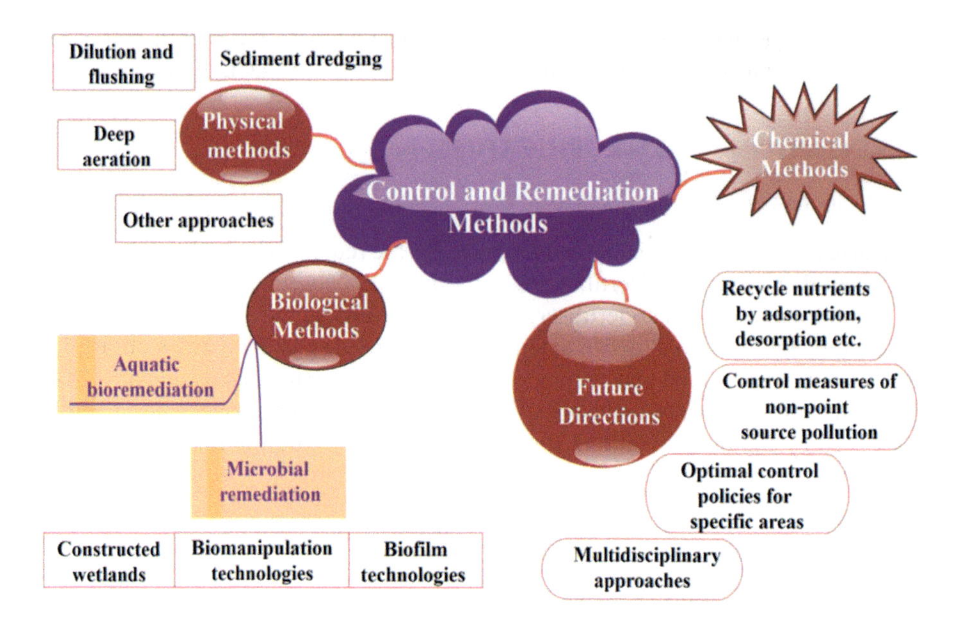

FIGURE 14.6 Strategies employed to tackle and reduce the problems associated with eutrophic lakes during the last three decades. (From Zhan et al., 2020.) Zhang Y, Luo P, Zhao S, Kang S, Wang P, Zhou M, Lyu J. Control and remediation methods for eutrophic lakes in the past 30 years. Water Sci Technol. 2020 Mar;81(6):1099-1113. doi: 10.2166/wst.2020.218. PMID: 32597398

of plant species can minimise the excess nitrogen and phosphorus in aquatic systems. *Eichhornia crassipes* and *Salvinia auriculata* are two examples of aquatic macrophytes that significantly reduce the levels of nitrogen and phosphorus compounds in water. The development of an appropriate management strategy for aquatic macrophytes to stop the eutrophication process in Imboassica Lagoon was deemed to be aided by this information (Petrucio and Esteves, 2000). According to Jiang et al. (2004), *Phragmites communis* and *Zizania latifolia* are effective in absorbing N and P, and as a result, these two species were discovered to be crucial in the purification of wetlands receiving non-point source pollutants. *Z. latifolia* has a greater ability for absorption and breakdown than *P. communis* (Jiang et al., 2004). Duckweeds speed up the decomposition of organic matter (Körner et al., 2003). The ability and relative contribution of the roots and fronds of the floating macrophyte *Lemna minor* for N uptake were examined by Cedergreen and Madsen in 2004. They demonstrated that *L. minor* roots and leaves can absorb a sizable amount of inorganic N from both roots and fronds (Figure 14.7).

The oyster aquaculture sector in Connecticut reduces nutrient loads to the tune of $8.5 to $23 million annually, according to a ground-breaking modelling experiment in Long Island Sound. The experiment also shows that a reasonable increase in oyster aquaculture may reduce nutrient levels to some extent. Using aquaculture modelling tools, the scientists showed that shellfish aquaculture compares favourably with existing nutrient management options in terms of nutrient removal effectiveness and implementation cost.

14.8.4 Improving Legislation and Rules Against Non-Point Source Pollution

Eutrophication can be effectively managed through the strengthening of rules and regulations against non-point water source pollution. The management of nutrient infiltration into water systems is most severely hampered, according to EPA (1986), by non-point pollution. Hence, reducing eutrophication is achieved by managing nutrient supplies. High requirements for water quality and a zero-tolerance policy for non-point solutions should be the focus of the laws. Eutrophication is

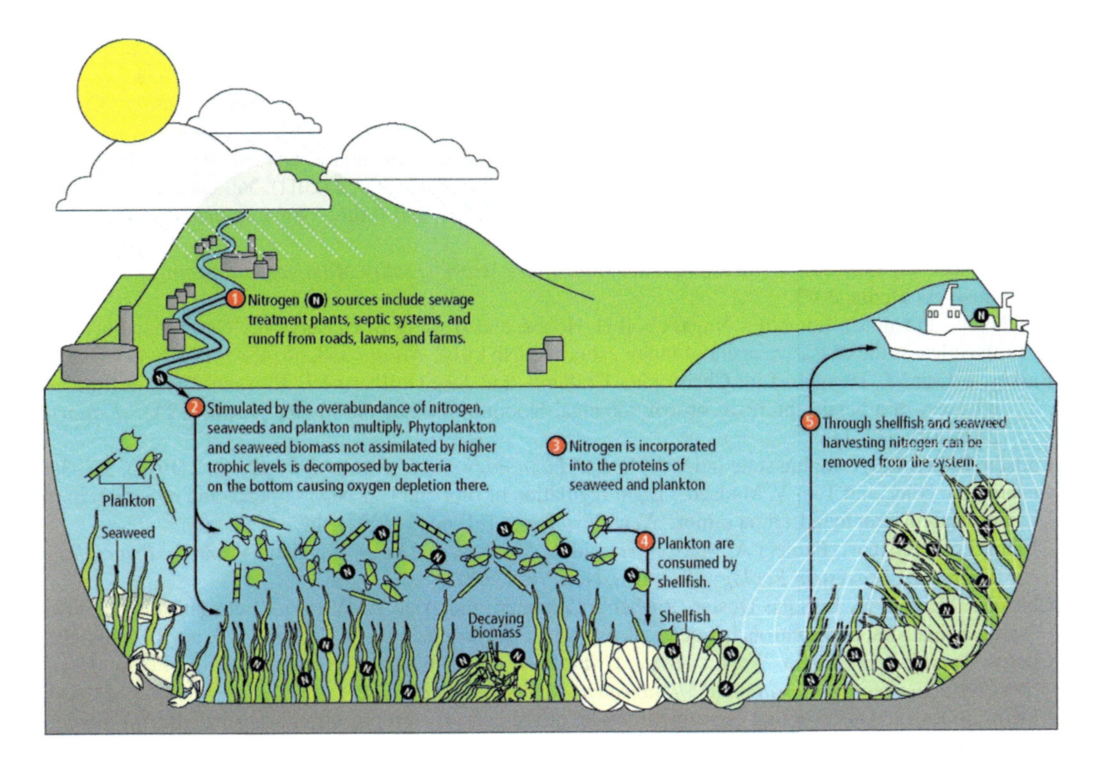

FIGURE 14.7 An illustration demonstrates how shellfish and seaweed can be utilised to remove nitrogen from coastal waters.

simple to manage with the help of government officials, citizens, pollution regulation agencies, and the legal system. Only a concerted community effort can more effectively reduce nutrient inputs to water bodies, as in the case of Lake Washington, where a decrease in detergent use was the result of public awareness. Local people's awareness of their surrounding environment and water sources has an important and long-lasting impact on the environment. Thus, it is crucial to design an integrated plan to reduce eutrophication as well as to raise public awareness and environmental education among individuals (Jorgensen, 2001).

Eutrophication can be prevented and reversed by taking steps to reduce nutrient pollution from agriculture and other non-point sources as well as point source pollution from sewage. Other technologies and practices, some of which are still in the experimental stage for prevention of eutrophication include raising shellfish in estuaries, seaweed, geo-engineering in lakes (chemical removal of phosphorus), ultrasonic irradiation, phosphorus removal by chemical precipitation, treatment of hypolimnetic water, and oxygenation of water.

REFERENCES

Addy, Kelly, and Linda Taylor Green. *Phosphorus and Lake Aging.* University of Rhode Island, College of Resource Development, Department of Natural Resources Science, 1996.

Anderson, Donald M., Patricia M. Glibert, and Joann M. Burkholder. "Harmful algal blooms and eutrophication: Nutrient sources, composition, and consequences." *Estuaries* 25 (2002): 704–726.

Anderson, N. John, Robert H. Foy, Daniel R. Engstrom, Brian Rippey, and Farah Alamgir. "Climate forcing of diatom productivity in a lowland, eutrophic lake: White Lough revisited." *Freshwater Biology* 57, no. 10 (2012): 2030–2043.

Andersson, Lars, and Lars Rydberg. "Exchange of water and nutrients between the Skagerrak and the Kattegat." *Estuarine, Coastal and Shelf Science* 36, no. 2 (1993): 159–181.

Ansari, A. A., and Fareed A. Khan. "Remediation of eutrophied water using Spirodelapolyrrhiza L. Shleid in controlled environment." *Pan-American Journal of Aquatic Sciences* 4, no. 1 (2009): 52–54.
Ansari, Abid A., Gill Sarvajeet Singh, Guy R. Lanza, and Walter Rast, eds. *Eutrophication: Causes, Consequences and Control.* Vol. 1. Springer Science & Business Media, 2010.
Arend, Kristin K., Dmitry Beletsky, Joseph V. DePinto, et al. "Seasonal and interannual effects of hypoxia on fish habitat quality in central Lake Erie." *Freshwater Biology* 56, no. 2 (2011): 366–383.
Beeton, Alfred M. "Large freshwater lakes: Present state, trends, and future." *Environmental Conservation* 29, no. 1 (2002): 21–38.
Boyd, Claude E., and Craig S. Tucker. *Pond Aquaculture Water Quality Management.* Springer Science & Business Media, 2012.
Burkholder, JoAnn M., Edward J. Noga, Cecil H. Hobbs, and Howard B. Glasgow Jr. "New 'phantom' dinoflagellate is the causative agent of major estuarine fish kills." *Nature* 358, no. 6385 (1992): 407–410.
Buskey, Edward J., Hongbin Liu, Christopher Collumb, and Jose Guilherme F. Bersano. "The decline and recovery of a persistent Texas brown tide algal bloom in the Laguna Madre (Texas, USA)." *Estuaries* 24 (2001): 337–346.
Carlson, Robert E. "A trophic state index for lakes." *Limnology and Oceanography* 22, no. 2 (1977): 361–369.
Cedergreen, Nina, and Tom V. Madsen. "Light regulation of root and leaf NO3− uptake and reduction in the floating macrophyte Lemna minor." *New Phytologist* 161, no. 2 (2004): 449–457.
Chapin, Francis Stuart, Pamela A. Matson, Harold A. Mooney, and Peter Morrison Vitousek. *Principles of Terrestrial Ecosystem Ecology.* Springer, New York, 2002.
Chislock, Michael F., Enrique Doster, Rachel A. Zitomer, and Alan E. Wilson. "Eutrophication: Causes, consequences, and controls in aquatic ecosystems." *Nature Education Knowledge* 4, no. 4 (2013): 10.
Chorus, Ingrid, and Martin Welker. *Toxic Cyanobacteria in Water: A Guide to Their Public Health Consequences, Monitoring and Management.* Taylor & Francis, 2021.
Cloern, James E. "Our evolving conceptual model of the coastal eutrophication problem." *Marine Ecology Progress Series* 210 (2001): 223–253.
Codd, G. A., K. A. Beattie, and S. L. Raggett. "The evaluation of Envirogard Microcystin plate and tube kits." *Environment Agency* (1997): 47.
Cox, Evelyn F., Marta Ribes, and Robert A. Kinzie III. "Temporal and spatial scaling of planktonic responses to nutrient inputs into a subtropical embayment." *Marine Ecology Progress Series* 324 (2006): 19–35.
Crisman, Thomas L., and John R. Beaver. "Applicability of planktonic biomanipulation for managing eutrophication in the subtropics." In *Biomanipulation Tool for Water Management: Proceedings of an International Conference held in Amsterdam*, The Netherlands, 8–11 August, 1989, pp. 177–185. Springer Netherlands, 1990.
Diaz, Robert J., and Rutger Rosenberg. "Spreading dead zones and consequences for marine ecosystems." *Science* 321, no. 5891 (2008): 926–929.
Dodds, Walter K., Wes W. Bouska, Jeffrey L. Eitzmann, et al. "Eutrophication of US freshwaters: Analysis of potential economic damages." *Environmental Science & Technology* 43 (2009): 12–19.
Downing, John A., Susan B. Watson, and Edward McCauley. "Predicting cyanobacteria dominance in lakes." *Canadian Journal of Fisheries and Aquatic Sciences* 58, no. 10 (2001): 1905–1908.
Elser, James J., Matthew E. S. Bracken, Elsa E. Cleland, et al. "Global analysis of nitrogen and phosphorus limitation of primary producers in freshwater, marine and terrestrial ecosystems." *Ecology Letters* 10, no. 12 (2007): 1135–1142.
Flemer, David A., Gail B. Mackiernan, Willa Nehlsen, Virginia K. Tippie, Robert B. Biggs, Dewey Blaylock, Ned H. Burger, et al. "Chesapeake Bay: A profile of environmental change." US Environmental Protection Agency, Washington, DC, 1983.
North Sea Task Force. *North Sea Subregion 6: Assessment Report 1993.* State Pollution Control Authority, 1993.
Fowler, David, Mhairi Coyle, Ute Skiba, Mark A. Sutton, J. Neil Cape, Stefan Reis, Lucy J. Sheppard, et al. "The global nitrogen cycle in the twenty-first century." *Philosophical Transactions of the Royal Society B: Biological Sciences* 368, no. 1621 (2013): 20130164.
Francis, George. "Poisonous Australian lake." *Nature* 18, no. 444 (1878): 11–12.
Furnas, Miles, Alan Mitchell, Michele Skuza, and Jon Brodie. "In the other 90%: Phytoplankton responses to enhanced nutrient availability in the Great Barrier Reef Lagoon." *Marine Pollution Bulletin* 51, no. 1–4 (2005): 253–265.
Gin, Karina Yew-Hoong, Xiaohua Lin, and Sheng Zhang. "Dynamics and size structure of phytoplankton in the coastal waters of Singapore." *Journal of Plankton Research* 22, no. 8 (2000): 1465–1484.

Glibert, Patricia M., and Michele A. Burford. "Globally changing nutrient loads and harmful algal blooms: Recent advances, new paradigms, and continuing challenges." *Oceanography* 30, no. 1 (2017): 58–69.

Glibert, P. M., Cynthia A. Heil, D. Hollander, M. Revilla, A. Hoare, J. Alexander, and S. Murasko. "Evidence for dissolved organic nitrogen and phosphorus uptake during a cyanobacterial bloom in Florida Bay." *Marine Ecology Progress Series* 280 (2004): 73–83.

Hammer, Mark J. *Water and Wastewater Technology*. John Wiley, New York, 1986.

Harrison, Paul. *World Agriculture: Towards 2015/2030*. Summary report. FAO, 2002.

Jacquet, Séverine, Bruno Delesalle, Jean-Pascal Torréton, and Jean Blanchot. "Response of phytoplankton communities to increased anthropogenic influences (southwestern lagoon, New Caledonia)." *Marine Ecology Progress Series* 320 (2006): 65–78.

Jeppesen, Erik, Martin Søndergaard, Brian Kronvang, Jens P. Jensen, Lars M. Svendsen, and Torben L. Lauridsen. "Lake and catchment management in Denmark." In *The Ecological Bases for Lake and Reservoir Management: Proceedings of the Ecological Bases for Management of Lakes and Reservoirs Symposium*, held 19–22 March 1996, Leicester, pp. 419–432. Springer Netherlands, 1999.

Ji, Bin, Jiechao Liang, Yingqun Ma, Lin Zhu, and Yu Liu. "Bacterial community and eutrophic index analysis of the East Lake." *Environmental Pollution* 252 (2019): 682–688.

Jiang, C. L., G. B. Cui, X. Q. Fan, and Y. B. Zhang. "Purification capacity of ditch wetland to agricultural non-point pollutants." *Huan Jing ke Xue = HuanjingKexue* 25, no. 2 (2004): 125–128.

Khan, Fareed A., and Abid Ali Ansari. "Eutrophication: An ecological vision." *The Botanical Review* 71, no. 4 (2005): 449–482.

Khan, M. Nasir, and Firoz Mohammad. "Eutrophication: Challenges and solutions." *Eutrophication: Causes, Consequences and Control* 2 (2014): 1–15.

Kiersztyn, Bartosz, Ryszard Chróst, Tomasz Kaliński, Waldemar Siuda, Aleksandra Bukowska, Grzegorz Kowalczyk, and Karolina Grabowska. "Structural and functional microbial diversity along a eutrophication gradient of interconnected lakes undergoing anthropopressure." *Scientific Reports* 9, no. 1 (2019): 11144.

Kim, Jang K., George P. Kraemer, and Charles Yarish. "Use of sugar kelp aquaculture in Long Island Sound and the Bronx River Estuary for nutrient extraction." *Marine Ecology Progress Series* 531 (2015): 155–166.

Kim, H. G., J. S. Park, S. G. Lee, and K. H. An. "Population cell volume and carbon content in monospecific dinoflagellate blooms." In *Toxic Phytoplankton Blooms in the Sea* (eds. Smayda, T.J. and Shimizu, Y.). Elsevier, 1993, pp. 769–773.

Kirby, Michael X., and Henry M. Miller. "Response of a benthic suspension feeder (Crassostrea virginica Gmelin) to three centuries of anthropogenic eutrophication in Chesapeake Bay." *Estuarine, Coastal and Shelf Science* 62, no. 4 (2005): 679–689.

Körner, Sabine, Jan E. Vermaat, and Siemen Veenstra. "The capacity of duckweed to treat wastewater: Ecological considerations for a sound design." *Journal of Environmental Quality* 32, no. 5 (2003): 1583–1590.

Lehman, P. W., Greg Boyer, Catherine Hall, Scott Waller, and Karen Gehrts. "Distribution and toxicity of a new colonial Microcystis aeruginosa bloom in the San Francisco Bay Estuary, California." *Hydrobiologia* 541 (2005): 87–99.

Lehtiniemi, Maiju, Jonna Engström-Öst, and Markku Viitasalo. "Turbidity decreases anti-predator behaviour in pike larvae, Esox lucius." *Environmental Biology of Fishes* 73 (2005): 1–8.

Lu, Xiaotian, Yonglong Lu, Deliang Chen, et al. "Climate change induced eutrophication of cold-water lake in an ecologically fragile nature reserve." *Journal of Environmental Sciences* 75 (2019): 359–369.

Li, Yaqin, and Theodore J. Smayda. "Heterosigmaakashiwo (Raphidophyceae): On prediction of the week of bloom initiation and maximum during the initial pulse of its bimodal bloom cycle in Narragansett Bay." *Plankton Biology and Ecology* 47, no. 2 (2000): 80–84.

Lindeman, Raymond L. "The trophic-dynamic aspect of ecology." *Ecology* 23, no. 4 (1942): 399–417.

Meyer-Reil, Lutz-Arend, and Marion Köster. "Eutrophication of marine waters: Effects on benthic microbial communities." *Marine Pollution Bulletin* 41, no. 1–6 (2000): 255–263.

Miralto, A., A. Ianora, S. A. Poulet, G. Romano, I. Buttino, and S. Scala. "Embryonic development in invertebrates is arrested by inhibitory compounds in diatoms." *Marine Biotechnology* 1, no. 4 (1999): 401.

Morris Jr., J. Glenn. "Harmful algal blooms: An emerging public health problem with possible links to human stress on the environment." *Annual Review of Energy and the Environment* 24, no. 1 (1999): 367–390.

Moss, Brian. "The Norfolk Broadland: Experiments in the restoration of a complex wetland." *Biological Reviews* 58, no. 4 (1983): 521–561.

Mukhopadhay, K., Ranjan Bera, and Ratneshwar Roy. "Modern agriculture and environmental pollution." *Everyman's Science* (2005): 186–192.
Namsaraev, Zorigto, Anna Melnikova, Vasiliy Ivanov, Anastasia Komova, and Anton Teslyuk. "Cyanobacterial bloom in the world largest freshwater lake Baikal." In *IOP Conference Series: Earth and Environmental Science*, vol. 121, p. 032039. IOP Publishing, 2018.
Newton, Ryan J., Stuart E. Jones, Alexander Eiler, Katherine D. McMahon, and Stefan Bertilsson. "A guide to the natural history of freshwater lake bacteria." *Microbiology and Molecular Biology Reviews* 75, no. 1 (2011): 14–49.
Paerl, Hans W. "Coastal eutrophication and harmful algal blooms: Importance of atmospheric deposition and groundwater as 'new' nitrogen and other nutrient sources." *Limnology and Oceanography* 42, no. 5part2 (1997): 1154–1165.
Paerl, Hans W. "Nuisance phytoplankton blooms in coastal, estuarine, and inland waters." *Limnology and Oceanography* 33, no. 4part2 (1988): 823–843.
Paerl, Hans W., and Jef Huisman. "Climate change: A catalyst for global expansion of harmful cyanobacterial blooms." *Environmental Microbiology Reports* 1, no. 1 (2009): 27–37.
Paerl, Hans W., and Valerie J. Paul. "Climate change: Links to global expansion of harmful cyanobacteria." *Water Research* 46, no. 5 (2012): 1349–1363.
Parry, Roberta. "Agricultural phosphorus and water quality: A US Environmental Protection Agency perspective." *Journal of Environmental Quality* 27, no. 2 (1998): 258–261.
Petrucio, M. M., and F. A. Esteves. "Uptake rates of nitrogen and phosphorus in the water by Eichhornia crassipes and Salvinia auriculata." *Revista Brasileira de Biologia* 60 (2000): 229–236.
Porter, Karen Glaus. "The plant-animal interface in freshwater ecosystems: Microscopic grazers feed differentially on planktonic algae and can influence their community structure and succession in ways that are analogous to the effects of herbivores on terrestrial plant communities." *American Scientist* 65, no. 2 (1977): 159–170.
Ptacnik, R., L. Lepistö, E. Willén, P. Brettum, T. Andersen, S. Rekolainen, A. Lyche Solheim, and Laurence Carvalho. "Quantitative responses of lake phytoplankton to eutrophication in Northern Europe." *Aquatic Ecology* 42 (2008): 227–236.
Qin, Boqiang, Liuyan Yang, Feizhou Chen, Guangwei Zhu, Lu Zhang, and Yiyu Chen. "Mechanism and control of lake eutrophication." *Chinese Science Bulletin* 51 (2006): 2401–2412.
Rabalais, Nancy N. "Nitrogen in aquatic ecosystems." *AMBIO: A Journal of the Human Environment* 31, no. 2 (2002): 102–112.
Ramachandra, T. V., and Malvikaa Solanki. "Ecological assessment of lentic water bodies of Bangalore." *The Ministry of Science and Technology* 25 (2007): 96.
Romdhane, Mohamed S., Hans C. Eilertsen, O. Kefi Daly Yahia, and M. N. Daly Yahia. "Toxic dinoflagellate blooms in Tunisian lagoons: Causes and consequences for aquaculture." In *Harmful Algae* (eds Reguera, B., Blanco, J., Fernândez, L. and Wyatt, T.), 1998, pp. 80–83.
Ruenglertpanyakul, W., S. Attasat, and P. Wanichpongpan. "Nutrient removal from shrimp farm effluent by aquatic plants." *Water Science and Technology* 50, no. 6 (2004): 321–330.
Sachoemar, Suhendar I., and Tetsuo Yanagi. "Seasonal variation in water quality at the northern coast of Karawang – West Java, Indonesia." *Umi/la mer. Tokyo* 37, no. 3 (1999): 91–101.
Sawyer, Clair N. "Basic concepts of eutrophication." *Journal (Water Pollution Control Federation)* 38 (1966): 737–744.
Scheuerell, Mark D., Daniel E. Schindler, Arni H. Litt, and W. T. Edmondson. "Environmental and algal forcing of Daphnia production dynamics." *Limnology and Oceanography* 47, no. 5 (2002): 1477–1485.
Schindler, David W. "The dilemma of controlling cultural eutrophication of lakes." *Proceedings of the Royal Society B: Biological Sciences* 279, no. 1746 (2012): 4322–4333.
Schindler, David W., and John R. Vallentyne. *The Algal Bowl: Overfertilization of the World's Freshwaters and Estuaries*. University of Alberta, 2008. ISBN 0-88864-484-1
Sellner, K. G., M. M. Olson, and K. Olli. "Copepod interactions with toxic and non-toxic cyanobacteria from the Gulf of Finland." *Phycologia* 35, no. sup6 (1996): 177–182.
Smith, Val H., and David W. Schindler. "Eutrophication science: Where do we go from here?." *Trends in Ecology & Evolution* 24, no. 4 (2009): 201–207.
Smith, Val H., G. David Tilman, and Jeffery C. Nekola. "Eutrophication: Impacts of excess nutrient inputs on freshwater, marine, and terrestrial ecosystems." *Environmental pollution* 100, no. 1–3 (1999): 179–196.
Smith, Val H., Samantha B. Joye, and Robert W. Howarth. "Eutrophication of freshwater and marine ecosystems." *Limnology and Oceanography* 51, no. 1part2 (2006): 351–355.

Tada, Kuninao, Kazuhiko Sakai, Yoshikatsu Nakano, Akihiro Takemura, and Shigeru Montani. "Size-fractionated phytoplankton biomass in coral reef waters off Sesoko Island, Okinawa, Japan." *Journal of Plankton Research* 25, no. 8 (2003): 991–997.

Tomec, M., I. Ternjej, M. Kerovec, E. Teskeredzic, and M. Mestrov. "Plankton in the oligotrophic lake Vrana (Croatia)." *Biologia* 57, no. 5 (2002): 579–588.

Turner, Andrew M., and Michael F. Chislock. "Blinded by the stink: Nutrient enrichment impairs the perception of predation risk by freshwater snails." *Ecological Applications* 20, no. 8 (2010): 2089–2095.

Van Duyl, F., G. Gast, W. Steinhoff, S. Kloff, M. Veldhuis, and R. Bak. "Factors influencing the short-term variation in phytoplankton composition and biomass in coral reef waters." *Coral Reefs* 21 (2002): 293–306.

van Leeuwen, Cornelis Johannes, and Theodorus Gabriel Vermeire, eds. *Risk Assessment of Chemicals: An Introduction*. Vol. 94. Dordrecht: Springer, 2007.

Walker, I. R. R. Chironomid. In *Encyclopedia of Quaternary Science*, 1, no. 1. S. A. Elias (Ed.), Elsevier: Amsterdam, The Netherlands, 2006, pp. 360–366.

Wen, Zhidan, Kaishan Song, Ge Liu, Yingxin Shang, Chong Fang, Jia Du, and Lili Lyu. "Quantifying the trophic status of lakes using total light absorption of optically active components." *Environmental Pollution* 245 (2019): 684–693.

Wetzel, R. G. *Limnology*. Philadelphia: WB Sunders Company Pub 740 (1975): 65. ISBN 0-7216-9240-0.

Whiteside, M. C. "The mythical concept of eutrophication." In *Paleolimnology: Proceedings of the Third International Symposium on Paleolimnology*, held at Joensuu, Finland, pp. 107–111. Springer Netherlands, 1983.

Widuto, J. "Zooplankton in an artificially aerated Lake Mutek in the Period 1977–1980." *Roczn. Nauk. roln. H, Rybactwo* 101 (1988): 173–186.

Young, Hannah. "Removal and reuse of phosphorus as a fertilizer from CAFO runoff." Chemical Engineering Undergraduate Honors Theses. 146, 2019.

Zhang, Yuan, Pingping Luo, Shuangfeng Zhao, Shuxin Kang, Pengbo Wang, Meimei Zhou, and Jiqiang Lyu. "Control and remediation methods for eutrophic lakes in the past 30 years." *Water Science and Technology* 81, no. 6 (2020): 1099–1113.

Zhou, Chuanqiao, Yu Peng, Li Chen, Miaotong Yu, Muchun Zhou, Runze Xu, Lanqing Zhang, et al. "Rapidly increasing sulfate concentration: A hidden promoter of eutrophication in shallow lakes." *Biogeosciences* 19, no. 17 (2022): 4351–4360.

Zohary, Tamar, and Richard D. Robarts. "Diurnal mixed layers and the long-term dominance of Microcystis aeruginosa." *Journal of Plankton Research* 11, no. 1 (1989): 25–48.

15 Bacterial Biofilm in the Aquatic Environment and Its Impact

Pramod Kumar Pandey

15.1 INTRODUCTION

Aquaculture is the most important food-producing sector, with 40% of the world's population relying on it for protein. Aquaculture accounts for roughly 46% of global fish output (179 million tonnes) and 52% of human fish consumption. Capture fish production has dropped dramatically in recent decades, and many marine fish stocks are already in decline, making aquaculture an attractive option for meeting rising protein demand, particularly in developing countries (FAO, 2020). When land and water resources are limited, intensive aquaculture operations are the only way to fulfil rising protein demands. However, as a result of intensification, problems such as pollution, sickness, and environmental consequences have shrunk (Beveridge et al., 1997; Priedahitra, 2003). As a result, a sustainable aquaculture technology is required to increase fish production without affecting the environment. At the same time, the technology should be economically viable, especially one that reduces feed usage. Feed makes up a large portion of the operating costs in intensive aquaculture, and excessive feeding results in increased pollution, disease incidence, and financial loss to the industry (Priedahitra, 2003). Biofilm-based aquaculture, which is applicable to both freshwater and brackish water aquaculture, has gained popularity in recent years. It has proven to be a successful nitrogen control strategy for long-term aquaculture (Crab et al., 2007). It effectively recycles waste nutrients into fish feed, lowering feed costs and reducing both in-situ and ex-situ nutrient pollution (Kumar et al., 2017). The well-being of the cultivated animals depends heavily on maintaining optimal water quality. Biofilm is a microbial community linked to submerged substrates and coupled with a matrix of extracellular polymeric molecules actively involved in nutrient recycling of the aquatic ecosystem. The bacteria present in the biofilm were found to be effective in recycling the nutrients and removing unwanted nutrients from the system (Meyer-Reil, 1994). The system also uses nitrification and heterotrophic microorganisms to minimise the hazardous inorganic nitrogen in the culture system (Avnimelech, 2007). The installation of an Aquamat™/ substrate for biofilm formation also promotes fish/shrimp growth and survival. It's also a great place for freshwater prawns and shrimp to hide when they're moulting. The presence of microalgae in the biofilm provides a good source of nutrition for the cultured animals, lowering the fiber carpet (FC) (Anand et al., 2013a,b; Kumar et al., 2017). When compared with a conventional aquaculture system, the economic study of biofilm-based aquaculture revealed a higher profit margin (Kumar et al., 2019). Furthermore, manipulating the C:N ratio and biofilm growth in substrate improves the well-balanced growth of both heterotrophic and autotrophic communities in the biofilm (Azim and Asaeda, 2005). Despite the advantages of biofilm-based aquaculture, it has yet to be commercialised, with a few exceptions. As a result, the current chapter attempts to compile studies on biofilm-based aquaculture (both freshwater and brackish water) in India and other areas of the world over the previous few decades, as well as viable candidate species and the role of biofilm in sustainable aquaculture production.

 DOI: 10.1201/9781003408543-15

15.2 BIOFILM

Organisms adhering to submerged substrates are referred to by a variety of terms. The organisms discovered clinging to submerged substrates, which might be natural or manufactured materials, are referred to as biofilm. Algae, heterotrophic and autotrophic bacteria, and other flora and fauna clinging to the substrates are among these organisms. In aquatic ecosystems, it is crucial for nutrient cycling, energy transport, and trophic transmission (Azim et al., 2005). "Aufwuchs" and "Periphyton" are two other terms. The only distinction between Aufwuchs and periphyton/ biofilm is that the former includes non-attached fauna associated with the biofilm, while the latter does not (Van Dam et al., 2002). The microfloral community associated with submerged substrates in water is referred to as periphyton in limnology (Wetzel, 1983). All creatures that are adhered to or move on submerged substrates in water are referred to as "Aufwuchs" in German. Algae, fungi, bacteria, protozoa, and other animal components are included (Azim et al., 2005). Attached algae, bacteria, protozoa, zooplankton, and other invertebrates make up the biofilm community. Biofilm is a term that is used in a variety of industries, including medicine, food processing, wastewater treatment, and drinking water technology. Biofilm formation includes the following steps. Electrostatic forces will cause the deposition of dissolved organic molecules such as mucopolysaccharides on the surface of the substrates at first. After that, bacteria begin to adsorb onto the substrate and attach utilising mucus layers, forming aggregates, cell division, and colony formation across the substrate. Algae begin to grow on top of the bacterial layer after a few days to a week. Bacillariophyceae (14 genera), Chlorophyceae (12 genera), Cyanophyceae (10 genera), and Euglenophyceae (10 genera) are the 4 groups of algae (3 genera). Zooplankton will find their way into the substrate after algal settlement and consume the algal population. Rotifers, copepods, ciliates, daphnia, and moina are the most common zooplankton in periphyton-based aquaculture.

15.2.1 Organisms Associated with Biofilm

The taxonomic makeup of the biofilm is mostly determined by factors such as grazing pressure, water quality, and substrate kinds (Azim and Asaeda, 2005). Most studies on biofilm-based aquaculture have focused on the algal makeup of the biofilm, leaving heterotrophic species out of the equation. The taxonomic composition of diverse aquaculture systems with varying substrate types has been examined. Both algae and zooplankton make up the biofilm in *Litopenaeus vannamei* culture using Aquamat™ as a substrate. Bacillariophyceae (14 genera), Chlorophyceae (12 genera), Cyanophyceae (10 genera), and Euglenophyceae (3 genera) were discovered among the algae. The Cyanophyceae were the most common periphytic algae found near the Aquamat™. Amphora, Coscinodiscus, Cyclotella, Cymbella, Diatoma, Fragellaria, Melosira, Navicula, Nitzschia, Pinnularia, Pleurosigma, Skeletonema, Surirella, Synedra, Anabaena, Anacystis, Aphanothece, Gloeocapsa, Gomphosphaeria, Lyngbya, Microcystis, Nostoc, Oscillatoria, Phormidium, Spirulina, Chaetophora, Chlamydomonas, Chlorella, Closterium, Enteromorpha, Gonatozygon, Mougeotia, Oedogonium, Scenedesmus, Sphaerocystis, Volvox, Ulothrix, Euglena, and Gyrodinium were among the algal taxa identified. Zooplankton community include rotifers, copepods, ciliates, daphnia, and moina. Similar observations were made by Anand et al. (2019)in biofilm (bamboo as a substrate)-based *Penaeus monodon* culture. Similarly, in mullet culture, Bacillariophyceae was the most abundant group present in the substrate. Audelo-Naranjo et al. (2010) used artificial substrates (Aquamats™) for the intensive farming of shrimp at two densities and found that the periphyton was composed of primary producers such as diatoms and cyanobacteria; primary consumers such as rhizopods, heliozoans, ciliates, flagellates, foraminifera, copepods, rotifers, and gastrotriches; and some detritivorous metazoans such as nematodes and amphipods. Pennate diatoms (*Bacillaria paradoxa*, *Surirella* sp., and *Pleurosigma* sp.), centric diatoms (*Melosira dubia*), filamentous cyanobacteria (*Oscillatoria* sp.), ciliates (mainly from the vorticellid and tintinnid groups), and nematodes dominated the taxonomic composition of biofilm present in a pink shrimp nursery (Ballester et al., 2007). Similar observations were found during sea cucumber culture in the installed system. Apart from diatoms, crustose coralline algae, *Gracilaria* sp., *Sargassum* sp., three species of sea slugs, *Stylocheilus striatus*, *S. longicauda*, *Bursatella leachii*, and many macroalgae propagules were also found (Gorospe et al., 2019). Bacillariophyceae, Chlorophyceae, Cyanophyceae, and Euglenophyceae, as well as two zooplankton groups, Crustacea and Rotifera, make up the biofilm communities in freshwater aquaculture ponds. Chlorophyceae, on the other hand, dominated the other algae groupings (Azim et al., 2002a). Haque et al. (2015) discovered that Chlorophyceae were the most dominant group in biofilm in a bamboo-installed polyculture system (*Macrobrachium rosenbergii* + *Oreochromis niloticus*). Chlorella, Pediastrum, Scenedesmus, and Ceratium were the most common chlorophyceae algae. Other benthic macroinvertebrates found in the biofilm included chironomids, oligochaetes, and molluscs, in addition to algae. The dominance of algal groups in the biofilm was clearly demonstrated in previous research, and it changed depending on the nutritional content of the culture water as well as the salinity of the water.

15.2.2 Biochemical Composition of Biofilm

Both cellular biomass and extracellular polymeric molecules make up biofilm. Organic matter (10–90%) and other molecules such as protein, nucleic acids, and heteropolymers account for the majority of both components (Nielsen et al., 1997). Biofilm's nutritional quality is largely determined by its taxonomic composition, environment, water quality, grazing pressure, and the substrates on which it grows. Biofilm developed in confined systems was shown to be of higher quality than biofilm grown in open water habitats. When employing different types of substrates, the nutritional composition of the biofilm varies greatly (Azim et al., 2005). Table 15.1 shows the proximate composition of biofilm formed on various substrates in various ecosystems. Various writers have well documented the nutritional content, which has also been proven to be suitable for the dietary requirements of fish (Azim et al., 2005). Biofilm not only covers all of the dietary needs of fish and shrimp, but also provides a high-quality protein supply for cultured animals. When compared with normal feed, an experimental diet made from biofilm resulted in improved shrimp growth and survival (Anand et al., 2013). Biofilm is a rich source of growth promoters (Kuhn et al., 2010), immunological stimulants (Supamattaya et al., 2005), bioactive substances (Ju et al., 2008), and vital fatty acids, in addition to nutrition sources. The presence of bacteria and phytoplankton in the biofilm was primarily responsible for the biofilm's above-mentioned features. Bacteria contained in biofilms were discovered to be an essential food source for penaeid shrimp, promoting their development and survival in pond aquaculture systems (Keshavanath and Gangadhar, 2005). Due to their high protein to energy conversion ratio, other biofilm-associated organisms, such as protozoans, nematodes, and polychaetes, also serve as a quality food supply for the fish (Ballester et al., 2007).

15.3 QUORUM SENSING

Quorum sensing (QS) is a bacterial intercellular communication mechanism that allows bacteria to coordinate individual and collective actions, such as biofilm development and dispersal (Liu et al., 2007; Muller et al., 2009). QS is dependent on groups of bacteria with high cell densities accumulating, releasing, and detecting extracellular signal molecules called autoinducers (AIs). AIs control a variety of processes, including biofilm formation, virulence factor accumulation, antibiotic synthesis, and bioluminescence, through regulating gene expression (Rutherford et al., 2011; Ng and Bassler, 2009; Verma and Miyashiro, 2013). Furthermore, these systems are involved in the synthesis of siderophores, the accumulation of virulence factors such as exotoxins, the development of exoproteases, and the production of a variety of secondary metabolites (Winson et al., 1995;

TABLE 15.1
Biochemical Composition of Biofilm Grown in Different Substrates Installed in Aquaculture System

Substrate Type	Habitat/culture System	Protein (% DM)	Lipid (% DM)	Carbohydrate (% DM)	Ash (% DM)	Energy (kJ/g DM)	References
Sugarcane bagasse	Mud bottom cemented tank	9.4	0.33	38	23	–	Mridula et al. (2003)
Bamboo	Earthen pond/carp culture	23–32	2–5	33	15–29	14–20	Azim et al. (2002a, b, c)
Hizol branch	Earthen pond/carp culture	15	6	38	41	12	Azim et al. (2002b)
Jute stick	Earthen pond/carp culture	13	1.5	–	31	14	Azim et al. (2002c)
Bamboo	*Penaeus monodon* culture/FRP (Fibre-reinforced plastic) tank	25.96	2.65	67.25	32.75	298.58	Anand et al. (2013)
Aquamat™	*L. vannamei*/HDPE lined pond; earthen bottom	24.97	2.67	–	37.5	278.11	Kumar et al. (2017)
Nylon mesh	Earthen pond/Nile tilapia	25.75	4.54	–	11.4–12.69	10.51	Tammam et al. (2020)

DM: dry matter

Hentzer et al., 2003). Thin-layer chromatography, gas chromatography, and liquid chromatography coupled with electrospray ionisation, hybrid quadrupole linear ion trap, and Fourier-transform ion-cyclotron-resonance mass spectrometry have all been used to detect and identify quorum-sensing molecules using bacterial biosensors (Ravn et al., 2001; Morin et al., 2003; Yang et al., 2005; Steindler and Venturi, 2007; Cataldi et al., 2008).

Vibrio species are common in aquatic environments (Igbinosa and Okoh, 2008). QS mediates a variety of physiological processes and controls, and influences the virulence system of a variety of infectious bacteria. It is widely employed and highly conserved in these species (Liu et al., 2013). Several *Vibrio* species, particularly *Vibrio fischeri* and *Vibrio harveyi*, have been studied for their model QS regulatory systems, which mediate bioluminescence (Stevens and Greenberg, 1997; Freeman and Bassler, 1999; Camara et al., 2002; Kim et al., 2003). When these bacteria infect the light organs of squid and certain fish species, they produce bioluminescence. In contrast, during planktonic growth in aquatic habitats, these organisms suppress bioluminescence (reviewed by Sitnikov et al., 1995 and Dunlap, 1999).

QS was identified by looking at the processes that control the luminescence induced within growing cultures of the Gram-negative bacterium *V. fischeri* (Nealson et al., 1970). Because higher qrr genes exist in the genomes of other Vibrionaceae members and because the LuxIR autoinducer (AI) system functions downstream from the transcriptional activator LitR in *V. fischeri*, the *V. fischeri* QS network is radically different from those found in all other Vibrionaceae members (Milton, 2006; Miyashiro et al., 2010). Meanwhile, *V. harveyi* uses two distinct AI response systems to govern luminescence, colony shape, and siderophore assembly, and creates three AIs that are recognised by membrane-bound two-component sensors (Bassler et al., 1994; Ng and Bassler, 2009). Using the *V. harveyi* bioassay (which can detect AI production by different bacterial species), *Salmonella typhimurium* produced AI-2 (Surette et al., 1999). The LuxS gene, which is important for QS in *V. harveyi* and is thought to be a vehicle for internal communication in bacteria, produces AI-2 (Cloak et al., 2002; Belval et al., 2006). The population density affects bacterial communication mediated by QS. Bassler (1999) previously established that AIs accumulate and bind to specific receptors as bacterial population densities rise. All pathogens govern the expression of particular phenotypes associated with low and high cell densities based on their unique cellular biology.

15.4 FACTORS INFLUENCING BIOFILM FORMATION

15.4.1 Nutrients

In an aquaculture environment, nutrients have a significant impact on biofilm productivity. The main sources of nutrients in aquaculture are feed and input water. Increased nutrition availability in the system results in increased biofilm mat density and thickness, as well as cyanobacteria dominance. The availability of silica in the water determines diatom dominance in the biofilm community. Diatoms in the biofilm community produced in reservoirs (Baffico and Pedrozo, 1996) and the Baltic Sea favoured high Si:P and N:P ratios (Sommer, 1996). Another essential ingredient that affects biofilm productivity is inorganic carbon. When compared with the treatment without carbon addition, the 2 mM concentration of inorganic carbon boosted biofilm productivity fourfold (Jones et al., 2002). Phosphorus has been discovered to be the limiting nutrient in freshwater ecosystems, and it has an impact on biofilm formation (Vymazal et al., 1994). In streams, the effect of various nutrients on biofilm growth has been widely documented (Azim et al., 2005). The increase in production was ascribed primarily to the proliferation of phytoplankton on the biofilm. Nitrogen is the limiting nutrient in both marine environments and shrimp farming, making shrimp development and survival difficult. With the manipulation of carbon, biofilm was discovered to absorb waste nitrogen present in aquaculture systems and transform it into useful biomass (Kumar et al., 2017). Asaduzzaman et al. (2010) used substrate-installed ponds manipulated with carbon to conduct the experiment on *Macrobrachium rosenbergii*. Anand et al. (2012) used bamboo as a substrate and added rice flour as a carbon source in an outdoor experiment with *P. monodon*. To encourage the expansion of the microbial population in biofilm, C:N ratios of 10:1 and 20:1 were maintained.

In comparison to other treatments, they discovered that adding carbon in a 20:1 ratio to the substrate-installed pond reduced inorganic nitrogen NH_3–N by 48.2%, NO_3–N by 41.6%, and NO_2–N by 42.7%. Kumar et al. (2017)cultivated *L. vannamei* in an Aquamat™-installed pond with a C:N ratio of 15:1, which significantly reduced total nitrogen (ammonia, nitrite, and nitrate) by 25% compared with the control pond. Uneaten feed and expelled faeces are the primary sources of nutritional waste accumulation in feed-based aquaculture systems. It has been discovered that between 20 and 40% of the feed nutrient is absorbed by the fish, with the remainder excreted (Funge-Smith and Briggs, 1998). As a result, the most effective ways to reduce nutrient waste in effluents are to reduce its creation or recycle it into harvestable biomass. Improved nutrient retention ability of fish helps reduce waste output (Amirkolaie, 2011). In shrimp, periphyton has the potential to activate digestive enzymes such as amylase, cellulase, protease, lipase, trypsin, and chymotrypsin, which enhances feed digestibility and minimises feed waste (Anand et al., 2013). Periphyton has the ability to transform waste nitrogen into biomass that may be harvested (Audelo-naranjo et al., 2010; Kumar et al., 2017). Furthermore, adding carbon to the periphyton system boosts the conversion of hazardous nitrogen to less toxic forms, improving pond water quality. Toxic nitrogen compounds are immobilised significantly faster by heterotrophic assimilation than by nitrifying bacteria. These carbon source applications in the substrate (Aquamat™)-placed pond offer an environment with high organic matter concentrations, which supports both heterotrophs and autotrophs in the aphotic and photic zones, respectively, while eliminating nitrogen compounds from the culture water (Suryakumar and Avnimelech, 2017; Kumar et al., 2017). Organic matter adsorption on the substrate also minimises sediment accumulation in the pond's bottom. When compared with control ponds, Azim et al. (2003)found that adding substrate for biofilm adherence to fertilised Tilapia culture ponds greatly boosted growth and doubled nitrogen retention. The installation of an Aquamat in a vertical position in a shrimp culture system reduced nitrogen content in earthen ponds (Azim et al., 2003; Kumar et al., 2017) and experimental tanks (Audelo-naranjo et al., 2010), as it was consumed by both periphyton and shrimp biomass in the system (Audelo-naranjo et al., 2010). This clearly shows that integrating culture procedures with substrate installation decreases in-situ nutrient waste generation, serves as a feed source for cultured animals, and eliminates the need for a separate effluent treatment system (ETS) in an aquaculture farm.

15.4.2 Grazing

The majority of the aquatic creatures in the ecosystem graze on the biofilm community. Gastropods, crustaceans (freshwater prawns, *P. monodon*, *L. vannamei*), and vertebrates (rohu, mullets, rabbit fish, Nile tilapia) all efficiently graze the biofilm community, although not all components of the biofilm community are grazed equally. The algal component of the biofilm is generally grazed by invertebrate snails. The rate at which animals graze on biofilms is mostly determined by the size of the grazers, their density, and the availability of other food sources. The rate of biofilm grazing was calculated to be between 0.03 and 0.9 mg ash-free dry weight (AFDW)/individual/day (Vermaat, 2005). Huchette and Beveridge (2005) grew Nile tilapia in cages containing biofilm and evaluated their feeding preferences on the biofilm community in an aquaculture system. The Nile tilapia grazed on big diatoms and filamentous algae found in the biofilm ecosystem, according to the study. Rohu is a promising candidate species for biofilm-based aquaculture because it grazes on the plankton present in the sub-periphytic layer of periphytic substrate (Majumder and Saikia, 2020). In aquaculture, *L. vannamei* has also been demonstrated to be an excellent grazer of biofilm communities (Kumar et al., 2017).

Mullet have gained weight and survival as a result of their successful grazing on biofilm, according to Biswas et al. (2017). Grazing also has a favourable impact on biofilm growth because it renews algal cells by eliminating dead and senescent cells, keeping the algal assemblage in a productive, early-successional stage. In comparison to ungrazed controls, studies have found a fourfold increase in periphyton specific productivity in grazed periphyton (Norberg, 1999).

15.4.3 Substrate

PVC pipes (Keshavanath et al., 2001), plastic sheets (Shresta and Knud-Hansen, 1994; Tidwell et al., 1998), and custom-designed materials like Aquamats™ can all be utilised to foster biofilm growth in biofilm-based aquaculture (Bratvold and Browdy, 2001). Biofilm has grown successfully on organic substrates such as bamboo, hizol, coconut husks, jute sticks, sugarcane bagasses, paddy stem, and various tree branches. Biofilm development, growth, composition, nutrient quality, and water quality are all heavily influenced by the substrate. The protein content of the biofilm formed in bamboo trees was 50% higher than that of the other substrates investigated, but the community of biofilm grown on different substrates remained unchanged. Another study evaluated the biofilm formation of synthetic (PVC) and natural (sugarcane bagasse) substrates. When compared with PVC substrates, sugarcane bagasse was demonstrated to be the best substrate for biofilm formation and enhanced culture water quality. When compared with various synthetic substrates including PVC, ceramic tiles, plastic sheets, and fibre scrubbers, Khatoon et al. (2007) found that bamboo produced greater biofilm biomass. The biodegradability of the substrates, texture, nutrient leaching, and the presence of hazardous compounds on the substrates all influence the biomass formation of biofilm on diverse substrates. Artificial substrate, such as Aquamats™, laid vertically inside the pond increased the shrimp's growth, health, and survival. This is due to their function as hideouts during the moulting season (Kumar et al., 2017). Therefore, the choice of substrates will be based upon the types of species cultivated, cost of substrates, and availability of raw material for the substrates.

15.4.4 Light and Temperature

For the formation of plankton populations in biofilm, sunlight is a primary source of illumination in the aquaculture system. Autotrophic and heterotrophic organisms make up the majority of the biofilm population, and their dominance is largely determined by the quantity and quality of accessible light. The impact of light on biofilm productivity has been demonstrated by the variation in biofilm assemblages at various depths of water bodies. Furthermore, algal components promote non-algal biofilm components by providing nutritional exudates and attracting more biofilm assemblage organisms (Azim et al., 2005). This demonstrates that algae are an essential component of biofilm and determine its productivity. As a result, irradiance is critical in improving the biofilm community. In recent research on *L. vannamei* culture systems, it was discovered that the algal community dominated the upper part of the substrate that was exposed to sunlight, while the heterotrophic community dominated the lower part of the substrate (Kumar et al., 2017). In an aquatic ecosystem, a shift in dominance from diatoms to anaerobic photosynthetic green and purple bacteria has been seen (Goldsborough, 1993). The composition of the periphyton community is also largely determined by temperature. Scenedesmus flourished at high temperatures, while Navicula thrived at low temperatures, according to previous research (Vermaat and Hootsmans, 1994; Azim et al., 2005). In temperate areas, the temperature-influenced succession of periphyton communities (i.e., from diatom dominance in the spring season to green algae/cyanobacteria dominance in the summer) was also observed earlier. There has been very little research looking at the effect of temperature and light incidence on biofilm formation and composition. More research is needed to discover the optimum temperature and light for biofilm growth, as well as their impact on biofilm quality.

15.5 FISHERIES AND BIOFILM

Bacterial biofilms colonise the tissue of fish, molluscs, and crustaceans, and can reveal information about the animal's health (Geesey et al., 1992). Biofilms on fish surfaces, as well as bacteria from the sea, can infect seafood processing plants. *Pseudomonas*, *Vibrio*, *Staphylococcus*, *Bacillus*, and *Aeromonas* are just a few of the bacteria that can cause food poisoning and build biofilms (Cahill, 1990). Traditional wild-caught species, as well as farm-raised species, are being added to the seafood business. As a result of this increase, new growing and processing technologies have emerged. From the farm (preharvest) to the processing plant, biofilm formation has been an issue in the aquaculture sector (post-harvest). The three product classes, mollusc, crustacean, and finfish, are discussed in this chapter in terms of probable biofilm formation. *Listeria monocytogenes*, *Salmonella* spp., *Vibrio* spp., *Bacillus* spp., *Aeromonas*, and *Pseudomonas* spp. are among the dangerous microbes that could be found in the biofilm. These microorganisms are known to create biofilms, and the fact that they are involved in biofilm formation in the seafood sector is a major concern. Future research will focus on enhancing the effluent water system for ponds, recirculating system filtration/disinfection processes, and processing plant sanitisation.

15.6 ROLE OF BIOFILM IN AQUACULTURE

15.6.1 Biofilm in Nursery System

Finfish nursery rearing entails raising spawn to the fingerling stage. The nursery phase for shrimp entails developing wild-caught or hatchery-produced post-larvae to the point where they can be stocked in grow-out ponds. The use of biofilm in both shrimp and fish spawning facilities is extensively recognised. The inclusion of sugarcane bagasse as a substrate for biofilm development considerably improved the growth, survival, and weight of *Labeo fimbriatus* fry, and it was also discovered that the stocking density could be doubled (Gangadhar et al., 2015). The influence of the biofilm on digestive enzyme activity in fry was mostly responsible for the better growth (Gangadhar et al., 2016). Sakr et al. (2015) investigated the impact of decreasing protein meals in Nile tilapia juvenile rearing systems using substrates. When compared with high-protein diet treatments without substrates, the data revealed that the diet containing 15% protein in substrate constructed ponds greatly improved juvenile growth. Biofilm development was observed to be directly associated with the development of sea cucumber juveniles in an ocean nursery system by Gorospe et al. (2019). Compared with the treatment without substrates, the installation of substrate (HDPE sheet) in the nursery rearing of *L. vannamei* in a recirculating aquaculture system (RAS)-based culture system doubled the shrimp biomass (Tierney et al., 2020). The mullet fry's development and survival were greatly increased (by 95%) when they were raised in bamboo substrates (10% of the total pond surface area). The addition of substrates (meshed Happa) to marine cages, on the other hand, had no effect on the development and survival of Mullet stocked at a density of 1 m^2 per cage. Similarly, biofilm had no effect on sea bream growth (Richard et al., 2010). Thompson et al. (2002) discovered that biofilm alone did not enhance shrimp development and that feed supplementation in combination with biofilm boosted shrimp growth. The differences in stocking density, substrate surface area, culture environment, and associated organisms observed in previous studies on the growth and production of cultured animals could be attributed to the differences in stocking density, substrate surface area, culture environment, and associated organisms (Tidwell and Bratvold, 2005).

15.6.2 Biofilm in Fish Stock Management

15.6.2.1 As a Feed Ingredient

The presence of high protein content, organic matter, and other nutrients in biofilm makes it an ideal feed element for the cultured animals. Aside from direct ingestion by grazers in the aquaculture system, biofilm has also been successfully tested as a feed additive. Periphyton grown in natural bamboo substrate was dried and supplied in shrimp baseline diet in a previous study by Anand et al. (2013). The addition of periphyton to the basal diet at a concentration of 6% considerably boosted the digestive enzyme activity of shrimp, as well as their growth and production.

Individuals' weight grew significantly when they ate a periphyton-supplemented meal. According to the findings, including periphyton in the feed at an optimal amount will boost the cultured animals' production by greatly improving the animals' digestive ability. Another study on *Farfantepenaeus paulensis* found that eating biofilm developed on the substrate resulted in enhanced growth (Abreu et al., 2007). In addition, the algal species responsible for the biofilm was identified and used as a feed supplement in the *P. monodon* nursery system. The results showed that marine periphytic diatoms considerably increased *P. monodon* post-larvae growth and survival (Khatoon et al., 2009).

15.6.2.2 As an Immunostimulant and Vaccine

Innate and adaptive immunity are both present in the immune system of fish. Shrimps only have a general immunological response. Biofilm has immunostimulant properties and boosts fish and shellfish immunity significantly. The use of biofilm as a feed additive as well as direct exposure to farmed fish increases their immune responses. Biofilm increases non-specific immune response in *L. vannamei* and *P. monodon*, according to research (Anand et al., 2013, 2015; Zhang et al., 2010). Tammam et al. (2020)found that biofilm could help researchers understand both particular and non-specific immune responses in Nile tilapia. In fish and shrimp, the biofilm that forms on the substrate might induce an immunological response (Verdegem et al., 2005). Shrimp's immune response and illness resistance are enhanced when biofilm is used as a food element (Anand et al., 2015). Furthermore, developing vaccines using biofilm is a novel avenue for enhancing aquaculture animal health. The efficacy of a biofilm-based *Aeromonas hydrophila* oral vaccination produced by Azad et al. (1997)was tested in Indian major carp (IMC). When compared with free cell vaccination, biofilm vaccinated IMC had higher antibody titres and immunological protection. Furthermore, a longer vaccine retention period was noted (Azad et al., 2000).

15.7 WATER QUALITY MANAGEMENT BY BIOFILM

Water quality control is critical for long-term fish production. Several studies have previously shown that a biofilm-based aquaculture system can improve water quality. The key process in the aquaculture system that determines the water quality is nitrification. In intensive or semi-intensive aquaculture, feed is the primary supply of nitrogen, which disrupts the natural nitrogen cycle (Kumar et al., 2017). Excess feed-based nitrogen is transformed to inorganic nitrogen species like ammonia, which has a negative impact on the health of the cultured animal. However, the nitrifying bacteria in the biofilm reduce ammonia levels by converting it to nitrate via nitrite. In addition, additional heterotrophic communities linked with the biofilm use inorganic nitrogen directly for their own growth. The substrate-installed treatment with a balanced C:N ratio greatly improves the water quality of the system and lowers the ammonia nitrogen in the system (Azim et al., 2002a; Kumar et al., 2017). The use of biofilms in the rearing of *Catla catla* fingerlings improved the water quality (Pradeep et al., 2003). The use of biodegradable substrate in freshwater ponds improved the water quality, allowing fish to thrive. When compared with a standard aquaculture system, the biofilm treatment produces much less ammonia (Keshavanath et al., 2012). Installation of non-biodegradable substrates reduces ammonia nitrogen and nitrite nitrogen in brackish water ponds, promoting the growth of *L. vannamei* (Kumar et al., 2017). The presence of both nitrifying bacteria and microalgae in the water could help to reduce hazardous nitrogen levels. This element is also used by microalgae to grow (Thompson et al., 2002). Other critical water quality criteria to manage for the welfare of cultural animals include alkalinity and pH. During the culture phase, substrate installation in *L. vannamei* culture significantly enhances the pH (7.4–8.4) and alkalinity (88–230 mgL^{-1} as $CaCO_3$) of the culture system. The autotrophic community's photosynthesis raises the pH and alkalinity of the culture system (Kumar et al., 2017). Several studies have shown that employing various types of substrates in biofilm-based eco systems can improve water quality (Anand et al., 2013; Asaduzzaman et al., 2010). The biofilm also reduces the turbidity of the water, which is due to the trapping of organic debris or the accumulation of suspended matter over the submerged substrate (Van Dam et al., 2002). Various studies have reported a decrease in dissolved oxygen (DO) levels in substrate-installed tanks, which they attribute to shading effects of substrates as well as increased biochemical oxygen demand (BOD) (Anand et al., 2019).

15.8 BIOFILM IN EFFLUENT TREATMENT

Researchers have been using microalgae to remediate aquaculture effluents for several decades (Oswald, 2003; Levy et al., 2017). The cost of extracting algae from the system further limits its use. Periphyton grown in vertical plastic net substrates and algal turf scrubbers has been discovered to be a viable option for reducing nutrient pollution caused by aquaculture effluents (Erler et al., 2004; Valeta and Verdegem, 2009). Fish mariculture effluents are contained in marine periphyton-based biofilters that allow the establishment of marine biofilm on a plastic net-based substrate in a bioreactor. Total ammonia nitrogen (TAN) and dissolved inorganic nitrogen (DIN) are removed by the periphyton in the range of 0.11 to 1.2 g N m^{-2} (substrate area) day^{-1}. They also discovered that the vertical integration of the plastic net, as well as biomass weight and effluent retention duration, had a substantial impact on the nutrient removal effectiveness of marine periphyton (Levy et al., 2017). In a RAS, Sereti et al. (2004)investigated the importance of periphyton in maintaining water quality. It has been discovered that using periphyton mat instead of biological filters in the RAS enhances Nile tilapia growth and nutrient (N, P) retention. Scraping periphyton from the periphyton mat at regular intervals and feeding it to the culture organisms improves nutrient retention capacity directly or indirectly. When finfish and periphyton are used together in the wastewater treatment process, the nutrient retention of farm effluents is improved when compared with using periphyton alone as an effluent treatment method (Erler et al., 2004). As a tertiary nutrient removal system, the fish–periphyton system mesocosm has proven to be effective. It is made up of a series of 375-L tanks, each with vertical plastic mesh to support periphytic algal growth and algal-munching tilapia fish (*Tilapia mossambica*), and into which wastewater has been pumped and analysed. c mg of total phosphorus (TP) m^{-2} d^{-1} and 108 mg of total nitrogen (TN) m^{-2} d^{-1}, this system eliminates TP and TN from wastewater by 82% and 23%, respectively. The patent number for this system is 5254252 in the United States (Rectenwald and Drenner, 2000). Natural substrates, such as sugarcane bagasse, have also been utilised to treat aquaculture effluents in addition to artificial substrates. Bagasse, a natural highly fibrous lignocellulosic byproduct of sugarcane, was used in the study to support the formation of periphyton. When administered at a dose of 1 to 6 g L^{-1}, the ammonia concentration in the shrimp wastewater was considerably reduced within 24 hours (Krishnani et al., 2006). It paves the way for further research into the use of environmentally friendly agriculture byproducts in the treatment of aquaculture effluents.

15.9 ROLE OF BIOFILM IN FISH GROWTH

Fish growth in aquaculture ponds is primarily influenced by the culture environment and feed availability in the system. Biofilm development in the culture system has been shown to improve water quality and act as a feed element, resulting in increased fish growth. In freshwater aquaculture systems, natural biodegradable substrates are used for supporting the growth of biofilm, and it has been found that differences exist in the growth performance of animals grown in different substrates. When compared with other substrates such as bamboo leaf, Eichhornia, paddy straw, and palm leaf, the growth of rohu and common carp in the treatment employing sugarcane bagasses as a substrate was found to be two to three times greater (Ramesh et al., 1999). Another study discovered that freshwater fish such as *Cirrhinus mrigala* performed better on natural biodegradable substrate than in non-biodegradable substrates such as plastic sheet and tiles (Pandey et al., 2014). Tidwell et al. (2010) discovered a 20% improvement in freshwater prawn output in a biofilm-based pond. In brackish water shrimp aquaculture, a similar boost in growth was observed. In comparison to the normal pond, the growth and survival of *L. vannamei* cultivated in a biofilm pond were extremely high. The majority of investigations on *L. vannamei* used non-biodegradable substrates to foster biofilm formation (Kumar et al., 2017). Other species, including *Fenneropenaeus paulensis* (Thompson et al., 2002; Ballester et al., 2007), *Penaeus esculentus* (Burford et al., 2003; Arnold et al., 2005), and *Penaeus monodon* (Anand et al., 2013), grew, survived, and produced more in substrate-installed ponds. The following factors contribute to the cultured animals' improved growth, survival, and production in a biofilm-based culture system: (1) increased food availability, (2) effective nutrient cycling, (3) conversion of toxic nutrients into useful forms, (4) reduction of pathogenic bacteria, (5) improved culture water quality, and (6) provision of special nutrients (Azim et al., 2005).

15.9.1 Suitable Fish Species for Biofilm-Based Aquaculture

Biofilm is an important component of primary production and the food web in the aquatic environment. It also provides a food source for fish existing in both natural and cultivated settings. Biofilm has been widely used in freshwater aquaculture for polyculture systems, including substrates such as bamboo, palm leaf, coconut leaf, and sugarcane bagasses. Table 15.2 lists the species cultivated in the biofilm-based aquaculture system. Freshwater fish species such as *Oreochromis niloticus*, *Catla catla*, *Cirrhinus mrigala*, *Labeo rohita*, *and L. calbasu* were cultivated together in a biofilm-based polyculture system, and it was discovered that *L. rohita* grew faster than the other species. When *L. rohita* and *L. gonius* were compared in another study, the former showed 77% higher growth than the latter. The consumption of organisms connected with the biofilm subsurface zone, which may eliminate competition from other planktonic feeders and grazers, is primarily responsible for rohu's accelerated development in the biofilm-based polyculture system (Majumder and Saikia, 2020). Freshwater prawns have been cultivated widely in biofilm-based aquaculture systems. In the instance of freshwater prawns cultivated in a biofilm-based aquaculture system, production, animal welfare, and mortality have all improved (Tidwell et al., 2000). Artificial substrate, such as Aquamat™, has been frequently used for biofilm growth in *L. vannamei* culture systems in order to increase shrimp biomass, health, and moulting behaviour. Several studies have found that shrimp raised in a biofilm-based aquaculture system develop and survive better (Kumar et al., 2017).

15.9.2 Biofilm Development in Aquaculture Ponds

Aquamats™ are horticultural black shade netting that may be purchased commercially. They cover an area of 20 × 0.5 m and have a mesh size of 1 mm. They can be used as a vertical gill net inside the pond, with plastic soft drink bottles as floats and sand-filled bottles as sinkers. A mat is suspended vertically 10 cm below the water level and 50 cm above the pond floor, fastened to the pole on both sides. Aquamats are positioned in front of the aerator to supply adequate oxygen to connected organisms. They are kept parallel to each other, 30 cm apart, to form a battery. In addition, a carbon source (molasses) can be placed in front of the aerator to allow heterotrophic organisms in the Aquamat to employ ammonia uptake (Suryakumar and Avnimelech, 2017; Kumar et al., 2017). This shows that both autotrophic and heterotrophic biomass are in charge of waste nutrient production. Scraping is done on a regular basis to remove extra suspended materials that accumulate on the Aquamat. When the substrate is submerged in the ponds, biofilm formation begins.

15.10 ECONOMICS OF BIOFILM-BASED AQUACULTURE

The economics of a substrate-based polyculture system were calculated by Gangadhar and Keshavanath (2002). When compared with a standard fish pond, the installation of substrate, transportation, and labour costs all result in higher production costs. However, the increased fish output and subsequent selling amount offset the production costs, increasing the polyculture pond's profit. The economics of substrate-based aquaculture were assessed in another study by Azim et al. (2002a), and the study indicated the effect of substrate on the profitability of the culture.

TABLE 15.2
Growth Performance of Animals Cultured on Biofilm-Based Aquaculture Using Different Substrates

Aquaculture System	Substrate Installed	Periphyton Productivity (DM Basis mg/cm²)	Culture Species	Weight Gain/Final Weight (g)	FCR	Survival (%)	Reference
Freshwater aquaculture	Bamboo + carbohydrate source addition	2.02–2.56	Rohu + tilapia	150^{R}, 294^{T}	–	85.8	Asaduzzaman et al. (2010)
	Sugarcane bagasse	–	*Labeo fimbriatus* fry	2.2–5.47	–	59.32–88.8	Gangadhar et al. (2015)
	Bamboo side shoots	1.7–2.9	*M. rosenbergii* + *Oreochromis niloticus*	12.78^{M}, 96.08^{T}	2.02	66^{M}, 90.5^{T}	Haque et al. (2015)
	Bamboo bagasse	$0.14–0.17^{Bam}0.11–0.17^{Bag}$	*Labeo rohita*	$28.63–37.67^{Bam}30.1–34.75^{Bag}$		$77.78–88.89^{Bam}66.67–100^{Bag}$	Keshavanath et al. (2012)
	Bamboo branches (*Kanchi)*	1.85	*M. rosenbergii* + GIFT tilapia	40.46^{M}, 199.35^{T}	2.27	$78.97^{M}91.57^{T}$	Haque et al. (2016)
Brackish water aquaculture	Polyethylene screens	–	*Farfantepenaeus paulensis*	0.6	–	95	Ballester et al. (2007)
	Bamboo	1.62–2.85	Milkfish (*Chanos chanos*)	13.3–25.5	–	93–96	Garg (2016)
	Bamboo poles	15.06 ± 2.78	Mullet fry	28.39 ± 1.94	–	94.3	Biswas et al. (2017)
	Aquamat™	4.81	*L. vannamei*	–	1.2–1.5	93	Kumar et al. (2017)
	Bamboo + C:N ratio of 10:1	12.77 ± 1.47 (3.4–22.7)	*Penaeus monodon*	4.85 ± 0.49	2.37	70	Anand et al. (2019)
	High density polyethylene mesh	–	*Litopenaeus vannamei*	1.52 ± 0.19	0.99	50.3	Tierney et al. (2020)
	Nylon mesh	–	*L. vannamei*	20.70 ± 2.18	1.53	92	Sofyani and Sambhu (2020)

FCR: feed conversion ratio

REFERENCES

Abreu, Paulo Cesar, Eduardo L. C. Ballester, Clarisse Odebrecht, Wilson Wasielesky Jr, Ronaldo O. Cavalli, Wilhelm Granéli, and Alexandre M. Anesio. "Importance of biofilm as food source for shrimp (Farfantepenaeuspaulensis) evaluated by stable isotopes (δ13C and δ15N)." *Journal of Experimental Marine Biology and Ecology* 347, no. 1–2 (2007): 88–96.

Agriculture Organization of the United Nations. Fisheries Department. *The State of World Fisheries and Aquaculture, 2000*. Vol. 3. Food & Agriculture Org., 2000.

Amirkolaie, Abdolsamad K. "Reduction in the environmental impact of waste discharged by fish farms through feed and feeding." *Reviews in Aquaculture* 3, no. 1 (2011): 19–26.

Anand, P. S. Shyne, C. P. Balasubramanian, L. Christina, Sujeet Kumar, G. Biswas, Debasis De, T. K. Ghoshal, and K. K. Vijayan. "Substrate based black tiger shrimp, Penaeus monodon culture: Stocking density, aeration and their effect on growth performance, water quality and periphyton development." *Aquaculture* 507 (2019): 411–418.

Anand, P. S. Shyne, M. P. S. Kohli, S. Dam Roy, J. K. Sundaray, Sujeet Kumar, Archana Sinha, G. H. Pailan, and Munilkumar Sukham. "Effect of dietary supplementation of periphyton on growth performance and digestive enzyme activities in Penaeus monodon." *Aquaculture* 392 (2013a): 59–68.

Anand, P. S. Shyne, Sujeet Kumar, A. Panigrahi, T. K. Ghoshal, J. Syama Dayal, G. Biswas, J. K. Sundaray et al. "Effects of C: N ratio and substrate integration on periphyton biomass, microbial dynamics and growth of Penaeus monodon juveniles." *Aquaculture International* 21 (2013b): 511–524.

Anand, P. S., Kumar, S., Panigrahi, A., Ghoshal, T.K., Syama Dayal, J., Biswas, G., ... and Ravichandran, P. "Effects of C: N ratio and substrate integration on periphyton biomass, microbial dynamics and growth of Penaeus monodon juveniles." *Aquaculture International* 21, (2013): 511–524.

Anand, Panantharayil S. Shyne, Mahinder P. S. Kohli, Sibnarayan Dam Roy, Jitendra K. Sundaray, Sujeet Kumar, Archana Sinha, Gour H. Pailan, and Munilkumar Sukham. "Effect of dietary supplementation of periphyton on growth, immune response and metabolic enzyme activities in P enaeus monodon." *Aquaculture Research* 46, no. 9 (2015): 2277–2288.

Arnold, Stuart J., Melony J. Sellars, Peter J. Crocos, and Greg J. Coman. "Response of juvenile brown tiger shrimp (Penaeus esculentus) to intensive culture conditions in a flow through tank system with three-dimensional artificial substrate." *Aquaculture* 246, no. 1–4 (2005): 231–238.

Asaduzzaman, M., M. M. Rahman, M. E. Azim, M. Ashraful Islam, M. A. Wahab, M. C. J. Verdegem, and J. A. J. Verreth. "Effects of C/N ratio and substrate addition on natural food communities in freshwater prawn monoculture ponds." *Aquaculture* 306, no. 1–4 (2010): 127–136.

Audelo-Naranjo, Juan M., Luis R. Martínez-Córdova, and Domenico Voltolina. "Nitrogen budget in intensive cultures of Litopenaeusvannamei in mesocosms, with zero water exchange and artificial substrates." *Revista de biología marina y oceanografía* 45, no. 3 (2010): 519–524.

Avnimelech, Yoram. "Feeding with microbial flocs by tilapia in minimal discharge bio-flocs technology ponds." *Aquaculture* 264, no. 1–4 (2007): 140–147.

Azad, I. S., K. M. Shankar, and C. V. Mohan. "Evaluation of an Aeromonas hydrophila biofilm for oral vaccination of carp." *Diseases in Asian Aquaculture* 9, no. 7 (1997): 181–186.

Azad, I. S., K. M. Shankar, C. V. Mohan, and B. Kalita. "Uptake and processing of biofilm and free-cell vaccines of Aeromonas hydrophila in Indian major carps and common carp following oral vaccination antigen localization by a monoclonal antibody." *Diseases of Aquatic Organisms* 43, no. 2 (2000): 103–108.

Azim, M. E., and T. Asaeda. "Periphyton structure, diversity and colonization." *Periphyton: Ecology, Exploitation and Management* (eds M. E. Azim, Marc C. J. Verdegem, Anne A. van Dam, and Malcolm C. M. Beveridge) 1 (2005). CABI Publishing.

Azim, M. E., A. Milstein, M. A. Wahab, and M. C. J. Verdegam. "Periphyton–water quality relationships in fertilized fishponds with artificial substrates." *Aquaculture* 228, no. 1–4 (2003): 169–187.

Azim, M. Ekram, Marc C. J. Verdegem, Anne A. van Dam, and Malcolm C. M. Beveridge, eds. *Periphyton: Ecology, Exploitation and Management*. CABI, 2005.

Azim, M. E., M. C. J. Verdegem, H. Khatoon, M. A. Wahab, A. A. Van Dam, and M. C. M. Beveridge. "A comparison of fertilization, feeding and three periphyton substrates for increasing fish production in freshwater pond aquaculture in Bangladesh." *Aquaculture* 212, no. 1–4 (2002): 227–243.

Azim, M. E., M. C. J. Verdegem, M. M. Rahman, M. A. Wahab, A. A. Van Dam, and M. C. M. Beveridge. "Evaluation of polyculture of Indian major carps in periphyton-based ponds." *Aquaculture* 213, no. 1–4 (2002a): 131–149.

Azim, Mohammed Ekram, Mohammed Abdul Wahab, Marc C. J. Verdegem, Anne A. van Dam, Jules M. van Rooij, and Malcolm C. M. Beveridge. "The effects of artificial substrates on freshwater pond productivity and water quality and the implications for periphyton-based aquaculture." *Aquatic Living Resources* 15, no. 4 (2002b): 231–241.

Azim, M. E., Verdegem, M. C., Khatoon, H., Wahab, M. A., Van Dam, A. A., andBeveridge, M. C. "A comparison of fertilization, feeding and three periphyton substrates for increasing fish production in freshwater pond aquaculture in Bangladesh." *Aquaculture* 212, no. 1–4 (2002c): 227–243.

Baffico, Gustavo D., and Fernando L. Pedrozo. "Growth factors controlling periphyton production in a temperate reservoir in Patagonia used for fish farming." *Lakes & Reservoirs: Research & Management* 2, no. 3–4 (1996): 243–249.

Ballester, Eduardo Luis Cupertino, Wilson Wasielesky Jr, Ronaldo Olivera Cavalli, and Paulo César Abreu. "Nursery of the pink shrimp Farfantepenaeuspaulensis in cages with artificial substrates: Biofilm composition and shrimp performance." *Aquaculture* 269, no. 1–4 (2007): 355–362.

Bassler, Bonnie L. "How bacteria talk to each other: Regulation of gene expression by quorum sensing." *Current Opinion in Microbiology* 2, no. 6 (1999): 582–587.

Bassler, Bonnie L., Miriam Wright, and Michael R. Silverman. "Multiple signalling systems controlling expression of luminescence in Vibrio harveyi: Sequence and function of genes encoding a second sensory pathway." *Molecular Microbiology* 13, no. 2 (1994): 273–286.

Beveridge, M. C. M., M. J. Phillips, and D. J. Macintosh. "Aquaculture and the environment: The supply of and demand for environmental goods and services by Asian aquaculture and the implications for sustainability." *Aquaculture Research* 28, no. 10 (1997): 797–807.

Biswas, G., J. K. Sundaray, S. B. Bhattacharyya, P. S. Shyne Anand, T. K. Ghoshal, Debasis De, Prem Kumar et al. "Influence of feeding, periphyton and compost application on the performances of striped grey mullet (Mugil cephalus L.) fingerlings in fertilized brackishwater ponds." *Aquaculture* 481 (2017): 64–71.

Bratvold, Delma, and Craig L. Browdy. "Effects of sand sediment and vertical surfaces (AquaMatsTM) on production, water quality, and microbial ecology in an intensive Litopenaeusvannamei culture system." *Aquaculture* 195, no. 1–2 (2001): 81–94.

Burford, Michele A., Peter J. Thompson, Robins P. McIntosh, Robert H. Bauman, and Doug C. Pearson. "Nutrient and microbial dynamics in high-intensity, zero-exchange shrimp ponds in Belize." *Aquaculture* 219, no. 1–4 (2003): 393–411.

Cahill, Marian M. "Bacterial flora of fishes: A review." *Microbial Ecology* 19 (1990): 21–41.

Cámara, Miguel, Paul Williams, and Andrea Hardman. "Controlling infection by tuning in and turning down the volume of bacterial small-talk." *The Lancet Infectious Diseases* 2, no. 11 (2002): 667–676.

Cataldi, Tommaso R. I., Giuliana Bianco, and Salvatore Abate. "Profiling of N-acyl-homoserine lactones by liquid chromatography coupled with electrospray ionization and a hybrid quadrupole linear ion-trap and Fourier-transform ion-cyclotron-resonance mass spectrometry (LC-ESI-LTQ-FTICR-MS)." *Journal of Mass Spectrometry* 43, no. 1 (2008): 82–96.

Challan Belval, Sylvain, Laurent Gal, Sylvain Margiewes, Dominique Garmyn, Pascal Piveteau, and Jean Guzzo. "Assessment of the roles of LuxS, S-ribosyl homocysteine, and autoinducer 2 in cell attachment during biofilm formation by Listeria monocytogenes EGD-e." *Applied and Environmental Microbiology* 72, no. 4 (2006): 2644–2650.

Cloak, Orla M., Barbara T. Solow, Connie E. Briggs, Chin-Yi Chen, and Pina M. Fratamico. "Quorum sensing and production of autoinducer-2 in Campylobacter spp., Escherichia coli O157: H7, and Salmonella enterica serovar Typhimurium in foods." *Applied and Environmental Microbiology* 68, no. 9 (2002): 4666–4671.

Crab, Roselien, Yoram Avnimelech, Tom Defoirdt, Peter Bossier, and Willy Verstraete. "Nitrogen removal techniques in aquaculture for a sustainable production." *Aquaculture* 270, no. 1–4 (2007): 1–14.

Dunlap, Paul V. "Quorum regulation of luminescence in Vibrio fischeri." *Journal of Molecular Microbiology and Biotechnology* 1, no. 1 (1999): 5–12.

Erler, Dirk, Peter Pollard, Peter Duncan, and Wayne Knibb. "Treatment of shrimp farm effluent with omnivorous finfish and artificial substrates." *Aquaculture Research* 35, no. 9 (2004): 816–827.

Freeman, Jeremy A., and Bonnie L. Bassler. "A genetic analysis of the function of LuxO, a two-component response regulator involved in quorum sensing in Vibrio harveyi." *Molecular Microbiology* 31, no. 2 (1999): 665–677.

Funge-Smith, Simon J., and Matthew R. P. Briggs. "Nutrient budgets in intensive shrimp ponds: Implications for sustainability." *Aquaculture* 164, no. 1–4 (1998): 117–133.

Gangadhar, B., and P. Keshavanath. "Substrates used in aquauculture—Economical and ecological implications." *Fish Chimes* 22, no. 8 (2002): 14–16.

Gangadhar, B., N. Sridhar, C. H. Raghavendra, H. J. Santhosh, and P. Jayasankar. "Growth performance and digestive enzyme activities of fringe-lipped carp Labeofimbriatus (Bloch, 1795) in periphyton based nursery rearing system." *Indian Journal of Fisheries* 63, no. 1 (2016): 125–131.

Gangadhar, B., N. Sridhar, S. Saurabh, C. H. Raghavendra, and M. R. Raghunath. "Influence of periphyton based culture systems on growth performance of fringe-lipped carp Labeofimbriatus (Bloch, 1795) during fry to fingerling rearing." *Indian Journal of Fisheries* 62, no. 3 (2015): 118–123.

Garg, Sudhir Krishan. "Impacts of grazing by milkfish (ChanoschanosForsskal) on periphyton growth and its nutritional quality in inland saline ground water: Fish growth and pond ecology." *Ecology and Evolutionary Biology* 1 (2016): 41–52.

Geesey, Gill G., M. W. Stupy, and Philip J. Bremer. "The dynamics of biofilms." *International Biodeterioration & Biodegradation* 30, no. 2–3 (1992): 135–154.

Goldsborough, L. Gordon. "Diatom ecology in the phyllosphere of the common duckweed (Lemna minor L.)." *Hydrobiologia* 269 (1993): 463–471.

Gorospe, JayR C., Marie Antonette Juinio-Menez, and Paul C. Southgate. "Effects of shading on periphyton characteristics and performance of sandfish, Holothuriascabra Jaeger 1833, juveniles." *Aquaculture* 512 (2019): 734307.

Haque, M. Rezoanul, M. Ashraful Islam, M. Mojibar Rahman, Mst Farzana Shirin, M. Abdul Wahab, and M. Ekram Azim. "Effects of C/N ratio and periphyton substrates on pond ecology and production performance in giant freshwater prawn M acrobrachiumrosenbergii (De Man, 1879) and tilapia O reochromisniloticus (Linnaeus, 1758) polyculture system." *Aquaculture Research* 46, no. 5 (2015): 1139–1155.

Haque, M. Rezoanul, M. Ashraful Islam, M. Abdul Wahab, Md Enamul Hoq, M. Mojibar Rahman, and M. Ekram Azim. "Evaluation of production performance and profitability of hybrid red tilapia and genetically improved farmed tilapia (GIFT) strains in the carbon/nitrogen controlled periphyton-based (C/N-CP) on-farm prawn culture system in Bangladesh." *Aquaculture Reports* 4 (2016): 101–111.

Hentzer, Morten, Hong Wu, Jens Bo Andersen, Kathrin Riedel, Thomas B. Rasmussen, Niels Bagge, Naresh Kumar et al. "Attenuation of Pseudomonas aeruginosa virulence by quorum sensing inhibitors." *The EMBO Journal* 22, no. 15 (2003): 3803–3815.

Huchette, and Beveridge. "Periphyton-based cage aquaculture." In *Periphyton. Ecology, Exploitation and Management* (eds E. Azim, Marc C. J. Verdegem, Anne A. van Dam, and Malcolm C. M. Beveridge). CABI Publishing, Wallingford (2005), p. 237.

Igbinosa, Etinosa O., and Anthony I. Okoh. "Emerging Vibrio species: An unending threat to public health in developing countries." *Research in Microbiology* 159, no. 7–8 (2008): 495–506.

Jones, P. L., Thanongsak Thanuthong, and P. Kerr. "Preliminary study on the use of synthetic substrate for juvenile stage production of the yabby, Cherax destructor (Clark)(Decapoda: Parastacidae)." *Aquaculture Research* 33, no. 10 (2002): 811–818.

Ju, Z. Y., I. Forster, L. Conquest, and W. Dominy. "Enhanced growth effects on shrimp (Litopenaeusvannamei) from inclusion of whole shrimp floc or floc fractions to a formulated diet." *Aquaculture Nutrition* 14, no. 6 (2008): 533–543.

Keshavanath, P., and B. Gangadhar. "13 research on Periphyton-based aquaculture in India." In *Periphyton: Ecology, Exploitation and Management* (eds M. E. Azim, Marc C. J. Verdegem, Anne A. van Dam, and Malcolm C. M. Beveridge), CABI Publishing (2005): 223.

Keshavanath, P., B. Gangadhar, T. J. Ramesh, J. M. Van Rooij, M. C. M. Beveridge, D. J. Baird, M. C. J. Verdegem, and A. A. Van Dam. "Use of artificial substrates to enhance production of freshwater herbivorous fish in pond culture." *Aquaculture Research* 32, no. 3 (2001): 189–197.

Keshavanath, P., J. K. Manissery, A. Ganapathi Bhat, and B. Gangadhara. "Evaluation of four biodegradable substrates for periphyton and fish production." *Journal of Applied Aquaculture* 24, no. 1 (2012): 60–68.

Khatoon, H., S. Banerjee, F. M. Yusoff, and M. Shariff. "Evaluation of indigenous marine periphytic Amphora, Navicula and Cymbella grown on substrate as feed supplement in Penaeus monodon postlarval hatchery system." *Aquaculture Nutrition* 15, no. 2 (2009): 186–193.

Khatoon, Helena, Fatimah Yusoff, Sanjoy Banerjee, Mohamed Shariff, and Japar Sidik Bujang. "Formation of periphyton biofilm and subsequent biofouling on different substrates in nutrient enriched brackishwater shrimp ponds." *Aquaculture* 273, no. 4 (2007): 470–477.

Kim, Soo Young, Shee Eun Lee, Young Ran Kim, Choon Mee Kim, Phil Youl Ryu, Hyon E. Choy, Sun Sik Chung, and Joon Haeng Rhee. "Regulation of Vibrio vulnificus virulence by the LuxS quorum-sensing system." *Molecular Microbiology* 48, no. 6 (2003): 1647–1664.

Krishnani, Kishore K., V. Parimala, B. P. Gupta, I. S. Azad, Xiaoguang Meng, and Mathew Abraham. "Bagasse-assisted bioremediation of ammonia from shrimp farm wastewater." *Water Environment Research* 78, no. 9 (2006): 938–950.

Kuhn, David D., Addison L. Lawrence, Gregory D. Boardman, Susmita Patnaik, Lori Marsh, and George J. Flick Jr. "Evaluation of two types of bioflocs derived from biological treatment of fish effluent as feed ingredients for Pacific white shrimp, Litopenaeusvannamei." *Aquaculture* 303, no. 1–4 (2010): 28–33.

Kumar, Santhana, P. K. Pandey, Saurav Kumar, T. Anand, Boriah Suryakumar, and Rathi Bhuvaneswari. "Effect of periphyton (aquamat installation) in the profitability of semi-intensive shrimp culture systems." *Indian Journal of Economics and Development* 7, no. 1 (2019): 1–9.

Kumar, V. Santhana, P. K. Pandey, T. Anand, Rathi Bhuvaneswari, and Saurav Kumar. "Effect of periphyton (aquamat) on water quality, nitrogen budget, microbial ecology, and growth parameters of Litopenaeusvannamei in a semi-intensive culture system." *Aquaculture* 479 (2017): 240–249.

Levy, Alon, Ana Milstein, Amir Neori, Sheenan Harpaz, Muki Shpigel, and Lior Guttman. "Marine periphyton biofilters in mariculture effluents: Nutrient uptake and biomass development." *Aquaculture* 473 (2017): 513–520.

Liu, Huan, S. Srinivas, H. He, G. Gong, C. Dai, Y. Feng, X. Chen, and S. Wang. "Quorum sensing in Vibrio and its relevance to bacterial virulence." *Journal of Bacteriology & Parasitology* 4, no. 172 (2013): 3.

Liu, Xiaoguang, Mohammed Bimerew, Yingxin Ma, Henry Müller, Marianna Ovadis, Leo Eberl, Gabriele Berg, and Leonid Chernin. "Quorum-sensing signaling is required for production of the antibiotic pyrrolnitrin in a rhizospheric biocontrol strain of Serratia plymuthica." *FEMS Microbiology Letters* 270, no. 2 (2007): 299–305.

Majumder, Sandip, and Surjya K. Saikia. "Ecological intensification for feeding rohu Labeorohita (Hamilton, 1822): A review and proposed steps towards an efficient resource fishery." *Aquaculture Research* 51, no. 8 (2020): 3072–3078.

Meyer-Reil, Lutz-Arend. "Microbial life in sedimentary biofilms—the challenge to microbial ecologists." *Marine Ecology Progress Series* (1994): 303–311.

Milton, Debra L. "Quorum sensing in vibrios: Complexity for diversification." *International Journal of Medical Microbiology* 296, no. 2–3 (2006): 61–71.

Miyashiro, Tim, Michael S. Wollenberg, Xiaodan Cao, Dane Oehlert, and Edward G. Ruby. "A single qrr gene is necessary and sufficient for LuxO-mediated regulation in Vibrio fischeri." *Molecular Microbiology* 77, no. 6 (2010): 1556–1567.

Morin, Danièle, Béatrice Grasland, Karine Vallée-Réhel, Chrystèle Dufau, and Dominique Haras. "On-line high-performance liquid chromatography–mass spectrometric detection and quantification of N-acylhomoserine lactones, quorum sensing signal molecules, in the presence of biological matrices." *Journal of Chromatography A* 1002, no. 1–2 (2003): 79–92.

Mridula, R. M., J. K. Manissery, P. Keshavanath, K. M. Shankar, M. C. Nandeesha, and K. M. Rajesh. "Water quality, biofilm production and growth of fringe-lipped carp (Labeofimbriatus) in tanks provided with two solid substrates." *Bioresource Technology* 87, no. 3 (2003): 263–267.

Müller, Henry, Christian Westendorf, Erich Leitner, Leonid Chernin, Kathrin Riedel, Silvia Schmidt, Leo Eberl, and Gabriele Berg. "Quorum-sensing effects in the antagonistic rhizosphere bacterium Serratia plymuthica HRO-C48." *FEMS Microbiology Ecology* 67, no. 3 (2009): 468–478.

Nealson, Kenneth H., Terry Platt, and J. Woodland Hastings. "Cellular control of the synthesis and activity of the bacterial luminescent system." *Journal of Bacteriology* 104, no. 1 (1970): 313–322.

Ng, Wai-Leung, and Bonnie L. Bassler. "Bacterial quorum-sensing network architectures." *Annual Review of Genetics* 43 (2009): 197–222.

Nielsen, Per Halkjær, Andreas Jahn, and Rikke Palmgren. "Conceptual model for production and composition of exopolymers in biofilms." *Water Science and Technology* 36, no. 1 (1997): 11–19.

Norberg, J. "Periphyton fouling as a marginal energy source in tropical tilapia cage farming." *Aquaculture Research* 30, no. 6 (1999): 427–430.

Oswald, William J. "My sixty years in applied algology." *Journal of Applied Phycology* 15 (2003): 99–106.

Pandey, P.K., Bharti, V., and Kumar, K. "Biofilm in aquaculture production." *Afr J Microbiol Res* 8, no. 13 (2014): 1434–1443.

Piedrahita, Raul H. "Reducing the potential environmental impact of tank aquaculture effluents through intensification and recirculation." *Aquaculture* 226, no. 1–4 (2003): 35–44.

Pradeep, B., P. K. Pandey, and S. Ayyappan. "Effect of probiotic and antibiotics on water quality and bacterial flora". *Journal of the Inland Fisheries Society of India* 35, no. 2 (2003): 68–72.

Ramesh, M. R., K. M. Shankar, C. V. Mohan, and T. J. Varghese. "Comparison of three plant substrates for enhancing carp growth through bacterial biofilm." *Aquacultural Engineering* 19, no. 2 (1999): 119–131.

Ravn, Lars, Allan Beck Christensen, Søren Molin, Michael Givskov, and Lone Gram. "Methods for detecting acylated homoserine lactones produced by Gram-negative bacteria and their application in studies of AHL-production kinetics." *Journal of Microbiological Methods* 44, no. 3 (2001): 239–251.
Rectenwald, Laura L., and Ray W. Drenner. "Nutrient removal from wastewater effluent using an ecological water treatment system." *Environmental Science & Technology* 34, no. 3 (2000): 522–526.
Richard, Marion, Julien-Thomas Maurice, Aurore Anginot, Francois Paticat, M. C. J. Verdegem, and J. M. E. Hussenot. "Influence of periphyton substrates and rearing density on Liza aurata growth and production in marine nursery ponds." *Aquaculture* 310, no. 1–2 (2010): 106–111.
Rutherford, Steven T., Julia C. Van Kessel, Yi Shao, and Bonnie L. Bassler. "AphA and LuxR/HapR reciprocally control quorum sensing in vibrios." *Genes & Development* 25, no. 4 (2011): 397–408.
Sakr, Eman M., Shymaa M. Shalaby, Elham A. Wassef, Abdel-Fattah M. El-Sayed, and Asmaa I. Abdel Moneim. "Evaluation of periphyton as a food source for Nile Tilapia (Oreochromis niloticus) juveniles fed reduced protein levels in cages." *Journal of Applied Aquaculture* 27, no. 1 (2015): 50–60.
Sereti, V., M. C. J. Verdegem, E. H. Eding, and J. A. J. Verreth. "Periphyton as water treatment in fresh water recirculation system." In *Conference on 'Biotechnologies for quality'*, Barcelona, Spanje, 20–23 October 2004, pp. 741–742, 2004.
Shrestha, Madhav K., and Christopher F. Knud-Hansen. "Increasing attached microorganism biomass as a management strategy for Nile tilapia (Oreochromis niloticus) production." *Aquacultural Engineering* 13, no. 2 (1994): 101–108.
Sitnikov, Dmitry M., Jeffrey B. Schineller, and Thomas O. Baldwin. "Transcriptional regulation of bioluminesence genes from Vibrio fischeri." *Molecular Microbiology* 17, no. 5 (1995): 801–812.
Sofyani, A., and C. Sambhu. "Role of artificial substrates on growth performance of pacific white shrimp, Litopenaeusvannamei (Boone, 1931) in semi-arid lands." (2020).
Sommer, Ulrich. "Nutrient competition experiments with periphyton from the Baltic Sea." *Marine Ecology Progress Series* 140 (1996): 161–167.
Steindler, Laura, and Vittorio Venturi. "Detection of quorum-sensing N-acyl homoserine lactone signal molecules by bacterial biosensors." *FEMS Microbiology Letters* 266, no. 1 (2007): 1–9.
Stevens, Ann M., and E. P. Greenberg. "Quorum sensing in Vibrio fischeri: Essential elements for activation of the luminescence genes." *Journal of Bacteriology* 179, no. 2 (1997): 557–562.
Supamattaya, Kidchakan, Suphada Kiriratnikom, Mali Boonyaratpalin, and Lesley Borowitzka. "Effect of a Dunaliella extract on growth performance, health condition, immune response and disease resistance in black tiger shrimp (Penaeus monodon)." *Aquaculture* 248, no. 1–4 (2005): 207–216.
Surette, Michael G., Melissa B. Miller, and Bonnie L. Bassler. "Quorum sensing in Escherichia coli, Salmonella typhimurium, and Vibrio harveyi: A new family of genes responsible for autoinducer production." *Proceedings of the National Academy of Sciences* 96, no. 4 (1999): 1639–1644.
Suryakumar, Boriah, and Yoram Avnimelech. "Adapting biofloc technology for use in small-scale ponds with vertical substrate." *World Aquaculture* (2017): 1–5.
Tammam, Marwa S., Elham A. Wassef, Mohamed M. Toutou, and Abdel-Fattah M. El-Sayed. "Combined effects of surface area of periphyton substrates and stocking density on growth performance, health status, and immune response of Nile tilapia (Oreochromis niloticus) produced in cages." *Journal of Applied Phycology* 32 (2020): 3419–3428.
Thompson, Fabiano Lopes, Paulo Cesar Abreu, and Wilson Wasielesky. "Importance of biofilm for water quality and nourishment in intensive shrimp culture." *Aquaculture* 203, no. 3–4 (2002): 263–278.
Tidwell, J. H., and D. Bratvold. "15 utility of added substrates in shrimp culture." In *Periphyton: Ecology, Exploitation and Management*, edited by M. E. Azim, Marc C. J. Verdegem, Anne A. van Dam, Malcolm C. M. Beveridge, CABIPublishing (2005): 247.
Tidwell, James H., Shawn Coyle, Aaron Van Arnum, and Charles Weibel. "Production response of freshwater prawns Macrobrachiumrosenbergii to increasing amounts of artificial substrate in ponds." *Journal of the World Aquaculture Society* 31, no. 3 (2000): 452–458.
Tidwell, James H., Shawn D. Coyle, and Greg Schulmeister. "Effects of added substrate on the production and population characteristics of freshwater prawns Macrobrachiumrosenbergii in ponds." *Journal of the World Aquaculture Society* 29, no. 1 (1998): 17–22.
Tierney, Thomas W., Leo J. Fleckenstein, and Andrew J. Ray. "The effects of density and artificial substrate on intensive shrimp Litopenaeusvannamei nursery production." *Aquacultural Engineering* 89 (2020): 102063.
Valeta, J. S., and M. C. M. Verdegem. "Phosphate removal from aquaculture effluent by algal turf scrubber technology." *Malawi Journal of Aquaculture and Fisheries (MJAF)* (2009): 19.

van Dam, Anne A., Malcolm C. M. Beveridge, M. Ekram Azim, and Marc C. J. Verdegem. "The potential of fish production based on periphyton." *Reviews in Fish Biology and Fisheries* 12 (2002): 1–31.

Verdegem, M. C. J., E. H. Eding, V. Sereti, R. N. Munubi, R. A. Santacruz-Reyes, and A. A. Van Dam. "Similarities between microbial and periphytic biofilms in aquaculture systems." In *Periphyton: Ecology, Exploitation and Management*, edited by Azim ME, Verdegem MCJ, van Dam AA, Berveridge MCM, CABI, Wallingford, (2005): 191–205.

Verma, Subhash C., and Tim Miyashiro. "Quorum sensing in the squid-Vibrio symbiosis." *International Journal of Molecular Sciences* 14, no. 8 (2013): 16386–16401.

Vermaat, J. E., and M. J. M. Hootsmans. "Periphyton dynamics in a temperature-light gradient." In *Lake Veluwe* a *Macrophyte-Dominated System under Eutrophication Stress*, pp. 193–212. Dordrecht: Springer Netherlands, 1994.

Vermaat, J. E. "Periphyton dynamics and influencing factors." In *Periphyton. Ecology, Exploitation and Management*. London: CABI Publishing (2005): 35–49.

Vymazal, Jan, Christopher B. Craft, and Curtis J. Richardson. "Periphyton response to nitrogen and phosphorus additions in Florida Everglades." *Algological Studies/Archiv für Hydrobiologie, Supplement Volumes* (1994): 75–97.

Wetzel, Robert G. "Attached algal-substrata interactions: Fact or myth, and when and how?." In *Periphyton of Freshwater Ecosystems: Proceedings of the First International Workshop on Periphyton of Freshwater Ecosystems held in Växjö*, Sweden, 14–17 September 1982, pp. 207–215. Dordrecht: Springer Netherlands, 1983.

Winson, Michael K., Miguel Camara, Amel Latifi, Maryline Foglino, Siri Ram Chhabra, Mavis Daykin, Marc Bally, Virginie Chapon, G. P. Salmond, and Barrie W. Bycroft. "Multiple N-acyl-L-homoserine lactone signal molecules regulate production of virulence determinants and secondary metabolites in Pseudomonas aeruginosa." *Proceedings of the National Academy of Sciences* 92, no. 20 (1995): 9427–9431.

Yang, B., Q. S. Li, J. G. Yang, X. F. Li, and W. T. Lin. "Analysis of diglyceride concentration by HPLC-RID." *China Oils Fats* 30, no. 9 (2005): 45–47.

Zhang, Bo, Wenhui Lin, Yajun Wang, and Runlin Xu. "Effects of artificial substrates on growth, spatial distribution and non-specific immunity factors of Litopenaeusvannamei in the intensive culture condition." *Turkish Journal of Fisheries and Aquatic Sciences* 10, no. 4 (2010): 491–497.

16 Microbial Remediation in an Aquatic Environment

Ritesh Shantilal Tandel, Sanjay Rathod, Raja Aadil Hussain Bhat, and Pragyan Dash

16.1 INTRODUCTION

Microorganism-based bioremediation technologies have aroused the interest of scientists due to their extraordinary benefits, which include high efficiency, low cost, and ecological sustainability. Because of rapid population growth, accompanying development, and globalization leading to more anthropogenic activities and the indiscriminate use of heavy metals, the problem of pollution has increased. The aquatic environment, including freshwater bodies, estuaries, and oceans, is vulnerable to pollution from various sources, such as industrial discharges, agricultural runoff, and municipal waste. Heavy metal contamination in aquatic environments is a significant concern due to its persistence and toxicity. Several microorganisms possess metal-resistant genes and enzymatic systems that enable them to transform or immobilize heavy metals. Toxic heavy metals can have an impact on the variety of microbial species and metabolic activity (Tchounwou et al., 2012). As a result, bacteria can develop defense mechanisms to tolerate the stress of toxic heavy metal ions. Heavy metals can be broken down using various techniques, such as chemical, physical, biological, or a mix of these, although many of these techniques are not both environmentally beneficial and practical from an economic standpoint (Liu et al., 2019). Most heavy metals are water soluble and cannot be physically separated. The presence and concentration of metal smelters, mining, coloring agents, batteries, fertilizers, and other necessary industrial/domestic commodities have altered dramatically. This change can result in the accumulation of one or more heavy metals in a site in excess of the permissible natural limits, contaminating the air, water, and soil.

Most heavy metals are hazardous, carcinogenic, and mutagenic by nature, even at extremely low doses (Jaishankar et al., 2014). When present in the soil, air, or water, heavy metals, especially the non-essential elements, have a significant negative impact on the biological variety of the surrounding environment (Ali et al., 2019). Metals are very persistent, remain in the environment, and have a tendency to collect and multiply in the food chain because of their natural ability to change from one form to another rather than to be eliminated (Wuana and Okieimen, 2011). Some metals are necessary for healthy growth and development (nutrition), but when consumed in excess, they can be dangerous (Singh et al., 2011; Hejna et al., 2018). According to several studies, when humans and other living forms are exposed to heavy metals for an extended period of time through cutaneous contact, inhalation, and intake of foods containing heavy metals, they develop a variety of disorders (Anyanwu et al., 2018; Sharma et al., 2014; Alissa and Ferns, 2011). Surface water, rivers, and soil get contaminated when heavy metal–laden effluents from domestic, agricultural, and industrial sources are dumped into them (Anyanwu et al., 2018).

Physicochemical processes, including filtration, reverse osmosis, chemical precipitation, membrane technology, oxidation or reduction, evaporation, and ion exchange, cause secondary pollution at the treatment site, according to Barakat (2011). Since microorganisms and plants may change heavy metals into a less hazardous or harmless form, their use in heavy metal remediation and degradation has been studied for decades. Ayangbenro and Babalola (2017) stated that it

 DOI: 10.1201/9781003408543-16

was comparatively more efficient, economical, and environment friendly. We have discussed the principle, mechanism, types, and factors affecting microbial remediation in aquatic ecosystems, which will help researchers design new bioremediation techniques and more effective bioremediation methods suitable for aquatic environments.

16.2 PRINCIPLE OF BIOREMEDIATION PROCESS

Bioremediation employs biological interventions through biodiversity to minimize, and ultimately eradicate, the harmful impacts of environmental pollutants in targeted areas. The use of microorganisms to biodegrade pollutants in contaminated environments is part of the bioremediation process (Sarma, 2012; Azubuike et al., 2016). Many microorganisms, such as bacteria, fungi, and yeast, decompose hazardous substances into less harmful or non-toxic forms. Microorganisms can use pollutants as food or energy sources (Tang et al., 2007; Mbhele, 2007). This process includes decomposing and metabolizing chemicals by microorganisms to improve environmental quality.

16.3 TYPES OF MICROBIAL BIOREMEDIATION

16.3.1 Augmentation

Pre-grown microbe cultures increase the population of microorganisms at the site, lowering cleaning time and costs; nevertheless, directly adding pre-grown microorganisms can destroy contaminants and accelerate their removal (Tyagi et al., 2011; Azubuike et al., 2016). If biodegrading microbial populations are not present due to pollutant toxicity, certain microorganisms can be added to the ecosystem as "introduced organisms" to complement existing populations. For example, during phytoremediation of metal-contaminated estuaries, bioaugmentation with indigenous rhizobacteria using *Spartina maritima* resulted in increased plant subsurface biomass, metal accumulation, and enhanced metal removal (Mesa et al., 2015). The procedure is known as bioaugmentation. Natural microorganisms are frequently found in small numbers. They may be unable to reduce the amount of pollution or digest a specific chemical in a single contaminated place.

16.3.2 Biostimulation

This improves the environment by increasing the ability of naturally occurring microbes to break down hazardous substances or pollutants by providing nutrients such as phosphorus (P) and nitrogen (N) (Sharma, 2012). Following an oil spill, for example, an increase in carbon content promotes the growth of bacteria that are already capable of breaking down oil, and the addition of extra nutrients in the proper proportions accelerates this process. This happens because the microbes reach their maximum growth rate and as a result, their maximum capacity for pollution absorption (Boufadel et al., 2006; Zahed et al., 2010). The proper nutrient concentration for optimum microbe proliferation, as well as maintaining that concentration for as long as possible, is critical for achieving optimal biostimulation.

16.3.3 Heavy Metal Bioremediation

Live or dead microorganisms can remove heavy metals from the environment with significant advantages like ease of use, low cost, high adsorption capacity, and widespread availability. The three types of microorganisms that are used the most frequently are bacteria, fungi, and algae. The most common type of microbes on the planet, bacteria, are adapted to various environmental conditions. Because of their advantages, including small size, quick rate of multiplication, and simplicity of cultivation, bacteria are commonly used to eliminate heavy metal pollutants from the environment. Several bacteria-based heavy metal remediation techniques, including those using

Escherichia, *Pseudomonas*, *Bacillus*, and *Micrococcus*, have been developed. Heavy metal ions are typically adsorbed on the polysaccharide slime layers of bacteria by functional groups such as carboxyl, amino, phosphate, and sulfate groups (Anirudhan et al., 2012; Yue et al., 2015; Yin et al., 2016). These groups can bind to heavy metal ions, allowing the building of appropriate adsorption capability. Heavy metal ion absorption capabilities in bacteria typically range from 1 to 500 mg g^{-1}. *Pseudomonas aeruginosa* is an example of a mercury-resistant bacterium, with a maximum absorption capacity of about 180 mg g^{-1} (Yin et al., 2016). *Arthrobacter viscosus* biomass, both dead and living, has a high adsorption capacity, an acceptable recovery efficiency, and the ability to regenerate. The extracellular polymeric substances (EPS) on the surface of microbial cells can protect the organisms from harmful consequences by keeping heavy metals out of the internal environment. Because of the existence of cationic and anionic functional groups, EPS can effectively collect heavy metal ions such as mercury, cobalt, copper, and cadmium ions (Fang et al., 2011; Wang et al., 2014).

Fungi can use both micronutrients and heavy metals in situations with high amounts of heavy metals. As a result, fungi have a high capacity for metal absorption and have been widely used to adsorb heavy metal ions. The chitin-chitosan complex, glucuronic acid, phosphate, and polysaccharides in/on fungal cells play an important role in heavy metal adsorption via ion exchange and coordination (Purchase et al., 2009). Numerous functional groups, such as amine, carboxyl, hydroxyl, phosphate, and sulfhydryl groups, as well as different types of ionizable sites, influence the aptitude and selectivity of fungal strains to bind to heavy metal ions. *Aspergillus niger* has been proposed as a promising bio-sorbent to remove Pb(II) due to its high biosorption capacity (Iram et al., 2015). It has been observed that a native fungus isolate termed *A. fumigatus* eliminates Cr(VI) from mine drainage via bioremediation. Maximum absorption of 48.2 mg g^{-1} Cr(VI) may be achieved under ideal conditions, following the Freundlich isotherm (Dhal and Pandey, 2018). Undamaged yeast cell biomass from *Saccharomyces cerevisiae* has been investigated for its ability to extract Cu(II) from wastewater (Amirnia et al., 2015). Removal of Cd(II) by trichoma corresponds well to the Langmuir and Freundlich isotherm models (Bazrafshan et al., 2016). In high-salt conditions, *S. cerevisiae* can clean up copper, zinc, and cadmium contamination, and sodium chloride can increase adsorption capability (Li et al., 2013).

16.4 FACTORS AFFECTING BIOREMEDIATION OF MICROBES IN AQUATIC ECOSYSTEMS

Monitoring the chemical and physical properties of a waterbody takes time. The extent of pollution can be assessed by examining the microbiological and biochemical features of the ecosystem following contamination (Nannipieri et al., 1997). According to Guthrie and Pfaender (1998), it is essential to consider these factors to maximize the advantages of the bioremediation process. The intricacy of the process is increased by the toxicity and hydrophobicity of the hydrocarbons included in petrochemicals. Research suggests that the bioremediation process appears to be constrained by a combination of physical, chemical, and biological factors. Several parameters influence the success of the bioremediation process, including pH, temperature, nutritional status, dissolved oxygen concentration, the presence of electron providers and acceptors, dissolved oxygen content, contaminant load, and so on (Mohan et al., 2006). The degradation rate is influenced by interactions between the substrate and the contaminants (Wang et al., 2007). Bacterial metabolism is greatly influenced by interactions between substrates at different concentrations. It has been discovered that the synergistic impacts of catabolic enzymes can accelerate the degradation of many pollutants' components. Because of their complicated interactions, benzene, toluene, ethylbenzene and xylene (BTEX) compounds inhibit microbial activity above a particular concentration (Mathur and Majumder, 2010). Various factors that govern the bioremediation success are discussed in the following subsections.

16.4.1 Concentration of Contaminants on Site

The concentration of the pollutant affects the microbes. If the contaminant concentration is low, bacterial degradation enzymes are not triggered. Nonetheless, the hazardous consequences of the pollutants are only visible once exceptionally high pollutant concentrations are reached (Adams et al., 2015). Synergistic interactions between many constituents of a contaminant accelerate catabolic enzyme breakdown. The growth rate of *Pseudomonas putida* in batch culture was found to be lowered at high substrate concentrations (Abuhamed et al., 2004). As a result of complex microbiological interactions, BTEX compounds inhibited the biodegradation process (Mathur and Majumder, 2010).

16.4.2 Availability of Nutrients

Microbes require nutrients such as calcium, potassium, phosphorus, nitrogen, and carbon for development and reproduction. Furthermore, the relative concentrations of easily available nutrients influence how quickly the pollutant degrades. Biodegradation was hampered by an excess of nitrogen, potassium, and phosphorus during the decomposition of hydrocarbons (Van Hamme et al., 2003). The bioavailability of organic contaminants, also known as microorganism accessibility, impacts the biodegradation rate. When mass transfer is a limiting factor, increasing microbial conversion does not result in enhanced biotransformation (Boopathy, 2000). Despite rigorous mixing and slicing of the coarse soil particles, the biodegradation of contaminating explosives in soil was blocked even after 50 years (Mannig et al., 1995).

16.4.3 Characteristics of Contaminated Site

The bioremediation process is significantly influenced by the characteristics of the contaminated site. Factors like soil texture, permeability, pH, water-holding capacity, soil temperature, and nutrient and oxygen levels affect the bioremediation process.

16.4.3.1 pH

An optimal pH level is necessary for the bioremediation process, as the contaminants present at the contaminated site are influenced by it. The ideal pH range commonly cited is 6 to 8 (Adams et al., 2015). It is important to mention that bacteria have the ability to break down polyaromatic hydrocarbons at high pH levels and can withstand the adverse environmental conditions present at the contaminated site (Mishra et al., 2021). Neutral pH enhances the mineralization of petroleum hydrocarbons. Research has shown that a temperature range of 15–20 °C can affect the process of biodegradation (Mueller et al., 1989). The property of a substance to dissolve in a solvent is referred to as solubility. It has been discovered that the solubility of organic hydrocarbons in the medium increases as the temperature rises, which allows petrochemical hydrocarbons to become easily accessible with a slight change in pH. Stapleton et al. (1998) discovered that certain fungi and acidophilic bacteria exhibit an increased ability to break down substances in acidic environments (Figure 16.1).

16.4.3.2 Temperature

Temperature has an impact on the degradation of pollutants, particularly hydrocarbons, both *in situ* and *ex-situ* (Margesin and Schinner, 2001). High-temperature ranges, such as 30–40 C, have been found to accelerate soil biodegradation. This is also true for microorganisms living in aquatic or marine environments.

16.4.3.3 Oxygen Availability

The bioremediation procedure can be either aerobic or anaerobic, depending on the availability of oxygen. The short first stages of the aerobic metabolism of polycyclic aromatic hydrocarbons (PAH)

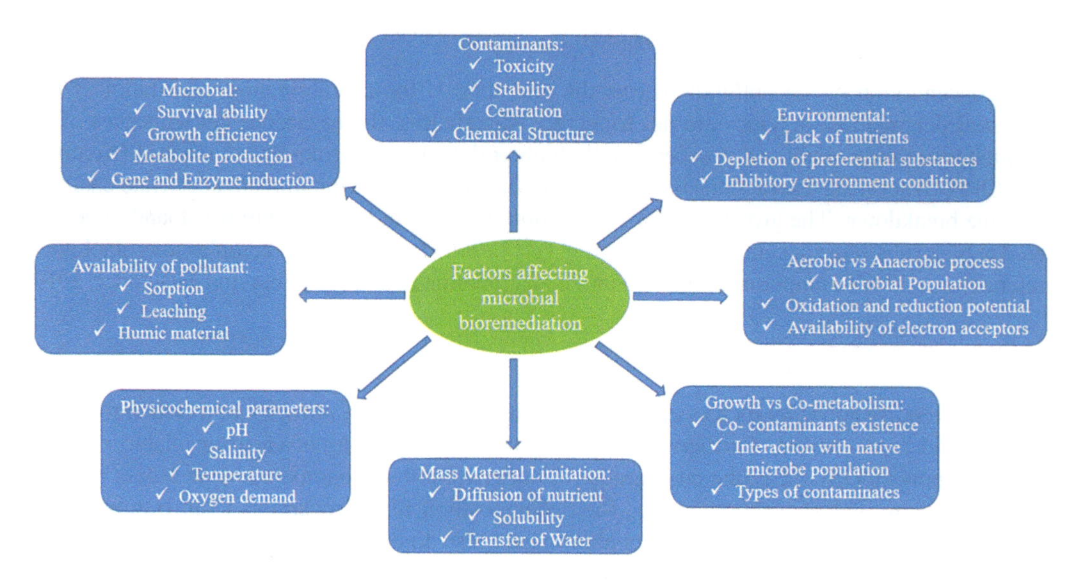

FIGURE 16.1 Factors affecting microbial bioremediation.

in oxygen determine how monooxygenase and dioxygenase enzymes work in the oxidation of aromatic rings (Mishra et al., 2021). Ferrous iron, nitrate, and sulfate in their substituted forms are necessary as electron acceptors for the anaerobic oxidation of aromatic molecules. Unfortunately, the procedure causes significant amounts of phosphorus and ferrous ions to be released, which harms the environment. Petrochemical hydrocarbons undergo anaerobic degradation, which results in the emission of methane and nitrogen dioxide, as well as an increase in pH (Bamforth and Singleton, 2005).

Surface soils exhibit elevated concentrations of organic materials. This is attributed to the increased diversity of microorganisms and their contentment. Bacteria that can use alternative electron acceptors grow increasingly as depth increases, outnumbering other microbial populations. Temperature, moisture content, and dissolved oxygen are the other factors that influence the spread of microbial populations.

16.4.3.4 Redox Potential

The amount of oxygen (O_2) in the gas and liquid phases directly affects the soil's redox potential (also known as the oxidation-reduction potential, Eh). For aerated soils, Eh varies from 0.8 to 0.4 V, for moderately reduced soils, from 0.4 to 0.1 V, for reduced soils, from 0.1 to 0.1 V, and for highly reduced soils, from 0.1 to 0.3 V. The loss of oxygen during respiration lowers the redox potential, resulting in anaerobic (i.e., reducing) circumstances. These conditions limit aerobic reactions while promoting anaerobic processes like fermentation, denitrification, and sulfate reduction. Because they are difficult to measure in the soil or groundwater, redox potentials are not widely used in the field (Pichtel, 2007).

16.5 OMICS

The term OMICs refers to the comprehensive determination and quantification of biological molecules' repositories associated with microbial function and metabolism, structure, and interactions, using techniques such as proteomics, genomics, and metabolomics (Rathoure and Dhatwalia, 2016; Kumari and Kumar, 2021).

16.6 GENOMICS AND METAGENOMICS

Microbiologists have benefited from the use of genomic techniques in the precise genetic analysis of the microbial communities that perform biodegradation and the identification of microorganisms that have prospective degradation capacity (Liu and Liu, 2013; Umar et al., 2017; Kostka et al., 2020).

The significance of marine oil snow in the decomposition of oil spills in aquatic habitats has also been highlighted by functional genomics. Marine oil snows originate when microbial populations, usually bacteria, gather over oil droplets and envelop them in the biosurfactant's clear exopolymers and biomass to create snows that ultimately fall to the seafloor (Ziervogel et al., 2012). During the deepwater spill, this technique was critical in cleaning oil through the water's surface (Vonk et al., 2015). Genomic techniques reveal the connection between certain elements, like phosphorus, nitrogen, iron, sulphur, and trace metals, and microbial degradation. The genes responsible for this metabolism have been found to be increased during biological breakdown (Bashir et al., 2014; Rodriguez-R et al., 2015).

16.7 PROTEOMICS

Proteomics provides a greater depth of understanding of a microbial population, produced phenotype than proteins; therefore, substantial changes can be recognized by monitoring these using proteomics (Singh and Nagaraj, 2006). When these peptides have been effectively extracted from the soil, they can be determined using Matrix-Assisted Laser Desorption Ionization Time-of-Flight Mass Spectrometry (MALDI-TOF), where the peptide patterns are converted into specific protein chains. However, the Liquid Chromatography Mass Spectrometry (LC-MS) technique has been extended to analyze the compounds indicating the biodegradation of surfactants, pesticides, pharmaceutical wastes, and other contaminants (Joo and Kim, 2005).

16.8 TRANSCRIPTOMICS

The transcriptome is a comprehensive description of a microorganism's transcription outputs, which serve as crucial links between genomes, proteomics, and intracellular traits. Transcriptomics is influenced by mutations in the genetic code caused by gene regulation expressed by the microbe to respond to various situations in the environment. The exploration of these is best done through DNA microarray studies, which permit investigations into specific messenger RNAs (mRNAs) (Dharmadi and Gonzalez, 2004; Diaz, 2004). The analysis of soils that have been experimentally poisoned with naphthalene using DNA microarray technology is one of the practical *in situ* uses of transcriptomics in bioremediation (Cho and Tiedje, 2002). The full genome can be replicated using these DNA microarray experiments. A schematic of the regulatory procedures governing biological remediation in bacteria can then be created using the information acquired (Seshadri et al., 2005; Wang et al., 2021).

16.9 METABOLOMICS

A wide range of microbes and their unrecognized potential for microbial breakdown can be comprehended through studies connected to metabolomics. These investigations may also clarify the functions of particular microbial groups or taxa that were earlier undiscovered. *Cycloclastus* sp. had been reported to break down polycyclic aromatic compounds at lower temperatures (Dubinsky et al., 2013). But, subsequent studies revealed that it was also capable of degrading propane, ethane, and butane through aerobic reactions both at the surface of the water and in particles, demonstrating a more flexible tolerance to fluctuations in temperature (Kostka et al., 2020).

Recent research has also made it possible to appreciate some species with diverse metabolomic identities. One such organism is *Marinobacter* sp. *M. hydro-carbonoclasticus* has recently been found to contain an array of metabolic functions for growth in both high- and low-salinity conditions, utilizing oxygen and nitrates or nitrites as electron acceptors and catalyzing both aromatics and alkanes (Huettel et al., 2018).

16.10 MULTI-OMICS

The multi-OMICs technique has proven useful in studying how microbial communities participate in biodegradation. These communities of microbes may already exist in small proportions before spills; then, spills cause their fast expansion, the dominance of the contaminated location's microbiome, and the expression of a variety of metabolic processes based on the kind of contamination along with the surrounding environment (Dubinsky et al., 2013; Rodriguez-R et al., 2015; Shin et al., 2019). Mason et al. (2012) used metagenome, meta-transcriptome, and single-cell genome sequencing to demonstrate the complete bacterial response of the many microbes to the spill, demonstrating the dissemination of metabolic processes among various individuals in the community and identifying the contribution and significance of every species in the metabolic process using meta-transcriptome and metagenome information.

A multi-omics approach is also employed for discovering biomarkers, such as anticipatory biomarkers, biomarkers for diagnosis that demonstrate the possibility of contamination in environments, and beneficial biomarkers that are employed in the remediation and restoration of spill-affected environments (Kostka et al., 2020). In conclusion, microbiologists were capable of using OMICs technologies to directly explore the complexities of bacterial connections in their ecosystems, using samples of water and sediment taken from oil-contaminated and free-of-contamination habitats. These allow analysts to gain unprecedented insight into the microbial responses to hydrocarbon pollution and the scientific basis of microbe-assisted environmental recovery (Kaster and Sobol, 2020; Kumari and Kumar, 2021; Santero and Diaz, 2020).

16.11 ADVANTAGES OF BIOREMEDIATION

1. It can be carried out on-site.
2. It is more affordable.
3. Site disruption during bioremediation is limited, in stark contrast to landfills, which require significant amounts of land.
4. Permanent waste elimination reduces long-term obligations.
5. Some regulatory incentives increase the public's acceptability of the technology.
6. It may also be combined with particular physical or chemical techniques.

16.12 DISADVANTAGES OF BIOREMEDIATION

1. Heavy metals, radionuclides, and some other chlorinated compounds cannot be eliminated by bioremediation.
2. Microbial metabolism may produce some harmful metabolites.
3. Since bioremediation is a highly scientific process, treatability tests must be completed before moving on to the actual cleanup of the site.

REFERENCES

Abuhamed, Tarık, Emine Bayraktar, Tanju Mehmetoğlu, and Ülkü Mehmetoğlu. "Kinetics model for growth of Pseudomonas putida F1 during benzene, toluene and phenol biodegradation." *Process Biochemistry* 39, no. 8 (2004): 983–988.

Adams, Godleads Omokhagbor, Prekeyi Tawari Fufeyin, Samson Eruke Okoro, and Igelenyah Ehinomen. "Bioremediation, biostimulation and bioaugmention: A review." *International Journal of Environmental Bioremediation & Biodegradation* 3, no. 1 (2015): 28–39.

Ali, Hazrat, Ezzat Khan, and Ikram Ilahi. "Environmental chemistry and ecotoxicology of hazardous heavy metals: Environmental persistence, toxicity, and bioaccumulation." *Journal of Chemistry* 2019 (2019): 14.

Alissa, Eman M., and Gordon A. Ferns. "Heavy metal poisoning and cardiovascular disease." *Journal of Toxicology* 1 (2011): 870125.

Amirnia, Shahram, Madhumita B. Ray, and Argyrios Margaritis. "Heavy metals removal from aqueous solutions using Saccharomyces cerevisiae in a novel continuous bioreactor–biosorption system." *Chemical Engineering Journal* 264 (2015): 863–872.

Anirudhan, T. S., S. Jalajamony, and S. S. Sreekumari. "Adsorption of heavy metal ions from aqueous solutions by amine and carboxylate functionalised bentonites." *Applied Clay Science* 65 (2012): 67–71.

Anyanwu, Brilliance Onyinyechi, Anthonet Ndidiamaka Ezejiofor, Zelinjo Nkeiruka Igweze, and Orish Ebere Orisakwe. "Heavy metal mixture exposure and effects in developing nations: An update." *Toxics* 6, no. 4 (2018): 65.

Ayangbenro, Ayansina Segun, and Olubukola Oluranti Babalola. "A new strategy for heavy metal polluted environments: A review of microbial biosorbents." *International Journal of Environmental Research and Public Health* 14, no. 1 (2017): 94.

Azubuike, Christopher Chibueze, Chioma Blaise Chikere, and Gideon Chijioke Okpokwasili. "Bioremediation techniques–classification based on site of application: Principles, advantages, limitations and prospects." *World Journal of Microbiology and Biotechnology* 32 (2016): 1–18.

Bamforth, S. M., and I. Singleton. "Bioremediation of polycyclic aromatic hydrocarbons: Current knowledge and future directions." *Journal of Chemical Technology & Biotechnology: International Research in Process, Environmental & Clean Technology* 80, no. 7 (2005): 723–736.

Barakat, M. A. "New trends in removing heavy metals from industrial wastewater." *Arabian Journal of Chemistry* 4, no. 4 (2011): 361–377.

Bashir, A., Z. D. Umar, and J. O. Steve. "Comparative analysis of Pneumosiderosis among different metal workers in Malumfashi, Katsina, Nigeria." *International Journal of Scientific and Engineering Research* 5, no. 6 (2014): 1429–1436.

Bazrafshan, Edris, Amin Allah Zarei, and Ferdos Kord Mostafapour. "Biosorption of cadmium from aqueous solutions by Trichoderma fungus: Kinetic, thermodynamic, and equilibrium study." *Desalination and Water Treatment* 57, no. 31 (2016): 14598–14608.

Boopathy, R. "Factors limiting bioremediation technologies." *Bioresource Technology* 74, no. 1 (2000): 63–67.

Boufadel, Michel C., Makram T. Suidan, and Albert D. Venosa. "Tracer studies in laboratory beach simulating tidal influences." *Journal of Environmental Engineering* 132, no. 6 (2006): 616–623.

Cho, Jae-Chang, and James M. Tiedje. "Quantitative detection of microbial genes by using DNA microarrays." *Applied and Environmental Microbiology* 68, no. 3 (2002): 1425–1430.

Dhal, Biswaranjan, and Banshi Dhar Pandey. "Mechanism elucidation and adsorbent characterization for removal of Cr (VI) by native fungal adsorbent." *Sustainable Environment Research* 28, no. 6 (2018): 289–297.

Dharmadi, Yandi, and Ramon Gonzalez. "DNA microarrays: Experimental issues, data analysis, and application to bacterial systems." *Biotechnology Progress* 20, no. 5 (2004): 1309–1324.

Díaz, Eduardo. "Bacterial degradation of aromatic pollutants: A paradigm of metabolic versatility." *Int. Microbiol.* 7 no. 3 (2004): 173–180.

Dubinsky, Eric A., Mark E. Conrad, Romy Chakraborty, Markus Bill, Sharon E. Borglin, James T. Hollibaugh, Olivia U. Mason et al. "Succession of hydrocarbon-degrading bacteria in the aftermath of the Deepwater Horizon oil spill in the Gulf of Mexico." *Environmental Science & Technology* 47, no. 19 (2013): 10860–10867.

Fang, Linchuan, Xing Wei, Peng Cai, Qiaoyun Huang, Hao Chen, Wei Liang, and XinmingRong. "Role of extracellular polymeric substances in Cu (II) adsorption on Bacillus subtilis and Pseudomonas putida." *Bioresource Technology* 102, no. 2 (2011): 1137–1141.

Guthrie, Elizabeth A., and Frederick K. Pfaender. "Reduced pyrene bioavailability in microbially active soils." *Environmental Science & Technology* 32, no. 4 (1998): 501–508.

Hejna, M., D. Gottardo, A. Baldi, V. Dell'Orto, F. Cheli, M. Zaninelli, and L. Rossi. "Nutritional ecology of heavy metals." *Animal* 12, no. 10 (2018): 2156–2170.

Huettel, Markus, Will A. Overholt, Joel E. Kostka, Christopher Hagan, John Kaba, Wm Brian Wells, and Stacia Dudley. "Degradation of Deepwater Horizon oil buried in a Florida beach influenced by tidal pumping." *Marine Pollution Bulletin* 126 (2018): 488–500.

Iram, Shazia, Rabia Shabbir, Hunnia Zafar, and Mehwish Javaid. "Biosorption and bioaccumulation of copper and lead by heavy metal-resistant fungal isolates." *Arabian Journal for Science and Engineering* 40 (2015): 1867–1873.

Jaishankar, Monisha, Tenzin Tseten, Naresh Anbalagan, Blessy B. Mathew, and Krishnamurthy N. Beeregowda. "Toxicity, mechanism and health effects of some heavy metals." *Interdisciplinary Toxicology* 7, no. 2 (2014): 60.

Joo, Won-A., and Chan-Wha Kim. "Proteomics of halophilicarchaea." *Journal of Chromatography B* 815, no. 1–2 (2005): 237–250.

Kaster, Anne-Kristin, and Morgan S. Sobol. "Microbial single-cell omics: the crux of the matter." *Applied Microbiology and Biotechnology* 104, no. 19 (2020): 8209–8220.

Kostka, Joel E., Samantha B. Joye, Will Overholt, Paul Bubenheim, Steffen Hackbusch, Stephen R. Larter, Andreas Liese et al. "Biodegradation of petroleum hydrocarbons in the deep sea." *Deep Oil Spills: Facts, Fate, and Effects* (2020): 107–124.

Kumari, Priyanka, and Yogesh Kumar. "Bioinformatics and computational tools in bioremediation and biodegradation of environmental pollutants." In *Bioremediation for Environmental Sustainability*, edited by: Vineet Kumar, Gaurav Saxena and Maulin P. Shah, pp. 421–444. Elsevier, 2021.

Li, Chunsheng, Ying Xu, Wei Jiang, Xiaoyan Dong, Dongfeng Wang, and Bingjie Liu. "Effect of NaCl on the heavy metal tolerance and bioaccumulation of Zygosaccharomycesrouxii and Saccharomyces cerevisiae." *Bioresource Technology* 143 (2013): 46–52.

Liu, Lingling, Xu-Biao Luo, Lin Ding, and Sheng-Lian Luo. "Application of nanotechnology in the removal of heavy metal from water." In *Nanomaterials for the Removal of Pollutants and Resource Reutilization*, edited by: Xubiao Luo and Fang Deng, pp. 83–147. Elsevier, 2019.

Liu, Zhanfei, and Jiqing Liu. "Evaluating bacterial community structures in oil collected from the sea surface and sediment in the northern Gulf of Mexico after the Deepwater Horizon oil spill." *Microbiologyopen* 2, no. 3 (2013): 492–504.

Mannig Jr, J. F., R. Boopathy, and C. F. Kulpa. *A Laboratory Study in Support of the Pilot Demonstration of Biological Soil Slurry Reactor.* Argonne National Lab IL Environmental Research Div, 1995.

Margesin, R., and F. Schinner. "Biodegradation and bioremediation of hydrocarbons in extreme environments." *Applied Microbiology and Biotechnology* 56 (2001): 650–663.

Mason, Olivia U., Terry C. Hazen, Sharon Borglin, Patrick S. G. Chain, Eric A. Dubinsky, Julian L. Fortney, James Han et al. "Metagenome, metatranscriptome and single-cell sequencing reveal microbial response to Deepwater Horizon oil spill." *The ISME Journal* 6, no. 9 (2012): 1715–1727.

Mathur, A. K., and C. B. Majumder. "Kinetics modelling of the biodegradation of benzene, toluene and phenol as single substrate and mixed substrate by using Pseudomonas putida." *Chemical and Biochemical Engineering Quarterly* 24, no. 1 (2010): 101–109.

Mbhele, Phelelani Phetheni. "Remediation of soil and water contaminated by heavy metals and hydrocarbons using silica encapsulation." PhD diss., University of the Witwatersrand, 2007.

Mesa, Jennifer, Ignacio David Rodríguez-Llorente, Eloisa Pajuelo, José María Barcia Piedras, Miguel Angel Caviedes, Susana Redondo-Gómez, and Enrique Mateos-Naranjo. "Moving closer towards restoration of contaminated estuaries: Bioaugmentation with autochthonous rhizobacteria improves metal rhizoaccumulation in native Spartinamaritima." *Journal of Hazardous Materials* 300 (2015): 263–271.

Mishra, Manisha, Sandeep Kumar Singh, and Ajay Kumar. "Environmental factors affecting the bioremediation potential of microbes." In *Microbe Mediated Remediation of Environmental Contaminants*, edited by: Ajay Kumar, Vipin Kumar Singh, ... and Virendra Kumar Mishrapp, 47–58. Woodhead Publishing, 2021.

Mohan, S. Venkata, Takuro Kisa, Takeru Ohkuma, Robert A. Kanaly, and Yoshihisa Shimizu. "Bioremediation technologies for treatment of PAH-contaminated soil and strategies to enhance process efficiency." *Reviews in Environmental Science and Bio/Technology* 5 (2006): 347–374.

Mueller, James G., Peter J. Chapman, and P. Hap Pritchard. "Creosote-contaminated sites. Their potential for bioremediation." *Environmental Science & Technology* 23, no. 10 (1989): 1197–1201.

Nannipieri, Paolo, L. Badalucco, Loretta Landi, and Giacomo Pietramellara. "Measurement in assessing the risk of chemicals to the soil eco system." In *Ecotoxicology: Responses, Biomarkers and Risk Assessment* (eds J.T. Zelikoff, J.M. Lynch, and J. Shepers), pp. 507–534. OECD, 1997.

Pichtel, J. *Fundamentals of Site Remediation for Metal-and Hydrocarbon-Contaminated Soils*, 437 p. Rockville, MD: Government Institutes. Inc., 2007.

Purchase, Diane, Lian N. L. Scholes, D. Mike Revitt, and R. Brian E. Shutes. "Effects of temperature on metal tolerance and the accumulation of Zn and Pb by metal-tolerant fungi isolated from urban runoff treatment wetlands." *Journal of Applied Microbiology* 106, no. 4 (2009): 1163–1174.

Rathoure, Ashok K., and Dhatwalia, V.K., ed. *Toxicity and Waste Management using Bioremediation*. IGI Global, 2016.

Rodriguez-R, Luis M., Will A. Overholt, Christopher Hagan, Markus Huettel, Joel E. Kostka, and Konstantinos T. Konstantinidis. "Microbial community successional patterns in beach sands impacted by the Deepwater Horizon oil spill." *The ISME Journal* 9, no. 9 (2015): 1928–1940.

Santero, Eduardo, and Eduardo Díaz. "Genetics of biodegradation and bioremediation." *Genes* 11, no. 4 (2020): 441.

Seshadri, Rekha, Lorenz Adrian, Derrick E. Fouts, Jonathan A. Eisen, Adam M. Phillippy, Barbara A. Methe, Naomi L. Ward et al. "Genome sequence of the PCE-dechlorinating bacterium Dehalococcoidesethenogenes." *Science* 307, no. 5706 (2005): 105–108.

Sharma, B., S. Singh, and N. J. Siddiqi. "Biomedical implications of heavy metals induced imbalances in redox systems." *BioMed Research International* no. 1 (2014): 640754.

Sharma, Shilpi. "Bioremediation: Features, strategies and applications." *Asian Journal of Pharmacy and Life Science* 2231 (2012): 4423.

Shin, Boryoung, Minjae Kim, Karsten Zengler, Kuk-Jeong Chin, Will A. Overholt, Lisa M. Gieg, Konstantinos T. Konstantinidis, and Joel E. Kostka. "Anaerobic degradation of hexadecane and phenanthrene coupled to sulfate reduction by enriched consortia from northern Gulf of Mexico seafloor sediment." *Scientific Reports* 9, no. 1 (2019): 1239.

Singh, Om V., and Nagathihalli S. Nagaraj. "Transcriptomics, proteomics and interactomics: Unique approaches to track the insights of bioremediation." *Briefings in Functional Genomics* 4, no. 4 (2006): 355–362.

Singh, Reena, Neetu Gautam, Anurag Mishra, and Rajiv Gupta. "Heavy metals and living systems: An overview." *Indian Journal of Pharmacology* 43, no. 3 (2011): 246.

Stapleton, Raymond D., Dwayne C. Savage, Gary S. Sayler, and Gary Stacey. "Biodegradation of aromatic hydrocarbons in an extremely acidic environment." *Applied and Environmental Microbiology* 64, no. 11 (1998): 4180–4184.

Tang, Chuyang Y., Q. Shiang Fu, Craig S. Criddle, and James O. Leckie. "Effect of flux (transmembrane pressure) and membrane properties on fouling and rejection of reverse osmosis and nanofiltration membranes treating perfluorooctanesulfonate containing wastewater." *Environmental Science & Technology* 41, no. 6 (2007): 2008–2014.

Tchounwou, Paul B. Yedjou, C.G., Patlolla, A.K., and Sutton, D.J. "Heavy metal toxicity and the environment, in molecular, clinical and environmental toxicology." In *Experientia Supplementum*, vol 101, edited by Luch, A. Springer, Basel, (2012).

Tyagi, Meenu, M. Manuela R. da Fonseca, and Carla C. C. R. de Carvalho. "Bioaugmentation and biostimulation strategies to improve the effectiveness of bioremediation processes." *Biodegradation* 22 (2011): 231–241.

Umar, Zubairu Darma, Nor Azwady Abd Aziz, Syaizwan Zahmir Zulkifli, and Muskhazli Mustafa. "Rapid biodegradation of polycyclic aromatic hydrocarbons (PAHs) using effective Cronobactersakazakii MM045 (KT933253)." *MethodsX* 4 (2017): 104–117.

Van Hamme, Jonathan D., Ajay Singh, and Owen P. Ward. "Recent advances in petroleum microbiology." *Microbiology and Molecular Biology Reviews* 67, no. 4 (2003): 503–549.

Vonk, Sophie M., David J. Hollander, and Alber Tinka J. Murk. "Was the extreme and wide-spread marine oil-snow sedimentation and flocculent accumulation (MOSSFA) event during the Deepwater Horizon blow-out unique?." *Marine Pollution Bulletin* 100, no. 1 (2015): 5–12.

Wang, Jianqiao, Tomohiro Suzuki, Toshio Mori, Ru Yin, Hideo Dohra, Hirokazu Kawagishi, and Hirofumi Hirai. "Transcriptomics analysis reveals the high biodegradation efficiency of white-rot fungus Phanerochaetesordida YK-624 on native lignin." *Journal of Bioscience and Bioengineering* 132, no. 3 (2021): 253–257.

Wang, Jin, Qing Li, Ming-Ming Li, Tian-Hu Chen, Yue-Fei Zhou, and Zheng-Bo Yue. "Competitive adsorption of heavy metal by extracellular polymeric substances (EPS) extracted from sulfate reducing bacteria." *Bioresource Technology* 163 (2014): 374–376.

Wang, L., S. Barrington and J. W. Kim. "Biodegradation of pentyl amine and aniline from petrochemical wastewater." *Journal of Environmental Management* 83, no. 2 (2007): 191–197.

Wuana, Raymond A., and Felix E. Okieimen. "Heavy metals in contaminated soils: A review of sources, chemistry, risks and best available strategies for remediation." *International Scholarly Research Notices*, edited by B. Montuelle, and A. D. Steinman 2011 (2011), Article ID 402647.

Yin, Kun, Min Lv, Qiaoning Wang, Yixuan Wu, Chunyang Liao, Weiwei Zhang, and Lingxin Chen. "Simultaneous bioremediation and biodetection of mercury ion through surface display of carboxylesterase E2 from Pseudomonas aeruginosa PA1." *Water Research* 103 (2016): 383–390.

Yue, Zheng-Bo, Qing Li, Chuan-chuan Li, Tian-hu Chen, and Jin Wang. "Component analysis and heavy metal adsorption ability of extracellular polymeric substances (EPS) from sulfate reducing bacteria." *Bioresource Technology* 194 (2015): 399–402.

Zahed, Mohammad Ali, Hamidi Abdul Aziz, Mohamed Hasnain Isa, and Leila Mohajeri. "Effect of initial oil concentration and dispersant on crude oil biodegradation in contaminated seawater." *Bulletin of Environmental Contamination and Toxicology* 84 (2010): 438–442.

Ziervogel, Kai, Luke McKay, Benjamin Rhodes, Christopher L. Osburn, Jennifer Dickson-Brown, Carol Arnosti, and Andreas Teske. "Microbial activities and dissolved organic matter dynamics in oil-contaminated surface seawater from the Deepwater Horizon oil spill site." *PloS One* 7, no. 4 (2012): e34816.

17 Environmental DNA and Its Application in Microbial Biodiversity Assessment

Neetu Shahi, Bhupendra Singh, Aslah Mohamad, and Sumanta Kumar Mallik

17.1 INTRODUCTION

Environmental DNA (eDNA) has become a powerful tool for assessing biodiversity in various ecosystems. This technology relies on the detection of genetic material left behind by organisms in their environment. eDNA has been applied to a range of organisms, from macrofauna to microorganisms. In recent years, eDNA technology has been used to assess microbial biodiversity in soils, sediments, water, and air, revealing new insights into the microbial world. The use of eDNA technology in microbial biodiversity assessment is becoming increasingly common due to its non-invasive, cost-effective, and rapid nature. This chapter will provide an overview of eDNA technology, its advantages and limitations, and its application in microbial biodiversity assessment.

17.2 PRINCIPLES OF EDNA TECHNOLOGY

Environmental DNA (eDNA) technology is a molecular tool that uses DNA molecules found in environmental samples to identify the presence or absence of a target organism in the sampled environment. The basic principle of eDNA technology is that organisms release DNA into the environment through various means such as excretion, secretion, and decay. This DNA can be extracted from environmental samples such as water, sediment, soil, or air (Figure 17.1) and amplified using PCR (Polymerase Chain Reaction) techniques to detect the presence or absence of target species (Ruppert et al., 2019; Ficetola et al., 2008).

The extraction of eDNA from environmental samples involves several steps (Figure 17.2) including the collection of the samples, filtration, concentration, and DNA extraction (Ficetola et al., 2008; Rees et al., 2014). Filtration is used to separate the target DNA from other organic and inorganic materials present in the sample. To enhance the quality of the filtrate, the sample could be filtered through a series of filters, from a large pore size, i.e., 11 µm, down to the smallest desired pore size, in this case 0.2 µm (Taberlet et al., 2012; Taberlet et al., 2018) (Figures 17.3 and 17.4).

After filtration, the DNA is concentrated using various methods such as precipitation or column-based purification. DNA extraction involves breaking down the cell membranes of the target organisms, releasing their DNA into solution, and purifying the DNA using various extraction kits and protocols, depending on the target organism, using the phenol-chloroform method for bacterial genomic DNA extraction and the CTAB method for fungal and algal genomic DNA extraction (Ruppert et al., 2019) (Figure 17.5).

After DNA extraction, PCR amplification is carried out using specific primers designed to target regions of the DNA of interest. PCR amplification generates large amounts of DNA, which can be detected and quantified using various techniques such as quantitative PCR, real-time PCR,

DOI: 10.1201/9781003408543-17

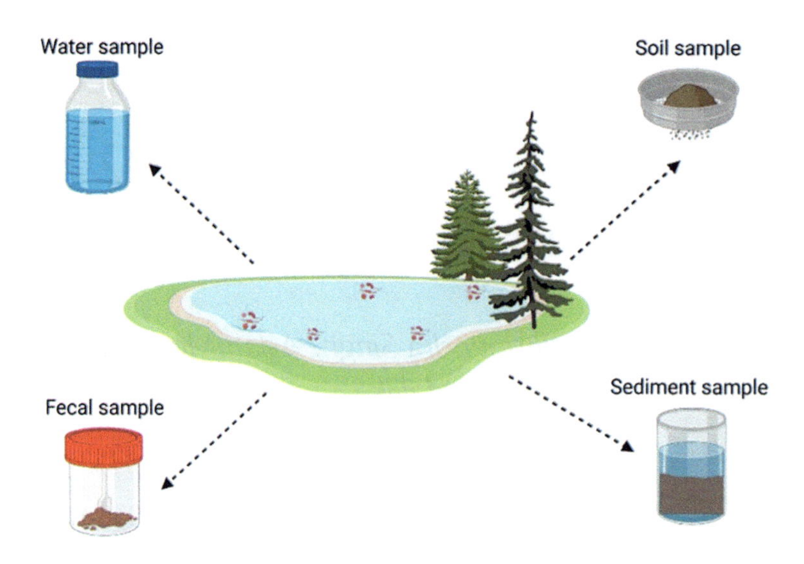

FIGURE 17.1 Possible sampling locations for eDNA study in aquatic habitat.

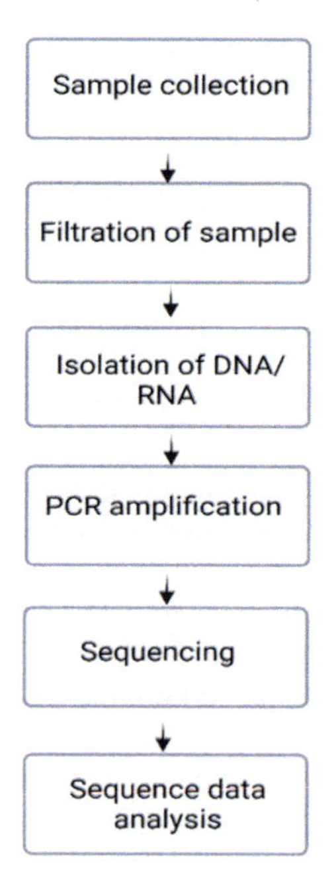

FIGURE 17.2 Workflow diagram of eDNA extraction and analysis.

multiplex PCR, or next-generation sequencing (NGS) (Igawa et al., 2019; Francioli et al., 2021) (Tables 17.1, 17.2, and 17.3 and Figure 17.6).

If we need to do metabarcoding, we should prepare the library and go for high-throughput sequencing (shotgun sequencing or NGS), then analyse the sequences through various bioinformatic tools (Figure 17.7).

The reliability of eDNA data is influenced by several factors, such as the quality of the sample, the concentration of the target DNA, and the choice of PCR primers (Mauvsseau et al., 2019). The reliability of eDNA data is also influenced by the sensitivity and specificity of the PCR assay, which

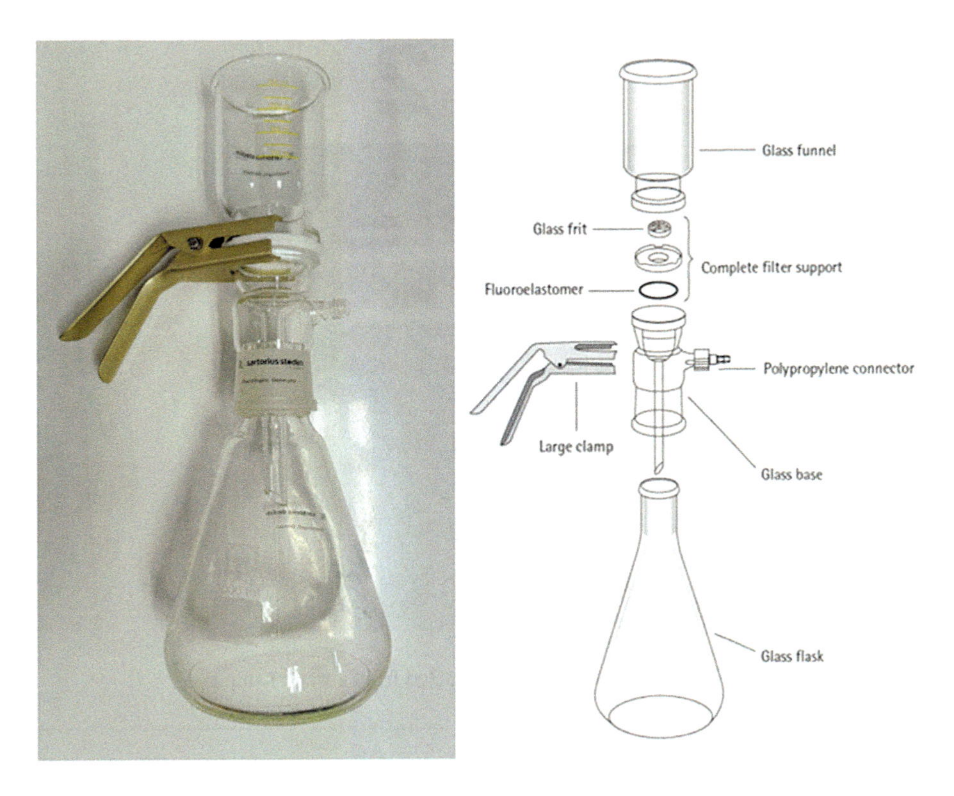

FIGURE 17.3 Filtration unit and description of parts.

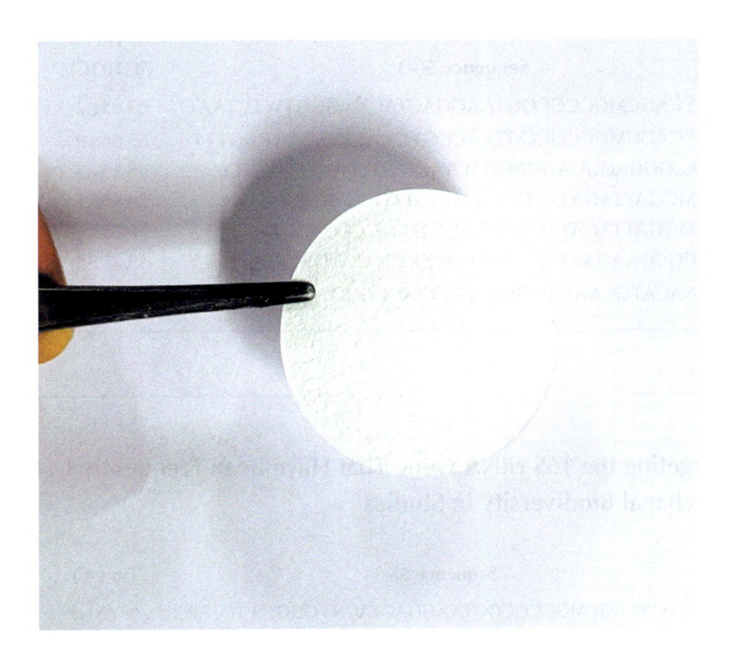

FIGURE 17.4 Cellulose nitrate ashless filter paper (0.2 μm pore size).

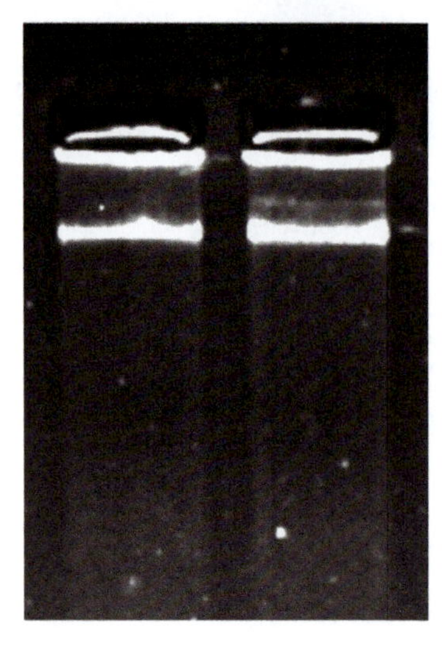

FIGURE 17.5 0.8% agarose gel containing eDNA extracted from a water sample.

TABLE 17.1

Primer Pairs Targeting the 16S rRNA Gene That Have Been Frequently Used to Characterize Bacterial Biodiversity in Studies

Primer Pair	Sequence 5′–3′	Tm (°C103C)	Amplified Region	Amplicon Length
515fB806rB	GTGYCAGCMGCCGCGGTAAGGACTACNVGGGTWTCTAAT	63.651.2	V4	253
515fB926r	GTGYCAGCMGCCGCGGTAACCGYCAATTYMTTTRAGTTT	63.648.9	V4–V5	394
341fB805r	CCTACGGGAGGCAGCAGGACTACHVGGGTATCTAATCC	58.251.3	V3–V4	418
799f1115r	AACMGGATTAGATACCCKGAGGGTTGCGCTCGTTG	50.956.1	V5–V6	301
799f1193r	AACMGGATTAGATACCCKGACGTCATCCCCACCTTCC	50.957.1	V5–V7	377
967f1391r	CAACGCGAAGAACCTTACCGACGGGCGGTGWGTRCA	53.859.5	V6–V8	405
68f518r	TNANACATGCAAGTCGRRCGWTTACCGCGGCTGCTG	55.556	V1–V3	438

TABLE 17.2

Primer Pairs Targeting the 16S rRNA Gene That Have Been Frequently Used to Characterize Archaeal Biodiversity in Studies

Primer Pair	Sequence 5′–3′	Tm (°C)	Amplified Region	Amplicon Length
515fB806rB	GTGYCAGCMGCCGCGGTAAGGACTACNVGGGTWTCTAAT	63.651.2	V4	253
340f806rB	CCCTAYGGGGYGCASCAGGGACTACNVGGGTWTCTAAT	61.351.2	V3–V4	388
SSU1ArFSSU520R	TCCGGTTGATCCYGCBRGGCTACGRRYGYTTTARRC	59.251	V1–V4	491
349f519r	GYGCASCAGKCGMGAAWTTACCGCGGCKGCTG	57.757.6	V3–V4	111
Parch519fArch915r	CAGCCGCCGCGGTAAGTGCTCCCCCGCCAATTCCT	59.462.9	V4–V5	386
1106F1378R	TTWAGTCAGGCAACGAGCTGTGCAAGGAGCAGGGAC	52.557.9	V7–V8	280

TABLE 17.3

Primer Pairs Targeting the ITS Region That Have Been Frequently Used to Characterize Fungal Biodiversity in Studies

Primer Pair	Sequence 5′–3′	Tm (°C)	Amplified Region	Amplicon Length
ITS1fITS2r	CTTGGTCATTTAGAGGAAGTAAGCTGCGTTCTTCATCGATGC	49.757	ITS1	357
ITS1F_KYO2ITS2_KYO2	TAGAGGAAGTAAAAGTCGTAATTYRCTRCGTTCTTCATC	4848.4	ITS1	358
ITS3ITS4	GCATCGATGAAGAACGCAGCTCCTCCGCTTATTGATATGC	5752.1	ITS2	306
gITS7ITS4ngs	GTGARTCATCGARTCTTTGTTCCTSCGCTTATTGATATGC	48.352.9	ITS2	288
fITS7ITS4	GTGARTCATCGAATCTTTGTCCTCCGCTTATTGATATGC	47.352.1	ITS2	292
ITS86fITS4	GTGAATCATCGAATCTTTGAATCCTCCGCTTATTGATATGC	48.652.1	ITS2	290

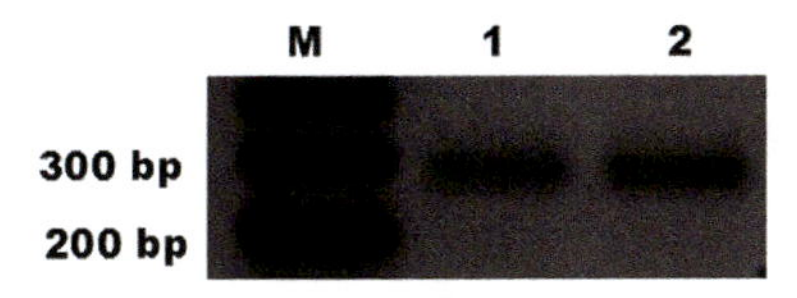

FIGURE 17.6 1.8% agarose gel with a band around 280 bp, showing the presence of the targeted organism in the water sample.

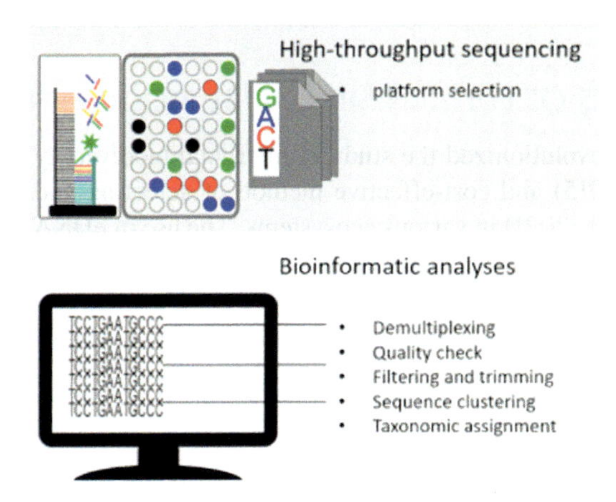

FIGURE 17.7 Metabarcoding analysis using high-throughput sequencing and later analysed with bioinformatic tools.

depends on the choice of PCR primers and the quality of the DNA template (Mauvsseau et al., 2019).

Overall, eDNA technology has several advantages over traditional methods of microbial assessment such as culturing and microscopy. eDNA technology is highly sensitive and can detect the presence of target species even when they are present in very low concentrations. eDNA technology is also highly specific, as it targets specific DNA sequences that are unique to the target species. Furthermore, eDNA technology is non-invasive and can be used to detect species that are difficult to sample using traditional methods (Thomsen et al., 2015).

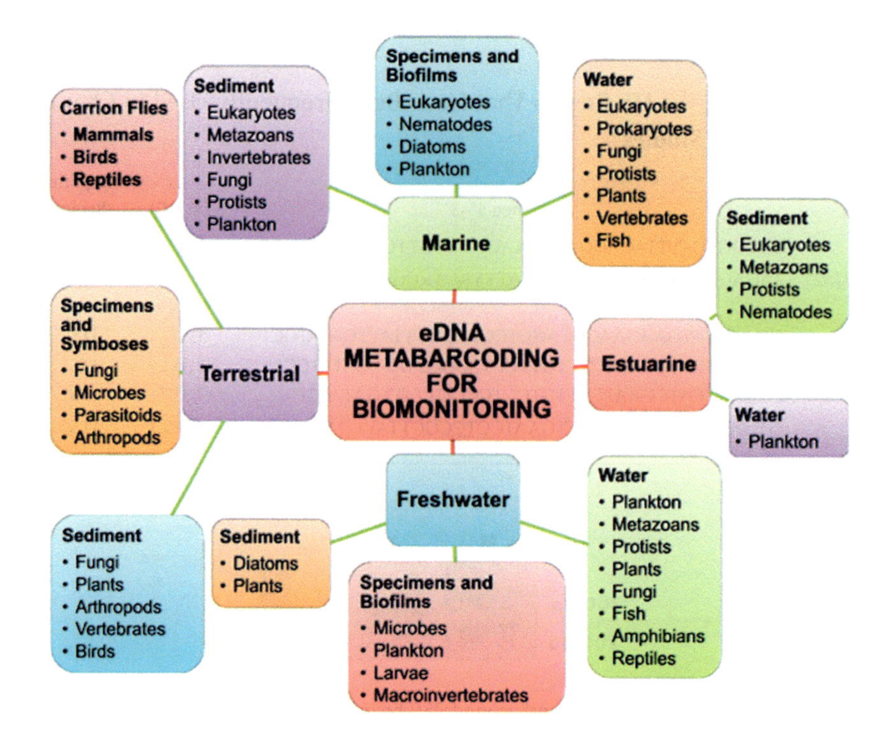

FIGURE 17.8 Schematic diagram of global ecosystem and biodiversity monitoring with environmental DNA metabarcoding.

17.3 APPLICATIONS OF EDNA IN MICROBIAL BIODIVERSITY ASSESSMENT

eDNA technology has revolutionized the study of microbial biodiversity by providing a non-invasive (Thomsen et al., 2015) and cost-effective method of detecting and characterizing microbial communities (Ladin et al., 2021) in various ecosystems. The use of eDNA technology has expanded rapidly in recent years, and it has become a powerful tool for biodiversity assessment and monitoring. The eDNA technology has application in microbial biodiversity assessment, with a focus on its use in studying microbial communities in soils, sediments, water, and air (Figures 17.8 and 17.9).

17.3.1 Soil Microbial Diversity

Soil is a complex and diverse ecosystem, hosting a wide variety of microorganisms that play crucial roles in nutrient cycling and other ecological processes. The study of soil microbial diversity is essential for understanding the functioning of terrestrial ecosystems, and eDNA technology provides a powerful tool for soil microbial assessment.

eDNA technology has been used to study soil microbial diversity in various ecosystems, including agricultural fields, forests, and grasslands. A recent study used eDNA sequencing to investigate the impact of land-use change on soil microbial diversity in grasslands (Skidmore et al., 2022; Ruppert et al., 2019). The study revealed significant differences in soil microbial diversity and composition between different land-use types, highlighting the importance of eDNA technology in understanding the impact of human activities on soil microbial communities.

17.3.2 Sediment Microbial Diversity

Sediments are an essential component of aquatic ecosystems, providing a habitat for diverse microbial communities that play important roles in nutrient cycling and other ecological processes. The

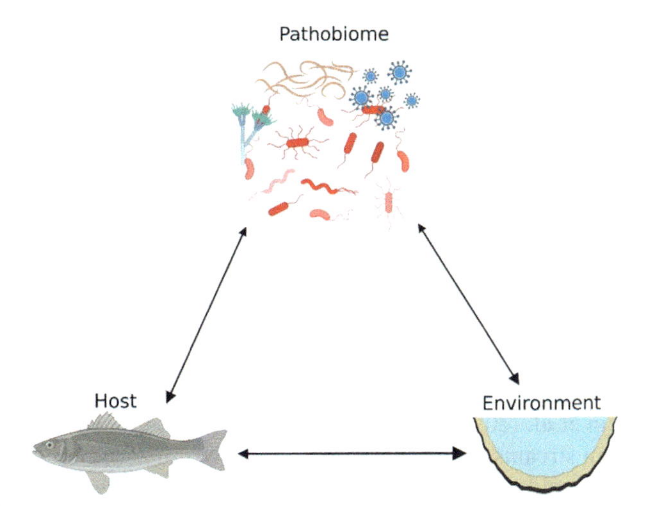

FIGURE 17.9 Pathobiome study by eDNA.

use of eDNA technology has enabled researchers to study sediment microbial diversity more comprehensively than by using traditional methods.

Several studies have used eDNA technology to investigate sediment microbial diversity in various aquatic ecosystems, including freshwater and marine environments. For example, a recent study used eDNA sequencing to assess the microbial diversity of sediments in a polluted river in China (Ruppert et al., 2019). The study revealed a significant decrease in microbial diversity and changes in microbial community composition in the polluted sediments compared with the reference site, demonstrating the potential of eDNA technology in monitoring the impact of anthropogenic activities on sediment microbial communities.

17.3.3 Water Microbial Diversity

Water is a critical component of many ecosystems, and the study of microbial diversity in aquatic environments is essential for understanding the functioning of these ecosystems. The use of eDNA technology has revolutionized the study of water microbial diversity, providing a non-invasive and cost-effective method for assessing microbial communities in different aquatic environments.

eDNA technology has been used to study microbial diversity in various aquatic environments, including lakes, rivers, and oceans. For example, a recent study used eDNA sequencing to investigate the microbial diversity of a eutrophic lake in China (Ruppert et al., 2019; Liu et al., 2022). The study revealed significant differences in microbial community composition between different sampling locations in the lake, highlighting the potential of eDNA technology in assessing the spatial distribution of microbial communities in aquatic environments.

17.3.4 Air Microbial Diversity

Airborne microorganisms play essential roles in various ecological processes, including nutrient cycling and disease transmission. The use of eDNA technology has enabled researchers to study air microbial diversity more comprehensively than traditional methods, providing a powerful tool for studying the ecology of airborne microorganisms.

Several studies have used eDNA technology to investigate air microbial diversity in various environments, including indoor and outdoor air. For example, a recent study used eDNA sequencing to assess the microbial diversity of indoor air in different environments, including hospitals, offices, and homes (Clare et al., 2022). The study revealed significant differences in microbial community composition between different indoor environments, demonstrating the potential of eDNA technology in assessing the impact of human activities on indoor air microbial communities.

17.4 RECENT STUDIES ON THE APPLICATION OF EDNA TECHNOLOGY IN MICROBIAL BIODIVERSITY ASSESSMENT

1. A study by Schenker (2023) used eDNA sequencing to investigate the bacterial communities in freshwater ecosystems. The study found that eDNA sequencing provided a more comprehensive picture of microbial diversity than traditional culturing methods.
2. A study by Nunez et al. (2021) used eDNA sequencing to investigate the microbial communities in soil samples from a variety of agricultural systems. The study found that eDNA analysis provided a more detailed understanding of soil microbial diversity and function.
3. In a study by Pin et al. (2021), eDNA sequencing was used to assess the microbial communities in urban streams. The study found that eDNA analysis could detect changes in microbial diversity and composition due to urbanization and other anthropogenic factors.
4. A study by Pawlowski et al. (2022) used eDNA technology to investigate the microbial communities in lake sediments. The study found that eDNA analysis provided a more comprehensive understanding of sediment microbial diversity and function than traditional methods.

17.5 USE OF EDNA TECHNOLOGY IN MONITORING AND MANAGING MICROBIAL BIODIVERSITY

eDNA technology can be used for the following purposes:

1. To monitor changes in microbial diversity and composition in response to anthropogenic activities such as pollution, land-use change, and climate change (Ruppert et al., 2019).
2. To monitor the success of microbial-based environmental remediation strategies, such as bioremediation (Gasparini, 2019).
3. To get information about the distribution and abundance of microbial communities in different ecosystems, which can be used to inform conservation and management efforts (Nagarajan et al., 2022).
4. To monitor the presence and spread of invasive microbial species, which can have significant ecological and economic impacts (Johnsen et al., 2020).

17.6 ADVANTAGES AND LIMITATIONS OF EDNA IN MICROBIAL BIODIVERSITY ASSESSMENT

Environmental DNA (eDNA) technology has emerged as a powerful tool for studying microbial communities in different ecosystems. The analysis of eDNA enables the identification of microorganisms present in the environment without the need for cultivation, which can be time-consuming and biased. However, as with any technology, eDNA has its advantages and limitations, which must be carefully considered when interpreting data and drawing conclusions.

17.6.1 Advantages of eDNA Technology

One of the main advantages of eDNA technology is that it allows the detection and identification of a wide range of microorganisms, including those that are difficult to culture, rare, or present at low concentrations. This is particularly important in complex microbial communities, where the vast majority of microorganisms are not yet described or cannot be easily cultured (Handelsman, 2004).

For example, eDNA analysis has been used to identify novel bacterial and fungal species in soil and water environments (Ruppert et al., 2019; Ladin et al., 2021).

Another advantage of eDNA technology is that it can provide a more comprehensive and representative view of microbial diversity compared with traditional cultivation-based methods. This is because eDNA analysis allows the detection of both living and dead microorganisms, as well as those that are present in a dormant state (Ruppert et al., 2019). Additionally, eDNA analysis can reveal the functional potential of microbial communities through the detection of genes involved in specific metabolic pathways (Wiseschart et al., 2019).

Moreover, eDNA analysis is a non-invasive and non-destructive method, which allows the sampling of microbial communities without disturbing the environment or altering the microbial community structure (Banerjee et al., 2022). This is particularly important in sensitive ecosystems where disturbances can have significant impacts on microbial diversity and ecosystem function.

17.6.2 Limitations of eDNA Technology

Despite its many advantages, eDNA technology also has several limitations that must be considered when interpreting data. One of the main limitations of eDNA analysis is the potential for false positives and false negatives. This can be caused by the presence of non-target DNA, contamination during sampling or laboratory processing, or the degradation of eDNA during transport and storage (De Souza et al., 2016). Therefore, careful quality control and validation of eDNA data are essential to ensure the accuracy and reliability of results.

Another limitation of eDNA analysis is that it may not provide information on the activity or physiological state of microorganisms. This is because eDNA can persist in the environment for long periods of time, even after the microorganisms that produced it have died or become dormant (Ruppert et al., 2019). Therefore, eDNA analysis should be combined with other methods, such as metatranscriptomics or metaproteomics, to obtain a more complete picture of microbial activity and function (Carvalhais et al., 2012).

Furthermore, eDNA analysis can be influenced by various factors, such as the type of ecosystem, the complexity of the microbial community, and the sensitivity and specificity of the eDNA markers used (Xia et al., 2021). Therefore, it is important to carefully select appropriate eDNA markers and use them in combination with other methods to validate and confirm results.

17.7 POTENTIAL APPLICATIONS OF EDNA TECHNOLOGY IN MICROBIAL CONSERVATION, RESTORATION, AND ECOLOGY

The use of eDNA technology in microbial conservation and restoration has the potential to revolutionize our ability to monitor and manage microbial biodiversity. One potential application of eDNA technology in microbial conservation is the identification and monitoring of rare or threatened microbial species. eDNA analysis can provide a non-invasive and cost-effective method to detect rare microbial species, which can be critical for conservation efforts. For example, a recent study used eDNA analysis to detect the presence of a rare and threatened species, a tardigrade, in Skråstadheia Nature Reserve in Norway (Topstad et al., 2021).

In addition to identifying rare species, eDNA analysis can also be used to monitor the effectiveness of conservation and restoration efforts. For example, it can be used to monitor the recovery of microbial communities in restored ecosystems. A recent study used it to monitor the recovery of microbial communities in a restored salt marsh, showing that microbial community composition had shifted towards a more natural state (Kim et al., 2023).

As eDNA technology continues to advance, there are many potential future applications in microbial ecology. It can be potentially applied to study the functional roles of microbial communities in ecosystem processes. By analysing eDNA from different microbial taxa, researchers can gain insights into the functional roles of these taxa in biogeochemical cycles, such as nutrient cycling and carbon sequestration (Ladin et al., 2021).

It also has potential to revolutionize our understanding and management of microbial biodiversity. From studying microbial communities in different ecosystems to monitoring and managing microbial conservation efforts, eDNA analysis has proven to be a powerful tool. As eDNA technology continues to advance, there are many emerging trends and technologies that are shaping the future of eDNA research, and there are many potential future applications in microbial ecology. However, it is important to be aware of the limitations and potential biases of eDNA analysis, and to validate eDNA data through independent methods.

17.8 EMERGING TRENDS AND TECHNOLOGIES IN EDNA RESEARCH

As eDNA technology continues to advance, there are emerging trends and technologies that are shaping the future of eDNA research. One of these trends is the use of nanopore sequencing technology, which allows rapid and accurate sequencing of DNA in real time. Nanopore sequencing has the potential to improve the accuracy and speed of eDNA analysis, making it an even more powerful tool for microbial biodiversity assessment (Hatfield et al., 2020).

Another emerging trend is the integration of eDNA data with other types of environmental data, such as remote sensing and in situ sensor data. By combining eDNA data with other types of environmental data, researchers can gain a more comprehensive understanding of microbial community dynamics and ecosystem processes (Thomsen et al., 2015).

eDNA technology can also be used to study the microbiomes of non-model organisms, such as insects and plants. eDNA analysis can provide a non-invasive and high-throughput method to study the microbiomes of these organisms, which can be important for understanding their ecology and evolution (Ruppert et al., 2019).

17.9 CONCLUSION

eDNA technology has revolutionized the study of microbial biodiversity and has led to the discovery of novel taxa in different ecosystems, including soils, sediments, water, and air. It has also been used to monitor and manage microbial biodiversity, including the assessment of the impact of anthropogenic activities on microbial communities. With the increasing availability of high-throughput sequencing technologies, eDNA technology will undoubtedly play a crucial role in advancing our understanding of microbial biodiversity and its role in ecosystem function and health.

REFERENCES

Banerjee, Pritam, Kathryn A. Stewart, Gobinda Dey, Caterina M. Antognazza, Raju Kumar Sharma, Jyoti Prakash Maity, Santanu Saha et al. "Environmental DNA analysis as an emerging non-destructive method for plant biodiversity monitoring: A review." *AoB Plants* 14, no. 4 (2022): plac031.

Carvalhais, Lilia C., Paul G. Dennis, Gene W. Tyson, and Peer M. Schenk. "Application of metatranscriptomics to soil environments." *Journal of Microbiological Methods* 91, no. 2 (2012): 246–251.

Clare, Elizabeth L., Chloe K. Economou, Frances J. Bennett, Caitlin E. Dyer, Katherine Adams, Benjamin McRobie, Rosie Drinkwater, and Joanne E. Littlefair. "Measuring biodiversity from DNA in the air." *Current Biology* 32, no. 3 (2022): 693–700.

De Souza, Lesley S., James C. Godwin, Mark A. Renshaw, and Eric Larson. "Environmental DNA (eDNA) detection probability is influenced by seasonal activity of organisms." *PloS One* 11, no. 10 (2016): e0165273.

Francioli, Davide, Guillaume Lentendu, Simon Lewin, and Steffen Kolb. "DNA metabarcoding for the characterization of terrestrial microbiota—Pitfalls and solutions." *Microorganisms* 9, no. 2 (2021): 361.

Ficetola, Gentile Francesco, Claude Miaud, François Pompanon, and Pierre Taberlet. "Species detection using environmental DNA from water samples." *Biology Letters* 4, no. 4 (2008): 423–425.

Gasparini, Louis. "Using environmental DNA can improve site remediation success." *Environmental Science & Engineering Magazin* (2019). (https://esemag.com/site-remediation/using-environmental-dna-can-improve-site-remediation-success)

Handelsman, Jo. "Metagenomics: Application of genomics to uncultured microorganisms." *Microbiology and Molecular Biology Reviews* 68, no. 4 (2004): 669–685.

Hatfield, Robert G., Frederico M. Batista, Timothy P. Bean, Vera G. Fonseca, Andres Santos, Andrew D. Turner, Adam Lewis, Karl J. Dean, and Jaime Martinez-Urtaza. "The application of nanopore sequencing technology to the study of dinoflagellates: A proof of concept study for rapid sequence-based discrimination of potentially harmful algae." *Frontiers in Microbiology* 11 (2020): 844.

Igawa, Takeshi, Teruhiko Takahara, Quintin Lau, and Shohei Komaki. "An application of PCR-RFLP species identification assay for environmental DNA detection." *PeerJ* 7 (2019): e7597.

Johnsen, Stein I., David A. Strand, Johannes C. Rusch, and Trude Vrålstad. "Environmental DNA (eDNA) monitoring of noble crayfish Astacus astacus in lentic environments offers reliable presence-absence surveillance–but fails to predict population density." *Frontiers in Environmental Science* 8 (2020): 612253.

Kim, Carol, Lorie W. Staver, Xuan Chen, Ashley Bulseco, Jeffrey C. Cornwell, and Sairah Y. Malkin. "Microbial community succession along a chronosequence in constructed salt marsh soils." *Microbial Ecology* (2023): 1–20.

Ladin, Zachary S., Barbra Ferrell, Jacob T. Dums, Ryan M. Moore, Delphis F. Levia, W. Gregory Shriver, Vincent D'Amico, Tara L. E. Trammell, João Carlos Setubal, and K. Eric Wommack. "Assessing the efficacy of eDNA metabarcoding for measuring microbial biodiversity within forest ecosystems." *Scientific Reports* 11, no. 1 (2021): 1629.

Liu, Qi, Hucai Zhang, Fengqin Chang, Ping Xie, Yun Zhang, Han Wu, Xiaonan Zhang, Wei Peng, and Fengwen Liu. "eDNA revealed in situ microbial community changes in response to Trapa japonica in Lake Qionghai and Lake Erhai, southwestern China." *Chemosphere* 288 (2022): 132605.

Mauvisseau, Quentin, Alfred Burian, Ceri Gibson, Rein Brys, Andrew Ramsey, and Michael Sweet. "Influence of accuracy, repeatability and detection probability in the reliability of species-specific eDNA based approaches." *Scientific Reports* 9, no. 1 (2019): 580.

Nagarajan, Raman P., Mallory Bedwell, Ann E. Holmes, Thiago Sanches, Shawn Acuña, Melinda Baerwald, Matthew A. Barnes et al. "Environmental DNA methods for ecological monitoring and biodiversity assessment in estuaries." *Estuaries and Coasts* 45, no. 7 (2022): 2254–2273.

Nunez, Nicolas Fernandez, Laurent Maggia, Pierre-Louis Stenger, Mélanie Lelievre, Kelly Letellier, Sarah Gigante, Aurore Manez, Pierre Mournet, Julie Ripoll, and Fabian Carriconde. "Potential of high-throughput eDNA sequencing of soil fungi and bacteria for monitoring ecological restoration in ultramafic substrates: The case study of the New Caledonian biodiversity hotspot." *Ecological Engineering* 173 (2021): 106416.

Pawlowski, J., K. Bruce, K. Panksep, F. I. Aguirre, S. Amalfitano, L. Apothéloz-Perret-Gentil, T. Baussant et al. "Environmental DNA metabarcoding for benthic monitoring: A review of sediment sampling and DNA extraction methods." *Science of the Total Environment* 818 (2022): 151783.

Pin, Lorenzo, Alexander Eiler, Stefano Fazi, and Nikolai Friberg. "Two different approaches of microbial community structure characterization in riverine epilithic biofilms under multiple stressors conditions: Developing molecular indicators." *Molecular Ecology Resources* 21, no. 4 (2021): 1200–1215.

Rees, Helen C., Ben C. Maddison, David J. Middleditch, James R. M. Patmore, and Kevin C. Gough. "The detection of aquatic animal species using environmental DNA–a review of eDNA as a survey tool in ecology." *Journal of Applied Ecology* 51, no. 5 (2014): 1450–1459.

Ruppert, Krista M., Richard J. Kline, and Md Saydur Rahman. "Past, present, and future perspectives of environmental DNA (eDNA) metabarcoding: A systematic review in methods, monitoring, and applications of global eDNA." *Global Ecology and Conservation* 17 (2019): e00547.

Schenekar, Tamara. "The current state of eDNA research in freshwater ecosystems: Are we shifting from the developmental phase to standard application in biomonitoring?." *Hydrobiologia* 850, no. 6 (2023): 1263–1282.

Skidmore, Andrew K., Andjin Siegenthaler, Tiejun Wang, Roshanak Darvishzadeh, Xi Zhu, Anthony Chariton, and G. Arjen De Groot. "Mapping the relative abundance of soil microbiome biodiversity from eDNA and remote sensing." *Science of Remote Sensing* 6 (2022): 100065.

Taberlet, Pierre, Aurélie Bonin, Lucie Zinger, and Eric Coissac. *Environmental DNA: For Biodiversity Research and Monitoring.* Oxford University Press, 2018.

Taberlet, Pierre, Eric Coissac, François Pompanon, Christian Brochmann, and Eske Willerslev. "Towards next-generation biodiversity assessment using DNA metabarcoding." *Molecular Ecology* 21, no. 8 (2012): 2045–2050.

Thomsen, Philip Francis, and Eske Willerslev. "Environmental DNA–An emerging tool in conservation for monitoring past and present biodiversity." *Biological Conservation* 183 (2015): 4–18.

Topstad, Lasse, Roberto Guidetti, Markus Majaneva, and Torbjørn Ekrem. "Multi-marker DNA metabarcoding reflects tardigrade diversity in different habitats." *Genome* 64, no. 3 (2021): 217–231.

Wiseschart, Apirak, Wuttichai Mhuantong, Sithichoke Tangphatsornruang, Duriya Chantasingh, and Kusol Pootanakit. "Shotgun metagenomic sequencing from Manao-Pee cave, Thailand, reveals insight into the microbial community structure and its metabolic potential." *BMC Microbiology* 19, no. 1 (2019): 1–14.

Xia, Zhiqiang, Aibin Zhan, Mattias L. Johansson, Emma DeRoy, Gordon Douglas Haffner, and Hugh J. MacIsaac. "Screening marker sensitivity: Optimizing eDNA-based rare species detection." *Diversity and Distributions* 27, no. 10 (2021): 1981–1988.

18 Metagenomic Approach to Unculturable Microbes of the Aquatic Environment

Mamta Singh

18.1 BACKGROUND

Microorganisms are as diverse as their habitats, and they are found in almost all known environments, including air, soil, and water. The immense microbial diversity and their presence in every environment are attributed to their ability to survive under a wide range of conditions. The microbial communities of aquatic environments, like ocean, sea, rivers, canals, wetlands, mud-plains, coast and estuaries, mangrove, hot water springs, aquatic flora and fauna, etc., are very diverse and constitute varied aquatic ecosystems. Their role in food web dynamics and biogeochemical processes is an essential component of all aquatic ecosystems (Grossart et al., 2020). These microorganisms are a treasure trove of many genes of economic importance, but more than 99% of such microorganisms are not cultivable or difficult to culture. The requirement for pure culture of microorganisms is a major limitation to understanding their genetic diversity, physiology, functionality, and possible economic use. To exploit economically important genes and/or gene products, it is necessary to sequence the genomes of microbial communities from diverse ecosystems. Recent advancements in DNA sequencing techniques have opened the door to rapid and cost-effective whole genome sequencing of microbial communities. Metagenomics is the culture-independent genomic analysis of microbial communities from environmental samples (Handelsman et al., 1998). Metagenomics is a cutting-edge molecular technology for identification of the entirety of microorganisms present in an environment by analysing their DNA, collected directly from the environment. This technique eliminates the requirement for pure culture for their isolation and identification. With the help of this technology, the DNA of microorganisms from a given environment can be analysed as a whole. Metagenomic study provides detailed insights about taxonomic identity, genetic diversity, and physiological and metabolic function that will help in the characterization of all the microorganisms present in a given environment (Coughlan et al., 2015; Langille et al., 2013).

Metagenomic data are mainly generated by two approaches: (1) whole genome shotgun sequencing of all the microbes presents in the microbial community or (2) targeting a single gene, such as the highly conserved and taxonomically important 16S ribosomal RNA (rRNA) gene (Gilbert and Dupont, 2011; Staley and Sadowsky, 2016). 16S rDNA sequencing does not include the sequencing of total environmental DNA (eDNA) and therefore, cannot be considered as a metagenomics approach, but it has been used by many researchers to identify previously unknown taxa from diverse environments and gained much popularity among the research group working with unculturable microorganisms. For whole genome sequencing of eDNA, three different strategies are mainly used by researchers: (1) construction of an eDNA library, followed by cloning microbial DNA into a high-capacity cloning vector like cosmid, fosmid, or bacterial artificial chromosome (BAC) and screening of particular genes or gene products of interest; (2) shotgun sequencing, in which eDNA clones are randomly sequenced, followed by assembly and annotation; and (3) next-generation sequencing, which utilizes high-throughput sequencing platforms like pyrosequencing, Illumina, Ion Torrent,

DOI: 10.1201/9781003408543-18

etc. to generate millions of sequence reads, which eliminates the need for cloning the eDNA in any type of vector (Kimura, 2006; Gilbert and Dupont, 2011; Staley and Sadowsky, 2016).

18.2 STEPS INVOLVED IN METAGENOMIC STUDY

To generate and analyse the metagenomic data of a microbial community, three steps are involved: (1) genome enrichment and eDNA extraction, (2) construction of an eDNA library, and (3) screening of the eDNA library. Details of these steps are given in the following subsections.

18.2.1 Genome Enrichment and eDNA Extraction

The very first step involves the genome enrichment to increase the portion of desired clones in an eDNA library to be cloned. Stable isotope probing is a technique that has been used by some researchers to concentrate the portion of desired DNA from environmental samples (Quaiser et al., 2003; Diaz-Torres et al., 2006). In this technique, C-13 label substrate is provided to microbes to label their DNA and RNA, followed by density gradient separation of labelled nucleic acid. Another method, known as multiple displacement amplification (MDA), was used by several researchers to amplify whole genome from various environments with a specific type of DNA polymerase, ϕ29. DNA polymerase ϕ29 is a DNA-dependent DNA polymerase that is highly sensitive and widely used for rolling circle amplification of circular DNA of up to 70 kilobases (kb) in size (Kimura, 2006). The development of a good-quality eDNA library, having the desired representation of the entire metagenome and clones with a good probability of a complete operon, needs isolation of high-quality and high-molecular-weight DNA. Several protocols have been developed to isolate the good-quality eDNA from different environments by direct and indirect DNA isolation methods (Berry et al., 2003; Bertrand et al., 2005). The direct DNA isolation method involves *in-situ* lysis of microorganisms prior to DNA isolation, whereas the indirect method involves separation of cells from the environmental sample before lysis. Gabor et al. (2003) compared the efficiency and suitability of both methods and concluded that the direct DNA isolation method has higher DNA recovery, but the indirect method is better when the microbial diversity of DNA is much greater.

18.2.2 Construction of eDNA Library

Initially, a small insert library (<10 kb insert size) in a standard cloning vector and *Escherichia coli* as a host were used by many researchers. But due to their inability to allow detection of large operons or gene clusters, this approach was found unsuitable for metagenomics. With this approach, sequencing of more than 10^7 clones (5 kb insert) will be required to represent the collective genome. To overcome the limitation of small insert size library, large-capacity vectors (carrying capacity of 40 kb insert) like cosmid and fosmid were used for metagenomic library preparation. Several researchers also used BAC vectors (carrying capacity of 200 kb) for the preparation of a metagenomic library from eDNA (Martinez et al., 2004). The most commonly used host for propagation of eDNA clones is *E. coli*, among other hosts like *Streptomyces lividans*, *Rhizobium leguminosarum*, and *Pseudomonas putida*.

18.2.3 Screening of eDNA Library

Two different approaches are generally used for screening of an eDNA library: sequence-based screening and function-based screening. Sequence-based screening mainly involves the sequencing of genes encoding microbial ribosomal RNAs (rRNAs). The bacterial 16S rRNA gene has conserved flanking regions, which were exploited for primer designing, that facilitate the amplification and sequencing of variable regions. These regions provide species-specific sequences useful for taxonomic identification of prokaryotes in an environmental sample (Lane et al., 1985; Coughlan et al.,

2015). In an indirect approach, the 16S rRNA gene sequence of a metagenome can also be used for functional profiling with the aid of comparative genomics, where information regarding the functional roles of an already characterized bacterial species or a close relative may be used to assign the functions. Phylogenetic Investigation of Communities by Reconstruction of Unobserved States (PICRUSt) is a computational program developed by Langille et al., (2013) to predict the functional properties of microorganisms in an eDNA sample. This program characterized the eDNA clones by searching for similarity with the 16S rRNA gene sequence of closely related species from publicly available databases. Another combinatorial approach by Keller et al., (2014) also used 16S rRNA gene sequences for determining the structural and functional diversity of a metagenomic library. The same strategies were also used for characterizing the eukaryotic microbial communities by using eukaryotic-specific 18S rRNA primers. This strategy was successfully used by Bates et al. (2012) to identify the eukaryotic microbial species present in three different lichens: *Alveolata*, *Metazoa*, and *Rhizaria*. Function-based screening of metagenome has been considered more powerful because random shotgun sequencing of eDNA, followed by assembly and annotation, helps in determining the functional characteristics of microbial communities at species level. In this approach, the microbial component of an environmental sample is identified by direct sequencing the eDNA sample instead of assigning function on the basis of 16S rDNA sequence similarity (Coughlan et al., 2015). Venter et al. (2004) used the shotgun sequencing technique for characterization of the microbial community of the Sargasso Sea. In this study, they identified about 1.2 million previously unknown genes and reported the first assignment of rhodopsin-like photoreceptors to bacterial species. Function-based analyses also involve identification of clones with desired traits, followed by characterization of isolated clones by sequencing and biochemical analysis. Several other researchers have also used the method of shotgun sequencing for functional characterization and assignments (Warnecke et al., 2007; Oh et al., 2014; Hess et al., 2011). This approach was also used for microbial population dynamics, genomic evolution, distribution, and redundancy of function (Chandler et al., 2014; Kay et al., 2014; Mendes et al., 2015; Coughlan et al., 2015). Screening of metagenomic libraries through phenotypic-based approach is another method of assigning functions to unknown genes. In this method, change in expression is detected in the host organism after inserting the DNA of an unknown gene. Screening is usually done with an appropriate indicator to reveal the presence of phenotypically significant clones (Rondon et al., 2000; Henne et al., 2000).

18.3 APPLICATION OF METAGENOMIC STUDY OF UNCULTURABLE MICROBES FROM THE AQUATIC ENVIRONMENT

Metagenomics enables the bioprospecting of the aquatic environment, which helps in the discovery of many novel biocatalysts/enzymes, bioactive compounds, and secondary metabolites with pharmaceutical and nutritional properties. Microbial enzymes, bioactive compounds, and secondary metabolites identified from the aquatic environment have great industrial importance. Several enzymes have been isolated from the aquatic environment, especially the marine environment. They have been used in almost all industries, such as food, pharmaceuticals, detergents, textiles, biofuel, leather, etc. The use of novel biocatalysts isolated from extreme environments with high process performance is always favoured due to their low cost, high efficiency, and better recovery. In the last two decades, several metagenomic studies have been conducted to identify various types of enzymes from aquatic environments that have great industrial importance. Novel lipases identified by metagenomic studies have important applications in the detergent industry, paper processing, food additives, and biofuel production (Jaeger and Eggert, 2002; Kennedy et al., 2008). Many novel metagenomically identified amylolytic enzymes, like β-glucosidase, endoglucanase, lignases and xylanase, have applications in bioethanol production. Chitinase, isolated from wetland environments, has applications in food, medicine, agriculture, and the cosmetic industry, and was patented by Chengdu Institute of Biology, Chinese Academy of Sciences (CN107475273B) in the year 2017. Another important enzyme, a protease identified by a metagenomic approach having antitumor, antifungal, antiviral, and antiparasitic

TABLE 18.1
Novel Biocatalysts of Commercial Importance Isolated from Aquatic Environments

S. No.	Enzyme Name	Aquatic Environment	Reference
1	Aldehyde dehydrogenase	Hot spring water	Chen et al. (2014)
		Antarctic seawater	Kazuoka et al. (2007)
2	Lipase	Hot water spring, hot spring sediment, marine sponge, deep sea	Yan et al. (2017); Sahoo et al. (2017); Su et al. (2015); Jeon et al. (2009a)
3	Lipolytic enzyme	Nieme River valley, Germany	Henne et al. (2000)
4	Low pH, thermostable α-amylase	Deep sea and acid soil	Richardson et al. (2002)
5	Esterases	Soil water, mudflat, beaches, surface seawater, South China Sea, brine:seawater interface, Uranian hypersaline basin, sea water sediment, seashore sediments, intertidal zone, Red Sea brine pool, tidal flat sediment	Kim et al. (2006); Elend et al. (2006); Ouyang et al. (2013); Chu et al. (2008); Ferrer et al. (2005a); De Santi et al. (2016); Jeon et al. (2009b); Fu et al. (2013); Mohamed et al. (2013), Jeon et al. (2012)
6	Fumarase	Marine water	Jiang et al. (2010)
7	Thermostable esterase	Mud sediment-rich water	Rhee et al. (2005)
8	Nitrilase genes	Soil water	Robertson et al. (2004)
9	Fibrinolytic metalloprotease (zinc-dependent)	Mud, Korean west coast	Lee et al. (2007)
10	Alkaline-stable family IV lipase	Marine sediment, South China Sea	Peng et al. (2014)
11	Two UDP glycotransferase (UGT) genes. One is a novel macroside glycotransferase (MGT)	Tidal flat sediment, Elbe River, Germany	Rabausch et al. (2013)
12	Polyketide synthase (PKS) gene	Marine sponge *Discodermia dissoluta*, Netherlands Antilles	Fujita et al. (2011)
13	Novel serine protease inhibitor (serpin) gene	Uncultured marine organisms	Jiang et al. (2011)
14	Subtilisin like serine protease	Underground water	Apolinar et al. (2016)
15	Xylanase	Sea-bottom	Hung et al. (2011)
16	Acidobacterial-laccase like copper oxidase	Marsh bog soil	Ausec et al. (2017)
17	Laccase	Water from South China Sea	Fang et al. (2011)
18	β-Glucosidase	Hydrothermal spring water	Schroder et al. (2014)
18	β-Glucosidase	Subsea floor sediment	Klippel et al. (2014)
20	β-Galactosidase	Marine sediment	Li et al. (2015)
21	α-Galactosidase	Hot spring water and sediment	Schroder et al. (2017)
22	Hydrolase	Marine environment, Sargasso Sea	Cottrell et al. (2005)
23	Protease (alkaline)	Deep-sea sediment	Zeng et al. (2003)
24	Chitinase deacetylase	Deep-sea sediment	Liu et al. (2016)
25	Carboxylesterase	Marine mud	Gao et al. (2016)
26	Endoglucanase	Hot spring sediment	Zhao et al. (2017)
27	Alkaline phosphatase	Tidal flat sediments	Lee et al. (2015)
28	Mercuric reductase	Red Sea brine pool	Sayed et al. (2014)
29	Amine transferase	Hot water spring	Ferrandi et al. (2017)
30	Epoxide hydrolase	Hot water spring	Ferrandi et al. (2015)
31	Catalase	Fish microbiota	Lorentzen et al. (2006)
32	Alanine dehydrogenase	Sea urchin	Irwin et al. (2003)

properties, was isolated from mangrove sediment and patented by Universidade Estadualhed De Santa Cruz, Brazil (BR102016000771A2) in the year 2016. Honda Motor Co., Ltd., Kazusa DNA Research Institute of Japan patented an endoglucanase identified from hot spring soil for bioethanol production in 2016 (JP6552098B2). l-Methionine γ-lyase, isolated from deep sea sediment, was also patented in 2010 (CN101962651 B); it has applications in clinical detection, food flavour production, and cancer therapy. Metagenomics also has roles in medical or forensic investigations. The Human Microbiome project was initiated in 2009 to identify the microbial communities associated with the human gut, mouth, skin, or vagina, and information generated from this project helps in the prognosis as well as the diagnosis of several human disorders. Extinct human and animal species, like woolly mammoth (Poinar et al., 2006) and Neanderthals (Noonan et al., 2006), have also been studied by the metagenomic approach. Fujita et al. (2011) identified a bioactive compound vibrioferrin (a siderophore) from tidal-flat sediment of the Ariake Sea using a metagenomic approach. A novel carboxylesterase, having antimicrobial activity similar to β-lactamase, was identified from soil of Upo wetland, South Korea (Jeon et al., 2011). Biosynthesis of an antitumor polyketide by an unculturable bacterial symbiont of a marine sponge is also reported (Piel et al., 2004). Some of the studies related to the identification of novel enzymes identified through metagenomic studies are listed in Table 18.1.

18.4 CONCLUSION

Metagenomics has revolutionized the identification of unculturable microorganisms from diverse and difficult aquatic environments. It also paves the way to exploiting various biocatalytic and bioactive compounds for many industries, such as food, pharmaceuticals, detergents, textiles, biofuel, leather, etc. Due to the paradigm shift in high-throughput sequencing techniques, the ever-increasing gene and genomic databases, and the availability of highly efficient bioinformatics tools, a fast-developing era of genomics has opened new avenues for microbiologists and molecular biologists to uncover the remarkable diversity of marine microorganisms. Metagenomics is a powerful technique with great potential for novel gene discovery and provides an opportunity for the industrial use of identified novel biocatalysts and bioactive compounds for humankind.

REFERENCES

Apolinar–Hernández, Max M., Yuri J. Peña–Ramírez, Ernesto Pérez-Rueda, Blondy B. Canto-Canché, César De los Santos-Briones, and Aileen O'Connor-Sánchez. "Identification and in silico characterization of two novel genes encoding peptidases S8 found by functional screening in a metagenomic library of Yucatán underground water." *Gene* 593, no. 1 (2016): 154–161.

Ausec, Luka, Francesca Berini, Carmine Casciello, Mariana Silvia Cretoiu, Jan Dirk van Elsas, Flavia Marinelli, and Ines Mandic-Mulec. "The first acidobacterial laccase-like multicopper oxidase revealed by metagenomics shows high salt and thermo-tolerance." *Applied Microbiology and Biotechnology* 101 (2017): 6261–6276.

Bates, Scott T., Berg-Lyons Donna, Christian L. Lauber, William A. Walters, Rob Knight, and Noah Fierer. "A preliminary survey of lichen associated eukaryotes using pyrosequencing." *The Lichenologist* 44, no. 1 (2012): 137–146.

Berry, Andrew E., Claudia Chiocchini, Tina Selby, Margherita Sosio, and Elizabeth M. H. Wellington. "Isolation of high molecular weight DNA from soil for cloning into BAC vectors." *FEMS Microbiology Letters* 223, no. 1 (2003): 15–20.

Bertrand, Hélène, Franck Poly, Nathalie Lombard, Renaud Nalin, Timothy M. Vogel, and Pascal Simonet. "High molecular weight DNA recovery from soils prerequisite for biotechnological metagenomic library construction." *Journal of Microbiological Methods* 62, no. 1 (2005): 1–11.

Chandler, James Angus, Panpim Thongsripong, Amy Green, Pattamaporn Kittayapong, Bruce A. Wilcox, Gary P. Schroth, Durrell D. Kapan, and Shannon N. Bennett. "Metagenomic shotgun sequencing of a Bunyavirus in wild-caught Aedes aegypti from Thailand informs the evolutionary and genomic history of the Phleboviruses." *Virology* 464 (2014): 312–319.

Chen, Rong, Chenglu Li, Xiaolin Pei, Qiuyan Wang, Xiaopu Yin, and Tian Xie. "Isolation an aldehyde dehydrogenase gene from metagenomics based on semi-nest touch-down PCR." *Indian Journal of Microbiology* 54 (2014): 74–79.

Chu, Xinmin, Haoze He, Changquan Guo, and Baolin Sun. "Identification of two novel esterases from a marine metagenomic library derived from South China Sea." *Applied Microbiology and Biotechnology* 80 (2008): 615–625.

Cottrell, Matthew T., Liying Yu, and David L. Kirchman. "Sequence and expression analyses of Cytophaga-like hydrolases in a Western arctic metagenomic library and the Sargasso Sea." *Applied and Environmental Microbiology* 71, no. 12 (2005): 8506–8513.

Coughlan, Laura M., Paul D. Cotter, Colin Hill, and Avelino Alvarez-Ordóñez. "Biotechnological applications of functional metagenomics in the food and pharmaceutical industries." *Frontiers in Microbiology* 6 (2015): 672.

De Santi, Concetta, Bjørn Altermark, Marcin Miroslaw Pierechod, Luca Ambrosino, Donatella de Pascale, and Nils-Peder Willassen. "Characterization of a cold-active and salt tolerant esterase identified by functional screening of Arctic metagenomic libraries." *BMC Biochemistry* 17 (2016): 1–13.

Diaz-Torres, Martha L., Aurelie Villedieu, Nigel Hunt, Rod McNab, David A. Spratt, Elaine Allan, Peter Mullany, and Michael Wilson. "Determining the antibiotic resistance potential of the indigenous oral microbiota of humans using a metagenomic approach." *FEMS Microbiology Letters* 258, no. 2 (2006): 257–262.

Elend, C., C. Schmeisser, C. Leggewie, Peter Babiak, José Daniel Carballeira, H. L. Steele, J.-L. Reymond, K.-E. Jaeger, and W. R. Streit. "Isolation and biochemical characterization of two novel metagenome-derived esterases." *Applied and Environmental Microbiology* 72, no. 5 (2006): 3637–3645.

Fang, Zemin, Tongliang Li, Quan Wang, Xuecheng Zhang, Hui Peng, Wei Fang, Yuzhi Hong, Honghua Ge, and Yazhong Xiao. "A bacterial laccase from marine microbial metagenome exhibiting chloride tolerance and dye decolorization ability." *Applied Microbiology and Biotechnology* 89 (2011): 1103–1110.

Ferrandi, Erica Elisa, Alessandra Previdi, Ivan Bassanini, Sergio Riva, Xu Peng, and Daniela Monti. "Novel thermostable amine transferases from hot spring metagenomes." *Applied Microbiology and Biotechnology* 101 (2017): 4963–4979.

Ferrandi, Erica Elisa, Christopher Sayer, Michail N. Isupov, Celeste Annovazzi, Carlotta Marchesi, Gianluca Iacobone, Xu Peng et al. "Discovery and characterization of thermophilic limonene-1, 2-epoxide hydrolases from hot spring metagenomic libraries." *The FEBS Journal* 282, no. 15 (2015): 2879–2894.

Ferrer, Manuel, Olga V. Golyshina, Tatyana N. Chernikova, Amit N. Khachane, Vitor A. P. Martins Dos Santos, Michail M. Yakimov, Kenneth N. Timmis, and Peter N. Golyshin. "Microbial enzymes mined from the Urania deep-sea hypersaline anoxic basin." *Chemistry & Biology* 12, no. 8 (2005): 895–904.

Fu, Juan, Hanna-Kirsti S. Leiros, Donatella de Pascale, Kenneth A. Johnson, Hans-Matti Blencke, and Bjarne Landfald. "Functional and structural studies of a novel cold-adapted esterase from an Arctic intertidal metagenomic library." *Applied Microbiology and Biotechnology* 97 (2013): 3965–3978.

Fujita, Masaki J., Nobutada Kimura, Atsushi Sakai, Yoichi Ichikawa, Tomohiro Hanyu, and Masami Otsuka. "Cloning and heterologous expression of the vibrioferrin biosynthetic gene cluster from a marine metagenomic library." *Bioscience, Biotechnology, and Biochemistry* 75, no. 12 (2011): 2283–2287.

Gabor, Esther M., Erik J. de Vries, and Dick B. Janssen. "Efficient recovery of environmental DNA for expression cloning by indirect extraction methods." *FEMS Microbiology Ecology* 44, no. 2 (2003): 153–163.

Gao, Wenyuan, Kai Wu, Lifeng Chen, Haiyang Fan, Zhiqiang Zhao, Bei Gao, Hualei Wang, and Dongzhi Wei. "A novel esterase from a marine mud metagenomic library for biocatalytic synthesis of short-chain flavor esters." *Microbial Cell Factories* 15 (2016): 1–12.

Gilbert, Jack A., and Christopher L. Dupont. "Microbial metagenomics: Beyond the genome." *Annual Review of Marine Science* 3 (2011): 347–371.

Grossart, Hans-Peter, Ramon Massana, Katherine D. McMahon, and David A. Walsh. "Linking metagenomics to aquatic microbial ecology and biogeochemical cycles." *Limnology and Oceanography* 65 (2020): S2–S20.

Handelsman, Jo, Michelle R. Rondon, Sean F. Brady, Jon Clardy, and Robert M. Goodman. "Molecular biological access to the chemistry of unknown soil microbes: A new frontier for natural products." *Chemistry & Biology* 5, no. 10 (1998): R245–R249.

Henne, Anke, Ruth A. Schmitz, Mechthild Bömeke, Gerhard Gottschalk, and Rolf Daniel. "Screening of environmental DNA libraries for the presence of genes conferring lipolytic activity on Escherichia coli." *Applied and Environmental Microbiology* 66, no. 7 (2000): 3113–3116.

Hess, Matthias, Alexander Sczyrba, Rob Egan, Tae-Wan Kim, Harshal Chokhawala, Gary Schroth, Shujun Luo et al. "Metagenomic discovery of biomass-degrading genes and genomes from cow rumen." *Science* 331, no. 6016 (2011): 463–467.

Hung, Kuo-Sheng, Shiu-Mei Liu, Wen-Shyong Tzou, Fu-Pang Lin, Chorng-Liang Pan, Tsuei-Yun Fang, Kuang-Hui Sun, and Shye-Jye Tang. "Characterization of a novel GH10 thermostable, halophilic xylanase from the marine bacterium Thermoanaerobacterium saccharolyticum NTOU1." *Process Biochemistry* 46, no. 6 (2011): 1257–1263.

Irwin, Jane A., Susan V. Lynch, Suzie Coughlan, Patrick J. Baker, Haflidi M. Gudmundsson, Gudni A. Alfredsson, David W. Rice, and Paul C. Engel. "Alanine dehydrogenase from the psychrophilic bacterium strain PA-43: Overexpression, molecular characterization, and sequence analysis." *Extremophiles* 7 (2003): 135–143.

Jaeger, Karl-Erich, and Thorsten Eggert. "Lipases for biotechnology." *Current Opinion in Biotechnology* 13, no. 4 (2002): 390–397.

Jeon, Jeong Ho, Jun-Tae Kim, Yun Jae Kim, Hyung-Kwoun Kim, Hyun Sook Lee, Sung Gyun Kang, Sang-Jin Kim, and Jung-Hyun Lee. "Cloning and characterization of a new cold-active lipase from a deep-sea sediment metagenome." *Applied Microbiology and Biotechnology* 81 (2009a): 865–874.

Jeon, Jeong Ho, Jun-Tae Kim, Sung Gyun Kang, Jung-Hyun Lee, and Sang-Jin Kim. "Characterization and its potential application of two esterases derived from the arctic sediment metagenome." *Marine Biotechnology* 11 (2009b): 307–316.

Jeon, Jeong Ho, Soo-Jin Kim, Hyun Sook Lee, Sun-Shin Cha, Jung Hun Lee, Sang-Hong Yoon, Bon-Sung Koo et al. "Novel metagenome-derived carboxylesterase that hydrolyzes β-lactam antibiotics." *Applied and Environmental Microbiology* 77, no. 21 (2011): 7830–7836.

Jeon, Jeong Ho, Hyun Sook Lee, Jun Tae Kim, Sang-Jin Kim, Sang Ho Choi, Sung Gyun Kang, and Jung-Hyun Lee. "Identification of a new subfamily of salt-tolerant esterases from a metagenomic library of tidal flat sediment." *Applied Microbiology and Biotechnology* 93 (2012): 623–631.

Jiang, Cheng-Jian, Zhen-Yu Hao, Rong Zeng, Pei-Hong Shen, Jun-Fang Li, and Bo Wu. "Characterization of a novel serine protease inhibitor gene from a marine metagenome." *Marine Drugs* 9, no. 9 (2011): 1487–1501.

Jiang, Chengjian, Lan-Lan Wu, Gao-Chao Zhao, Pei-Hong Shen, Ke Jin, Zhen-Yu Hao, Shuang-Xi Li et al. "Identification and characterization of a novel fumarase gene by metagenome expression cloning from marine microorganisms." *Microbial Cell Factories* 9 (2010): 1–9.

Kay, Gemma L., Martin J. Sergeant, Valentina Giuffra, Pasquale Bandiera, Marco Milanese, Barbara Bramanti, Raffaella Bianucci, and Mark J. Pallen. "Recovery of a medieval Brucella melitensis genome using shotgun metagenomics." *MBio* 5, no. 4 (2014): e01337–14.

Kazuoka, Takayuki, Tadao Oikawa, Ikuo Muraoka, Shun'ichi Kuroda, and Kenji Soda. "A cold-active and thermostable alcohol dehydrogenase of a psychrotorelant from Antarctic seawater, Flavobacterium frigidimaris KUC-1." *Extremophiles* 11 (2007): 257–267.

Keller, Alexander, Hannes Horn, Frank Foerster, and Joerg Schultz. "Computational integration of genomic traits into 16S rDNA microbiota sequencing studies." *Gene* 549, no. 1 (2014): 186–191.

Kennedy, Jonathan, Julian R. Marchesi, and Alan D. W. Dobson. "Marine metagenomics: Strategies for the discovery of novel enzymes with biotechnological applications from marine environments." *Microbial Cell Factories* 7, no. 1 (2008): 1–8.

Kim, Yun-Jung, Gi-Sub Choi, Seung-Bum Kim, Gee-Sun Yoon, Yong-Sung Kim, and Yeon-Woo Ryu. "Screening and characterization of a novel esterase from a metagenomic library." *Protein Expression and Purification* 45, no. 2 (2006): 315–323.

Kimura, Nobutada. "Metagenomics: Access to unculturable microbes in the environment." *Microbes and Environments* 21, no. 4 (2006): 201–215.

Klippel, Barbara, Kerstin Sahm, Alexander Basner, Sigrid Wiebusch, Patrick John, Ute Lorenz, Anke Peters et al. "Carbohydrate-active enzymes identified by metagenomic analysis of deep-sea sediment bacteria." *Extremophiles* 18 (2014): 853–863.

Lane, David J., Bernadette Pace, Gary J. Olsen, David A. Stahl, Mitchell L. Sogin, and Norman R. Pace. "Rapid determination of 16S ribosomal RNA sequences for phylogenetic analyses." *Proceedings of the National Academy of Sciences* 82, no. 20 (1985): 6955–6959.

Langille, Morgan G. I., Jesse Zaneveld, J. Gregory Caporaso, Daniel McDonald, Dan Knights, Joshua A. Reyes, Jose C. Clemente et al. "Predictive functional profiling of microbial communities using 16S rRNA marker gene sequences." *Nature Biotechnology* 31, no. 9 (2013): 814–821.

Lee, Dae-Hee, Su-Lim Choi, Eugene Rha, Soo Jin Kim, Soo-Jin Yeom, Jae-Hee Moon, and Seung-Goo Lee. "A novel psychrophilic alkaline phosphatase from the metagenome of tidal flat sediments." *BMC Biotechnology* 15 (2015): 1–13.

Lee, Dong-Geun, Jeong Ho Jeon, Min Kyung Jang, Nam Young Kim, Jong Hyun Lee, Jung-Hyun Lee, Sang-Jin Kim, Gun-Do Kim, and Sang-Hyeon Lee. "Screening and characterization of a novel fibrinolytic metalloprotease from a metagenomic library." *Biotechnology Letters* 29 (2007): 465–472.

Li, Liang, Gang Li, Li-chuang Cao, Guang-hui Ren, Wei Kong, Si-di Wang, Geng-shan Guo, and Yu-Huan Liu. "Characterization of the cross-linked enzyme aggregates of a novel β-galactosidase, a potential catalyst for the synthesis of galacto-oligosaccharides." *Journal of Agricultural and Food Chemistry* 63, no. 3 (2015): 894–901.

Liu, Jinlin, Zhijuan Jia, Sha Li, Yan Li, Qiang You, Chunyan Zhang, Xiaotong Zheng et al. "Identification and characterization of a chitin deacetylase from a metagenomic library of deep-sea sediments of the Arctic Ocean." *Gene* 590, no. 1 (2016): 79–84.

Lorentzen, Marit Sjo, Elin Moe, Hélène Marie Jouve, and Nils Peder Willassen. "Cold adapted features of Vibrio salmonicida catalase: Characterisation and comparison to the mesophilic counterpart from Proteus mirabilis." *Extremophiles* 10 (2006): 427–440.

Martinez, Asuncion, Steven J. Kolvek, Choi Lai Tiong Yip, Joern Hopke, Kara A. Brown, Ian A. MacNeil, and Marcia S. Osburne. "Genetically modified bacterial strains and novel bacterial artificial chromosome shuttle vectors for constructing environmental libraries and detecting heterologous natural products in multiple expression hosts." *Applied and Environmental Microbiology* 70, no. 4 (2004): 2452–2463.

Mendes, Lucas W., Siu M. Tsai, Acácio A. Navarrete, Mattias De Hollander, Johannes A. van Veen, and Eiko E. Kuramae. "Soil-borne microbiome: Linking diversity to function." *Microbial Ecology* 70 (2015): 255–265.

Mohamed, Yasmine M., Mohamed A. Ghazy, Ahmed Sayed, Amged Ouf, Hamza El-Dorry, and Rania Siam. "Isolation and characterization of a heavy metal-resistant, thermophilic esterase from a Red Sea Brine Pool." *Scientific Reports* 3, no. 1 (2013): 3358.

Noonan, James P., Graham Coop, Sridhar Kudaravalli, Doug Smith, Johannes Krause, Joe Alessi, Feng Chen et al. "Sequencing and analysis of Neanderthal genomic DNA." *Science* 314, no. 5802 (2006): 1113–1118.

Oh, Julia, Allyson L. Byrd, Clay Deming, Sean Conlan, Heidi H. Kong, and Julia A. Segre. "Biogeography and individuality shape function in the human skin metagenome." *Nature* 514, no. 7520 (2014): 59–64.

Ouyang, Li-Ming, Jia-Ying Liu, Ming Qiao, and Jian-He Xu. "Isolation and biochemical characterization of two novel metagenome-derived esterases." *Applied Biochemistry and Biotechnology* 169 (2013): 15–28.

Peng, Qing, Xu Wang, Meng Shang, Jinjin Huang, Guohua Guan, Ying Li, and Bo Shi. "Isolation of a novel alkaline-stable lipase from a metagenomic library and its specific application for milkfat flavor production." *Microbial Cell Factories* 13, no. 1 (2014): 1–9.

Piel, Jörn, Dequan Hui, Gaiping Wen, Daniel Butzke, Matthias Platzer, Nobuhiro Fusetani, and Shigeki Matsunaga. "Antitumor polyketide biosynthesis by an uncultivated bacterial symbiont of the marine sponge Theonella swinhoei." *Proceedings of the National Academy of Sciences* 101, no. 46 (2004): 16222–16227.

Poinar, Hendrik N., Carsten Schwarz, Ji Qi, Beth Shapiro, Ross D. E. MacPhee, Bernard Buigues, Alexei Tikhonov et al. "Metagenomics to paleogenomics: large-scale sequencing of mammoth DNA." *Science* 311, no. 5759 (2006): 392–394.

Quaiser, Achim, Torsten Ochsenreiter, Christa Lanz, Stephan C. Schuster, Alexander H. Treusch, Jürgen Eck, and Christa Schleper. "Acidobacteria form a coherent but highly diverse group within the bacterial domain: Evidence from environmental genomics." *Molecular Microbiology* 50, no. 2 (2003): 563–575.

Rabausch, U., J. Juergensen, N. Ilmberger, S. Böhnke, S. Fischer, B. Schubach, M. Schulte, and W. R. Streit. "Functional screening of metagenome and genome libraries for detection of novel flavonoid-modifying enzymes." *Applied and Environmental Microbiology* 79, no. 15 (2013): 4551–4563.

Rhee, Jin-Kyu, Dae-Gyun Ahn, Yeon-Gu Kim, and Jong-Won Oh. "New thermophilic and thermostable esterase with sequence similarity to the hormone-sensitive lipase family, cloned from a metagenomic library." *Applied and Environmental Microbiology* 71, no. 2 (2005): 817–825.

Richardson, Toby H., Xuqiu Tan, Gerhard Frey, Walter Callen, Mark Cabell, David Lam, John Macomber, Jay M. Short, Dan E. Robertson, and Carl Miller. "A novel, high performance enzyme for starch liquefaction: Discovery and optimization of a low pH, thermostable α-amylase." *Journal of Biological Chemistry* 277, no. 29 (2002): 26501–26507.

Robertson, Dan E., Jennifer A. Chaplin, Grace DeSantis, Mircea Podar, Mark Madden, Ellen Chi, Toby Richardson et al. "Exploring nitrilase sequence space for enantioselective catalysis." *Applied and Environmental Microbiology* 70, no. 4 (2004): 2429–2436.

Rondon, Michelle R., Paul R. August, Alan D. Bettermann, Sean F. Brady, Trudy H. Grossman, Mark R. Liles, Kara A. Loiacono et al. "Cloning the soil metagenome: A strategy for accessing the genetic and functional diversity of uncultured microorganisms." *Applied and Environmental Microbiology* 66, no. 6 (2000): 2541–2547.

Sahoo, Rajesh Kumar, Mohit Kumar, Lala Behari Sukla, and Enketeswara Subudhi. "Bioprospecting hot spring metagenome: Lipase for the production of biodiesel." *Environmental Science and Pollution Research* 24 (2017): 3802–3809.

Sayed, Ahmed, Mohamed A. Ghazy, Ari J. S. Ferreira, João C. Setubal, Felipe S. Chambergo, Amged Ouf, Mustafa Adel et al. "A novel mercuric reductase from the unique deep brine environment of Atlantis II in the Red Sea." *Journal of Biological Chemistry* 289, no. 3 (2014): 1675–1687.

Schröder, Carola, Viktoria-Astrid Janzer, Georg Schirrmacher, Jörg Claren, and Garabed Antranikian. "Characterization of two novel heat-active α-galactosidases from thermophilic bacteria." *Extremophiles* 21 (2017): 85–94.

Schröder, Carola, Skander Elleuche, Saskia Blank, and Garabed Antranikian. "Characterization of a heat-active archaeal β-glucosidase from a hydrothermal spring metagenome." *Enzyme and Microbial Technology* 57 (2014): 48–54.

Staley, Christopher, and Michael J. Sadowsky. "Application of metagenomics to assess microbial communities in water and other environmental matrices." *Journal of the Marine Biological Association of the United Kingdom* 96, no. 1 (2016): 121–129.

Su, Jing, Fengli Zhang, Wei Sun, Valliappan Karuppiah, Guangya Zhang, Zhiyong Li, and Qun Jiang. "A new alkaline lipase obtained from the metagenome of marine sponge Ircinia sp." *World Journal of Microbiology and Biotechnology* 31 (2015): 1093–1102.

Venter, J. Craig, Karin Remington, John F. Heidelberg, Aaron L. Halpern, Doug Rusch, Jonathan A. Eisen, Dongying Wu et al. "Environmental genome shotgun sequencing of the Sargasso Sea." *Science* 304, no. 5667 (2004): 66–74.

Warnecke, Falk, Peter Luginbühl, Natalia Ivanova, Majid Ghassemian, Toby H. Richardson, Justin T. Stege, Michelle Cayouette et al. "Metagenomic and functional analysis of hindgut microbiota of a wood-feeding higher termite." *Nature* 450, no. 7169 (2007): 560–565.

Yan, Wei, Furong Li, Li Wang, Yaxin Zhu, Zhiyang Dong, and Linhan Bai. "Discovery and characterizaton of a novel lipase with transesterification activity from hot spring metagenomic library." *Biotechnology Reports* 14 (2017): 27–33.

Zeng, Runying, Rui Zhang, Jing Zhao, and Nianwei Lin. "Cold-active serine alkaline protease from the psychrophilic bacterium Pseudomonas strain DY-A: Enzyme purification and characterization." *Extremophiles* 7 (2003): 335–337.

Zhao, Chao, Yanan Chu, Yanhong Li, Chengfeng Yang, Yuqing Chen, Xumin Wang, and Bin Liu. "High-throughput pyrosequencing used for the discovery of a novel cellulase from a thermophilic cellulose-degrading microbial consortium." *Biotechnology Letters* 39 (2017): 123–131.

19 Microalgal Remediation in the Aquatic Environment

V. Santhana Kumar, Dhruba Jyoti Sarkar, Soma Das Sarkar, and Basanta Kumar Das

19.1 INTRODUCTION

Aquatic pollution is one of the major hindrances to attaining the global sustainable development goals, especially "clean and safe water" and "life below the water". The discharge of pollutants from untreated effluents, agriculture runoffs, and industrial and sewage effluents into the aquatic environment has a toxic effect on the organisms that live there. Pollutants such as heavy metals, pesticides, nutrients and some emerging contaminants like antibiotics, microplastics, cosmetics products and flame retardants are commonly found in the aquatic environment due to increased anthropogenic activities. This also affects human beings through the food chain via bioaccumulation and biomagnification (Pacheco et al., 2020). There is a need for a technology to remove pollutants from the environment or from the pollution source for the sustainable management of aquatic resources. The existing physicochemical processes, i.e., precipitation, ion exchange, reverse osmosis and filtration, are not efficient in removing pollutants because of the complex nature of the different pollutants present in the effluents or the environment. The other disadvantages are high cost and energy requirements, disposal of the recovered pollutants and incomplete pollutant removal (Fu and Wang, 2011; Nguyen et al., 2021). Hence, the biological process is one of the most promising technologies for remediation, since many microbes are involved in the biogeochemical cycling of toxic metals or other substances in the environment (Pacheco et al., 2020).

Microalgae-based bioremediation is an emerging technology that has gained momentum in recent years (Leong and Chang, 2020). Microalgae are photosynthetic microbes found in both fresh water and marine waters, containing chlorophyll pigment, which is involved in photosynthesis, and a thallus with no differentiated root, stem or leaves (Andersen, 2013). They are the primary producers of the aquatic ecosystem and form the base of the food chain. They are also involved in the biogeochemical cycling of toxic pollutants and have the ability to detoxify the pollutants using their storage products, such as lipids and carbohydrates, for stress amelioration (Kumar et al., 2022a). For most of the microalgae, heavy metals are required in small quantities for their growth and photosynthesis. Hence, they accumulate the pollutants and convert them into less/non-toxic forms, which they utilize for growth, and the remainder will be excreted into the system (Chugh et al., 2022). This method has enormous potential to remediate organic and inorganic pollutants from the waters due to the following advantages: (1) low capital and operational costs, (2) increased removal efficiency, (3) resource recovery, (4) removal of multiple pollutants and (5) production of high-value products after remediation (Forján et al., 2015).

The present chapter discusses the process of bioremediation, including the selection, isolation and domestication of microalgae; the use of different microalgae for the remediation of organic, inorganic and emerging pollutants, and their mechanisms of remediation; and emerging techniques in microalgal bioremediation for aquatic environment management.

DOI: 10.1201/9781003408543-19

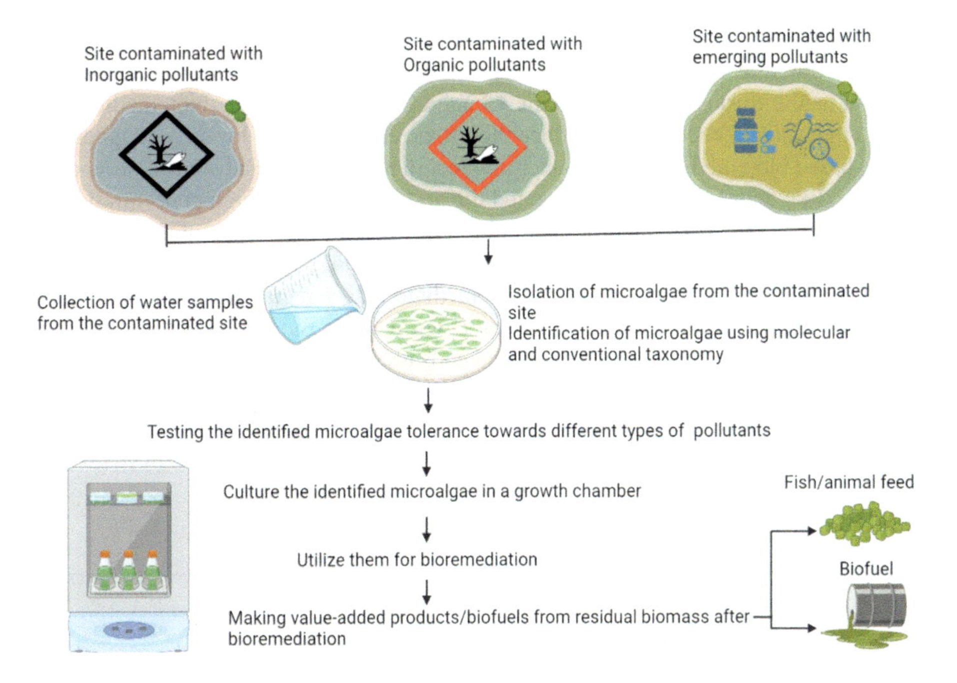

FIGURE 19.1 Steps involved in microalgal bioremediation process.

19.2 MICROALGAL BIOREMEDIATION PROCESS

Microalgal bioremediation is defined as the process by which microalgae are isolated, cultured and used for removing pollutants or reducing their toxic effects. The initial step in any bioremediation process is the isolation of specific or a mixture of microalgae from the contaminated site, after which they are acclimatized under controlled conditions, screened for tolerance towards targeted pollutants, and utilized for the bioremediation process besides the production of valuable bioproducts (Pandey et al., 2019) (Figure 19.1). Likewise, researchers isolated native microalgae strains from Lake Massaciucoli, Italy, and exposed them to various contaminants present in the lake. They identified the four most tolerant strains of *Chlorella sorokiniana*, which could be used for further waste water remediation purposes. Microalgae were also acclimatized to a polluted environment under controlled culture conditions and utilized for bioremediation (Chiellini et al., 2020). A study evaluated the potential of microalgae stains such as *Chlorella* sp., SL7A, *Chlorococcum* sp., SL7B and *Neochloris* sp. SK57 acclimatized in river water contaminated with pharmaceutical wastes and utilized for waste removal. It was found that *Neochloris* sp. SK57 grew well in the contaminated water and removed the contaminants as well as increasing the saturated fatty acid content (Singh et al., 2020). This suggests that screening and tolerance testing is one of the important steps for successful bioremediation.

19.3 BIOREMEDIATION OF INORGANIC POLLUTANTS USING MICROALGAE

Inorganic pollutants originate from minerals, including salts, heavy metals, and radioactive substances (Mondal et al., 2019). Among these, heavy metals are the most important pollutants, which are released into the ecosystem via both natural processes (volcanic eruptions) and anthropogenic activities (industrial sewage) (Chugh et al., 2022). Heavy metals like copper, zinc, iron and cobalt are required by organisms in small quantities for normal physiological functions, but in high amounts, they have a toxic effect on organisms. Other heavy metals, like mercury, lead, arsenic, cadmium and chromium, are toxic to organisms even at low concentrations and are not required by organisms for essential body functions (Mondal et al., 2019). These metals forcibly enter aquatic organisms via anthropogenic pressure and are accumulated inside the system, where they are biomagnified and affect human health through the food chain (Chugh et al., 2022). Hence, bioremediation of heavy metals from the water is necessary to reduce their toxic effect on both lower and higher organisms. Microalgae-based bioremediation is an emerging technique due to the advantageous characteristics of microalgae, like widespread occurrence and the ability to grow in high concentrations of heavy metals (Kumar et al., 2022b). Numerous studies have shown the utilization of microalgae for heavy metal remediation and reported significant removal (Table 19.1).

Microalgae employ different strategies to remediate heavy metals from the cell: (1) biosorption and (2) bioaccumulation and detoxification (Figure 19.2). Biosorption is a rapid and passive extracellular process in which both live and dead cells are involved in the removal of heavy metals. Biosorption of microalgae occurs either by binding with the heavy metals, using extracellular proteins and polysaccharides or using functional groups in the cell wall like sulfate, amino, hydroxyl and carboxyl, or exchanging ions like calcium, magnesium, sodium and potassium on the cell surface

TABLE 19.1
Microalgae Species Used for Inorganic Pollutant Removal

Heavy Metals	Microalgae Species	Biomass Conc. (g/L)	Removal Efficiency (%)	Duration (min)	References
Arsenic	Immobilized microalgal biofilm (mixture of algae)	–	96.7	5 days	Kumar et al., 2022
	Chlamydomonas reinhardtii	1	38.6	180	Saavedra et al., 2018
	Maugeotia genuflexa	4	96	60	Sari et al., 2011
	Chlorophyceae algae mixture	10	70	180	Sulaymon et al., 2013
Boron	*Scenedesmus almeriensisis*	–	38.6	10	Saavedra et al., 2018
Cadmium	*Scenedesmus* sp.	1.5	60.5	–	Jena et al., 2015
	Chlorella minutissima	4	–	20	Yang et al., 2015
	Chlorella sp. immobilized in biochar	1.3	92.5	–	Shen et al., 2018
	Desmodesmus sp.	–	>58%	16 days	Abinandan et al., 2019
	Chlorococcum humicola	–	17	6 days	Borah et al., 2020
Chromium	*Chlorella vulgaris*	1	43	240	Sibi, 2016
	Scenedesmus quadricauda	2	98.3 for Cr(III), 47.6 for Cr(V)	120	Shokri Khoubestani et al., 2015
	Pseudochlorella pringsheimii and *Chlorella vulgaris*	–	80	1 day	Saranya and Shanthakumar, 2019
	Oedogonium westi	–	93	7 days	Shamsad et al., 2016
Lead	*Chaetoceros* sp.	1.5	180	60	Molazedah et al., 2015
	Chlorella sp.	1.5	–	78	Molazedah et al., 2015
	Phormidium sp.	4	40	92.2	Das et al., 2016
	Oedogonium westi	–	61–96	7 days	Shamsad et al., 2016

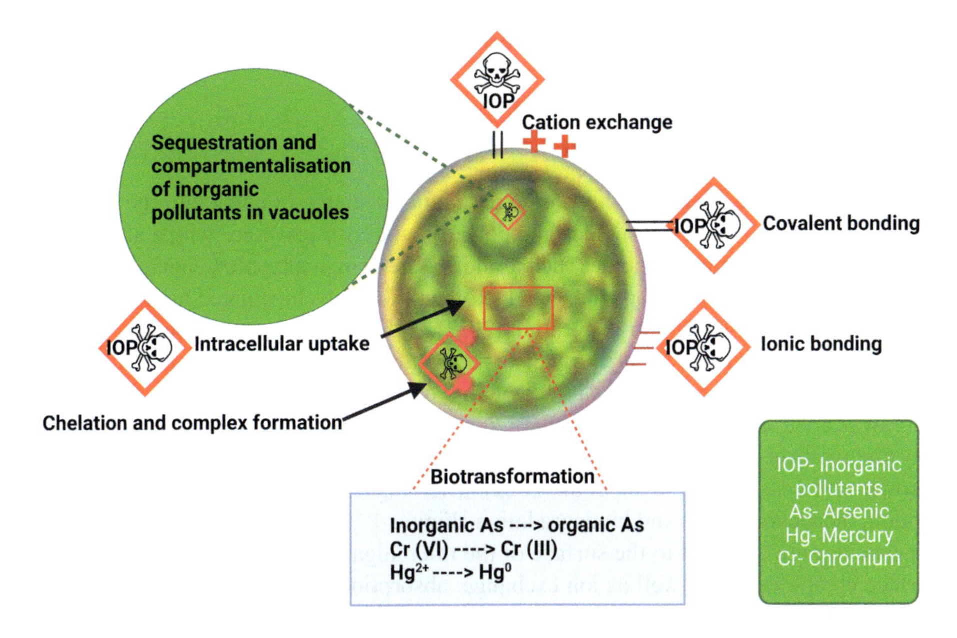

FIGURE 19.2 Mechanism of inorganic pollutant removal by microalgae. (Modified from Leong, Y.K. and Chang, J.-S., *Bioresource Technology* 303, 122886, 2020.)

with heavy metals (Leong and Chang, 2020; Chugh et al., 2022). Earlier studies have reported biosorption of heavy metals using microalgae and found significantly higher removal of heavy metals. For instance, immobilized *Chlorella* sp. has the ability to remove 92.5% of cadmium within a stipulated time period (Shen et al., 2018), and a ~40% removal of arsenic by *Chlamydomonas reinhardtii* within 180 minutes has been reported (Saavedra et al., 2018).

Bioaccumulation of heavy metals is an active, slow and energy-intensive process in which only live microalgal biomass is involved in accumulation and detoxification. Heavy metals enter the cytoplasm via the cell membrane transport system, and microalgae defend themselves against the toxicity of heavy metals through the following mechanism. Once the heavy metal enters the cytoplasm, it induces the production of more reactive oxygen species in the cell. To counteract this, microalgae produce both enzymatic and non-enzymatic antioxidants such as superoxide dismutase (SOD), catalase, glutathione reductase, peroxidase and ascorbic acids, carotenoids, cysteine and glutathione (Upadhdhay et al., 2016). Microalgae also produce peptides that chelate heavy metals and form organometallic complexes. These complexes are sequestered inside the vacuoles, reducing the toxic effect. Phytochelatins are small metal-binding peptides that play a major role in metal binding (Balaji et al., 2016). Earlier studies reported the activation of phytochelatins after exposing the microalgae to heavy metals like cadmium, gold, silver, lead, copper, zinc and arsenic (Torres et al., 2008). Microalgae have also the capacity to biotransform the toxic forms of heavy metals into non-toxic forms. For instance, the toxic forms of arsenic, like arsenite and arsenate, are methylated into organic forms like mono- and di-methyl arsinic acid in microalgae and excreted from the microalgae. The arsenate present in the water is transported into microalgae via the phosphate transport pathway due to the structural similarity between phosphate and arsenate, and is biotransformed in the cell. Direct oxidation of arsenite into arsenate in the cell surface has also been noticed. Microalgae such as *Chlorella* sp. and *Scenedesmus* sp. are reported to oxidize the arsenite into arsenate, using functional groups as binding agents. Besides detoxification, microalgae have the ability to produce lipids to overcome the stress induced by the heavy metals, and these quality lipids could be used for producing biodiesel (Kumar et al., 2022a). The use of microalgae for bioremediation and the grown biomass for the production of quality lipid and other by-products has also been reported to be an eco-friendly and economical option (Vassilev et al., 2016).

19.4 BIOREMEDIATION OF ORGANIC POLLUTANTS USING MICROALGAE

Industrial discharge and agricultural runoffs are the major source of entry into the aquatic ecosystem for organic pollutants, i.e., pesticides and polyaromatic hydrocarbons, which are also known as persistent organic pollutants. Pesticides such as aldrin, dieldrin, heptachlor, endrin, dichlorodiphenyltrichloroethane, chlordane, toxophene, etc. and others like polychlorinated biphenyls, dioxins and furans are listed in the Stockholm convention (2012) as the most toxic compounds. They are more readily bioaccumulated inside aquatic organisms due to their low water and high lipid solubility, and are biomagnified and affect human health through the food chain (Mondal et al., 2019).

Microalgae are a suitable option for the bioremediation of pesticides due to their high biosorption capacity and removal efficiency (Table 19.2) and the utilization of pesticides as a nutrient source for their growth and energy (Touliabah et al., 2022). Microalgae remove pesticides from the water via biosorption, bioaccumulation and biodegradation (Figure 19.3). Biosorption is a passive process in which pesticides are attached to the surface of the microalgal cell through electrostatic interaction and surface complexation as well as ion exchange, absorption and precipitation (Nie et al., 2020). It is the most efficient process of removing pesticides from the system, since more than 90% of the pesticides in the aqueous phase were removed by the microalgae through biosorption (Mishaqa, 2017). The adsorption of pesticides in the microalgal cell wall is mainly because of the functional groups, like carboxyl, hydroxyl and amine, and also due to the sulfated polysaccharides, carbohydrates, intercellular spaces and fibrous matrix in the cell wall of microalgae (Hammed et al., 2016; Qiu et al., 2017).

Bioaccumulation is an active metabolic energy-dependent process in which pollutants enter the cell via the membrane transport system and are biodegraded into simple substances. It has been noticed that the bioaccumulation capacity of microalgae is mainly dependent on the biodegradation process (Xu and Huang, 2017). During bioaccumulation of pesticides, there is huge production of reactive oxygen species, which leads to damage to DNA, and cell death may also occur. Microalgae produce antioxidant enzymes such as SOD and catalase to overcome the oxidative stress and cell damage caused by the pesticides. Biodegradation is the process of degrading organic compounds

TABLE 19.2
Microalgae Species Used for Organic Pollutant Removal

Pesticides	Microalgae Species	Biomass Conc. (g/L)	Removal Efficiency (%)	Duration (days)	References
Chlorpyrifos	*Chlorella sorokiniana*	–	97.86-99.85	120 h	Habbibah et al., 2020
Diazinon	*Chlorella vulgaris*	0.4	94	20	Weis et al., 2020
Fluroxypyr	*Chlamydomonas reinhardtii*	–	57	–	Philippat et al., 2018
Imidacloprid	*Nannochloropsis* sp.	0.02	50	12	Encarnacao et al., 2021
Isoproturan	*Platymonas subcordiformis*	188.4 μg/L	70.75	120 h	Wang et al., 2019
Mesotrione	*Scenedesmus quadricauda*	–	15.2	7	Ni et al., 2014
Trichlorfon	*Chlamydomonas reinhardtii*	–	96.2	–	Wan et al., 2020

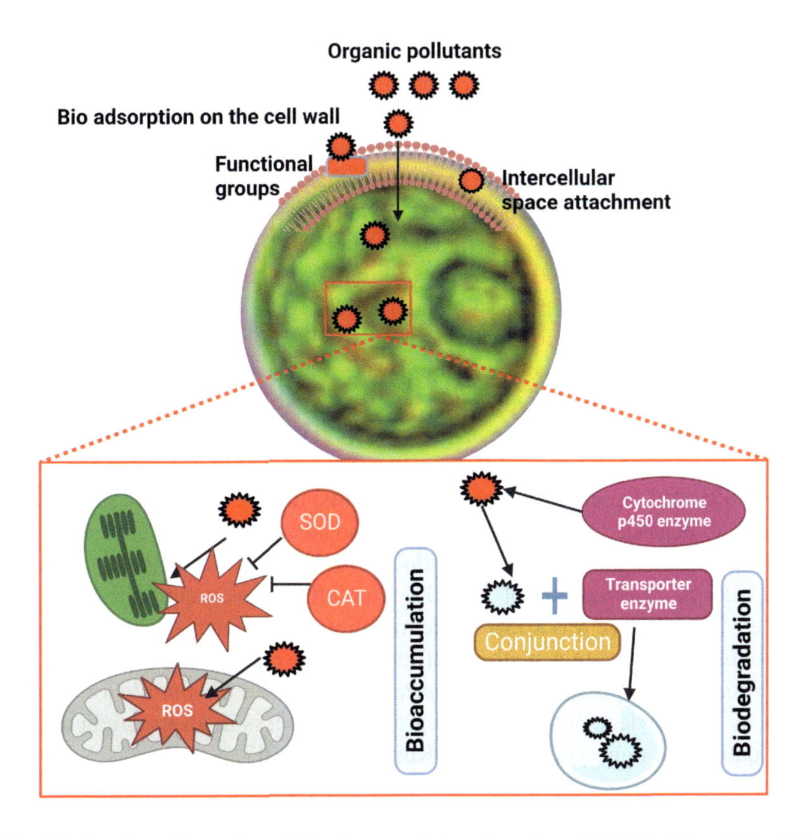

FIGURE 19.3 Mechanism of organic pollutant removal by microalgae. (Modified from Nie et al., 2020.)

into smaller compounds by microalgae and utilizing them as nutrients for their growth (Pérez-Legaspi et al., 2016). The degradation of pesticides involves the activation of pesticides by cytochrome p450 enzyme. Then, enzymes are transferred from the cytosol to the activated pesticides for degradation. The enzymes involved in the pesticide degradation process include esterase, transferase, cytochrome P450, hydrolase, phosphatase, phosphotriesterase, oxygenase, and oxidoreductases. Pesticides also form conjugates with functional groups and are activated and transported by glutathione transporters into the vacuoles, where they are sequestered (Nie et al., 2020).

19.5 BIOREMEDIATION OF EMERGING POLLUTANTS USING MICROALGAE

Emerging pollutants are synthetic organic chemicals, which include antibiotics and drugs, flame retardants, cosmetic products like triclosan, microplastics, etc. and pesticides. These have been recently discovered in the aquatic environment and drinking waters and are of emerging concern worldwide due to their increased concentration in the environment and the associated health effects (Rempel et al., 2021). Microalgae have been increasingly used for the bioremediation of emerging pollutants via one or more of the following mechanisms: bio-adsorption, intracellular uptake and biodegradation. Bio-adsorption is the process of adherence of emerging pollutants to the surface of the microalgal cell wall. The attraction towards the cell wall occurs either due to electrostatic interaction or due to the functional groups present in the microalgae (Norvill et al., 2016; Xiong et al., 2016). The rate of the adsorption process depends on other external factors like temperature, pH and redox. Previously, many studies have utilized the microalgal single cell or consortium for bioremediation of emerging pollutants, and a detailed list is given in the review by Sutherland and Ralph (2019). It has been noticed that algal cells like *Chlorella* sp., *Chlamydomonas* sp. and *Mychonastes* sp. have the capacity to remove 100% of 7-amino cephalosporanic acid from water (Guo et al., 2016). *Nanochloris* sp. removed 100% of the triclosan from culture medium (Bai and Acharya, 2016), and *C. reinhardtii* removed 100% of added hormones like β-estradiol and 17a-ethinylestradiol from culture medium. Microalgae such as *Scenedesmus abudans* were used for the removal of microplastics, i.e., polystyrene, polymethyl methacrylate and polylactide, from water. It has been found that more than 70% of the removal of microplastics added to water was due to hetero-aggregation, i.e., secretion of extracellular polysaccharides (Cheng and Wang, 2022). Bioremediation of flame retardants such as polybrominated diphenyl ethers using *Chlorella* sp. was successfully demonstrated with a removal efficiency of more than 80% (Deng and Tam, 2016).

The intracellular uptake of microalgae is an active metabolic process in which emerging pollutants enter the cells through the cell wall membrane, either due to its permeability or due to the transporter proteins present in microalgae or directly using energy (Sutherland and Ralph, 2019). The biodegradation or biotransformation of emerging pollutants includes the breakdown/conversion of complex compounds into simpler molecules through metabolic processes or by co-metabolism (Tiwari et al., 2017). Among all emerging contaminants, most studies have found that microalgae have the capacity to degrade pharmaceutical and personal care products from a contaminated system. For instance, microalgae such as *Navicula* sp. biodegraded drugs such as ibuprofen, carbamazepine and caffeine to the extent of 30–80% from effluent treatment plants (Ding et al., 2017; Matamoros et al., 2016). There was a 95% biodegradation of hormones like progesterone when using *Scenedesmus obliquus* for treating synthetic effluents (Peng et al., 2014). A study using microalgae such as *Scenedesmus dimorphus*, *Anabaena spiroides*, and *Navicula pupula* for the biodegradation of plastic such as polyethylene noticed a biodegradation process in all the studied microalgae with the help of oxidative and lignolytic enzymes (Kumar et al., 2017).

19.6 EMERGING TECHNIQUES IN MICROALGAL BIOREMEDIATION

Microalgae have immense potential to bioremediate or degrade contaminants from water or effluents. However, the commercial-scale operation of wastewater treatment using microalgae has important consequences, such as its high cost, laborious nature, and inefficient harvesting techniques. Therefore, as an alternative, two methods have emerged as advantageous options for microalgal bioremediation purposes, namely, co-cultivation of microalgae with other microbes and immobilization of microalgae on substrates, which may be organic or inorganic. Microalgal co-cultivation is defined as the process of cultivating microalgae either with other microbes such as bacteria, fungi and yeasts or with other microalgal species in the same medium. The microbes survive with the help of a mutualistic relationship and could increase the phyco-remediation and biomass yield of microalgae. Co-cultivation of microalgae and fungi was found more efficient than the monoculture of microalgae for wastewater treatment (Muradov et al., 2015). Earlier, many studies reported the utilization of co-cultivated algae and fungi for the bioremediation of organic and inorganic pollutants with positive results. For instance, co-cultivation of cyanobacteria *Leptolyngbya tenuis* and microalgae *Chlorella* sp. increased the biomass production and CO_2 fixation ability twofold compared with a monoculture of either species. Besides, the cadmium removal efficiency of the co-cultured species was 20–25% higher than the removal efficiency of either monocultured species (Satpati and Pal, 2021). Co-culture of fungi *Aspergillus oryzae* and microalgae *C. pyrenoidosa* was used for the treatment of organic pollutants from starch wastewater, and removed 92.08, 83.56, and 96.58% of chemical oxygen demand (COD), total nitrogen (TN) and total phosphorus (TP), respectively from the wastewater. Moreover, the protein and lipid content of the grown biomass also increased (Wang et al., 2022).

Immobilization of microalgae on either organic or synthetic substrates is also known as a microalgal biofilm. It is an emerging technique for bioremediation due to its ease of operation, survivability in harsh environments and increased tolerance towards pollutants (Mishra et al., 2022). A study utilized both organic and synthetic substrate-grown microalgal biofilm for the removal of arsenic from contaminated water. It was found that organic and inorganic substrate-grown biofilm removed more than 70 and 90% of arsenic, respectively, from the water within 5 days with little toxic effect on the microalgal biomass (Kumar et al., 2022). Another study utilized microalgal biofilm for the removal of copper from water and found 99% removal of copper within 12 h of hydraulic retention time; high tolerance among the biofilm community was also noticed (Ma et al., 2017). Hence, biofilm-based bioremediation is one of the effective options for the removal of pollutants from contaminated water due to the ease of harvesting.

19.7 CONCLUSION

Microalgal bioremediation is one of the most important techniques for the removal of organic, inorganic and emerging pollutants from contaminated water and is widely applied for waste water treatment. A wide range of publications are available on the utilization of microalgae for bioremediation purposes, and the microalgal biomass was used for the production of value-added products after bioremediation. However, on-site application of microalgae-based bioremediation is still scarce due to its disadvantages, like difficulty in harvesting and toxicity of pollutants to microalgal growth. New methods like co-cultivation and immobilization of microalgae are promising options for remediating toxic heavy metals, pesticides and other emerging contaminants from water. Most of the studies have utilized these technologies at laboratory scale. There is an urgent need for exploring the potential of these technologies on a large scale; a field study is required to find the exact potential of microalgae in a dynamic and diverse system, and their effect on removal efficiency also needs to be explored.

REFERENCES

Abinandan, Sudharsanam, Suresh R. Subashchandrabose, Kadiyala Venkateswarlu, Isiri Adhiwarie Perera, and Mallavarapu Megharaj. "Acid-tolerant microalgae can withstand higher concentrations of invasive cadmium and produce sustainable biomass and biodiesel at pH 3.5." *Bioresource Technology* 281 (2019): 469–473.

Andersen, Robert A. "The microalgal cell." In *Handbook of Microalgal Culture: Applied Phycology and Biotechnology* (eds Richmond and Hu). Wiley Blackwell (2013): 1–20.

Bai, Xuelian, and Kumud Acharya. "Removal of trimethoprim, sulfamethoxazole, and triclosan by the green alga Nannochloris sp." *Journal of Hazardous Materials* 315 (2016): 70–75.

Balaji, S., T. Kalaivani, B. Sushma, C. Varneetha Pillai, M. Shalini, and C. Rajasekaran. "Characterization of sorption sites and differential stress response of microalgae isolates against tannery effluents from Ranipet industrial area—An application towards phycoremediation." *International Journal of Phytoremediation* 18, no. 8 (2016): 747–753.

Borah, Dharitri, Bervin Kennedy, Subramanian Gopalakrishnan, Arutselvan Chithonirai, and Thajuddin Nooruddin. "Bioremediation and biomass production with the green microalga Chlorococcumhumicola and textile mill effluent (TE)." *Proceedings of the National Academy of Sciences, India Section B: Biological Sciences* 90 (2020): 415–423.

Cheng, Yu-Ru, and Hsiang-Yu Wang. "Highly effective removal of microplastics by microalgae *Scenedesmus abundans*." *Chemical Engineering Journal* 435 (2022): 135079.

Chiellini, Carolina, Lorenzo Guglielminetti, Sabrina Sarrocco, and Adriana Ciurli. "Isolation of four microalgal strains from the Lake Massaciuccoli: Screening of common pollutants tolerance pattern and perspectives for their use in biotechnological applications." *Frontiers in Plant Science* 11 (2020): 607651.

Chugh, Mohita, Lakhan Kumar, Maulin P. Shah, and Navneeta Bharadvaja. "Algal Bioremediation of heavy metals: An insight into removal mechanisms, recovery of by-products, challenges, and future opportunities." *Energy Nexus* (2022): 100129.

Das, Debayan, Sudip Chakraborty, Chiranjib Bhattacharjee, and Ranjana Chowdhury. "Biosorption of lead ions (Pb2+) from simulated wastewater using residual biomass of microalgae." *Desalination and Water Treatment* 57, no. 10 (2016): 4576–4586.

Deng, D. and Tam, N.F. "Adsorption-uptake-metabolism kinetic model on the removal of BDE-47 by a Chlorella isolate." *Environmental Pollution* 212 (2016): 290–298.

Ding, Tengda, Kunde Lin, Bo Yang, Mengting Yang, Juying Li, Wenying Li, and Jay Gan. "Biodegradation of naproxen by freshwater algae *Cymbella* sp. and *Scenedesmus quadricauda* and the comparative toxicity." *Bioresource Technology* 238 (2017): 164–173.

Encarnação, Telma, Daniel Santos, Simone Ferreira, Artur J. M. Valente, J. C. Pereira, M. G. Campos, Hugh D. Burrows, and Alberto A. C. C. Pais. "Removal of imidacloprid from water by microalgae Nannochloropsis sp. and its determination by a validated RP-HPLC method." *Bulletin of Environmental Contamination and Toxicology* 107 (2021): 131–139.

Forján, Eduardo, Francisco Navarro, María Cuaresma, Isabel Vaquero, María C. Ruíz-Domínguez, Živan Gojkovic, María Vázquez et al. "Microalgae: Fast-growth sustainable green factories." *Critical Reviews in Environmental Science and Technology* 45, no. 16 (2015): 1705–1755.

Fu, Fenglian, and Qi Wang. "Removal of heavy metal ions from wastewaters: A review." *Journal of Environmental Management* 92, no. 3 (2011): 407–418.

Guo, Wan-Qian, He-Shan Zheng, Shuo Li, Juan-Shan Du, Xiao-Chi Feng, Ren-Li Yin, Qing-Lian Wu, Nan-Qi Ren, and Jo-Shu Chang. "Removal of cephalosporin antibiotics 7-ACA from wastewater during the cultivation of lipid-accumulating microalgae." *Bioresource Technology* 221 (2016): 284–290.

Habibah, Raudhatul, B. Iswanto, and A. Rinanti. "The significance of tropical microalgae chlorella sorokiniana as a remediate of polluted water caused by chlorpyrifos." *International Journal of Scientific & Technology Research* 9 (2020): 4460–4463.

Hammed, Ademola Monsur, Sanjeev Kumar Prajapati, Senay Simsek, and Halis Simsek. "Growth regime and environmental remediation of microalgae." *Algae* 31, no. 3 (2016): 189–204.

Jena, Jayashree, Nilotpala Pradhan, V. Aishvarya, Rati Ranjan Nayak, Bisnu Prasad Dash, Lala Behari Sukla, Prasanna Kumar Panda, and BaradaKanta Mishra. "Biological sequestration and retention of cadmium as CdS nanoparticles by the microalga Scenedesmus-24." *Journal of Applied Phycology* 27 (2015): 2251–2260.

Kumar, Lakhan, Mohita Chugh, Saroj Kumar, Krishna Kumar, Jaigopal Sharma, and Navneeta Bharadvaja. "Remediation of petrorefinery wastewater contaminants: A review on physicochemical and bioremediation strategies." *Process Safety and Environmental Protection* 159 (2022b): 362–375.

Kumar, R. Vimal, G. R. Kanna, and Sanniyasi Elumalai. "Biodegradation of polyethylene by green photosynthetic microalgae." *Journal of Bioremediation & Biodegradation* 8, no. 381 (2017): 2.

Kumar, V. Santhana, Dhruba Jyoti Sarkar, Basanta Kumar Das, Srikanta Samanta, Gayatri Tripathi, B. K. Behera, and Soma Das Sarkar. "Recycling banana pseudostem waste as a substrate for microalgae biofilm and their potential in arsenic removal." *Journal of Cleaner Production* 367 (2022a): 132772.

Leong, Yoong Kit, and Jo-Shu Chang. "Bioremediation of heavy metals using microalgae: Recent advances and mechanisms." *Bioresource Technology* 303 (2020): 122886.

Ma, Lan, Fengwu Wang, Yuanchun Yu, Junzhuo Liu, and Yonghong Wu. 2017. "Cu removal and response mechanisms of Periphytic biofilms in a tubular bioreactor." *Bioresource Technology* 248: 61–67. https://doi.org/10.1016/j.biortech.2017.07.014.

Matamoros, V., E. Uggetti, J. García, and J. M. Bayona. 2016. "Assessment of the mechanisms involved in the removal of emerging contaminants by microalgae from wastewater: A laboratory scale study." *Journal of Hazardous Materials* 301: 197e205.

Mishaqa, E. S. I. "Biosorption potential of the microchlorophyte *Chlorella vulgaris* for some pesticides." *Journal of Fertilizers & Pesticides* 8 (2017): 177.

Mishra, Sandhya, Yaohua Huang, Jiayi Li, Xiaozhen Wu, Zhe Zhou, Qiqi Lei, Pankaj Bhatt, and Shaohua Chen. "Biofilm-mediated bioremediation is a powerful tool for the removal of environmental pollutants." *Chemosphere* (2022): 133609.

Molazadeh, Parvin, Narges Khanjani, Mohammad Reza Rahimi, and Alireza Nasiri. "Adsorption of lead by microalgae Chaetoceros sp. and Chlorella sp. from aqueous solution." *Journal of Community Health Research* 4, no. 2 (2015): 114–127.

Mondal, Madhumanti, Gopinath Halder, Gunapati Oinam, Thingujam Indrama, and Onkar Nath Tiwari. "Bioremediation of organic and inorganic pollutants using microalgae." In *New and Future Developments in Microbial Biotechnology and Bioengineering* (eds Gupta and Pandey), pp. 223–235. Elsevier, 2019.

Muradov, Nazim, Mohamed Taha, Ana F. Miranda, Digby Wrede, Krishna Kadali, Amit Gujar, Trevor Stevenson, Andrew S. Ball, and Aidyn Mouradov. "Fungal-assisted algal flocculation: Application in wastewater treatment and biofuel production." *Biotechnology for Biofuels* 8, no. 1 (2015): 1–23.

Nguyen, Hoang Tam, Yeomin Yoon, Huu Hao Ngo, and Am Jang. "The application of microalgae in removing organic micropollutants in wastewater." *Critical Reviews in Environmental Science and Technology* 51, no. 12 (2021): 1187–1220.

Ni, Yan, Jinhu Lai, Jinbao Wan, and Lianshui Chen. "Photosynthetic responses and accumulation of mesotrione in two freshwater algae." *Environmental Science: Processes & Impacts* 16, no. 10 (2014): 2288–2294.

Nie, Jing, Yuqing Sun, Yaoyu Zhou, Manish Kumar, Muhammad Usman, Jiangshan Li, Jihai Shao, Lei Wang, and Daniel C. W. Tsang. "Bioremediation of water containing pesticides by microalgae: Mechanisms, methods, and prospects for future research." *Science of the Total Environment* 707 (2020): 136080.

Norvill, Zane N., Andy Shilton, and Benoit Guieysse. "Emerging contaminant degradation and removal in algal wastewater treatment ponds: Identifying the research gaps." *Journal of Hazardous Materials* 313 (2016): 291–309.

Pacheco, Diana, Ana Cristina Rocha, Leonel Pereira, and Tiago Verdelhos. "Microalgae water bioremediation: Trends and hot topics." *Applied Sciences* 10, no. 5 (2020): 1886.

Pandey, Ashutosh, Sameer Srivastava, and Sanjay Kumar. "Isolation, screening and comprehensive characterization of candidate microalgae for biofuel feedstock production and dairy effluent treatment: A sustainable approach." *Bioresource Technology* 293 (2019): 121998.

Peng, F. Q., G. G. Ying, B. Yang, S. Liu H. J. Lai, Y. S. Liu, Z. F. Chen & G. J. Zhou. (2014). "Biotransformation of progesterone and norgestrel by two freshwater microalgae (*Scenedesmus obliquus* and *Chlorella pyrenoidosa*): Transformation kinetics and products identification." *Chemosphere* 95: 581e588.

Pérez-Legaspi, Ignacio Alejandro, Luis Alfredo Ortega-Clemente, Jesús David Moha-León, Elvira Ríos-Leal, Sergio Curiel-Ramírez Gutiérrez, and Isidoro Rubio-Franchini. "Effect of the pesticide lindane on the biomass of the microalgae *Nannochlorisoculata*." *Journal of Environmental Science and Health, Part B* 51, no. 2 (2016): 103–106.

Philippat, Claire, Jacqueline Barkoski, Daniel J. Tancredi, Bill Elms, Dana Boyd Barr, Sally Ozonoff, Deborah H. Bennett, and Irva Hertz-Picciotto. "Prenatal exposure to organophosphate pesticides and risk of autism spectrum disorders and other non-typical development at 3 years in a high-risk cohort." *International Journal of Hygiene and Environmental Health* 221, no. 3 (2018): 548–555.

Qiu, Yao-Wen, Eddy Y. Zeng, Hanlin Qiu, Kefu Yu, and Shuqun Cai. "Bioconcentration of polybrominated diphenyl ethers and organochlorine pesticides in algae is an important contaminant route to higher trophic levels." *Science of the Total Environment* 579 (2017): 1885–1893.

Rempel, Alan, Julia Pedó Gutkoski, Mateus Torres Nazari, Gabrielle Nadal Biolchi, Vítor Augusto Farina Cavanhi, Helen Treichel, and Luciane Maria Colla. "Current advances in microalgae-based bioremediation and other technologies for emerging contaminants treatment." *Science of the Total Environment* 772 (2021): 144918.

Saavedra, Ricardo, Raúl Muñoz, María Elisa Taboada, Marisol Vega, and Silvia Bolado. "Comparative uptake study of arsenic, boron, copper, manganese and zinc from water by different green microalgae." *Bioresource Technology* 263 (2018): 49–57.

Saranya, D. and Shanthakumar, S. "Green microalgae for combined sewage and tannery effluent treatment: Performance and lipid accumulation potential." *Journal of Environmental Management* 241 (2019): 167–178.

Sarı, Ahmet, Özgür Doğan Uluozlü, and Mustafa Tüzen. "Equilibrium, thermodynamic and kinetic investigations on biosorption of arsenic from aqueous solution by algae (Maugeotiagenuflexa) biomass." *Chemical Engineering Journal* 167, no. 1 (2011): 155–161.

Satpati, Gour Gopal, and Ruma Pal. "Co-Cultivation of *Leptolyngbya tenuis* (Cyanobacteria) and *Chlorella ellipsoidea* (green alga) for biodiesel production, carbon sequestration, and cadmium accumulation." *Current Microbiology* 78 (2021): 1466–1481.

Shamshad, Isha, Sardar Khan, Muhammad Waqas, Maliha Asma, Javed Nawab, Nayab Gul, Arjumand Raiz, and Gang Li. "Heavy metal uptake capacity of fresh water algae (Oedogonium westti) from aqueous solution: A mesocosm research." *International Journal of Phytoremediation* 18, no. 4 (2016): 393–398.

Shen, Ying, Wenzhe Zhu, Huan Li, Shih-Hsin Ho, Jianfeng Chen, Youping Xie, and Xinguo Shi. "Enhancing cadmium bioremediation by a complex of water-hyacinth derived pellets immobilized with *Chlorella* sp." *Bioresource Technology* 257 (2018a): 157–163.

Shokri Khoubestani, Roghayeh, Nourollah Mirghaffari, and Omidvar Farhadian. "Removal of three and hexavalent chromium from aqueous solutions using a microalgae biomass-derived biosorbent." *Environmental Progress & Sustainable Energy* 34, no. 4 (2015): 949–956.

Sibi, G. "Biosorption of chromium from electroplating and galvanizing industrial effluents under extreme conditions using Chlorella vulgaris." *Green Energy & Environment* 1, no. 2 (2016): 172–177.

Singh, Anamika, Sabeela Beevi Ummalyma, and Dinabandhu Sahoo. "Bioremediation and biomass production of microalgae cultivation in river water contaminated with pharmaceutical effluent." *Bioresource Technology* 307 (2020): 123233.

Sulaymon, Abbas H., Ahmed A. Mohammed, and Tariq J. Al-Musawi. "Competitive biosorption of lead, cadmium, copper, and arsenic ions using algae." *Environmental Science and Pollution Research* 20 (2013): 3011–3023.

Sutherland, Donna L., and Peter J. Ralph. "Microalgal bioremediation of emerging contaminants-Opportunities and challenges." *Water Research* 164 (2019): 114921.

Tiwari, Bhagyashree, Balasubramanian Sellamuthu, Yassine Ouarda, Patrick Drogui, Rajeshwar D. Tyagi, and Gerardo Buelna. "Review on fate and mechanism of removal of pharmaceutical pollutants from wastewater using biological approach." *Bioresource Technology* 224 (2017): 1–12.

Torres, Moacir A., Marcelo P. Barros, Sara C. G. Campos, Ernani Pinto, Satish Rajamani, Richard T. Sayre, and Pio Colepicolo. "Biochemical biomarkers in algae and marine pollution: A review." *Ecotoxicology and Environmental Safety* 71, no. 1 (2008): 1–15.

Touliabah, Hussein El-Sayed, Mostafa M. El-Sheekh, Mona M. Ismail, and Hala El-Kassas. "A review of microalgae-and cyanobacteria-based biodegradation of organic pollutants." *Molecules* 27, no. 3 (2022): 1141.

Upadhyay, A. K., S. K. Mandotra, N. Kumar, N. K. Singh, Lav Singh, and U. N. Rai. "Augmentation of arsenic enhances lipid yield and defense responses in alga *Nannochloropsis* sp." *Bioresource Technology* 221 (2016): 430–437.

Vassilev, Stanislav V., and Christina G. Vassileva. "Composition, properties and challenges of algae biomass for biofuel application: An overview." *Fuel* 181 (2016): 1–33.

Wan, Liang, Yixiao Wu, Huijun Ding, and Weihao Zhang. "Toxicity, biodegradation, and metabolic fate of organophosphorus pesticide trichlorfon on the freshwater algae Chlamydomonas reinhardtii." *Journal of Agricultural and Food Chemistry* 68, no. 6 (2020): 1645–1653.

Wang, Luyun, Han Xiao, Ning He, Dong Sun, and Shunshan Duan. "Biosorption and biodegradation of the environmental hormone nonylphenol by four marine microalgae." *Scientific Reports* 9, no. 1 (2019): 5277.

Wang, Shi-Kai, Kun-Xiao Yang, Yu-Rong Zhu, Xin-Yu Zhu, Da-Fang Nie, Ning Jiao, and Irini Angelidaki. "One-step co-cultivation and flocculation of microalgae with filamentous fungi to valorize starch wastewater into high-value biomass." *Bioresource Technology* 361 (2022): 127625.

Weis, Leticia, Rosana de Cassia de Souza Schneider, Michele Hoeltz, Alexandre Rieger, Schirley Tostes, and Eduardo A. Lobo. "Potential for bifenthrin removal using microalgae from a natural source." *Water Science and Technology* 82, no. 6 (2020): 1131–1141.

Xiong, Jiu-Qiang, Mayur B. Kurade, Reda A. I. Abou-Shanab, Min-Kyu Ji, Jaeyoung Choi, Jong Oh Kim, and Byong-Hun Jeon. "Biodegradation of carbamazepine using freshwater microalgae *Chlamydomonas mexicana* and *Scenedesmus obliquus* and the determination of its metabolic fate." *Bioresource Technology* 205 (2016): 183–190.

Xu, Peng, and Ledan Huang. "Stereoselective bioaccumulation, transformation, and toxicity of triadimefon in *Scenedesmus obliquus*." *Chirality* 29, no. 2 (2017): 61–69.

Yang, JinShui, Jing Cao, GuanLan Xing, and HongLi Yuan. "Lipid production combined with biosorption and bioaccumulation of cadmium, copper, manganese and zinc by oleaginous microalgae Chlorella minutissima UTEX2341." *Bioresource Technology* 175 (2015): 537–544.

20 Role of Microorganisms as Biocontrol Agents in Aquatic Environments

Sumanta Kumar Mallik, Richa Pathak, Neetu Shahi, and Mohan Singh

20.1 INTRODUCTION

Biological control, often known as biocontrol, is the use of live organisms to suppress pest and pathogen populations. In biocontrol, organisms take advantage of fundamental ecological interactions such as predation, parasitism, antagonism, herbivory, pathogenicity, and competition for common resources. Most biological control agents (BCAs) in use today are invertebrate predators, parasitoids, or herbivores, which are sometimes referred to as "macrobials" or "invertebrate biocontrol agents" (IBCAs) (Sundh and Goettel, 2013). However, antagonistic or pathogenic microorganisms (bacteria, fungi, or viruses) are also utilized in biocontrol.

The biocontrol agents or microbial antagonists prevent pathogen infection of the host or pathogen establishment in the host. Currently, several microorganisms are being researched and used as BCAs or biopesticides. *Richoderma* spp., *Pseudomonas fluorescens*, *Bacillus* spp., *Ampelomyce squisqualis*, *Agrobacterium radiobacter*, non-pathogenic *Fusarium*, *Coniothyrium*, and atoxigenic *Aspergillus niger* are all common BCAs (Singh, 2006; Keswani, 2015; Mishra et al., 2015).

Aquatic ecosystems represent a dynamic and intricate web of life, where a delicate balance between various organisms is essential for maintaining ecological harmony. Within these ecosystems, microorganisms play a pivotal role as biocontrol agents, exerting significant influence over the population dynamics of aquatic organisms. Their presence and activities can have profound implications for the health and sustainability of these environments.

Microorganisms, including bacteria, viruses, fungi, and protists, are omnipresent in aquatic habitats, from freshwater lakes and rivers to marine oceans and estuaries. While they are often associated with disease-causing agents, it is increasingly evident that many microorganisms serve as natural regulators of aquatic ecosystems. Their interactions with other organisms, both beneficial and detrimental, can have far-reaching consequences for aquatic biota and the overall ecological balance. The role of microorganisms as biocontrol agents in aquatic environments is a multifaceted and fascinating area of study, encompassing a wide range of ecological and physiological processes. Aquatic environments, whether freshwater or marine, are among the most diverse and dynamic ecosystems on Earth. They are home to an array of organisms, from microscopic phytoplankton to colossal whales, all interconnected in a delicate web of life. Within this complex tapestry, microorganisms play a pivotal role as biocontrol agents, influencing the balance and health of aquatic ecosystems. This chapter explores the multifaceted roles of microorganisms in controlling various aspects of aquatic environments, from mitigating pathogens to participating in nutrient cycling and maintaining ecosystem stability. It also looks into the intricate mechanisms by which microorganisms influence the dynamics of aquatic populations, addressing their roles in controlling pathogens, nutrient cycling, and overall ecosystem stability.

DOI: 10.1201/9781003408543-20

20.2 BIOLOGICAL CONTROL PRINCIPLES IN THE ENVIRONMENT

20.2.1 Biological Control Methods

Biological control is classified as "classical", "augmentation," or "conservation," depending on whether a BCA is exotic or indigenous to the location to which it is used or introduced, and the aims about its persistence and proliferation (Eilenberg et al., 2001; Bailey et al., 2010). Classical biological control is the introduction of a foreign natural enemy to manage the (typically exotic) pest organism. In the case of augmentation biological control, the agent may or may not already be present in the region of application. There has been a tendency in the last decade toward commercialization of more indigenous than foreign species for augmentation of biocontrol (van Lenteren, 2012). The injected agent is predicted to multiply if the augmentation is inoculative. Conservation biological control attempts to create conditions that favor resident species with intrinsic biocontrol activity, through either manipulation of the environment or crop or pest management approaches.

20.2.2 Mode of Action

In general, IBCAs are utilized to manage other invertebrates or weeds. In general, predators/herbivores consume the prey/host plant, while parasitoids deposit an egg or their progeny within or near the host, following which the immatures consume the host. Most entomopathogenic nematodes operate as a vector for a symbiotic bacterium, which colonizes and kills the host once the worm enters the host body cavity and supplies a food source for the nematode (Vega and Kaya, 2012).

Microbial biological control agents (MBCAs) can be used to kill a wide range of species, including insects, mites, nematodes, weeds, various microbes that cause plant disease, and even vertebrates. The method of action of MBCAs varies greatly. Some microbial agents are pathogens that infect and kill their hosts. Entomopathogenic fungi, for example, can penetrate the exterior cuticle of insects, whereas entomopathogenic viruses and bacteria must be consumed (Vega and Kaya, 2012). Many MBCAs that control fungal or bacterial plant pathogens, on the other hand, rely on one or more of the following mechanisms: competitive displacement through competition for nutrients or space; antibiosis through production of antagonistic metabolites or specific enzymes (e.g., cell-wall degrading); direct interaction as parasitism (e.g., mycoparasitism in *Trichoderma*); and strengthening of the plant's innate resistance against the pathogen (Cook, 2007).

It is frequently difficult to tell which of these mechanisms (or mechanisms) is dominant. It may be more difficult to pinpoint the precise route of action in conservation biocontrol, since this sort of biocontrol frequently relies on indirect stimulation of resident organisms or additions of entire communities. The antagonistic qualities of resident microorganisms, for example, are responsible for the "crop rotation effect" as a disease management technique (Mathre et al., 1999; Weller et al., 2002) Compost addition to soils can be viewed as a means of stimulating or introducing certain microbial biocontrol agents, resulting in long-term disease prevention due to the activities of microbial communities (Hoitink and Boehm, 1999). Certain hostile MBCAs may be identical to those involved in soil's "natural" disease suppression (Chet and Baker, 1981)

20.3 TYPES OF BIOLOGICAL CONTROL AGENT IN AQUACULTURE

20.3.1 Bacteria as a Biocontrol Agent

Bacteria can play important roles as BCAs in aquaculture systems. These beneficial bacteria are often referred to as probiotics or bioaugmentation agents and are used to maintain water quality, enhance the health of aquatic organisms, and improve overall aquaculture system performance. Several important methods by which bacteria serve as BCAs in aquaculture are listed here:

a) Water Quality Management: Bacteria can help maintain proper water quality parameters in aquaculture systems by participating in the nitrogen cycle. Beneficial bacteria, particularly nitrifying bacteria, convert toxic ammonia (NH_3) and nitrite (NO_2^-) into less harmful nitrate (NO_3^-). This process, called nitrification, is crucial for preventing ammonia and nitrite toxicity in aquatic organisms.
b) Pathogen Control: Certain probiotic bacteria can outcompete and inhibit the growth of harmful pathogens such as *Vibrio* spp. and *Aeromonas* spp. in the aquatic environment. By creating a competitive advantage for beneficial bacteria, the overall disease pressure in the aquaculture system can be reduced.
c) Digestive Health: Probiotic bacteria can be added to the diet of aquatic organisms to improve their digestive health and nutrient absorption. These bacteria can help break down complex organic matter and enhance the efficiency of feed utilization, leading to improved growth rates and feed conversion ratios.
d) Biofloc Technology: Biofloc technology is a recirculating aquaculture system that relies on the development of dense microbial communities, including bacteria, to maintain water quality and provide a natural food source for cultured organisms. Bacteria within the biofloc contribute to nutrient cycling and waste management.
e) Algae Control: Beneficial bacteria can help control excessive algal growth in aquaculture ponds by competing for nutrients essential for algal growth, such as nitrogen and phosphorus. This can prevent the depletion of dissolved oxygen in the water caused by algal blooms.
f) Improving Larval Rearing: Probiotic bacteria are sometimes used in the rearing of larval aquatic organisms to improve survival rates and reduce disease susceptibility. They can help establish a healthy microbial community in larval tanks, enhancing the development of the larvae's immune system.
g) Biosecurity: Beneficial bacteria can contribute to the overall biosecurity of aquaculture systems by reducing the prevalence of harmful pathogens. This can help reduce the need for antibiotics and other chemical treatments.
h) Waste Reduction: Some bacteria can assist in breaking down organic waste, including uneaten feed and feces, reducing the accumulation of organic matter in aquaculture systems and minimizing the risk of water quality issues.

The Japan Sea Farming Association, which has been generating millions of larvae of the crab *Portunus trituberculatus* each year since the beginning of the 1980s, began operations at Tamano Station in June 1985. A significant majority of larvae older than a few days were infected with *Vibrio* spp. in the 1940s. Three-quarters of the larvae died within a day as a result of the bacteria multiplying inside the larval body. It was futile to use any antibiotics to stop bacterial development; on the morning after treatment, practically all the larvae were floating close to the surface and perished later that day. Mycelia of the fungus *Haliphehorns* sp. were abundant inside the deceased bodies (Nogami and Maeda, 1992).

20.3.2 Marine Yeasts as Biocontrol Agents and Producers of Bio-Products

20.3.2.1 Industrial Enzymes

It has been demonstrated that diverse species of marine-derived yeasts may generate amylase, cellulase, inulinase, lipase, alkaline protease, acid protease, and phytase (Chi et al., 2009b). All the enzymes have been purified and identified. Inulinase, lipase, alkaline protease, acid protease, and phytase genes have been cloned and described. Several of them have been expressed heterologously in other hosts. Several possible applications for enzymes and their encoding genes have been demonstrated. The inulinase gene, for example, was expressed in *Pichia pastoris* X-33 after being cloned from the marine-derived yeast *P. guilliermondii* strain 1 (collection number 2E00005 at MCCC). After optimizing the conditions for producing recombinant inulinase, 286.85.4 U/ml of recombinant inulinase activity in the supernatant of a 2-l fermentor culture was obtained after 120 hours of fermentation (Zhang et al., 2009a). Exo-inulinase activity was likewise high in pure recombinant inulinase (Zhang et al., 2009b). After the recombinant inulinase thoroughly hydrolyzed the inulin (over 20% w/v), the hydrolysate produced could be converted into ethanol by a high ethanol–generating yeast, *Saccharomyces* sp. W0, and the ethanol concentration in fermented medium may reach over 15% (v/v). The inulinase gene was recently cloned and expressed in high ethanol–generating yeast. In one stage, the recombinant yeast can make more than 15% (v/v) ethanol from inulin and 12% (v/v) ethanol from powdered Jerusalem artichoke tubers. This suggests that recombinant inulinase and the gene cloned from marine yeast have great potential for use in the biofuel business (Chi et al., 2009a).

Chitinase, agarase, and carrageenase from marine bacteria or other sources could be expressed in marine yeasts to overproduce recombinant chitinase, agarase, and carrageenase for chitin, agar, alginate, and carrageenan hydrolysis and synthesis of bioactive oligosaccharides (Liu et al., 2010; Chi et al., 2009b). Because some of the enzyme-producing marine bacteria are pathogens to marine algae and animals, the genes in the GRAS marine yeasts must be overexpressed. For example, the alginate lyase structural gene (AlyVI gene) was amplified from the plasmid pET24-ALYVI, which carried the alginate lyase gene from the marine bacterium *Vibrio* sp. QY101, a pathogen of *Laminaria* sp. (Han et al., 2004), and expressed in *Yarrowia lipolytica* cells. To create varied lengths of oligosaccharides, recombinant alginate lyase can hydrolyze poly-D-mannuronate (M), poly-L-guluronate (G), and sodium alginate (more than pentasaccharides) (Liu et al., 2010). As illustrated, *Y. lipolytica* is also common in maritime habitats. The expression vector pINA1317 can also be utilized to clone and express genes from various sources in the marine-derived *Y. lipolytica* (Wang et al., 2009a).

Several marine fungi, including *Laminaria* sp., manufacture β-glucans. Non-cellulosic β-glucans are important biopolymers due to their unique biological activity, which includes anti-tumor, anti-inflammatory, and immunomodulatory properties.

Using microbial **β-glu**canases to fragment the polysaccharide into shorter chain lengths is one method of enhancing the water solubility of **β-glu**cans. These enzymes could be used in biotechnology for cell fusion, transformation, and protoplast preparation, as well as in the food, feed, agricultural, pharmaceutical, and fermentation industries, and during the clarification of slimy material (Martin et al., 2007). It was recently shown that crude enzymes from numerous marine-derived yeasts may hydrolyze laminarin, -1,3-glucan isolated from *Laminaria* sp. (Wang et al., 2008).

Purified -1,3-glucanase from the marine yeast *Williopsis saturnus* WC91-2 with a molecular mass of 47.5 kDa was able to convert laminarin into monosaccharides and disaccharides, indicating that the enzyme is an exo-1,3-glucanase (Peng et al., 2009). The pure killer toxin from the marine killer yeast YF07b actively hydrolyzes laminarin into monosaccharides and disaccharides (Wang et al., 2007). This demonstrates that the marine-derived yeasts' -1,3-glucanases can be employed to actively hydrolyze -1,3-glucan from *Laminaria* sp. The *W. saturnus* WC91-2 gene (WsEXG1) has been cloned and described (Peng et al., 2009). The amino acids inferred from the WsEXG1 gene contain the conserved amino acid sequence NLCGEWSAA, where the Glu (E) has been demonstrated to be the catalytic nucleophile for -1,3-glucan hydrolysis (Martin et al., 2007). In *Y. lipolytica*, the gene can be overexpressed, and the released -1,3-glucanase can aggressively hydrolyze laminarin into solo monosaccharides and minor disaccharides. This could imply that the altered *Y. lipolytica*'s recombinant -1,3-glucanase can be employed to hydrolyze -1,3-glucan (Peng et al., 2009) (Table 20.1 and Figure 20.1).

TABLE 20.1
Various Applications of Marine Yeasts and Their Products

Products	Mechanisms	Marine Yeasts	Applications	References
Probiotics	Strong antagonism, induction of intestinal maturation, and improvement of growth and survival	*Rhodotorula rubra*, *R. glutinis*, *Candida zeylanoides*, *Saccharomyces cerevisiae*, *Debaryomyces hansenii*, *S. cerevisiae* var. *boulardii*	Inhibition of pathogens	Gatesoupe (2007)
Immunostimulants	Immuno-stimulation of fish	*S. cerevisiae*, *C. sake*	Stop bacterial and viral infection	Sajeevan et al. (2006)
Siderophore	High iron-chelating ability	*A. pullulans* HN6.2	Inhibition of pathogen	Wang et al. (2009b)
Killer toxins	Inhibition and hydrolysis of cell wall of sensitive cells	*W. saturnus* WC91-2, *P. anomala* YF07b	Killing of pathogens	Peng et al. (2010)
Vaccine	Antibody	*S. cerevisiae*, *Y. lipolytica*	Enhanced immunity	Zhu et al. (2006)
Amylase	Hydrolysis of starch	*A. pullulans*	Fermentation	Chi et al. (2009b)
Alkaline protease	Hydrolysis of protein	*A. pullulans*	Food and pharmaceutical industries	Chi et al. (2009b)
Acid protease	Hydrolysis of protein	*Metschnikowia reukaufii*	Food and pharmaceutical industries	Chi et al. (2009b)
Lipase	Hydrolysis of lipid	*A. pullulans*	Chemical industry	Chi et al. (2009b)
Cellulase	Hydrolysis of cellulose	*A. pullulans*	Chemical industry	Chi et al. (2009b)
Inulinase	Hydrolysis of inulin	*P. guilliermondii*	Food and fuel industry	Zhang et al. (2009b)
Phytase	Hydrolysis of phytate	*K. ohmeri*	Feed industry	Chi et al. (2009b)
Alginate lyase	Hydrolysis of alginate	*Y. lipolytica*	Pharmaceutical industries	Liu et al. (2009)
Glucanase	Hydrolysis of β-glucan	*W. saturnus*	Pharmaceutical industry	Peng et al. (2009)
Riboflavin	Nutrition	*C. membranifaciens* subsp. *flavinogenie*	Food and pharmaceutical industry	Wang et al. (2008)
Single-cell protein	Nutrition	*C. aureus*, *Y. lipolytica*	Food and feed industry	Zhang et al. (2009b)
Single-cell oil	Bioenergy	*R. mucilaginosa*	Biodiesel industry	Li et al. (2010)
Nanoparticles	Materials	*Y. lipolytica*	Biomaterial industry	Agnihotri et al. (2009)
Degrader of pollutants	Degradation of pollutants	*Y. lipolytica*	Bioremediation	Bankar et al. (2009)

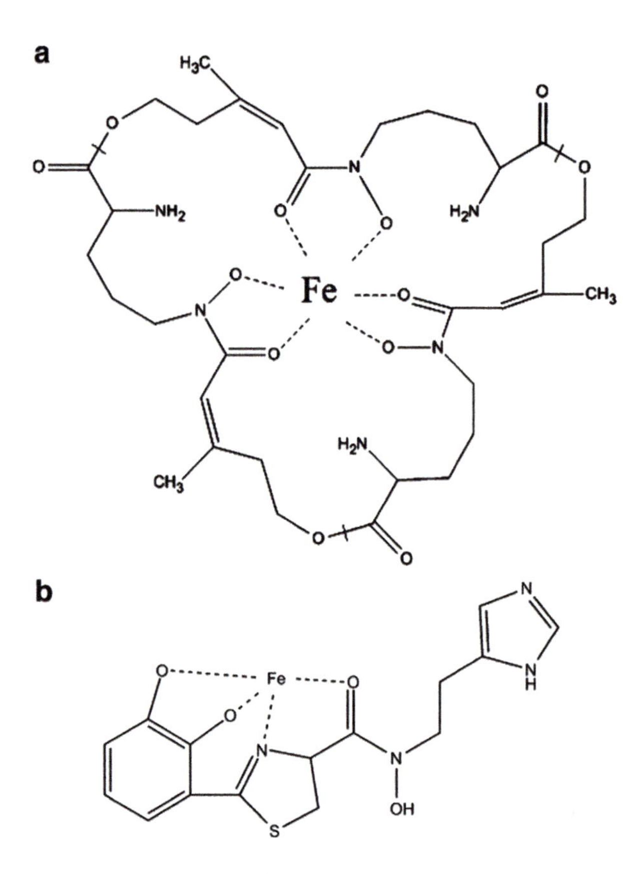

FIGURE 20.1 The chemical structures of fusigen (a) produced by *Aureobasidium pullulans* HN6.2 and anguibactin (b) produced by *Vibrio anguillarum*. (From Chi, Z.-M. et al., *Applied Microbiology and Biotechnology*, 86, 1227–1241, 2010.)

20.3.3 Probiotic Bacteria as Biological Control Agents in Aquaculture

Many definitions of probiotics have been proposed as new data have emerged. Fuller (1989) defined them precisely. Probiotics are a live microbial feed additive that benefits the host animal by enhancing its intestinal equilibrium.

Historically, the focus has been on terrestrial species, and the term "probiotic" has invariably referred to gram-positive bacteria from the genus *Lactobacillus*. However, the implementation of Fuller's (1989) definition in aquaculture necessitates several modifications. It is assumed in aquaculture that the gut microbiota does not exist as a separate entity but is always in interaction with the environment and host processes (Holzapfel et al., 1998). Various researchers have already looked into the interaction between the gut microbiota and the aquatic environment or food. Cahill (1990) synthesized the findings of these fish studies, demonstrating that bacteria in the aquatic environment alter the composition of the gut microbiota and vice versa. The genera found in the intestine appear to be those from the environment or diet that can survive and multiply in the intestine. Nonetheless, it may be argued that in aquaculture systems, the immediate surrounding environment has a considerably greater influence on health than in terrestrial animals or people.

Indeed, host–microbe interactions differ qualitatively and quantitatively across aquatic and terrestrial species. Hosts and microbes coexist in the water. In most of the terrestrial systems, however, the gut serves as a wet habitat in an otherwise water-stressed environment. In certain ways, microorganisms in an aquatic environment have the option of coexisting with the possible host (intestinal tract, gills, or skin), whereas in the terrestrial environment, appreciable activity may be confined to aquatic niches such as those given by host animal intestines (Harris, 1993).

Aquatic farmed animals are surrounded by an environment that sustains their infections independently of the host animals, allowing opportunistic pathogens to reach high numbers around the animal (Moriarty, 1998). Bacteria from the environment are constantly consumed, either with the meal or when the host drinks. This is especially true for filter feeders, which consume bacteria at a rapid pace from the surrounding environment, resulting in a natural interplay between ambient microbiota and live food.

While probiotic research in aquaculture initially focused on fish juveniles, greater attention has since been paid to fish and shellfish larvae, as well as live food organisms. By interaction with the mother, terrestrial animals (mammals) inherit an important portion of the initially colonizing bacteria, whereas aquatic species often spawn axenic eggs in the water, with no further contact with the parents. This permits germs from the environment to colonize the egg surface. Additionally, newly hatched larvae or newborn animals lack a completely established intestinal system and have no microbial community in the intestine, gills, or skin. Because the early stages of aquatic larvae rely on the water in which they are reared (Cahill, 1990; Hagiwara et al., 1994; Ringoe and Birkbeck, 1999) for their major microbiota, the qualities of the bacteria in the ambient water are critical (Skjermo et al., 1997).

As previously noted, the interaction between microbiota, including probiotics, and the host is not restricted to the gastrointestinal system. Probiotic bacteria may be active not just on the host's gills or skin, but also in its surrounding environment. Because of the intense interaction between the culture environment and the host in aquaculture, many probiotics are received through the culture environment rather than directly from the feed.

As a result, the following updated definition is proposed, which allows a broader application of the term "probiotic" and addresses the preceding objections. A probiotic is a live microbial supplement that has a favorable effect on the host by affecting the host-associated or ambient microbial population, boosting feed utilization or nutritional value, enhancing the host's reaction to disease, or improving the quality of its ambient environment. Probiotics, according to this definition, are microbial adjuncts that prevent pathogens from multiplying in the intestinal tract, on the superficial structures, and in the culture environment of the cultured species, that ensure optimal use of the feed by assisting in its digestion, that improve water quality, or that stimulate the host's immune system. Bacteria that give vital nutrients to the host (single-cell protein) without being active in the host or interacting with other bacteria, the host's environment, or the host itself are excluded from the definition. Although probiotics may also contribute significantly to nutritional health and zootechnical performance, and although it is sometimes impossible to separate feeding of aquatic organisms from environmental control, this review is limited to the use of probiotics as BCAs in aquaculture (Table 20.2).

20.3.4 Chitinase-Producing Bacteria and Their Role in Biocontrol

Chitin is plentiful and pervasive in nature. It is, in fact, one of the most abundant biopolymers on the planet, second only to cellulose. Chitin can be found in a variety of creatures, including arthropod (crustaceans and insects) shells, exoskeletons, and gut linings. It is also found in the cell walls of various fungi, including some yeasts, as well as the structural frameworks of certain protista and nematode eggs. Many microbial genomes contain diverse genes that encode chitinolytic enzymes, which have been extensively studied; however, studies on the utilization of microbes that use insoluble chitin as a carbon source in biotechnology are limited (Stoykov et al., 2015). Fungicide leaching

TABLE 20.2
Overview of Literature Reports Dealing with Probiotics as Biological Control Agents in Aquaculture

Putative Probiotic	Origin	Observations	Method of Administration	Mode of Action	Reference
Fish eggs and larvae					
Several strains (unidentified)	Eggs of cod and halibut	Failure of the strains to prevent the adherence of environmental bacteria to cod eggs	Bathing in bacterial suspension	Antagonism	Hansen et al. (1989)
Vibrio salmonicida-like and *Lactobacillus plantarum*	Fish	Increase of survival of halibut larvae 2 weeks after hatching	Addition to culture water	Immunostimulation	Olafsen (1998)
Bacillus strain IP5832 spores (Paciflor 9)	–	Increase of weight of turbot larvae when fed spore-fed rotifers; decrease of mortality when challenged with an opportunistic *Vibrionaceae* member	Addition to rotifer diet	Antagonism and/or improved nutritional value of the rotifers	Gatesoupe (2007)
Streptococcus lactis and *Lactobacillus bulgaricus*	–	Increase of survival of turbot larvae 17 days after hatching	Enrichment of rotifers and Artemia	–	García et al. (1992)
Lactobacillus or *Carnobacterium*	Rotifers (*Brachionus plicatilis*)	Decrease of mortality of turbot larvae challenged with a pathogenic *Vibrio* sp.	Enrichment of rotifers	Antagonism and/or improved nutritional value of the rotifers	Gatesoupe (1994)
Vibrio pelagius	Copepod-fed turbot larvae	Decrease of mortality of turbot larvae challenged with *Aeromonas caviae*	Addition to culture water	–	Ringø and Vadstein (1998)
Strain E (*Vibrio alginolyticus*-alike)	Healthy turbot larvae	Decrease of mortality of turbot larvae challenged with a pathogenic *Vibrio* strain P; also, the growth rate of the larvae may be increased	Enrichment of rotifers	Competition for iron	Gatesoupe (1997)
Microbially matured water	–	Increase of initial growth rate of turbot and halibut larvae	As culture water	–	Skjermo et al. (1997); Vadstein et al. (2020)

(Continued)

TABLE 20.2 (CONTINUED)
Overview of Literature Reports Dealing with Probiotics as Biological Control Agents in Aquaculture

Putative Probiotic	Origin	Observations	Method of Administration	Mode of Action	Reference
Fish juveniles and adults					
Lyophilized *Carnobacterium divergens*	Atlantic salmon intestines	Increase of mortality of Atlantic salmon fry challenged with cohabitants infected with *A. salmonicida*	Addition to diet	–	Gildberg et al. (1995)
Lyophilized *Carnobacterium divergens*	Atlantic salmon intestines	Decrease of mortality of Atlantic cod fry when challenged with a pathogenic *V. anguillarum* strain	Addition to diet	–	Gildberg et al. (1997)
Lyophilized *Carnobacterium divergens*	Atlantic salmon intestines	Decrease of mortality of Atlantic cod fry challenged with a pathogenic *V. anguillarum* strain 12 days after the infection; 4 weeks after the infection, however, the same mortality as in the control was reached	Addition to diet	Antagonism	Gildberg and Mikkelsen (1998)
Carnobacterium strain K1	Atlantic salmon intestines	Growth inhibition of *V. anguillarum* and *A. salmonicida* in fish intestinal mucus and fecal extracts (no in vivo test)	–	Antagonism	Jöborn et al. (1997)
Carnobacterium	Atlantic salmon intestines	Growth inhibition of *V. anguillarum* in turbot fecal extracts (no in vivo test)	–	Antagonism	Olsson et al. (1998)
Fluorescent pseudomonad F19/3	Fish mucus	Decrease of mortality of Atlantic salmon presmolts challenged with stress-inducible *A. salmonicida* infection	Bathing in bacterial suspension	Competition for iron	Smith and Davey (1993)
Pseudomonas fluorescens AH2	Iced Lake Victorian Nile perch	Decrease of mortality of rainbow trout juveniles challenged with a pathogenic *V. anguillarum*	Addition to culture water and/or bathing in bacterial suspension	Competition for iron	Gram et al. (1999)
Vibrio alginolyticus	Commercial shrimp hatchery in Ecuador	Decrease of mortality of Atlantic salmon juveniles challenged with a pathogenic *A. salmonicida*, *V. anguillarum*, and *V. ordalii*	Bathing in bacterial suspension	Antagonism	Austin and Day (1990)

(Continued)

TABLE 20.2 (CONTINUED)
Overview of Literature Reports Dealing with Probiotics as Biological Control Agents in Aquaculture

Putative Probiotic	Origin	Observations	Method of Administration	Mode of Action	Reference
Bacillus megaterium, *B. polymyxa*, *B. licheniformis*, two strains of *B. subtilis* (Biostart)	–	Increase of survival and net production of channel catfish	Addition to pond water	–	Queiroz and Boyd (1998)
Spray-dried *Tetraselmis suecica* (unicellular alga)	–	Decrease of mortality of Atlantic salmon juveniles challenged with several pathogens; alga was effective prophylactically as well as therapeutically	Addition to diet	Antagonism	Austin et al. (1992)
Crustaceans					
Bacillus strain S11	*Penaeus monodon* or mud and water from shrimp ponds	Increase of mean weight and survival of *P. monodon* larvae and post larvae; decrease of mortality after challenge with the pathogen *V. harveyi* D331	Addition to diet	Antagonism	Rengpipat et al. (1998)
Vibrio alginolyticus	Pacific Ocean seawater	Increase of survival and weight of *L. vannamei* post larvae; decreased observation of *V. parahaemolyticus* in the shrimps	Addition to culture water	Antagonism	Garrique (1995)
Bacillus	–	Increase of survival of penaeid shrimps; decrease of luminous *Vibrio* densities	Addition to pond water	Antagonism	Moriarty (1998); Moriarty and Body (1995)
Strain PM-4 and/or NS-110	Soil	Increase of survival of *P. monodon* and *P. trituberculatus* larvae; decrease of *Vibrio* densities	Addition to culture water	Antagonism/food source for larvae	Maeda (1994); Maeda and Liao (1994); Nogami and Maeda (1992)
Strain BY-9	Coastal seawater	Increase of survival of *P. monodon* larvae; decrease of *Vibrio* densities	Addition to culture water	–	Sugama and Tsumura, abstract
Bivalve mollusks					
Vibrio strain 11	Microalgae in a scallop hatchery	Decrease of mortality of scallop larvae challenged with a pathogenic *V. anguillarum*-like strain	Bathing in bacterial suspension	Antagonism	Riquelme et al. (1997)

(Continued)

TABLE 20.2 (CONTINUED)
Overview of Literature Reports Dealing with Probiotics as Biological Control Agents in Aquaculture

Putative Probiotic	Origin	Observations	Method of Administration	Mode of Action	Reference
Aeromonas media A 199	–	Decrease of mortality and suppression of the pathogen of Pacific oyster larvae when challenged with a pathogenic *V. tubiashii*	Addition to culture water	Antagonism	Gibson et al. (1998)
Live food—Algae					
Several strains	Turbot larvae	Growth stimulation of *P. lutheri*	Addition to culture water	–	Munro et al. (1995)
Flavobacterium sp.	*Chaetoceros gracilis* mass culture	Improved growth characteristics of *C. gracilis*, *I. galbana*, and *P. lutheri*	Addition to culture water	–	Suminto and Hirayama (1997)
Strain SK-05	*Skeleton emacostatum* culture	Inhibition of the growth of *V. alginolyticus* in *Skeleton emacostatum* culture	Addition to culture water	Competition for resources	Rico-Mora et al. (1998)
Live food—Rotifers					
Lactobacillus plantarum	–	Inhibition of the growth of a fortuitous *A. salmonicida* strain in rotifer culture	Addition to diet	Antagonism	Gatesoupe (2007)
Lactococcus lactis AR21	Rotifer culture	Counteraction of rotifer growth inhibition when challenged with *V. anguillarum*	Addition to culture water	–	Harzevili et al. (1998)
Several strains	Rotifer culture	Increase of the specific growth rate of rotifer culture	Addition to culture water		Rombaut et al. (1999)
Live food—*Artemia*					
Vibrio alginolyticus C14	–	Decrease of mortality of *Artemia* nauplii when challenged with *V. parahaemolyticus*	Addition to culture water	–	Gomez-Gil et al. (1998)
Several strains	*Artemia* culture	Decrease of mortality of *Artemia* juveniles when challenged with *V. proteolyticus*			Verschuere et al. (2000b)

Source: Verschuere et al. (2000a).

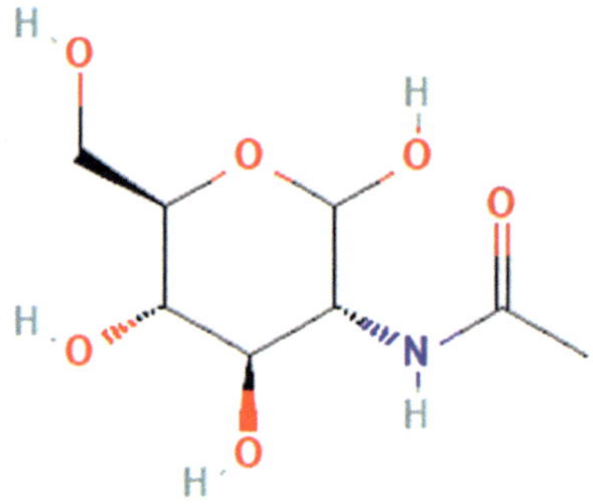

FIGURE 20.2 2D structure of N-acetylglucosamine.

into ground water harms both aquatic habitats and drinking water resources, and pesticide use has been related to a variety of human health issues. A recent investigation of honey discovered that the samples contained insecticides (Chiesa et al., 2016). Because of public concern about these chemical residues, as well as the development of fungicide resistance by pathogens, new ways of controlling both pre- and post-harvest illnesses have been developed (Russell, 2006). Chitin and cellulose are functionally and physically similar (Figure 20.2). Unlike chitin, which is a linear homopolymer of (14)-linked N-acetyl-D-glucosamine (GlcNAc) residues, cellulose is made up of glucose monomers. GlcNAc residues differ from glucose in that the hydroxyl at the C-2 position is replaced by an acetyl amino group. Chitin is also related to murein, a structural polymer composed of alternating GlcNAc and N-acetylmuramic acid monomers.

Chitinases are among the many hydrolytic enzymes produced by microorganisms. Actinobacteria, like *Firmicutes*, are widely known for their chitinolytic enzyme synthesis (Liu et al., 2010) and activity, as are certain Proteobacteria (Beier and Bertilsson, 2013). Microbial chitinases break down and weaken the cell walls of many pests and pathogens, displaying antibacterial, antifungal, insecticidal, or nematicidal activity. Chitinolytic enzymes will become a more visible and crucial answer to the environmental and human dangers caused by the use of synthetic pesticides and fungicides. Thus, chitinolytic bacteria hold potential as a safer alternative to the more hazardous practice of using insect- and fungal-killing pesticides.

Chitinases are classified into distinct glycoside hydrolase (GH) families, based on their amino acid sequence, such as GH18, GH19, and GH20. GH18 endo chitinases are present in bacteria, fungi, and mammals, as well as some plant chitinases. The GH18 family contains the majority of bacterial chitinases (Adrangi and Faramarzi, 2013). Based on amino acid sequence homology of the different catalytic domains, bacterial GH18 chitinases are classified into three primary subfamilies, A, B, and C (Horn et al., 2006). It is assumed that each chitinase has distinct, albeit complementary, processes (Kawase et al., 2004). Endo chitinases of the GH19 family are mostly found in plants, but they have also been found in bacteria, particularly actinobacteria, green non-sulfur (*Chloroflexi*), and purple (*Chromatiales* and *Rhodospirillaceae*) bacteria (Kawase et al., 2004; Udaya Prakash et al., 2010; Watanabe et al., 1999). Plant GH19 chitinases are hypothesized to be involved in defense against fungal infections (Kawase et al., 2004). Because GH families 18 and 19 have no sequence similarity, 3D structure, or molecular mechanisms in common, they are thought to have evolved as distinct lineages (Dahiya et al., 2006). Actinobacterial GH19 chitinases are related to plant class IV chitinases based on amino acid sequence analysis. They appear to have evolved from plant chitinases and may have been acquired by bacteria via horizontal gene transfer (Kawase et al., 2004; Udaya Prakash et al., 2010). GH19 chitinases have been found in several actinobacteria, typically in conjunction with GH18 chitinases (Udaya Prakash et al., 2010).

20.3.4.1 Chitin and Its Hydrolysis

Chitin is available in three crystalline forms, α-, β-, and γ-chitin, which differ in the arrangement of polymer chains and thus, have varying mechanical properties (Jang et al., 2004). To give rigidity and strength, the multiple chains cross-link to other structural polymers such as proteins and β-glucans (Gooday, 1994). Chitin is present in varying degrees of deacetylation in the environment, ranging from totally acetylated chitin to its wholly deacetylated form, known as chitosan (Beier and Bertilsson, 2013). Because chitin is so prevalent in nature, its breakdown has received substantial attention in general biochemistry, molecular biology, biogeochemistry, and microbial ecology (Beier and Bertilsson, 2013; Adrangi and Faramarzi, 2013; Cohen-Kupiec and Chet, 1998; Nagpure et al., 2014).

Chitinases, which are found in chitin-containing animals such as insects, crustaceans, and fungi, catalyze the destruction of chitin. These enzymes have also been discovered in viruses, bacteria, archaea, protista, higher plants, and mammals. Chitinases have several roles in these species, including morphogenesis, nutrition cycling, and defense against chitin-containing pests and parasites (Dahiya et al., 2006; Herrera-Estrella and Chet, 1999). Chitinolytic bacteria are found in a variety of settings and degrade chitin in both aerobic and anaerobic circumstances. They participate in the nutritional cycle of chitin generated from arthropod shells and other sources in marine environments (Souza et al., 2011). Bacteria utilize chitin from insects and fungi as a carbon and nitrogen source in the soil and rhizosphere (Cohen-Kupiec and Chet, 1998; Geisseler et al., 2010).

Because chitin and cellulose have structural similarities, the chitinolytic and cellulolytic pathways proceed in parallel. Chitin hydrolysis begins with cleaving the polymer into water-soluble oligomers, which are then split into dimers by another enzyme, which then divides the dimers into monomers. An endo-acting chitinase (EC 3.2.1.14) hydrolyzes chitin and the resultant oligomers at random, yielding a mixture of end products of various sizes. However, this enzyme is incapable of breaking down compounds other than diacetyl chitobiose. β-N-acetyl hexosaminidases (EC 3.2.1.52), on the other hand, are exo-acting enzymes that cleave chitin oligomers as well as chitin from the non-reducing end. They are the sole enzymes capable of cleaving diacetyl chitobiose. The nomenclature for the enzymes currently categorized as β-N-acetyl hexosaminidases, however, is not totally agreed upon (Brzezinska et al., 2013; Patil et al., 2000).

20.3.4.2 Chitinolytic Microorganisms as Chemical Pesticide Alternatives in Pathogen Control

The most common bacteria utilized as biocontrol agents are *Streptomyces*, *Bacillus*, and *Pseudomonas* species (Paulitz and Bélanger, 2001). Given the number of chitinolytic bacteria being examined, as well as the fact that current disease control tactics are detrimental, the idea of developing new pesticides based on biocontrol bacteria offers viable alternatives. Pesticides' harmful impacts on the environment are well documented and reported in the literature. These chemicals not only harm organisms in the environment other than the target organisms, but they also have the potential to migrate within and outside the application site (Kalia and Gosal, 2011).

Chitinases have also been shown to inhibit insect growth; when larvae come into contact with chitinases, their feeding rate and body weight drop, eventually leading to death. These symptoms are caused by a breakdown of the peritrophic membrane that lines the larvae's gut epithelium, the major component of which is chitin (Carozzi and Koziel, 1997). Brandt et al. 1978 discovered that chitinases destroyed the peritrophic membrane of *Orgyia pseudotsugata*, and this effect was later observed in vivo with *Spodoptera littoralis* and *Escherichia coli* that expressed the *Serratia marcescens* endochitinase ChiAII (Regev et al., 1996).

20.3.5 Bacteriophages as Biocontrol Agents

Antibiotic resistance in bacteria, as well as difficulties with antibiotic residues in aquatic habitats and aquaculture products, underlines the need for alternate therapies for pathogenic bacteria

TABLE 20.3
Possible Advantages and Disadvantages of Bacteriophages as Biocontrol Measures for Aquatic Bacterial Disease

S. No.	Advantages	Disadvantages
1.	Abundance in nature, including lytic and lysogenic bacteriophages.	Only strong lytic bacteriophages are needed for phage therapy.
2.	Treatment does not require repeated administration.	Difficult to extrapolate from in vitro treatment to in vivo expectation.
3.	Narrow host range can provide an effective treatment to targeted bacteria, without any effect on other bacteria.	Need to identify and isolate the bacterium causing the infection/disease.
4.	Rapid process to isolate and select new lytic bacteriophages.	Need expertise and experimental setting up and careful screening to determine the activity spectrum of phages.
5.	Administration though feeding, injection, and immersion.	There might be practical difficulties, e.g., injecting large numbers of aquaculture animals.
6.	High specificity of killing of pathogens, including antibiotic-resistant bacteria.	Phage resistance can be developed by bacteria.
7.	Phage-resistant colonies might not be pathogenic.	Newly isolated phages are required for phage-resistant bacteria.
8.	No side-effects on microbiota and environment during or after phage application.	Phages could transfer virulence factors and other genes coding for undesired traits.
9.	Phage cocktails can reduce the phage-resistant bacteria.	All infecting bacteria must be correctly identified, which might have time constraints.
10.	Phage therapy might be less expensive than antibiotics.	More studies in phage therapy might cause additional costs.

Source: Oliveira et al., *Aquaculture International*, 20, 879–910, 2012; Le and Kurtböke, *Microbiology Australia*, *40*(1), 37–41, 2019.

management in aquaculture. One of these possible techniques could be bacteriophage-mediated biocontrol (Rong et al., 2014; Wang et al., 2017).

Other considerations, such as the presence of phage-resistant bacteria, must be considered before broad application of bacteriophage therapy may occur. Phage-resistant *Streptococcus iniae*, for example, causes beta-hemolytic streptococcosis in the Japanese flounder *Paralichthys olivaceus* (Matsuoka et al., 2007). Bacteriophages can also mediate toxicity, such as that observed in *Penaeus monodon* after infection with *Vibrio harveyi* (Ruangpan et al., 1999). The accidental introduction of lysogenic phages was identified as a potential issue for shrimp growers (Flegel et al., 2005). After boiling for 10 minutes, *V. harveyi Siphophage* 1 (VHS1) lost its ability to lyse cells but preserved its ability to lysogenize. As a result, frying crustaceans may not be adequate to completely inactivate phages that may be present in the seafood (Flegel et al., 2005) (Table 20.3).

20.4 POTENTIAL HAZARDS AND RISKS OF BIOCONTROL AGENTS

The possible hazards and risks of BCAs vary depending on whether they are used for classical, augmentation, or conservation biocontrol. Risks should be assessed separately for each method of biocontrol, and how they differ for microorganisms and invertebrates.

20.4.1 Classical Biocontrol

Since the organism is brought into an area where it did not previously exist, traditional biocontrol entails distinct environmental risks. As a result, any negative consequences would be irreversible or extremely difficult to ameliorate. The risks are supposedly negligible when the targeted host is exotic in and of itself, which is usually the case, because the goal is to drastically reduce a population that has become invasive, probably due to lack of natural adversaries. If the targeted host is indigenous, the potential dangers increase, since the targeted host has a long history in the ecosystem. The effects of a catastrophic depopulation of an indigenous creature on the ecosystem are difficult to anticipate (Lockwood, 1993).

The greatest risk in traditional biocontrol is that the BCA has a negative impact on non-target host populations, leading to depopulation or in the worst-case scenario, extinction. In principle, this hazard is the same whether the BCA is a pathogenic microbe, a parasitoid, or a predatory macroorganism.

Nonetheless, present information of biogeographical distribution patterns for microorganisms in general is significantly weaker, making it impossible to apply alien or indigenous features. Recently developed molecular approaches (e.g., Enkerli and Widmer, 2010; Holmberg et al., 2009) are now giving more efficient tools for research into microorganism biogeography and population genetics.

20.4.2 Augmentation Biocontrol

If the biocontrol organisms are indigenous, augmentation biocontrol (inoculative or inundative) should provide fewer environmental risks than classical biocontrol. However, when BCAs are utilized in inundative techniques, distinct dangers may develop, since the BCA is injected at considerably higher densities than usual. According to research, augmentative additions with indigenous creatures rarely constitute a significant environmental concern. Any negative impacts would be expected to end once the augmentations were removed and the organisms returned to their natural population numbers.

A possible risk of using IBCAs for augmentative purposes is that they spread, develop enduring populations, and negatively affect non-target creatures outside the area where they are applied. Abiotic influences like climate, as well as biotic host/prey interactions, can have a significant impact on their ability to persist (Boivin et al., 2006). The vast majority of IBCA releases in augmentation biocontrol have had no negative side effects, and we have identified no cases of significant consequences when utilizing indigenous species (van van Lenteren et al., 2008). However, there are examples of non-target effects when the biocontrol agent is employed outside the organism's native region (Louda et al., 2003).

20.5 BIOCONTROL AS AN EXTENSIVE BACKGROUND

Biological pest management had its beginnings long before the establishment of BCA laws. Attempts have been made to boost natural enemies of pests and diseases. For ages, pesticides have been used to protect crops (Table 20.4).

20.6 MICROORGANISMS AS PATHOGEN BIOCONTROL AGENTS

A general overview of microorganisms as pathogen biocontrol agents is given below:

a) Bacterial Predation: Certain microorganisms, notably bacteriophages, act as natural predators of harmful bacteria in aquatic environments. They infect and lyse bacterial cells, regulating their populations and preventing disease outbreaks.
b) Competition for Resources: Microorganisms, including beneficial bacteria and protists, compete with pathogenic organisms for essential resources such as nutrients and space, thereby reducing the proliferation of harmful pathogens.
c) Microbial Biofilms: Biofilms formed by microorganisms can serve as a physical barrier against pathogens, preventing their attachment to host organisms and surfaces.

TABLE 20.4
Major Features of Chemical Pesticides and Microbial and Invertebrate Species Utilized in Augmented Biocontrol

Chemical Pesticides	Microbial Biocontrol Agents	Macrobial Biocontrol Agents
Non-living	Living organism	Living organism
Majority are anthropogenically produced xenobiotics	Naturally occurring	Naturally occurring
Affect biota by direct, toxic mechanisms	Affect biota by ecological or pathogenic (nematodes) interactions	Affect biota by ecological or pathogenic interactions
Many can contaminate groundwater	Extremely unlikely to spread to and grow in groundwater	Cannot contaminate groundwater
May bioaccumulate in tissue	Local increase in abundance may cause disease or sensitization in humans and affect other non-target organisms	Local increase in abundance may affect non-target organisms
High risk for induction of resistance	Lower risk for induction of resistance	Lower risk for induction of resistance

Source: Sundh and Goettel (2013).

d) Nutrient Cycling and Microbial Control: Nutrient cycling involves recycling essential elements by microorganisms, promoting soil fertility and plant health. Microbial control uses beneficial microbes to suppress pathogens, ensuring sustainable agriculture through natural means.
e) Biological Oxygen Demand (BOD): Microorganisms play a key role in decomposing organic matter, reducing BOD levels in aquatic systems, and preventing oxygen depletion, which can be harmful to aquatic life.
f) Nitrogen and Phosphorus Removal: Microbial communities assist in the removal of excess nitrogen and phosphorus, reducing the risk of eutrophication and algal blooms.
g) Symbiotic Nitrogen Fixation: Nitrogen-fixing bacteria in aquatic environments form symbiotic relationships with certain plants, contributing to nutrient availability and ecosystem productivity.

20.7 MICROBES IN ECOSYSTEM STABILITY AND RESILIENCE

a) Microbial diversity: High microbial diversity enhances ecosystem stability, as diverse communities can better adapt to environmental changes and disturbances.
b) Biogeochemical cycling: Microorganisms drive essential biogeochemical cycles, such as the carbon, nitrogen, and sulfur cycles, maintaining the chemical balance of aquatic ecosystems.
c) Microbial interactions: The intricate interactions among microorganisms, including mutualism, predation, and competition, contribute to ecosystem resilience and adaptability.

20.8 CHALLENGES AND APPLICATIONS

a) Antibiotic resistance: The overuse of antibiotics in aquaculture and agriculture can lead to the development of antibiotic-resistant microorganisms, posing challenges to biocontrol efforts.

b) Bioaugmentation: The deliberate introduction of beneficial microorganisms into aquatic systems can enhance their biocontrol capabilities, improving water quality and ecosystem health.
c) Bioremediation: Microorganisms can be harnessed for bioremediation purposes, assisting in the cleanup of polluted aquatic environments.

20.9 CONCLUSION

Microorganisms are the unsung heroes of aquatic ecosystems, exerting a profound influence on their health and stability. Their roles as biocontrol agents extend from pathogen regulation to nutrient cycling and overall ecosystem resilience. Understanding and harnessing the power of microorganisms in aquatic environments is not only crucial for preserving biodiversity but also for ensuring the sustainability of these vital ecosystems. As we continue to explore and appreciate the intricate relationships between microorganisms and aquatic life, we pave the way for innovative solutions to mitigate environmental challenges and safeguard the future of our oceans, lakes, and rivers.

REFERENCES

Adrangi, Sina, and Mohammad Ali Faramarzi. "From bacteria to human: A journey into the world of chitinases." *Biotechnology Advances* 31, no. 8 (2013): 1786–1795.

Agnihotri, Mithila, Swanand Joshi, Ameeta Ravi Kumar, Smita Zinjarde, and Sulabha Kulkarni. "Biosynthesis of gold nanoparticles by the tropical marine yeast Yarrowia lipolytica NCIM 3589." *Materials Letters* 63, no. 15 (2009): 1231–1234.

Austin, B., E. Baudet, and M. Stobie. "Inhibition of bacterial fish pathogens by Tetraselmis suecica." *Journal of Fish Diseases* 15, no. 1 (1992): 55–61.

Austin, B. A., and Day, J. G. (1990). "Inhibition of prawn pathogenic Vibrio spp. by a commercial spray-dried preparation of Tetraselmis suecica." *Aquaculture* 90, no. 3–4: 389–392.

Bailey, A., D. Chandler, W. P. Grant, J. Greaves, G. Prince, and M. Tatchell. *Biopesticides: Pest Management and Regulation*. Wallingford: CABI, 2010, pp. 1–232.

Bankar, Ashok V., Ameeta R. Kumar, and Smita S. Zinjarde. "Environmental and industrial applications of Yarrowia lipolytica." *Applied Microbiology and Biotechnology* 84 (2009): 847–865.

Beier, Sara, and Stefan Bertilsson. "Bacterial chitin degradation—Mechanisms and ecophysiological strategies." *Frontiers in Microbiology* 4 (2013): 149.

Boivin, Guy, Ursula M. Kölliker-Ott, Jeffrey Bale, and Franz Bigler. "Assessing the establishment potential of inundative biological control agents." In *Environmental Impact of Invertebrates for Biological Control of Arthropods: Methods and Risk Assessment*. Wallingford: CABI Publishing, 2006, pp. 98–113.

Brandt, Curtis R., Michael J. Adang, and Kemet D. Spence. "The peritrophic membrane: Ultrastructural analysis and function as a mechanical barrier to microbial infection in Orgyia pseudotsugata." *Journal of Invertebrate Pathology* 32, no. 1 (1978): 12–24.

Brzezinska, Maria Swiontek, Urszula Jankiewicz, and Maciej Walczak. "Biodegradation of chitinous substances and chitinase production by the soil actinomycete Streptomyces rimosus." *International Biodeterioration & Biodegradation* 84 (2013): 104–110.

Cahill, Marian M. "Bacterial flora of fishes: A review." *Microbial Ecology* 19 (1990): 21–41.

Carozzi, Nadine B., and Michael Koziel. *Advances in Insect Control: The Role of Transgenic Plants*. CRC Press, 1997.

Chet, Ilan, and Ralph Baker. "From soil naturally suppressive to Rhizoctonia solani." *Phytopathology* 71, no. 3 (1981): 286–290.

Chi, Zhenming, Zhe Chi, Tong Zhang, Guanglei Liu, and Lixi Yue. "Inulinase-expressing microorganisms and applications of inulinases." *Applied Microbiology and Biotechnology* 82 (2009a): 211–220.

Chi, Zhenming, Zhe Chi, Tong Zhang, Guanglei Liu, Jing Li, and Xianghong Wang. "Production, characterization and gene cloning of the extracellular enzymes from the marine-derived yeasts and their potential applications." *Biotechnology Advances* 27, no. 3 (2009b): 236–255.

Chi, Zhen-Ming, Guanglei Liu, Shoufeng Zhao, Jing Li, and Ying Peng. "Marine yeasts as biocontrol agents and producers of bio-products." *Applied Microbiology and Biotechnology* 86 (2010): 1227–1241.

Chiesa, L. M., G. F. Labella, A. Giorgi, S. Panseri, R. Pavlovic, S. Bonacci, and F. Arioli. "The occurrence of pesticides and persistent organic pollutants in Italian organic honeys from different productive areas in relation to potential environmental pollution." *Chemosphere* 154 (2016): 482–490.

Cohen-Kupiec, Rachel, and Ilan Chet. "The molecular biology of chitin digestion." *Current Opinion in Biotechnology* 9, no. 3 (1998): 270–277.

Cook, R. James. "Take-all decline: Model system in the science of biological control and clue to the success of intensive cropping." In *Biological Control a Global Perspective: Case Studies from Around the World*, C. Vincent, M. Goettel, and G. Lazarovits, eds. Wallingford: CAB International, 2007, pp. 399–414.

Dahiya, Neetu, Rupinder Tewari, and Gurinder Singh Hoondal. "Biotechnological aspects of chitinolytic enzymes: A review." *Applied Microbiology and Biotechnology* 71 (2006): 773–782.

Eilenberg, J., A. Hajek, and C. Lomer. "Suggestions for unifying the terminology in biological control." *BioControl* 46 (2001): 387–400.

Enkerli, Jürg, and Franco Widmer. "Molecular ecology of fungal entomopathogens: Molecular genetic tools and their applications in population and fate studies." *BioControl* 55 (2010): 17–37.

Flegel, T. W., T. Pasharawipas, L. Owens, and H. J. Oakey. "Evidence for phage-induced virulence in the shrimp pathogen Vibrio harveyi." *Diseases in Asian Aquaculture V* (2005): 329–337.

Fuller, Rachel. "Probiotics in man and animals." *The Journal of Applied Bacteriology* 66, no. 5 (1989): 365–378.

Garcia De La Banda, I., O. Chereguini, and I. Rasines. "Influence of lactic bacterial additives on turbot (Scophthalmus maximus L.) larvae culture." *Boletin Del Instituto Espanol De Oceanografia (Espana)* 8 (1992): 247–254.

Garriques, David. "An evaluation of the production and use of a live bacterial isolate to manipulate the microbial flora in the commercial production of Penaeus vennamei postlarvae in Ecuador." In *Swimming through Troubled Water. Proceedings of the Special Session on Shrimp Farming, Aquaculture'95. World Aquaculture Society* (1995): 53–59.

Gatesoupe, François-Joël. "Lactic acid bacteria increase the resistance of turbot larvae, Scophthalmus maximus, against pathogenic Vibrio." *Aquatic Living Resources* 7, no. 4 (1994): 277–282.

Gatesoupe, François-Joël. "Siderophore production and probiotic effect of Vibrio sp. associated with turbot larvae, Scophthalmus maximus." *Aquatic Living Resources* 10, no. 4 (1997): 239–246.

Gatesoupe, F. J. "Live yeasts in the gut: Natural occurrence, dietary introduction, and their effects on fish health and development." *Aquaculture* 267, no. 1–4 (2007): 20–30.

Geisseler, Daniel, William R. Horwath, Rainer Georg Joergensen, and Bernard Ludwig. "Pathways of nitrogen utilization by soil microorganisms–a review." *Soil Biology and Biochemistry* 42, no. 12 (2010): 2058–2067.

Gibson, L. F., J. Woodworth, and A. M. George. "Probiotic activity of Aeromonas media on the Pacific oyster, Crassostrea gigas, when challenged with Vibrio tubiashii." *Aquaculture* 169, no. 1–2 (1998): 111–120.

Gildberg, Asbjørn, and Helene Mikkelsen. "Effects of supplementing the feed to Atlantic cod (Gadus morhua) fry with lactic acid bacteria and immuno-stimulating peptides during a challenge trial with Vibrio anguillarum." *Aquaculture* 167, no. 1–2 (1998): 103–113.

Gildberg, Asbjørn, Audny Johansen, and Jarl Bøgwald. "Growth and survival of Atlantic salmon (Salmo salar) fry given diets supplemented with fish protein hydrolysate and lactic acid bacteria during a challenge trial with Aeromonas salmonicida." *Aquaculture* 138, no. 1–4 (1995): 23–34.

Gildberg, Asbjørn, Helene Mikkelsen, Elin Sandaker, and Einar Ringø. "Probiotic effect of lactic acid bacteria in the feed on growth and survival of fry of Atlantic cod (Gadus morhua)." *Hydrobiologia* 352 (1997): 279–285.

Gomez-Gil, B., A. Roque, J. F. Turnbull, and V. Inglis. "Reduction in Artemia nauplii mortality with the use of a probiotic bacterium." In *Proceedings of the Third International Symposium on Aquatic Animal Health*. APC Press, Baltimore, MD, 1998, p. 116.

Gooday, Graham W. "Physiology of microbial degradation of chitin and chitosan." In *Biochemistry of Microbial Degradation*, edited by Ratledge, C., Springer, Dordrecht (1994) . https://doi.org/10.1007/978-94-011-1687-9_9 279–312.

Gram, Lone, Jette Melchiorsen, Bettina Spanggaard, Ingrid Huber, and Torben F. Nielsen. "Inhibition of Vibrio anguillarum by Pseudomonas fluorescens AH2, a possible probiotic treatment of fish." *Applied and Environmental Microbiology* 65, no. 3 (1999): 969–973.

Hagiwara, Atsushi, Kenichiro Hamada, Sonoyo Hori, and Kazutsugu Hirayama. "Increased sexual reproduction in Brachionus plicatilis (Rotifera) with the addition of bacteria and rotifer extracts." *Journal of Experimental Marine Biology and Ecology* 181, no. 1 (1994): 1–8.

Han, Feng, Qian-hong Gong, Kai Song, Jing-bao Li, and Wen-gong Yu. "Cloning, sequence analysis and expression of gene alyVI encoding alginate lyase from marine bacterium Vibrio sp. QY101." *DNA Sequence* 15, no. 5–6 (2004): 344–350.

Hansen, Geir Høvik, and Jan A. Olafsen. "Bacterial colonization of cod (Gadus morhua L.) and halibut (Hippoglossus hippoglossus) eggs in marine aquaculture." *Applied and Environmental Microbiology* 55, no. 6 (1989): 1435–1446.

Harris, Jean M. "The presence, nature, and role of gut microflora in aquatic invertebrates: A synthesis." *Microbial Ecology* 25 (1993): 195–231.

Herrera-Estrella, Alfredo, and Ilan Chet. "Chitinases in biological control." *EXS-BASEL* 87 (1999): 171–184.

Hoitink, H. A. J., and M. J. Boehm. "Biocontrol within the context of soil microbial communities: A substrate-dependent phenomenon." *Annual Review of Phytopathology* 37, no. 1 (1999): 427–446.

Holmberg, Anna-Ida Johnsson, Petter Melin, Jens P. Levenfors, and Ingvar Sundh. "Development and evaluation of SCAR markers for a Pseudomonas brassicacearum strain used in biological control of snow mould." *Biological Control* 48, no. 2 (2009): 181–187.

Holzapfel, Wilhelm H., Petra Haberer, Johannes Snel, Ulrich Schillinger, and Jos H. J. Huis in't Veld. "Overview of gut flora and probiotics." *International Journal of Food Microbiology* 41, no. 2 (1998): 85–101.

Horn, Svein Jarle, Audun Sørbotten, Bjørnar Synstad, Pawel Sikorski, Morten Sørlie, Kjell M. Vårum, and Vincent G. H. Eijsink. "Endo/exo mechanism and processivity of family 18 chitinases produced by Serratia marcescens." *The FEBS Journal* 273, no. 3 (2006): 491–503.

Jang, Mi-Kyeong, Byeong-Gi Kong, Young-Il Jeong, Chang Hyung Lee, and Jae-Woon Nah. "Physicochemical characterization of α -chitin, β -chitin, and γ -chitin separated from natural resources." *Journal of Polymer Science Part A: Polymer Chemistry* 42, no. 14 (2004): 3423–3432.

Jöborn, A., J. C. Olsson, A. Westerdahl, P. L. Conway, and S. Kjelleberg. "Colonization in the fish intestinal tract and production of inhibitory substances in intestinal mucus and faecal extracts by Carnobacterium sp. strain K1." *Journal of Fish Diseases* 20, no. 5 (1997): 383–392.

Kalia, Anu, and S. K. Gosal. "Effect of pesticide application on soil microorganisms." *Archives of Agronomy and Soil Science* 57, no. 6 (2011): 569–596.

Kawase, T., A. Saito, T. Sato et al. (2004). "Distribution and phylogenetic analysis of family 19 chitinases in actinobacteria." *Applied and Environmental Microbiology* 70: 1135–1144.

Keswani, C. "Ecofriendly management of plant diseases by biosynthesized secondary metabolites of Trichoderma spp." *The Journal of Brief Ideas* 10.5281, (2015): zenodo.15571.

Le, Son Tuan, and İpek Kurtböke. "Bacteriophages as biocontrol agents in aquaculture." *Microbiology Australia* 40, no. 1 (2019): 37–41.

Li, Mei, Guang-Lei Liu, Zhe Chi, and Zhen-Ming Chi. "Single cell oil production from hydrolysate of cassava starch by marine-derived yeast Rhodotorula mucilaginosa TJY15a." *Biomass and Bioenergy* 34, no. 1 (2010): 101–107.

Liu, Guanglei, Lixi Yue, Zhe Chi, Wengong Yu, Zhenming Chi, and Catherine Madzak. "The surface display of the alginate lyase on the cells of Yarrowia lipolytica for hydrolysis of alginate." *Marine Biotechnology* 11 (2009): 619–626.

Lockwood, Jeffrey A. "Environmental issues involved in biological control of rangeland grasshoppers (Orthoptera: Acrididae) with exotic agents." *Environmental Entomology* 22, no. 3 (1993): 503–518.

Louda, Svata M., R. W. Pemberton, M. T. Johnson, and P. A. Follett. "Nontarget effects—The Achilles' heel of biological control? Retrospective analyses to reduce risk associated with biocontrol introductions." *Annual review of Entomology* 48, no. 1 (2003): 365–396.

Maeda, M. "Biocontrol of the larvae rearing biotope in aquaculture." *Bulletin of National Research Institute of Aquaculture* (1994): 71–74.

Maeda, M. "Effect of bacterial population on the growth of a prawn larva, *Penaeus monodon*." *Bulletin of National Research Institute of Aquaculture* 21 (1992): 25–29.

Maeda, M., and Liao, I.C. "Microbial processes in aquaculture environment and their importance for increasing crustacean production." *Japan Agricultural Research Quarterly* 28 (1994): 283–283.

Martin, Kirstee, Barbara M. McDougall, Simon McIlroy, Jayus, Jiezhong Chen, and Robert J. Seviour. "Biochemistry and molecular biology of exocellular fungal β-(1, 3)-and β-(1, 6)-glucanases." *FEMS Microbiology Reviews* 31, no. 2 (2007): 168–192.

Mathre, D. E., R. J. Cook, and Nancy W. Callan. "From discovery to use: Traversing the world of commercializing biocontrol agents for plant disease control." *Plant Disease* 83, no. 11 (1999): 972–983.

Matsuoka, Satoru, Takaya Hashizume, Hiroyuki Kanzaki, Emi Iwamoto, Park Chang, Terutoyo Yoshida, and Toshihiro Nakai. "Phage therapy against β-hemolytic Streptococcicosis of Japanese flounder Paralichthys olivaceus." *Fish Pathology* 42 (2007): 181–189.

Mishra, Sandhya, Akanksha Singh, Chetan Keswani, Amrita Saxena, B. K. Sarma, and H. B. Singh. "Harnessing plant-microbe interactions for enhanced protection against phytopathogens." In *Plant Microbes Symbiosis: Applied Facets*, edited by Arora, N., Springer, New Delhi. Vol 9, (2015): 111–125.

Moriarty, D. J. W. "Control of luminous Vibrio species in penaeid aquaculture ponds." *Aquaculture* 164, no. 1–4 (1998): 351–358.

Moriarty, D. J. W., and A. G. C. Body. "Modifying microbial ecology in ponds: The key to sustainable aquaculture." In *Proceedings of Fish Asia'95 Conference: 2nd Asian Aquaculture and Fisheries Exhibition and Conference*. RAI Exhibitions, Singapore, 1995, pp. 1–10.

Munro, P. D., A. Barbour, and T. H. Birkbeck. "Comparison of the growth and survival of larval turbot in the absence of culturable bacteria with those in the presence of Vibrio anguillarum, Vibrio alginolyticus, or a marine Aeromonas sp." *Applied and Environmental Microbiology* 61, no. 12 (1995): 4425–4428.

Nagpure, Anand, Bharti Choudhary, and Rajinder K. Gupta. "Chitinases: In agriculture and human healthcare." *Critical Reviews in Biotechnology* 34, no. 3 (2014): 215–232.

Nogami, Kinya, and Masachika Maeda. "Bacteria as biocontrol agents for rearing larvae of the crab Portunus trituberculatus." *Canadian Journal of Fisheries and Aquatic Sciences* 49, no. 11 (1992): 2373–2376.

Olafsen, J. A. "Interactions between hosts and bacteria in aquaculture." In *Proceedings from the US-EC Workshop on Marine Microorganisms: Research Issues for Biotechnology*. European Commission, Brussels, 1998, pp. 127–145.

Oliveira, J., F. Castilho, A. Cunha, and M. J. Pereira. "Bacteriophage therapy as a bacterial control strategy in aquaculture." *Aquaculture International* 20 (2012): 879–910.

Olsson, Jöborn, Blomberg Westerdahl, and Conway Kjelleberg. "Survival, persistence and proliferation of Vibrio anguillarum in juvenile turbot, Scophthalmus maximus (L.), intestine and faeces." *Journal of Fish Diseases* 21, no. 1 (1998): 1–9.

Patil, R. S., V. Ghormade, and M. V. Deshpande. "Chitinolytic enzymes: An exploration." *Enzyme and Microbial Technology* 26 (2000): 473–483.

Paulitz, Timothy C., and Richard R. Bélanger. "Biological control in greenhouse systems." *Annual Review of Phytopathology* 39, no. 1 (2001): 103–133.

Peng, Ying, Zhen-Ming Chi, Xiang-Hong Wang, and Jing Li. "Purification and molecular characterization of exo-β-1, 3-glucanases from the marine yeast Williopsis saturnus WC91-2." *Applied Microbiology and Biotechnology* 85 (2009): 85–94.

Peng, Ying, Zhenming Chi, Xianghong Wang, and Jing Li. "β-1, 3-Glucanase inhibits activity of the killer toxin produced by the marine-derived yeast Williopsis saturnus WC91-2." *Marine Biotechnology* 12 (2010): 479–485.

Queiroz, Julio F., and Claude E. Boyd. "Effects of a bacterial inoculum in channel catfish ponds." *Journal of the World Aquaculture Society* 29, no. 1 (1998): 67–73.

Regev, Avital, Menachem Keller, Nicolai Strizhov, Baruch Sneh, Evgenya Prudovsky, Ilan Chet, Idit Ginzberg et al. "Synergistic activity of a Bacillus thuringiensis delta-endotoxin and a bacterial endochitinase against Spodoptera littoralis larvae." *Applied and Environmental Microbiology* 62, no. 10 (1996): 3581–3586.

Rengpipat, Sirirat, Wannipa Phianphak, Somkiat Piyatiratitivorakul, and Piamsak Menasveta. "Effects of a probiotic bacterium on black tiger shrimp Penaeus monodon survival and growth." *Aquaculture* 167, no. 3–4 (1998): 301–313.

Rico-Mora, Roxana, Domenico Voltolina, and Julio A. Villaescusa-Celaya. "Biological control of Vibrio alginolyticus in Skeletonema costatum (Bacillariophyceae) cultures." *Aquacultural Engineering* 19, no. 1 (1998): 1–6.

Ringø, E., and O. Vadstein. "Colonization of Vibrio pelagius and Aeromonas caviae in early developing turbot (Scophthalmus maximus L.) larvae." *Journal of Applied Microbiology* 84, no. 2 (1998): 227–233.

Ringoe, Einar, and T. H. Birkbeck. "Intestinal microflora of fish larvae and fry." *Aquaculture Research* 30, no. 2 (1999): 73.

Riquelme, Carlos, Rubén Araya, Nelson Vergara, Alejandro Rojas, Mauricio Guaita, and Marcela Candia. "Potential probiotic strains in the culture of the Chilean scallop Argopecten purpuratus (Lamarck, 1819)." *Aquaculture* 154, no. 1 (1997): 17–26.

Rombaut, Geert, Ph Dhert, J. Vandenberghe, Laurent Verschuere, Patrick Sorgeloos, and Willy Verstraete. "Selection of bacteria enhancing the growth rate of axenically hatched rotifers (Brachionus plicatilis)." *Aquaculture* 176, no. 3–4 (1999): 195–207.

Rong, Rong, Hong Lin, Jingxue Wang, Muhammad Naseem Khan, and Meng Li. "Reductions of Vibrio parahaemolyticus in oysters after bacteriophage application during depuration." *Aquaculture* 418 (2014): 171–176.

Ruangpan, Lila, Yaowanit Danayadol, Sataporn Direkbusarakom, Siriporn Siurairatana, and T. W. Flegel. "Lethal toxicity of Vibrio harveyi to cultivated Penaeus monodon induced by a bacteriophage." *Diseases of Aquatic Organisms* 35, no. 3 (1999): 195–201.

Russell, P. E. "The development of commercial disease control." *Plant Pathology* 55, no. 5 (2006): 585–594.

Sajeevan, T. P., Rosamma Philip, and I. S. Bright Singh. "Immunostimulatory effect of a marine yeast Candida sake S165 in Fenneropenaeus indicus." *Aquaculture* 257, no. 1–4 (2006): 150–155.

Shiri Harzevili, A. R., H. Van Duffel, Ph Dhert, J. Swings, and P. Sorgeloos. "Use of a potential probiotic Lactococcus lactis AR21 strain for the enhancement of growth in the rotifer Brachionus plicatilis (Müller)." *Aquaculture Research* 29, no. 6 (1998): 411–417.

Singh, H. B. "Trichoderma: A boon for biopesticides industry." *Journal of Mycology and Plant Pathology* 36 (2006): 373–384.

Skjermo, J., I. Salvesen, G. Øie, Y. Olsen, and O. Vadstein. "Microbially matured water: A technique for selection of a non-opportunistic bacterial flora in water that may improve performance of marine larvae." *Aquaculture International* 5 (1997): 13–28.

Smith, P. R., and S. Davey. "Evidence for the competitive exclusion of Aeromonas salmonicida from fish with stress-inducible furunculosis by a fluorescent pseudomonad." *Journal of Fish Diseases* 16, no. 5 (1993): 521–524.

Souza, Claudiana P., Bianca C. Almeida, Rita R. Colwell, and Irma N. G. Rivera. "The importance of chitin in the marine environment." *Marine Biotechnology* 13 (2011): 823–830.

Stoykov, Yuriy Mihaylov, Atanas Ivanov Pavlov, and Albert Ivanov Krastanov. "Chitinase biotechnology: Production, purification, and application." *Engineering in Life Sciences* 15, no. 1 (2015): 30–38.

Suminto, and Kazutsugu Hirayama. "Application of a growth-promoting bacteria for stable mass culture of three marine microalgae." In *Live Food in Aquaculture: Proceedings of the Live Food and Marine Larviculture Symposium held* in Nagasaki, Japan, September 1–4, 1996. Netherlands: Springer, 1997, pp. 223–230.

Sundh, Ingvar, and Mark S. Goettel. "Regulating biocontrol agents: A historical perspective and a critical examination comparing microbial and macrobial agents." *BioControl* 58 (2013): 575–593.

Udaya Prakash, N. A., M. Jayanthi, R. Sabarinathan, P. Kangueane, Lazar Mathew, and Kanagaraj Sekar. "Evolution, homology conservation, and identification of unique sequence signatures in GH19 family chitinases." *Journal of Molecular Evolution* 70 (2010): 466–478.

Vadstein, O., G. Øie, Y. Olsen, I. Salvesen, J. Skjermo, and G. Skjåk-Bræk. "A strategy to obtain microbial control during larval development of marine fish." In *Fish Farming Technology*. CRC Press, 2020, pp. 69–75.

Van Lenteren, Joop C. "The state of commercial augmentative biological control: Plenty of natural enemies, but a frustrating lack of uptake." *BioControl* 57, no. 1 (2012): 1–20.

Van Lenteren, Joop C., Antoon J. M. Loomans, Dirk Babendreier, and Franz Bigler. "Harmonia axyridis: An environmental risk assessment for Northwest Europe." In *From Biological Control to Invasion: The Ladybird Harmonia axyridis as a Model Species*, edited by Roy, H.E., and Wajnberg, E., Springer, Dordrecht (2008): 37–54.

Vega, Fernando E., and Harry K. Kaya. *Insect Pathology*. Vol. 2. Academic Press, 2012.

Verschuere, L., H. Heang, G. Criel, S. Dafnis, P. Sorgeloos, and W. Verstraete. "Protection of Artemia against the pathogenic effects of Vibrio proteolyticus CW8T2 by selected bacterial strains." *Applied and Environmental Microbiology* 66 (2000a): 1139–1146.

Verschuere, Laurent, Geert Rombaut, Patrick Sorgeloos, and Willy Verstraete. "Probiotic bacteria as biological control agents in aquaculture." *Microbiology and Molecular Biology Reviews* 64, no. 4 (2000b): 655–671.

Wang, X., Z. Chi, L. Yue, and J. Li. "Purification and characterization of killer toxin from a marine yeast Pichia anomala YF07b against the pathogenic yeast in crab." *Current Microbiology* 55 (2007): 396–401.

Wang, Lin, Lixi Yue, Zhenming Chi, and Xianghong Wang. "Marine killer yeasts active against a yeast strain pathogenic to crab Portunus trituberculatus." *Diseases of Aquatic Organisms* 80, no. 3 (2008): 211–218.

Wang, Fang, Lixi Yue, Lin Wang, Catherine Madzak, Jing Li, Xianghong Wang, and Zhenming Chi. "Genetic modification of the marine-derived yeast Yarrowia lipolytica with high-protein content using a GPI-anchor-fusion expression system." *Biotechnology Progress* 25, no. 5 (2009a): 1297–1303.

Wang, W. L., Z. M. Chi, Z. Chi, J. Li, and X. H. Wang. "Siderophore production by the marine-derived Aureobasidium pullulans and its antimicrobial activity." *Bioresource Technology* 100, no. 9 (2009b): 2639–2641.

Wang, Yanhui, Mary Barton, Lisa Elliott, Xiaoxu Li, Sam Abraham, Mark O'Dea, and James Munro. "Bacteriophage therapy for the control of Vibrio harveyi in greenlip abalone (Haliotis laevigata)." *Aquaculture* 473 (2017): 251–258.

Watanabe, Takeshi, Ryo Kanai, Tomokazu Kawase, Toshiaki Tanabe, Masaru Mitsutomi, Shohei Sakuda, and Kiyotaka Miyashita. "Family 19 chitinases of Streptomyces species: Characterization and distribution." *Microbiology* 145, no. 12 (1999): 3353–3363.

Weller, David M., Jos M. Raaijmakers, Brian B. McSpadden Gardener, and Linda S. Thomashow. "Microbial populations responsible for specific soil suppressiveness to plant pathogens." *Annual Review of Phytopathology* 40, no. 1 (2002): 309–348.

Zhang, Tong, Fang Gong, Ying Peng, and Zhenming Chi. "Optimization for high-level expression of the Pichia guilliermondii recombinant inulinase in Pichia pastoris and characterization of the recombinant inulinase." *Process Biochemistry* 44, no. 12 (2009a): 1335–1339.

Zhang, Tong, Fang Gong, Zhe Chi, Guanglei Liu, Zhenming Chi, Jun Sheng, Jing Li, and Xianghong Wang. "Cloning and characterization of the inulinase gene from a marine yeast Pichia guilliermondii and its expression in Pichia pastoris." *Antonie van Leeuwenhoek* 95 (2009b): 13–22.

Zhu, Kailing, Zhenming Chi, Jing Li, Fengli Zhang, Meiju Li, Hirimuthugoda Nalini Yasoda, and Longfei Wu. "The surface display of haemolysin from Vibrio harveyi on yeast cells and their potential applications as live vaccine in marine fish." *Vaccine* 24, no. 35–36 (2006): 6046–6052.

21 Thermophilic Microbial Enzymes from Hot Springs and Their Role in Bioprocessing

Amit Seth

21.1 INTRODUCTION

Hot springs emanate from the earth's surface and are known for their highly thermal environment. The occurrence of high mineral contents at specific sites underneath the earth's surface is the main reason for their existence, as they also occur in areas prone to volcanic eruptions. Hot springs are also regarded as the sites for the origin of life on our planet. These hot springs are rich in both acid and alkaline-based compounds. The inherent high heat of the water stimulates a chemical interaction between the soil (surface and sub-surface) components, resulting in the accumulation of diverse mineral compounds. The extremophilic heat conditions of the hot springs are a major hindrance to the existence of any life form. Only prokaryotic species, having unique cellular mechanisms, can withstand and survive in these extreme and harsh environments (Marais and Walter, 2019).

Thermophilic microorganisms are inherently programmed to withstand high-temperature conditions. Consequently, their constituent enzymes are also resistant to denaturation at high temperatures. Thus, heat-stable enzymes can tolerate temperature fluctuations and display comparatively better applicability in scale-up operations. Thermophilic bacteria and fungi produce stable enzymes, valuable for the pharmaceutical, chemical, food, and biotechnology industries (Sharma et al., 2024) (Figure 21.1). Microorganisms are characterized by their rapid multiplication rate, thus facilitating their large-scale cultivation under *in vitro* conditions. Furthermore, huge quantities of industrial enzymes can be obtained from microbial sources. Although higher organisms also contain catalytic enzymes, their cellular machinery is very complex, thus hindering extensive growth and enzyme productivity (Bhatia et al., 2018). Microorganisms contain various constitutive and inducible enzymes that can participate in biochemical reactions and synthesize useful chemical compounds. The microbial enzymes must be identified and characterized for future industrial applications. In recent years, a shift in mindset to reduce the usage and production of toxic compounds in industrial processes has also provided an impetus to bio-based industries, wherein enzymes play a crucial role in product development. The process parameters for enzyme production must therefore be thoroughly investigated before their selection for industrial operations. Microbial bioprocess development is intricately linked with process optimization and downstream processing (Figure 21.2). Homologous and heterologous expression of industrial enzymes paves the way for enhanced productivity (Pratush et al., 2017). Enzyme-catalysed reactions are possible under normal conditions instead of chemical reactions that require harsh conditions. Enzymes display high substrate specificity; thus, only the desired compounds are generated, unaccompanied by unwanted by-products.

High product yields and purity are attained through enzyme-mediated bioprocesses. Microbial enzymes with high substrate and product tolerance are preferentially selected for industrial operations because such enzymes tend to maintain high activities for a longer duration and in fluctuating reaction conditions (Raj et al., 2007). Moreover, microbial enzymes are better suited for genetic alterations than their eukaryotic counterparts. The molecular changes in genes and proteins

DOI: 10.1201/9781003408543-21

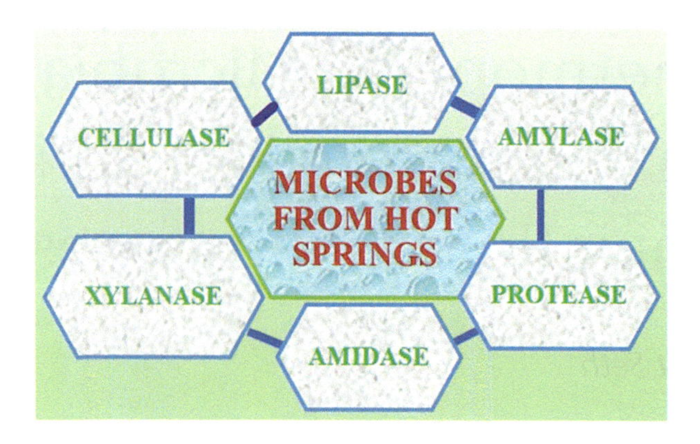

FIGURE 21.1 Various enzymes derived from microorganisms from thermal hot springs.

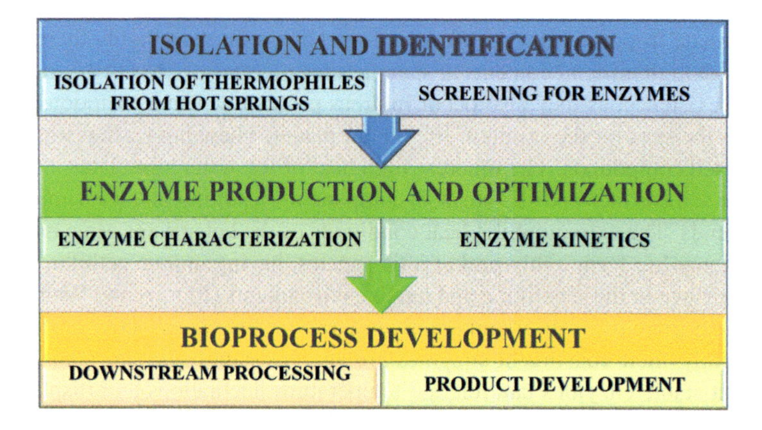

FIGURE 21.2 Scheme outlining isolation of enzymes from microbes of thermal hot springs and subsequent use in bioprocess development.

influence their performance in bioprocesses because the three-dimensional conformation deviations introduced in proteins improve substrate–enzyme interaction, boosting product formation. Genetic alterations are also possible in higher organisms, but attaining the desired results becomes limiting in complex organisms. Thus, microbial enzymes are highly suited to the commercial production of useful biochemicals (Pratush et al., 2010). The thermophilic enzymes, obtained from thermal hot springs, are more stable and are better suited to industrial operations.

21.2 THERMOPHILIC MICROBIAL LIPASE ENZYMES FROM HOT SPRINGS

Thermophilic enzymes from microorganisms have a robust nature that allows them to work in extreme environments. Lipase enzymes from various species of *Bacillus* have been thoroughly investigated to harness their industrial potential. Lipase enzymes remain stable in various solvents, thus enhancing their commercial utility. One such enzyme has been isolated from the hot springs of Malaysia. The microbial isolates have been thoroughly identified and characterized using modern standard molecular tools. There are numerous hot springs in Malaysia, and the principal ones are Gadek, Labis, Pedas, and Selayang. Most lipase-producing microorganisms from the Malaysian hot springs belong to two major genera, *Bacillus* and *Geobacillus*, whereas one strain belongs to the genus *Anoxybacillus* (Zuridah et al., 2011). Another site in Malaysia that has been explored for bacterial lipase enzymes is Setapak hot springs. The microbial isolates belonged to two genera,

Bacillus and *Ralstonia*, which displayed stable lipase activities from 50 to 80 C. Some strains from this site contained endospores, whereas others were gram-negative. Almost all the strains were rod-shaped bacteria (Sheikh Abdul Hamid et al., 2003).

Lipase enzyme-harbouring bacteria have also been detected from the hot springs in Spain. The lipases are extracellular, and these selected microorganisms can be cultured in liquid media to obtain high growth along with elevated production of lipase enzymes. Even at very high temperatures, these enzymes can cleave ester bonds in diverse conjugated glycerol molecules. Commercial large-scale operations require stable enzymes to achieve the desired targets. The cultures of thermophilic microbes also are less prone to contamination, as most of the intruding microbes are mesophilic and cannot survive the harsh high-temperature cultivation conditions of these extremophilic microbes. The media ingredients for cultivating thermophilic microorganisms should be heat stable and maintain their integrity in hot conditions of growth vessels. Otherwise, the degraded nutrients will hamper the cultivation and propagation of thermophilic microorganisms. The growth media should contain various types of lipid molecules that stimulate the activation of lipase enzymes. The lipase enzymes from the thermal springs of Spain were found to be stable at temperatures above 50 C (Deive et al., 2013).

Another instance of lipase from thermophiles is *Bacillus thermoleovorans* from the hot springs of Indonesia. This microbe was cultivated and isolated in a growth medium with olive oil as the chief carbon source, and the lipase enzyme was stable above 60 C. The thermophilic lipase enzyme was purified through liquid chromatography. Its molecular mass was determined by gel electrophoresis. The tendency of this strain to survive high temperatures is due to the presence of thick-walled endospores within the cell, enabling it to withstand extreme environments. This species of *Bacillus* is a fastidious microorganism capable of amassing high growth rates within a short interval of a few hours. Normally, fat molecules are preferred for cultivating lipase-producing microorganisms, as they show repressed growth in the presence of glucose molecules (Lee et al., 1999).

The same species, *B. thermoleovorans,* was also discovered from the hot springs in Iceland. Various reports from around the globe clearly illustrate the universal global distribution of thermophilic microorganisms in different parts of the world. It displayed similar characteristics to the strain isolated from Indonesian hot springs. The stability of lipase enzyme enables the utilization of these microbes in scale-up bioremediation operations to degrade lipid-rich waste materials. However, this strain could metabolize diverse carbon-containing compounds comprising lipid molecules and carbohydrate compounds, including glucose (Markossian et al., 2000).

21.3 THERMOPHILIC MICROBIAL PROTEASE ENZYMES FROM HOT SPRINGS

Proteases are one of the primary industrial enzymes. They have tremendous applications in various industries ranging from food to pharmaceutical interests. Protease enzymes are ubiquitously reported from various plant, animal, and microbial species. Most commercial proteases are obtained from bacteria and fungi, including yeasts. Protease enzymes from microbes can be obtained in ample quantities for large-scale industrial production. The production and downstream processing of proteases are relatively easy. Genetic analysis and mutational studies are far more amenable in microbes than in large eukaryotic organisms. The protease enzyme is extracellular and secreted from the microbial cells after production. The main consideration for selecting productive microbial strains should be safety, referred to as GRAS (Generally Regarded as Safe). The microorganisms designated as GRAS are non-toxic, non-pathogenic, and do not release toxins having adverse effects. Among all bacterial and fungal protease producers, *Bacillus* genus is the most reported for commercial large-scale enzyme production (Aksoy et al., 2012).

Tarabalo, a hot spring in eastern India, has an average mean temperature of 60 C, along with an alkaline pH. Bacterial samples have been isolated and screened for the presence of industrial enzymes like proteases. Like all thermophilic isolates, the bacterial strains show the presence of thick-walled endospores. One of the characteristic species isolated from Tarabalo hot springs was

Bacillus amyloliquefaciens. Protease enzymes obtained from this species displayed ample thermotolerance and could withstand temperatures above 50 C (Panda et al., 2013).

The presence of protease enzyme-containing thermophilic bacteria has been reported from Buranga hot springs in western Uganda in the African continent. Selective culture media were used for the isolation of the desired microorganisms. These microbes required high temperatures for growth under *in vitro* conditions. The principal microbes from Buranga hot springs belonged to two genera: *Bacillus* and *Geobacillus*. The highest protease activity was measured at 50–100 C (Hawumba et al., 2002).

In Iran, a hot spring called Gavmesh Goli has been explored for protease-producing microorganisms. Bacterial isolation experiments were carried out to obtain thermophilic proteases, cellulases, and lipases. The isolated samples underwent microbiological, biochemical, and molecular phylogenetic analysis. It was further concluded that the main thermophiles present in this hot spring are *Thermomonas* and *Bacillus*. The temperature tolerance ability was remarkable, and the highest enzyme activities were recorded above 50 C (Abdollahi et al., 2021).

Some other hot spring sites from which protease-producing microorganisms are reported include Nenehatun (Turkey), Pasinler (Turkey), Ilica (Turkey), Moinit (Indonesia), Erzurum (Turkey), Van (Turkey), Siirt (Turkey), Mardin Sirnak (Turkey), Tshipise (South Africa), Siloam (South Africa), Mphephu (South Africa), Lekkerrus (South Africa), Libertas (South Africa), Belhachani (Algeria), Guerfa (Algeria), Saida (Algeria), Guerdjuma (Algeria), Sidi el hadj (Algeria), and El Knif (Algeria) (Oztas and Gormez, 2020; Ginting et al., 2020; Ulucay et al., 2022; Jardine et al., 2018; Benammar et al., 2020).

21.4 THERMOPHILIC MICROBIAL NITRILE-DEGRADING ENZYMES FROM HOT SPRINGS

The three main nitrile-degrading enzymes are nitrile hydratase (NHase), nitrilase, and amidase. Mehta et al. (2013) isolated an amidase-containing microbe, *Geobacillus subterraneus* RL-2a, from hot springs in the western Himalayas in India. This enzyme did not require any specific inducer for activation. The enzyme was found to be active at high temperature. It displayed the catalytic ability to convert amide compounds into carboxylic acids. The enzyme from *G. subterraneus* RL-2a could biotransform a wide variety of amide compounds irrespective of their structural identity. Scale-up operations were also designed to establish the efficacy of the amidase enzyme from *Geobacillus*, and successful synthesis of nicotinic acid was achieved with a high product yield. Special enrichment culture media were used to isolate thermophilic bacteria from hot springs. These enrichment media must be supplemented with appropriate nitrile compounds to promote the selective growth of nitrile-degrading microorganisms. The constitutive nature of the amidase enzyme from *G. subterraneus* RL-2a makes it highly suitable for industrial applications. This enzyme was further purified, and its molecular mass was also determined. The purified enzyme was characterized and reported to display decreased activity in the presence of metal ions. The enzyme kinetic analysis revealed the K_m and V_{max} values of purified amidase enzyme (Mehta et al., 2013)

Nitrilase enzyme is another nitrile-degrading enzyme that produces acid compounds from nitriles. In India, five main hot springs, Kheerganga, Tattapani, Kasol, Vashisht, and Manikaran, are in the hilly state of Himachal Pradesh. They are considered biodiversity hotspots for the abundant occurrence of thermophilic microorganisms. Among these hot springs, Manikaran and Kheerganga were explored by Sharma et al. (2024) for nitrilase enzyme-synthesizing thermophilic bacteria. Traditional enrichment culture techniques were used to obtain the target microorganisms. A rigorous three-step screening approach was adopted to obtain the anticipated bacterial species. Typical microbiological, biochemical, and molecular tools were employed to establish the identity of the selected microorganisms. Detailed growth and activity profiling were undertaken to understand the nuances of enzyme production in thermophilic species. All the thermophilic nitrilase enzymes, obtained from various isolates, were inducible and required the presence of nitrile compounds in the growth medium for enzyme activation. The culture plate indicator method, containing phenol red dye, also assisted in the preliminary screening of bacterial samples. Eventually, five thermophilic bacteria were obtained from the two hot spring sites, including two species of *Geobacillus*, two species of *Bacillus*, and one species of *Lysinibacillus*. One species, *G. icigianus* MAC VI, was used for large-scale production of mandelic acid. The cells of *G. icigianus* were provided with multiple feedings of inducer molecules for raising the enzyme activity values to suitable levels for enhanced product yield. The enantioselectivity of the enzyme was also tested, and R-mandelic acid with a 74% ee value was obtained.

Due to the high specificity of nitrilase enzymes, the products are relatively free from contaminant molecules. This makes the downstream processing of the final product relatively easier, and the high purity of the final product is ensured. This is the first report confirming the presence of *G. icigianus* from the Manikaran hot springs of Himachal that can produce an enantioselective nitrilase enzyme. Likewise, an amidase-producing thermophilic *G. pallidus* BTP-5X species was obtained for the Tattapani hot springs (Sharma et al., 2013). This enzyme could tolerate temperatures as high as 70 C. The enzyme was purified, and N-terminal sequencing was undertaken to confirm its identity. The amidase from *G. pallidus* BTP-5X displayed a substrate preference for aliphatic compounds.

Sometimes during screening processes, mesophilic bacteria are also obtained from hot spring sites. Although these microbial types are not thermotolerant, they are also suitable for industrial applications (Jyoti et al., 2017; Singh et al., 2019; Singh et al., 2020). Another species of heat-tolerant *Geobacillus* with amide-degrading capability was reported from the Manikaran hot springs by Kumar et al. (2022). This species was designated as *G. thermoleovorans* MTCC 13131. Statistical optimization of growth parameters was undertaken to enable microbe culture under *in vitro* conditions. The substrate affinity displayed a broad-spectrum tendency, and as a result, the bacterium could metabolize a wide variety of amide compounds and enable their conversion to the corresponding acids. All these instances clearly illustrate the significance of diverse hot springs in the Himalayan region. These novel isolates are promising candidates for industrial applicability. The stable enzymes housed in these single-celled organisms can radically transform the industrial landscape vis-à-vis the biotech-based industries. Manikaran hot spring in the Kullu district of Himachal Pradesh has gained much fame over the years for microbiological research. Most species extracted from this site belong to either *Bacillus* or *Geobacillus*. Recently, Sharma et al. (2022) also described the existence of thermophilic amide-hydrolysing bacteria (*B. smithii* IIIMB2907). In this case, the amidase enzyme was also inducible and was activated by the caprolactam compound. Acyl transfer reactions were catalysed to convert benzamide into its acid counterpart benzohydroxamic acid. Detailed investigations also highlighted the usage of this microbe in large-scale elevated production of hydroxamic acids characterized by impressive conversion rates.

Bacillus pallidus Dac521, procured from a hot spring in New Zealand, is a thermophilic bacterial species showing nitrilase activity. The enzyme was initially precipitated and purified through liquid chromatography, including hydrophobic interaction and gel filtration chromatography. The purified enzyme's molecular mass and amino acid sequences were also determined. Most nitrile-degrading enzymes have an intracellular presence; thus, the microbial cells must be lysed to obtain the enzyme. Nitrilase of *B. pallidus* Dac521 showed the highest activity at 65 C (Almatawah et al., 1999).

Another hot spring site in India is Surajkund in Ranchi in Jharkhand. One thermophilic bacterium, *B. tequilensis* (BITNR004), has been obtained and explored to produce nitrile-degrading enzymes. The biotransformation process, catalysed by the amidase enzyme, was explored for the degradation of acrylamide into acrylic acid. Fourier transform infrared spectroscopy (FTIR) and high-performance liquid chromatography (HPLC) analysis confirmed the conversion of the substrate into the product. The enzyme was active even at 50 C and was utilized to degrade polluted water contaminated with acrylamide (Prabha and Nigam, 2020).

21.5 THERMOPHILIC MICROBIAL XYLANASE AND CELLULASE ENZYMES FROM HOT SPRINGS

Cellulase and xylanase are the principal lignocellulosic enzymes reported from various bacterial and fungal species. These enzymes can utilize plant biomass and specifically target cellulose and hemicellulose constituents of plant cell walls. Xylanases are of immense use in biofuel production and find utility as natural bleaching agents in pulp and paper industries. One such bacterial species, *B. aestuarii*, was extracted from the hot spring of Tattapani in India. The taxonomic identity of the selected isolate was established using standard methods and molecular phylogenetic analysis. The thermal stability of the enzyme was commendable, with the enzyme remaining active at around 50 C temperature. At these elevated temperatures, the mesophilic enzymes undergo denaturation and witness a complete loss of enzyme activity. Maximum enzyme activity was attained in 72 hours. Select carbon and nitrogen sources in the growth medium profoundly influence xylanase activity. Both the parameters need to be carefully optimized for maximal enzyme production. The thermophilic xylanases also display high enzyme activity in the presence of ferrous ions (Chauhan et al., 2015). This strain was subsequently subjected to random chemical mutagenesis to improve the stability and efficacy of the xylanase enzyme. The outcome was quite promising, indicating the success of the random mutagenesis approach (Chauhan et al., 2020).

Xylanase enzyme-producing *B. subtilis* was isolated from the Azores hot spring in Portugal. The isolate was further cultivated in a large bioreactor, and a greater yield in enzyme productivity was witnessed, signifying the value of this isolate for industrial usage (Sá-Pereira et al., 2002). Microbial samples from Huancarhuaz hot spring in Peru were processed to establish the presence of thermophilic *B. licheniformis* and *Cohnella laeviribosi*. Although many *Bacillus* species have been isolated from various sites worldwide, the second species mentioned in this research article is a relatively new microbial source for the xylanase enzyme (Tamariz-Angeles et al., 2014). Four hot spring sites in the Myagdi province of Nepal have also been explored for xylanase enzyme-producing bacteria. These hot springs in Nepal are Paudwar, Bhurung, Sinkosh, and Ratopani. The bacterial isolates were screened for xylanase, protease, cellulase, lipase, and gelatinase enzymes (Yadav et al., 2018). Some other hot spring sites explored for thermophilic xylanase enzymes are in Tunisia (Thebti et al., 2016) and Tibet (Liu et al., 2020).

Correspondingly, thermophilic cellulases have been reported from Grensdalur hot springs in Iceland (Zarafeta et al., 2016), Gazan hot springs in Saudi Arabia (Khalil, 2011), Debuk hot springs in Indonesia (Fachrial et al., 2020), Changbai hot springs in China (Bing et al., 2015), Great Basin hot springs in the United States (Zhao et al., 2014), and Unhale, Vajeshwari, and Bakra hot springs in India (Shajahan et al., 2017; Acharya and Chaudhary, 2012; Sarangthem et al., 2023). A few additional examples of thermophilic enzymes from hot springs are summarized in Table 21.1.

21.6 CONCLUSION

Thermophilic microorganisms have immense industrial potential. Hot springs are home to a diverse group of robust microorganisms. The ecological and economic importance of these sites is immense. A tangible improvement in various bioprocesses can be accomplished by incorporating enzymes derived from hot spring bacteria and fungi. The conservation of these sites assumes critical importance, and concerted steps should be undertaken to maintain the sanctity and viability of these precious ecological hotspots. The advent of metagenomics can further provide an impetus to discovering novel enzymes and proteins from thermal hot springs. A lot remains to be explored, and new-age technologies must be incorporated to reveal the hidden potential of thermophilic microbes.

TABLE 21.1

Summary of Various Microbial Enzymes and Their Properties from Diverse Hot Springs

S. No.	Thermophilic Enzyme	Microorganism	Hot Spring	Enzyme Properties	Reference
1.	Pectinase	*Bacillus licheniformis* UNP-1	Unapdev hot spring (Maharashtra, India)	Optimum pH 11.0Optimum temperature 80 C	Jadhav and Pathak 2019)
2.	Pectinase	*Bacillus subtilis* BK-3	Bakreshwar hot spring (West Bengal, India)	Optimum pH 5.0Optimum temperature 50 C	Prajapati et al. (2021)
3.	Amylase	*Bacillus subtilis*	Cermik Belkishatum hot springs (Turkey)	Enzymes produced by solid-state fermentation (SSF)	Baysal et al. (2003)
4.	Pectate lyase	*Aeribacillus pallidus* TD-1	Taodam hot springs (Thailand)	Optimum pH 7.0–8.0Optimum temperature 50-60°C	Yasawong et al. (2011)
5.	Glycosyl hydrolases	Various species of *Bacillus*	Manikaran and Vashishta hot springs (Himachal Pradesh, India)	Optimum temperature 50–70 C	Thankappan et al. (2018)
6.	Superoxide dismutase	*Anoxybacillus gonensis* KA-55	Manikaran hot springs (Himachal Pradesh, India)	Optimum pH 9.0Optimum temperature 70 C	Bhatia et al. (2018)
7.	Amylase	*Aspergillus niger* G2-1	Afyon, Usak, Ankara hot springs (Turkey)	Enzyme used in bread production	Unal et al. (2022)
8.	Laccase	*Anoxybacillus ayderensis* SK 3-4	Sungai Klah hot springs (Malaysia)	Optimum pH 7.0Optimum temperature 75 C	Wang et al. (2020)
9.	Cellulase	*Melanocarpus albomyces*	Nimu hot spring (Tibet)	Optimum pH 5.5Optimum temperature 58 C	Lu et al. (2019)
10.	Phytase	*Aspergillus tubingensis* TEM 37	Gediz hot springs (Turkey)	Optimum pH 2.0–5.5Optimum temperature 80 C	Çalışkan-Özdemir et al. (2021)

REFERENCES

Abdollahi, Pantea, Maryam Ghane, and Laleh Babaeekhou. "Isolation and characterization of Thermophilic bacteria from Gavmesh Goli hot spring in Sabalan geothermal field, Iran: *Thermomonas hydrothermalis* and *Bacillus altitudinis* isolates as a potential source of Thermostable Protease." *Geomicrobiology Journal* 38, no. 1 (2021): 87–95.

Acharya, Somen, and Anita Chaudhary. "Optimization of fermentation conditions for cellulases production by *Bacillus licheniformis* MVS1 and Bacillus sp. MVS3 isolated from Indian hot spring." *Brazilian Archives of Biology and Technology* 55 (2012): 497–503.

Aksoy, Semiha Çetinel, Ataç Uzel, and E. Esin Hameş Kocabaş. "Extracellular serine proteases produced by Thermoactinomyces strains from hot springs and soils of West Anatolia." *Annals of Microbiology* 62 (2012): 483–492.

Almatawah, Qadreyah A., Rebecca Cramp, and Don A. Cowan. "Characterization of an inducible nitrilase from a thermophilic bacillus." *Extremophiles* 3 (1999): 283–291.

Baysal, Zübeyde, Fikret Uyar, and Çetin Aytekin. "Solid state fermentation for production of α-amylase by a thermotolerant *Bacillus subtilis* from hot-spring water." *Process Biochemistry* 38, no. 12 (2003): 1665–1668.

Benammar, Leyla, K. İnan Bektaş, T. Menasria, A. O. Beldüz, H. I. Güler, I. K. Bedaida, J. M. Gonzalez, and A. Ayachi. "Diversity and enzymatic potential of thermophilic bacteria associated with terrestrial hot springs in Algeria." *Brazilian Journal of Microbiology* 51 (2020): 1987–2007.

Bhatia, Kavita, Gorakh Mal, Rasbihari Bhar, Chandrika Attri, and Amit Seth. "Purification and characterization of thermostable superoxide dismutase from *Anoxybacillus gonensis* KA 55 MTCC 12684." *International Journal of Biological Macromolecules* 117 (2018): 1133–1139.

Bing, Wei, Honglei Wang, Baisong Zheng, Feng Zhang, Guangshan Zhu, Yan Feng, and Zuoming Zhang. "*Caldicellulosiruptor changbaiensis* sp. nov., a cellulolytic and hydrogen-producing bacterium from a hot spring." *International Journal of Systematic and Evolutionary Microbiology* 65, no. Pt_1 (2015): 293–297.

Çalışkan-Özdemir, Sennur, Seçil Önal, and Ataç Uzel. "Partial purification and characterization of a Thermostable Phytase produced by thermotolerant *Aspergillus tubingensis* TEM 37 isolated from hot spring soil in Gediz Geothermal Field, Turkey." *Geomicrobiology Journal* 38, no. 10 (2021): 895–904.

Chauhan, Shweta, C. A. Seth, and Amit Seth. "Bioprospecting thermophilic microorganisms from hot springs of western Himalayas for xylanase production and its statistical optimization by using response surface methodology." *Journal of Pure and Applied Microbiology* 9, no. 2 (2015): 1417–1428.

Chauhan, Shweta, Varun Jaiswal, Chandrika Attri, and Amit Seth. "Random mutagenesis of thermophilic xylanase for enhanced stability and efficiency validated through molecular docking." *Recent Patents on Biotechnology* 14, no. 1 (2020): 5–15.

Deive, Francisco J., María S. Álvarez, M. Angeles Sanromán, and Maria A. Longo. "North Western Spain hot springs are a source of lipolytic enzyme-producing thermophilic microorganisms." *Bioprocess and Biosystems Engineering* 36 (2013): 239–250.

Des Marais, David J., and Malcolm R. Walter. "Terrestrial hot spring systems: Introduction." *Astrobiology* 19, no. 12 (2019): 1419–1432.

Fachrial, E. D. Y., Raden Roro Jenny Satyo Putri, I. Nyoman Ehrich Lister, Sari Anggraini, Harmileni Harmileni, Titania T. Saryono Saryono. "Molecular identification of cellulase and protease producing *Bacillus tequilensis* UTMSA14 isolated from the geothermal hot spring in Lau Sidebuk Debuk, North Sumatra, Indonesia." *Biodiversitas Journal of Biological Diversity* 21, no. 10 (2020): 4719–4725.

Ginting, Elvy Like, Kurniati Kemer, Stenly Wullur, and Agustinus R. Uria. "Identification of proteolytic thermophiles from Moinit Coastal hot-spring, North Sulawesi, Indonesia." *Geomicrobiology Journal* 37, no. 1 (2020): 50–58.

Hawumba, Joseph F., Jacques Theron, and Volker S. Brözel. "Thermophilic protease-producing *Geobacillus* from Buranga hot springs in Western Uganda." *Current Microbiology* 45 (2002): 144–150.

Jadhav, Swati Rangrao, and Anupama Prabhakarrao Pathak. "Production and characterization of a thermo-pH stable pectinase from *Bacillus licheniformis* UNP-1: A novel strain isolated from Unapdev hot spring." *Indian Journal of Geo-Marine Sciences* 48, no. 5 (2019): 670–677.

Jardine, J. L., S. Stoychev, V. Mavumengwana, and E. Ubomba-Jaswa. "Screening of potential bioremediation enzymes from hot spring bacteria using conventional plate assays and liquid chromatography-Tandem mass spectrometry (Lc-Ms/Ms)." *Journal of Environmental Management* 223 (2018): 787–796.

Jyoti, Bhatia Kavita, Kalpna Chauhan, Chandrika Attri, and Amit Seth. "Improving stability and reusability of Rhodococcus pyridinivorans NIT-36 nitrilase by whole cell immobilization using chitosan." *International Journal of Biological Macromolecules* 103 (2017): 8–15.

Khalil, Amjad. "Isolation and characterization of three thermophilic bacterial strains (lipase, cellulose and amylase producers) from hot springs in Saudi Arabia." *African Journal of Biotechnology* 10, no. 44 (2011): 8834–8839.

Kumar, Arun, Refana Shahul, Rajendra Singh, Sanjay Kumar, Ashok Kumar, and Praveen Kumar Mehta. "*Geobacillus thermoleovorans* MTCC 13131: An Amide-Hydrolyzing Thermophilic bacterium isolated from a hot spring of Manikaran." *Indian Journal of Microbiology* 62, no. 4 (2022): 618–626.

Lee, Dong-Woo, You-Seok Koh, Ki-Jun Kim, Byung-Chan Kim, Hak-Jong Choi, Doo-Sik Kim, Maggy T. Suhartono, and Yu-Ryang Pyun. "Isolation and characterization of a thermophilic lipase from *Bacillus thermoleovorans* ID-1." *FEMS Microbiology Letters* 179, no. 2 (1999): 393–400.

Liu, Lan, Jian-Yu Jiao, Bao-Zhu Fang, Ai-Ping Lv, Yu-Zhen Ming, Meng-Meng Li, Nimaichand Salam, and Wen-Jun Li. "Isolation of *Clostridium* from Yunnan-Tibet hot springs and description of *Clostridium thermarum* sp. nov. with lignocellulosic ethanol production." *Systematic and Applied Microbiology* 43, no. 5 (2020): 126104.

Lu, YuXin, HangKe Zhao, XiaoFei Tang, KaiHui Liu, BaiWan Deng, and XiaoWei Ding. "Identification of a cellulase producing fungus from Tibet hot spring and optimization of its enzyme production conditions." *Journal of Henan Agricultural Sciences* 48, no. 10 (2019): 77–83.

Markossian, Samson, Peter Becker, Herbert Märkl, and Garabed Antranikian. "Isolation and characterization of lipid-degrading *Bacillus thermoleovorans* IHI-91 from an icelandic hot spring." *Extremophiles* 4 (2000): 365–371.

Mehta, Praveen Kumar, Shashi Kant Bhatia, Ravi Kant Bhatia, and Tek Chand Bhalla. "Purification and characterization of a novel thermo-active amidase from *Geobacillus subterraneus* RL-2a." *Extremophiles* 17 (2013): 637–648.

Oztas Gulmus, Ebru, and Arzu Gormez. "Identification and characterization of novel thermophilic bacteria from hot springs, Erzurum, Turkey." *Current Microbiology* 77, no. 6 (2020): 979–987.

Panda, Mrunmaya Kumar, Mahesh Kumar Sahu, and Kumananda Tayung. "Isolation and characterization of a thermophilic *Bacillus* sp. with protease activity isolated from hot spring of Tarabalo, Odisha, India." *Iranian Journal of Microbiology* 5, no. 2 (2013): 159.

Prabha, Riddhi, and Vinod Kumar Nigam. "Biotransformation of acrylamide to acrylic acid carried through acrylamidase enzyme synthesized from whole cells of *Bacillus tequilensis* (BITNR004)." *Biocatalysis and Biotransformation* 38, no. 6 (2020): 445–456.

Prajapati, Jenika, Pravin Dudhagara, and Kartik Patel. "Production of thermal and acid-stable pectinase from *Bacillus subtilis* strain BK-3: Optimization, characterization, and application for fruit juice clarification." *Biocatalysis and Agricultural Biotechnology* 35 (2021): 102063.

Pratush, Amit, Amit Seth, and Tek Bhalla. "Generation of mutant of *Rhodococcus rhodochrous* PA-34 through chemical mutagenesis for hyperproduction of nitrile hydratase." *Acta Microbiologica et Immunologica Hungarica* 57, no. 2 (2010): 135–146.

Pratush, Amit, Amit Seth, and Tek Chand Bhalla. "Expression of nitrile hydratase gene of mutant 4D strain of *Rhodococcus rhodochrous* PA 34 in Pichia pastoris." *Biocatalysis and Biotransformation* 35, no. 1 (2017): 19–26.

Raj, J., A. Seth, S. Prasad, and T. C. Bhalla. "Bioconversion of butyronitrile to butyramide using whole cells of *Rhodococcus rhodochrous* PA-34." *Applied Microbiology and Biotechnology* 74 (2007): 535–539.

Sá-Pereira, Paula, Alexandra Mesquita, José C. Duarte, Maria Raquel Aires Barros, and Maria Costa-Ferreira. "Rapid production of thermostable cellulase-free xylanase by a strain of *Bacillus subtilis* and its properties." *Enzyme and Microbial Technology* 30, no. 7 (2002): 924–933.

Sarangthem, Indira, Lynda Rajkumari, Ng Ngashangva, Jusna Nandeibam, Randhir B. S. Yendrembam, and Pulok K. Mukherjee. "Isolation and characterization of bacteria from natural hot spring and insights into the thermophilic cellulase production." *Current Microbiology* 80, no. 2 (2023): 64.

Shajahan, S., I. Ganesh Moorthy, N. Sivakumar, and G. Selvakumar. "Statistical modeling and optimization of cellulase production by *Bacillus licheniformis* NCIM 5556 isolated from the hot spring, Maharashtra, India." *Journal of King Saud University-Science* 29, no. 3 (2017): 302–310.

Sharma, Hitesh, Rahul Vikram Singh, Ananta Ganjoo, Amit Kumar, Ravail Singh, and Vikash Babu. "Development of effective biotransformation process for benzohydroxamic acid production using *Bacillus smithii* IIIMB2907." *3 Biotech* 12, no. 2 (2022): 44.

Sharma, Mamta, Chandrika Attri, and Amit Seth. "Characterization of nitrilase-producing thermophilic bacteria and exploring their potential for mandelic acid synthesis." *Vegetos* 37, no. 1 (2024): 257-265.

Sharma, Monica, Nitya Nand Sharma, and Tek Chand Bhalla. "Purification studies on a thermo-active amidase of *Geobacillus pallidus* BTP-5x MTCC 9225 isolated from thermal springs of Tatapani (Himachal Pradesh)." *Applied Biochemistry and Biotechnology* 169 (2013): 1–14.

Sheikh Abdul Hamid, N., Hee B. Zen, Ong B. Tein, Yasin M. Halifah, Nazamid Saari, and Fatimah Abu Bakar. "Screening and identification of extracellular lipase-producing thermophilic bacteria from a Malaysian hot spring." *World Journal of Microbiology and Biotechnology* 19 (2003): 961–968.

Singh, Poonam, Ansu Kumari, Chandrika Attri, and Amit Seth. "Efficient lactamide synthesis by fed-batch method using nitrile hydratase of Rhodococcus pyridinivorans NIT-36." *The Journal of Microbiology, Biotechnology and Food Sciences* 9, no. 3 (2019): 567.

Singh, Poonam, Ansu Kumari, Kalpana Chauhan, Chandrika Attri, and Amit Seth. "Nitrile hydratase mediated green synthesis of lactamide by immobilizing *Rhodococcus pyridinivorans* NIT-36 cells on N, N′-Methylene bis-acrylamide activated chitosan." *International Journal of Biological Macromolecules* 161 (2020): 168–176.

Tamariz-Angeles, Carmen, Percy Olivera-Gonzales, Gretty K. Villena, and Marcel Gutiérrez-Correa. "Isolation and identification of cellulolytic and xylanolytic bacteria from huancarhuaz hot spring, Peru." *Annual Research & Review in Biology* (2014): 2920–2930.

Thankappan, Sugitha, Sujatha Kandasamy, Beslin Joshi, Ksenia N. Sorokina, Oxana P. Taran, and Sivakumar Uthandi. "Bioprospecting thermophilic glycosyl hydrolases, from hot springs of Himachal Pradesh, for biomass valorization." *AMB Express* 8, no. 1 (2018): 1–15.

Thebti, Wajdi, Yosra Riahi, Rawand Gharsalli, and Omrane Belhadj. "Screening and characterization of thermo-active enzymes of biotechnological interest produced by thermophilic Bacillus isolated from hot springs in Tunisia." *Acta Biochimica Polonica* 63, no. 3 (2016): 581–587.

Ulucay, Orhan, Arzu Gormez, and Cem Ozic. "Identification, characterization and hydrolase producing performance of thermophilic bacteria: Geothermal hot springs in the Eastern and Southeastern Anatolia Regions of Turkey." *Antonie van Leeuwenhoek* (2022): 1–18.

Ünal, Arzu, Asiye Seis Subaşı, Semra Malkoç, İjlal Ocak, S. Elif Korcan, Elif Yetilmezer, Seyhun Yurdugül, Hülya Yaman, Turgay Şanal, and Alaettin Keçeli. "Potential of fungal thermostable alpha amylase enzyme isolated from Hot springs of Central Anatolia (Turkey) in wheat bread quality." *Food Bioscience* 45 (2022): 101492.

Wang, Jingjing, Fei Chang, Xiaoqing Tang, Wei Li, Qiang Yin, Yang Yang, and Yang Hu. "Bacterial laccase of *Anoxybacillus ayderensis* SK3-4 from hot springs showing potential for industrial dye decolorization." *Annals of Microbiology* 70 (2020): 1–9.

Yadav, Punam, Suresh Korpole, Gandham S. Prasad, Girish Sahni, Jyoti Maharjan, Lakshmaiah Sreerama, and Tribikram Bhattarai. "Morphological, enzymatic screening, and phylogenetic analysis of thermophilic bacilli isolated from five hot springs of Myagdi, Nepal." *Journal of Applied Biology and Biotechnology* 6, no. 3 (2018): 1–8.

Yasawong, Montri, Supatra Areekit, Arda Pakpitchareon, Somchai Santiwatanakul, and Kosum Chansiri. "Characterization of thermophilic halotolerant *Aeribacillus pallidus* TD1 from Tao dam hot spring, Thailand." *International Journal of Molecular Sciences* 12, no. 8 (2011): 5294–5303.

Zarafeta, Dimitra, Dimitrios Kissas, Christopher Sayer, Sóley R. Gudbergsdottir, Efthymios Ladoukakis, Michail N. Isupov, Aristotelis Chatziioannou et al. "Discovery and characterization of a thermostable and highly halotolerant GH5 cellulase from an icelandic hot spring isolate." *PLoS One* 11, no. 1 (2016): e0146454.

Zhao, Chao, Yunjin Deng, Xingna Wang, Qiuzhe Li, Yifan Huang, and Bin Liu. "Identification and characterization of an anaerobic ethanol-producing cellulolytic bacterial consortium from great basin hot springs with agricultural residues and energy crops." *Journal of Microbiology and Biotechnology* 24 (2014): 1280–1290.

Zuridah, H., N. Norazwin, M. Siti Aisyah, M. N. A. Fakhruzzaman, and N. A. Zeenathul. "Identification of lipase producing thermophilic bacteria from Malaysian hot springs." *African Journal of Microbiology Research* 5 (2011): 3569–3573.

22 Overview of Microbial Toxins in the Aquatic Environment

Pramod Kumar Pandey, M. Junaid Sidiq, and Rameshori Yumnam

22.1 INTRODUCTION

Microbial toxins are toxic compounds produced by a range of microorganisms: bacteria, fungi, viruses, and algae. These toxins can have adverse effects on various organisms, encompassing plants, animals, and humans. The severity of their effects in humans can vary from mild symptoms to severe illness, and in some cases, can even be fatal (Valério and Tenreiro, 2010). Microorganisms, being present everywhere, are the key sources of water- and foodborne intoxications, which are critical issues in environmental safety, human health, and animal welfare. Such harmful microbes may exist in drinking and recreational waters. Also, these toxins have significant economic influence as a result of their detrimental effects (Valerio et al., 2010).

Bacterial toxins are some of the best-known microbial toxins. For example, the botulinum toxin produced by certain strains of *Clostridium botulinum* is one of the most lethal compounds known to humankind. It causes botulism, a severe disease that attacks the central nervous system. This can result in respiratory failure and ultimately, death. Other bacterial toxins include Shiga toxins, which are produced by certain strains of *Escherichia coli*, and Staphylococcal enterotoxins produced by *Staphylococcus aureus*. Both can induce a range of symptoms, such as diarrhoea and vomiting (Hernández-Cortez et al., 2017).

Fungal toxins are also important microbial toxins. Some common examples include aflatoxins produced by *Aspergillus flavus*, which can contaminate crops such as peanuts and maize, and can lead to serious liver damage. Likewise, T-2 toxin and deoxynivalenol are two important mycotoxins produced by *Fusarium* fungi that can contaminate grains and can cause a range of adverse effects such as vomiting, diarrhoea, and anaemia in humans (Zaki et al., 2012). Algal microorganisms can also produce toxins that can have significant effects on a range of organisms. For example, dinoflagellates such as *Alexandrium* sp., *Gymnodinium* sp., and *Dinophysis* sp. can produce toxins that have neurolytic effects on their consumers, which can result in respiratory and muscular paralysis, and in severe cases, can cause death. Similarly, microcystins (MCs) produced by certain species of cyanobacteria can contaminate freshwater ecosystems, causing adverse effects on wildlife and humans (Brown et al., 2020). Understanding and managing the risks from microbial toxins are important for ensuring public welfare and maintaining a healthy ecosystem (Figure 22.1).

22.2 MICROBIAL TOXINS IN AQUATIC ENVIRONMENTS

22.2.1 Cyanobacterial Toxins

Cyanobacterial strains produce some of the most prevalent microbial toxins found in water. Cyanobacteria represent prokaryotic microorganisms under the major bacterial phyla that exhibit the general characteristics of gram-negative bacteria and have successfully colonised freshwater, brackishwater, and marine waters, as well as hypersaline environments, nonacidic hot springs,

DOI: 10.1201/9781003408543-22

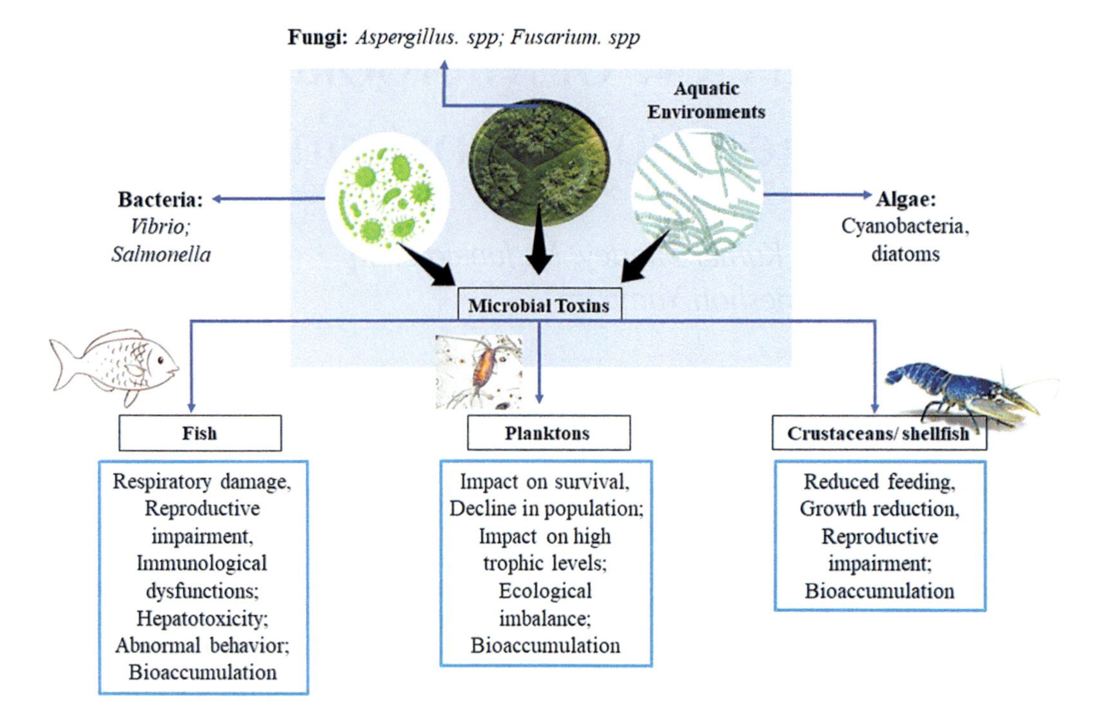

FIGURE 22.1 Origin and impacts of microbial toxins in aquatic environments.

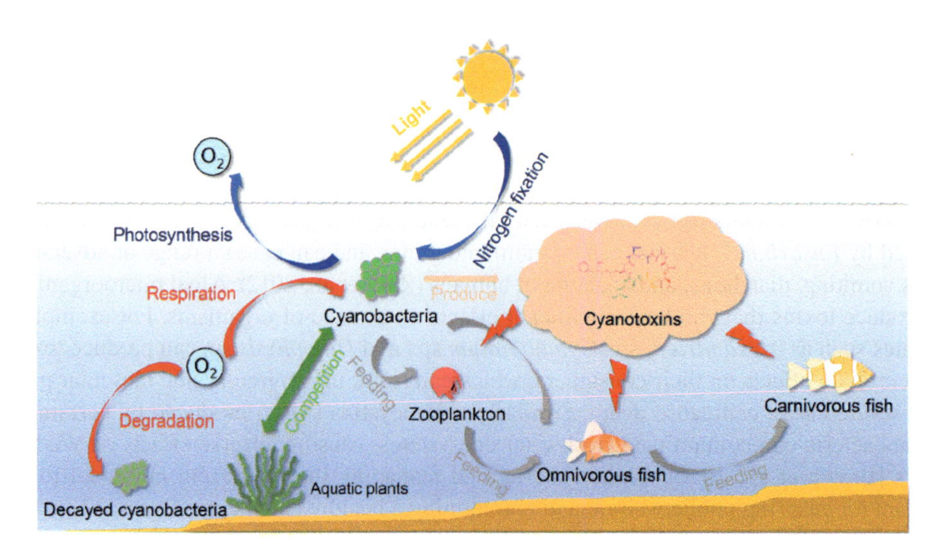

FIGURE 22.2 Impact of cyanobacteria and cyanotoxins on the aquatic food web. The respiration and degradation of cyano-HABs result in oxygen deprivation in the water. The cyanotoxins, produced and released by cyanobacteria, can harm aquatic animals through biological concentration or food chain transfer.

Antarctic regoliths, and even deserts (Falconer, 2004). Water managers are extremely concerned about the negative effects of such cyanobacterial blooms and the creation of toxins (Figure 22.2). They are now a growing issue on a global scale in artificial water storage reservoirs as well as aquatic environments (lakes, rivers, estuaries, and seas). The gradual eutrophication of the streams, constant sunlight and warmth, and the presence of nutrients such as phosphates and nitrates are all variables that may lead to these occurrences (Castenholz et al., 2001).

In contrast to some other aquatic microbial and toxicant health hazards that are undetectable to the human eye, cyanobacteria are frequently plainly visible to the human eye and occasionally identifiable by smell. They are identified by water discolouration, bloom formation, and smelly chemical creation in the residing waters. It is well known that a number of cyanobacterial toxins are capable of building up, moving up the food chain, and bioaccumulating. Mussels, crayfish, fish and crustaceans, and even plants that are irrigated with polluted water can absorb and contain cyanotoxins such as MCs. *Daphnia magna*, a cladoceran zooplankton, can collect paralytic shellfish poison

TABLE 22.1
Summary of Immunotoxicity of Pure Microcystin in Fish Studies in vivo

Species	Microbial Toxin/Exposure Type	Dosage	Critical Findings/Impacts
Silver carp	MC-LR/IP	104.9 μg/kg, 262.1 μg/kg	Increased levels in serum: ALT, AST, lysozyme, complement C3, TNF-α, IL-1β, IFN-γ
Common carp	MC-LR/IP	150 μg/kg BW	Increased levels in serum for CAT, SOD, GSH, GPx, and LPO; decreased complement C3 and lysozyme activity
Grass carp	MC-LR/IP	50 μg/kg	Induction of mitochondrial oedema, apoptotic lymphocytes, and chromatin condensation in kidney and spleen
Bighead carp	MC-LR/IP	50, 200, 500 μg MC-LR/kg BW	Temporal- and dose-related increment in IL-8 of kidney and liver
Grass carp	MC-LR/IP	25, 75, 100 μg/kg BW	Upregulation of complement and coagulation cascade pathways in liver, *soc3*, and *hsp70*
Gold fish	MC-RR/IP	50, 200 μg/kg BW	Decreased levels of T-AOC and GPx, increased SOD in kidney
Tilapia	MC-LR, MC-RR/IP	500 μg/kg MC-LR or MC-RR	Increased levels of SOD, CAT, GPx LPO in kidney
Zebrafish	MC-LR/Immersion	200, 800 μg/L	Upregulation of *Ragl*, *Rag2*, *Ikaros*, *GATA*, *Lck*, and *TCRα* in the fish larvae
Zebrafish	MC-LR/Immersion	1, 5, 20 μg/L	Ultrastructural damage in spleen and upregulation of *infl*, *il8*, *il1β*, and *tnfα*
Zebrafish	MC-LR/Immersion	0.3, 1, 3, 10, 30 μg/L	Histopathological lesions in spleen and changes in the complement C3 system
Zebrafish	MC-LR/Immersion	0.4, 2, 10 μg/L	Histopathological lesions and apoptosis in spleen along with upregulation of *tnf-α*, *il-1β*, *myd88*, and complement C3
Zebrafish	MC-LR/Immersion	20 μg/L	Increased levels of TBARS, GSH, LDH, while decrease in GST, CAT, in liver; upregulation of *cas3a* and *cas3b*

Source: adapted from Lin, W. et al., *Toxins*, 13, 11, 765, 2021.

ALT: alanine aminotransferase, AST: aspartate aminotransferase, IFN: interferon, BW: body weight, CAT: catalase, SOD: superoxide dimustase, GSH: glutathione, GPx: glutathione peroxidase, LPO: lipid peroxidase, T-AOC: total antioxidant capacity, TBARS: thiobarbituric acid reactive substances, GST: glutathione S-transferase, LDH: lactate dehydrogenase

(PSP) toxins and move them along the food chains and food webs dependent on them (Valério et al., 2010). Similarly, PSP toxins can build up in freshwater mussels, clams, crabs, and other cladocerans.

The bioactive compounds reported from cyanobacteria possess a wide range of structural variations and are reported to exist in the form of non-ribosomal peptides and polyketides, lipopeptides, and alkaloids. The effects that cyanobacteria-produced toxins have on mammals and other vertebrates are typically used to categorize them. Various toxins have been isolated and characterized from different cyanobacteria: hepatotoxins (liver-damaging), cytotoxins (cell-damaging), neurotoxins (nerve-damaging), and toxins that cause allergic reactions (dermatotoxins).

22.2.2 Microcystins (MCs)

Microcystins are the most prevalent types of cyanotoxin in the world and are the subject of extensive research. Structurally, the most prevalent microcystins are those with the L-amino acids leucine (L), arginine (R), or tyrosine (Y) at the X position, respectively. The most researched variation is MC-LR because of its ubiquity, abundance, and toxicity. Numerous cyanobacterial taxa, such as *Microcystis*, *Planktothrix*, *Oscillatoria*, *Anabaena*, *Anabaenopsis*, *Nostoc*, *Hapalosiphon*, *Snowella*, and *Woronichinia*, have been found to produce MCs (Valério et al., 2010).

Hepatocytes, the most commonly occurring cell type in the liver, are the primary target of MCs. MCs induce toxicity by activating phosphorylase-b enzyme and in turn, inhibit all the protein phosphatases, which causes excessive phosphorylation of cytoskeletal filaments and ultimately, their apoptosis. Hepatocyte death causes the liver's smaller blood arteries to burst, resulting in severe hepatic haemorrhage. Human exposure to these toxins can have an impact on organs like the kidney and colon (Campos and Vasconcelos, 2010). Nodularin (NOD), like MCs, is a powerful tumour promoter that inhibits serine/threonine protein phosphatase-1 and 2A and may also operate as a carcinogen or tumour initiator.

22.2.3 Cytotoxins: Cylindrospermopsin

Numerous studies have been conducted on cylindrospermopsin (CYN) because of its toxic effects, particularly in Australia. It is a potent alkaloid, consisting of a tricyclic guanidine moiety combined with hydroxymethyluracil CYN. *Cylindrospermopsis raciborskii*, *Umezakia natans*, *Aphanizomenon ovalisporum*, *Raphidiopsis curvata*, *Anabaena bergii*, and more recently, *Aphanizomenon flos-aquae* and *Lyngbya wollei* have all been found to produce analogues of CYN. Terao et al., 1994 found that the liver was the primary target of this cyanotoxin, but additional histopathological studies revealed that the kidneys, thymus, and heart are also affected. Kidney and liver failure are the initial signs of CYN consumption (WHO, 2003). CYN poisoning in the liver results in the inhibition of protein synthesis, proliferation of membranes, accumulation of fat droplets, and death of cells. Despite extensive research, the precise chemical interactions that trigger CYN-mediated toxicity remain unknown. But, it is known that this toxin is a general cytotoxin and a genotoxic and hepatotoxic in vivo toxin, and it inhibits protein synthesis.

22.2.4 Anatoxin-a (ANA) and Homoanatoxin (HNA)

These rare secondary amines having low molecular weights, anatoxin-a (ANA-a) (165 Da) and homoanatoxin-a (179 Da), are generated only by cyanobacteria. *Anabaena*, *Cylindrospermum*, *Microcystis*, *Oscillatoria*, *Raphidiopsis*, *Planktothrix*, and *Phanizomenon* are among the genera that produces anatoxin-a, while species belonging to the genera *Oscillatoria*, *Anabaena*, *Raphidiopsis*, and *Phormidium* generate homoanatoxin-a. Both the toxins are also produced concurrently by some strains (Namikoshi et al., 2003; Aráoz et al., 2005). Anatoxin-a functions as a post-synaptic neuromuscular blocker and was also referred to as "Very Fast Death Factor" in the past. Anatoxin-A and homoanatoxin-A are strongly agonistic to the nicotinic acetylcholine receptors in the muscles and neurons. Because of the toxin's irreversible binding to the nicotinic acetylcholine receptor, sodium channels open, allowing sodium ions to enter cells continually. Depolarization and desensitization of the membrane cause muscle cell overstimulation. Acute asphyxia, which results in animal death when respiratory muscles are compromised, may cause convulsions when there is a lack of oxygen in the brain (Herrero-Galán et al., 2009).

22.2.5 Saxitoxin

Tricyclic perhydropurine alkaloids, often known as saxitoxins (STXs) or paralytic shellfish poisons (PSPs), have a molecular weight of 299 Da (Figure 22.3). More than 30 structural variants can emerge from potential substitutions at different locations of the molecule, including non-sulfated (saxitoxins and neosaxitoxins), single sulfated (gonyautoxins), and doubly sulfated (C-toxins) (Mihali et al., 2009). The most dangerous compounds are gonyautoxins (GTX1-4), neosaxitoxin (NEO), and saxitoxin (STX), which vary in potency depending on the type of variation produced. Dinoflagellates and cyanobacteria found in marine environments create saxitoxins. Freshwater cyanobacteria belonging to the genera *Anabaena*, *Aphanizomenon*, *Cylindrospermopsis*, *Lyngbya*, and *Planktothrix* can produce these toxins. Other cyanobacterial toxins include jamaicamides, kalkitoxin, antillatoxins, and lipopeptides, produced by the marine cyanobacterium *Lyngbya majuscula*.

In addition to the major toxigenic prokaryotes, cyanobacteria, there are additional bacteria present in aquatic environments that can create toxins with high relevance for human and animal health. One of the most prevalent subgroups of aquatic toxin makers, *Vibrio* sp., are frequently linked to infections and toxicity brought on by seafood. The synthesis of toxins has also been attributed to *Aeromonas hydrophila*, which is likewise connected to aquatic habitats (Biscardi et al., 2002). Additionally, known organisms like *E. coli*, *Campylobacter* sp., and *Legionella pneumophila* have been identified as emergent toxin makers and are water pollutants.

22.2.6 Bacterial Toxins Contaminating Water

The environment has an abundance of additional toxic species, such as *Clostridium* sp. and *Pseudomonas* sp. Despite not being aquatic microorganisms, they can easily contaminate irrigation or drinking water and pose risks to both people and animals. *Clostridium* spp. bacteria are a genus of pleomorphic, gram-positive, anaerobic rods found in soils, marshes, lakes, and coastal waters, as well as fish gills, crabs, and other shellfish viscera. Because they can create endospores, they have a high level of resistance to unfavourable environmental circumstances. The cellular structures that make up these bacterial spores can withstand high temperatures, desiccation, chemicals, and radiation. The spores can germinate and produce viable vegetative cells when favourable growth circumstances are restored (Bergey, 1994). *Clostridium* sp. spores can pollute recreational and drinking water via sources like dirt, dust, and insects.

22.2.7 Plant Toxins (Phytotoxins)

The primary function of the plant-based secondary metabolites called phytotoxins is to defend the plant from external dangers. Alkaloids, terpenes, glycosides, proteinaceous chemicals, organic acids, and resinoid compounds are the main categories of plant poisons (Gunthardt et al., 2018). Each group is divided into many subgroups in accordance with its structure. Numerous phytotoxic chemicals have been described, and extensive lists have also been suggested (Quattrocchi, 2012). The regulations governing plant toxins in surface and drinking water have not yet been developed, despite the large number of plant toxins that could potentially end up in natural aquatic settings (distributed between sediments and water according to their octanol/water partition constants). In addition, only a small number of compounds, such as glycoalkaloids made from potatoes (*Solanum tuberosum*), ptaquiloside made from bracken (*Pteridium aquilinum*), or some isoflavones, have had their environmental

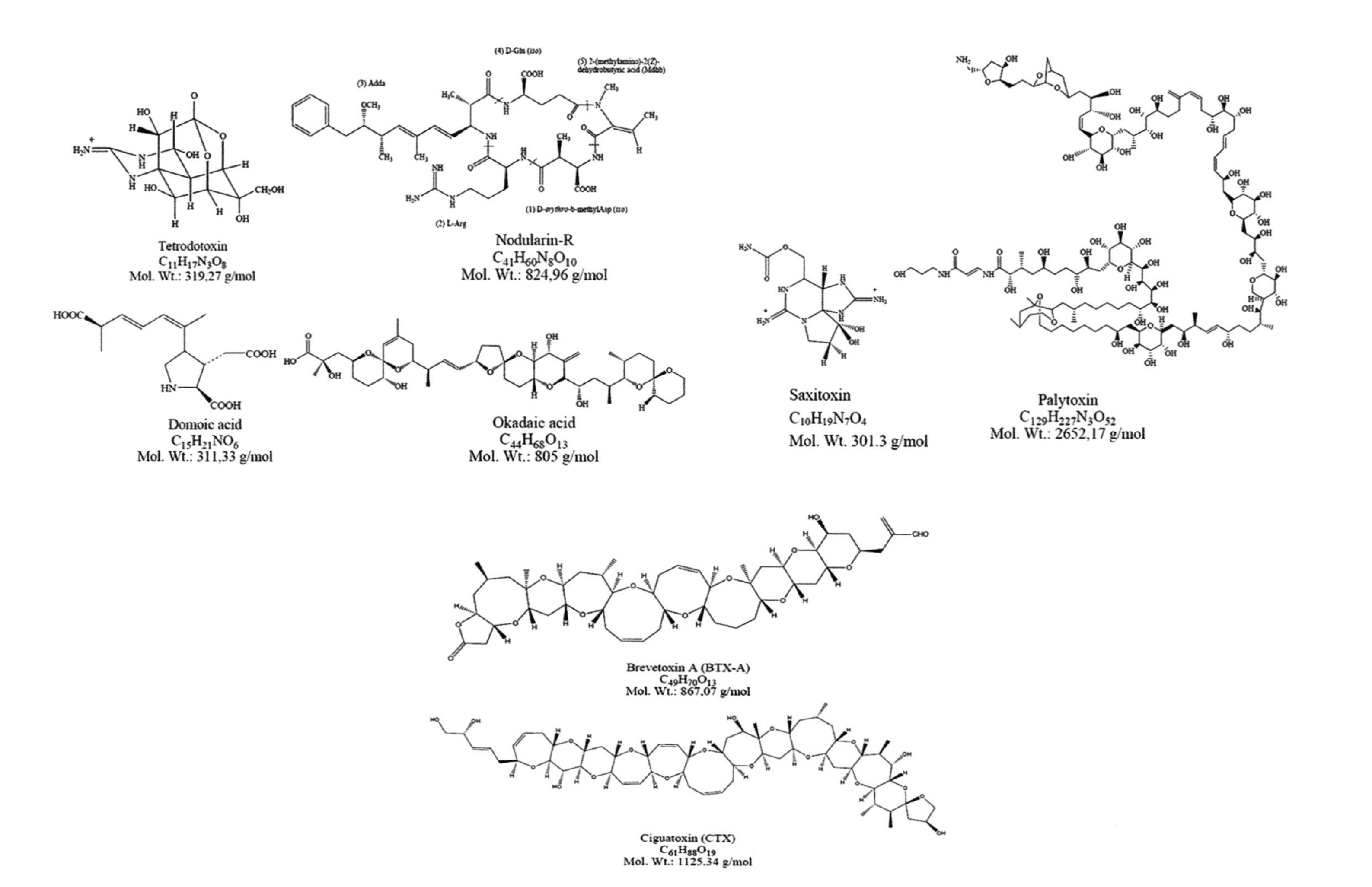

FIGURE 22.3 Molecular structures of popular microbial toxins: tetrodotoxin, nodularin, domoic acid, okadaic acid, saxitoxin palytoxin, brevetoxin, and ciguatoxin. (Adapted from Vasconcelos, V. et al., *Marine Drugs*, 8, 1, 59–79, 2010.)

fate and behaviour studied (Picardo et al., 2019). However, these compounds are highly relevant in terms of their capacity to generate biological consequences such as toxicity, carcinogenicity, or oestrogenicity. These toxins have been studied in relation to food and feed, but for the majority of them, data on their prevalence, fate, and behaviour in the environment is still insufficient. The last ten years have seen the development of many analytical approaches for the detection of isoflavones in wastewaters and aquatic habitats due to their estrogenic action and applications in pharmacy.

22.2.8 Mycotoxins

Mycotoxins are produced by fungi of the genera *Fusarium*, *Aspergillus*, and *Penicillium* under specific temperature and humidity conditions. They can enter the human food chain directly through polluted drinking water or plant-based dietary components, or indirectly through toxigenic contamination by fungal growth on food or feed. Mycotoxins can accumulate in cereals, corn, peanuts, soybeans, and spices, among other things, during maturation, storage, and transportation. Human and animal consumption of mycotoxin-contaminated food or feed can result in acute or chronic poisoning (Marijani et al., 2019). They are genotoxic, carcinogenic, and mutagenic, and some of them are immunosuppressive (Smith et al., 1995). For these reasons, the Regulation (EC) 1881/2006 established the maximum levels of mycotoxins in food, and the Commission Directive 2003/100/EC, which amended the Directive 2002/32/EC (Cheli et al., 2013), modified the maximum levels of mycotoxins in feed, but no limits are currently applied to surface water.

22.3 METHODS FOR DETECTION OF MICROBIAL TOXINS IN AQUATIC ENVIRONMENTS

22.3.1 Molecular Methods

The discovery of cyanobacteria-specific genes and those involved in the production of their toxins is the foundation of molecular approaches. These techniques include real-time polymerase chain reaction (PCR), multiplex PCR, conventional PCR, random amplified polymorphic DNA (RAPD), terminal restriction fragment length polymorphism (T-RFLP), and denaturing gradient gel electrophoresis (DGGE). Some non-PCR methods, like fluorescence in situ hybridization (FISH) and DNA microarrays, are also employed. PCR is the most popular type of analysis, which involves amplifying a DNA sequence in vitro using specialized primers that aim for the unique sequence present on DNA. Amplification of the 16S ribosomal RNA (rRNA) gene, the phycocyanin operon, the internal transcribed spacer (ITS) region, and the RNA polymerase b subunit gene (rpoB) using taxon-specific primers are only a few of the sequences employed in the study of cyanobacteria (Gaget et al., 2011).

Similarly, biotoxins from the major cyanotoxin groups, including MCs and their congeners, NOD, STX, ANA-a, and CYN, can be identified using molecular techniques by looking for the presence of the genes in the gene clusters that encode these biosynthetic enzymes. For other families of cyanobacteria, including MCs, STX, and ANA-a, various multiplexed PCR tests have also been devised. However, the real-time PCR method enables highly sensitive measurement of both cyanobacteria and cyanotoxins in environmental materials (Moreira et al., 2011).

22.3.2 Biochemical Methods

22.3.2.1 Methods Based on Enzyme Inhibition

Different techniques to determine the toxins and toxicity of a sample have been developed based on the enzyme inhibition capabilities of different families of cyanotoxins. The MCs, NOD, and ANA cyanotoxin groups can be identified using enzyme inhibition techniques. The ANA-a group are acetylcholinesterase (AChE) inhibitors, whereas MCs and NODs are both phosphatase (PP) inhibitors. Such methods were initially developed by employing a colorimetric protein phosphatase test to identify MC and NOD, and were modified with time (An and Carmichael, 1994). For instance, MC and NOD levels in drinking water were found using the phosphate released from a phosphorylated protein substrate. For the measurement of phosvitin, colorimetric and fluorometric reactions are used to detect several cyanotoxins. Moreover, the liquid chromatography and colorimetric processes often work well together to obtain high levels of sensitivity and specificity.

Various biosensors have already been created using enzyme inhibition. For instance, Campas et al. 2007 developed an electrochemical biosensor for the detection of MC based on the suppression of protein phosphatase 2A (PP2A). The enzyme was trapped inside a poly (vinyl alcohol) azide-unit pendant water-soluble photopolymer (PVA-AWP) to immobilize it. The protein phosphatase activity was measured by amperometry using the phosphorylated substrates catechol monophosphate (CMP), a-naphthyl phosphate (a-NP), and 4-methylumbelliferyl phosphate (4-MUP). The first of these provided the highest chronoamperometric currents at the proper working potentials (+450 mV versus Ag/AgCl), and a limit of detection of 37 mg L^{-1} was attained.

Some biosensor-based strategies have been developed for ANA-a as well: for instance, the development of a biosensor based on the amperometric measurement of the activity of electric eel acetylcholinesterase. The oxime reactivation was utilized to distinguish between the toxin and possible insecticides present in the sample, and the system demonstrated a limit of detection of 1 mg L^{-1} ANA-a(s) in natural ambient sample. Overall, the speed of reaction without sample preparation is the major benefit of enzyme inhibition-based approaches. However, the lack of specificity and lack of adaptability of these assays are their primary overall drawbacks.

22.3.2.2 Immunochemical Methods

In order to create detection techniques for naturally occurring poisons in water, the affinity qualities between antibodies and antigens have been thoroughly investigated. Brooks and Codd reported one of the earliest polyclonal rabbit antibodies against MCs (Brooks and Codd, 1987). Numerous polyclonal and monoclonal antibodies were developed following the first, leading to a diversity of immunoassays. Enzyme-linked immunosorbent tests (ELISAs), which are available in a variety of commercial kits, have been widely used. Some of these antibodies could have broad specificity against a variety of cyanotoxins, which is good for quick screening but might be bad for specificity among congeners or chemically related compounds. For instance, researchers have developed a direct competitive assay to detect MCs in waters and raised a polyclonal antibody to the MC group, which demonstrated good cross-reactivity with MC-LR, MC-RR, MC-YR, MC-LF, and NOD, and has a limit of detection (LOD) value for MC-LR of 0.12 mg L^{-1} (Sheng et al., 2006).

Similarly, in vitro designed monoclonal antibodies, with LOD between 0.16 and 0.10 mg L^{-1} and recoveries of 62–86%, were able to detect NOD and eight types of MCs (Yang et al., 2016). The authors have suggested that a significant selectivity can also be attained in some circumstances. Moreover, a disclosed method of immunoassay permits LODs of less than 4 ng L^{-1}. LODs of roughly 500 ng L^{-1} were reached by variations of this technique that included automated array biosensors (Picardo et al., 2019).

In general, high-throughput analysis, which results in fewer or even no sample preparation requirements, and quick results are the key benefits of immunoassays. The fundamental restrictions, on the other hand, are the cross-reactivity of structurally related compounds that are bound with low specificity. The sensitivity of the assays might also be significantly influenced by matrix effects. The challenge of developing antibodies against very dangerous compounds, which restricts their availability, is one further downside of the immunological detection assays.

22.3.3 Chemical Analysis

The analytical methods that are frequently used to identify natural poisons in water samples are based on separation-based methods like liquid or gas chromatography (LC or GC) connected to various detectors, including mass fluorescence (FL), spectrometry (MS), ultraviolet (UV), and spectroscopy. However, LC-MS is currently the method of choice because of its specificity and sensitivity. Also, prior to the analysis, preconcentration and cleaning steps are required.

22.3.3.1 Extraction and Clean-Up Strategies for Chemical Detection Methods

The most popular techniques include liquid-liquid extraction, solid phase microextraction (SPME), freeze-drying, and solid phase extraction (SPE) (Picardo et al., 2019). However, SPE is the most extensively used technique because of its adaptability, the large variety of possible stationary phases, the possibility of automation, and the reduced solvent usage. To distinguish between the levels of toxins present within and outside cells, a filtration step is sometimes required to first separate the cells from the water, and then the two fractions are treated simultaneously. The sample is put through ultrasonication, lyophilization, and freeze-thawing to break up the cells and release the toxins if the complete toxin content is needed.

SPME is another intriguing method for removing and cleaning up pollutants from water samples. This method provides reusable fibres using a range of materials, including carbopack-z, polydimethylsiloxane (PDMS), polystyrene (PS), divinylbenzene (DVB), and carboxen. SPME, however, is a relatively new technology that has not seen much use for the investigation of natural poisons. Only a few application examples for the extraction of MCs, NOD, and ANA-a in waters have been documented (Picardo et al., 2019).

22.3.3.2 Gas Chromatography Coupled to Mass Spectrometry

Researchers have created a regular procedure for ANA analysis in blue-green algae (Sano et al., 1992). The study described a technique based on a quantitative measurement of 2-methyl-3-methoxy-4-phenylbutyric acid (MMPB) as an oxidation product of MCs by GC linked to a single quadrupole MS (Q-MS). In this method, the MCs are first oxidized and then derivatized to their methyl esters. The method was later modified by suggesting the use of ozonolysis to obtain the MCs' oxidation product, 3-methoxy-2-methyl-4-phenylbutyric acid. This method resulted in a much shorter reaction time, including the elimination of sample preparation.

For the examination of endocrine disruptor chemicals (EDCs), including natural oestrogens and phytoestrogens such formononetin, biochanin A, genistein, and daidzein, the usage of GC-MS is gaining high popularity. For instance, the use of an ion trap mass spectrometer (ITQ) connected to a GC outfitted with a programmable temperature vaporizer (PTV) for the measurement of EDC following a SPE clean-up and preconcentration step of river water samples has been described (Rocha et al., 2016). The "method limit of detection" (MLOD) for this technique ranged from 5.5 to 0.9 ng L^{-1}. A similar investigation was conducted by using a GC linked to an ion trap mass spectrometer for the precise mass identification of phytoestrogens and other natural chemicals in estuary waters at ng/L levels. The study demonstrated the existence of natural poisons in these waters (Picardo et al., 2019). In contrast to approaches based on LC separation, the application of GC-MS methods for the analysis of natural poisons in freshwater has generally been negligible so far.

22.3.3.3 Liquid Chromatography Coupled to Mass Spectrometry

Toxins found in aqueous media possess moderate to high polarity. As a result, LC is a more practical method, since it offers separation without derivatization. Although LC in combination with detectors like UV/VIS has also been used, it is not possible to validate the toxin identification. Due to their ability to perform simultaneous identification and quantification, LC-MS techniques are the most used techniques. These methods typically rely on the use of LC linked to tandem mass spectrometry (MS/MS); however, it has recently come to light that certain methods also utilize the benefits of high-resolution MS (HRMS).

22.4 IMPACTS OF HARMFUL ALGAL BLOOMS (HABS) ON FISHERIES

The HAB species affect fish by clogging their gills or producing fish toxins (ichthyotoxins). Furthermore, as the blooms decay, microbes degrade the stored algal biomass, causing oxygen loss in aquatic ecosystems as a whole (Smayda et al., 2004; Svendsen et al., 2018). Around 300 HAB species have been discovered worldwide, and more than a third of them, mostly dinoflagellates, have been shown to generate toxins that are dangerous to aquatic organisms and people who consume them. For some HAB species, toxin production can change across genetic strains or even by environmental factors. These poisons may also produce other metabolites, many of which have not yet been thoroughly described in terms of chemical composition, potency, or relevance to human health (Brown et al., 2020).

22.5 ACUTE TOXICITY FROM HABS TO FINFISH AND SHELLFISH

HAB species from broad taxonomic groups with few similarities, such as dinoflagellates, dictyophytes, haptophytes, prymnesiophytes, and raphidophytes, have been linked to major finfish mortality in coastal fisheries and mariculture. In rare instances, the toxicity can be passed on to seabirds and marine animals further up the food chain. The Atlantic salmon (*Salmo salar*), rainbow trout (*Oncorhynchus mykiss*), and yellowtail amberjack/kingfish (*Seriola quinqueradiata*) are three commonly cultivated fish species that are impacted by HABs (Clément et al., 2016). However, the toxicity mechanisms for "fish-killing HABs" are not well elucidated.

HABs affect finfishes at molecular and physiological levels. *Heterosigma akashiwo* is used to illustrate the complexities involved with HAB toxicity in fish. Effects in this case might be brought on by the generation of reactive oxygen species, the presence of compound(s) that resemble brevetoxin, an excessive amount of mucus that prevents oxygen exchange, mucocyst-induced gill injury, and haemolytic activity. When the toxicities of wild HAB populations and laboratory cultures differ, uncertainties arise. For instance, it has been demonstrated that long-term cultivation of *H. akashiwo* results in decreased toxicity (Cochlan et al., 2013). Additionally, various strains of microalgae, such as *Pseudochattonella farcimen*, may produce mucocysts differently.

Major finfish and shellfish kills in the United Kingdom (1978, 1980) and Ireland (1976, 1978, 1979, and 2005) have been linked to *Karenia mikimotoi* (also known as *Gyrodinium/Gymnodinium aureolum*). Widespread fish deaths from these blooms have been attributed to acute toxicity from phycotoxins with neurotoxic, haemolytic, or cytotoxic effects, or to oxygen deprivation from decaying blooms. *Scombrus*, an Atlantic mackerel, and *Homarus americanus*, an American lobster, are two examples of commercially significant fish and shellfish species that may be fatally affected by the STX generated by *Alexandrium* sp. Moreover, considerable fish kills and mortality of seabirds and marine mammals have also been connected to biomagnification of STX in the marine food chain (Brown et al., 2020).

HAB risks are especially severe for finfish limited to protected inshore embankments, where onshore winds and currents might concentrate HABs. For example, the raphidophyte *Chatonella antiqua* was responsible for the deaths of at least 21.8 million farmed yellowtail amberjacks (*Seriola quinqueradiata*) in the Seto Inland Sea, Japan, between 1972 and 1982. The economic cost of the summer outbreak in 1972 was US$70 million. Annual losses have lessened since then, but the catastrophic consequences have continued. Another poisonous raphidophyte, *H. akashiwo*, has been linked to multiple risks, including finfish deaths in Chile, Iceland, Spain, British Columbia, and Iceland (Landsberg, 2002). It has been calculated that each episode of *H. akashiwo* outbreaks affecting wild and net-penned finfish near Puget Sound, Washington, results in losses of between US$2 million and US$6 million (Brown et al., 2020).

22.6 ENVIRONMENTAL FACTORS CONTRIBUTING TO HAB RISK

Aquatic environments naturally experience HABs as part of the seasonal cycles of planktonic microorganisms (Shumway et al., 2018). The number, length, and worldwide effect of hazardous occurrences appear to be rising in recent decades, and a top research objective is to verify them

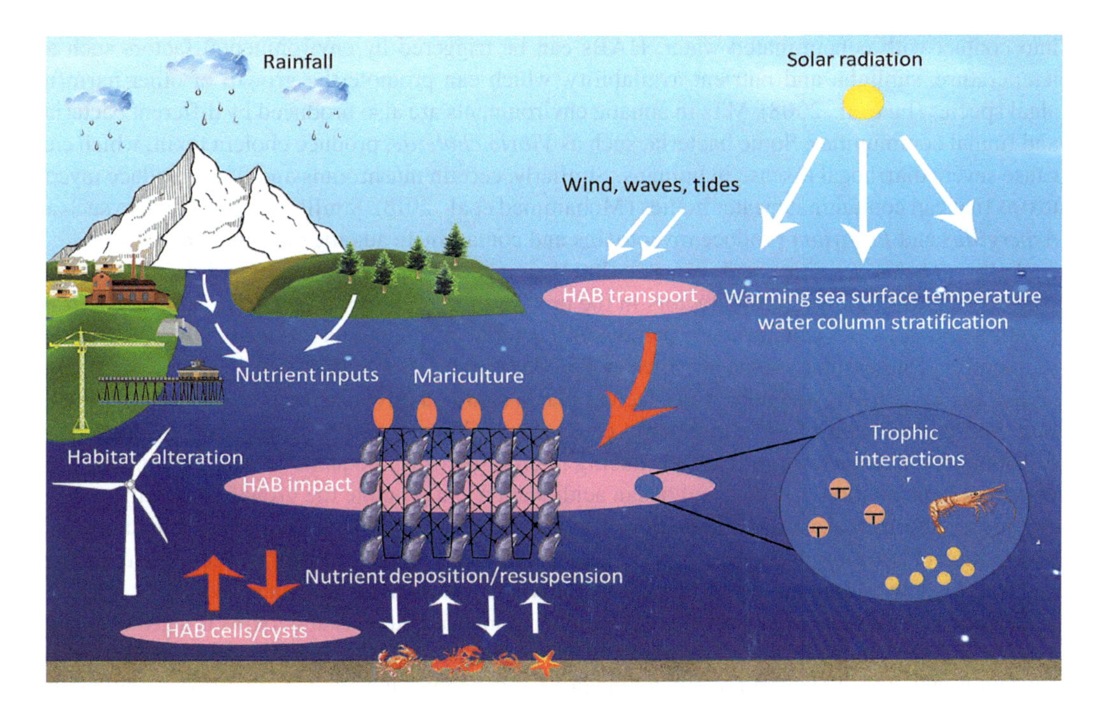

FIGURE 22.4 Environmental factors augmenting HABs. Complex relations among environmental indices (solar energy, waves, tides, rainfall, wind, nutrients), ecological and trophic interactions, and biological processes can facilitate the multiplication of phytoplankton along with harmful algal communities. (Adapted from Brown et al., *Reviews in Aquaculture* 12. 3: 1669, 2020.)

(GlobalHAB, 2017). Several environmental factors and their combinations are found to increase the occurrence and proliferation of HABs (Figure 22.4) and include:

a) increasing sea surface temperatures, water column stratification, and range expansion of toxic species of tropical organisms
b) frequent and intense storm events and flooding with associated increases in nutrient inputs, upwelling intensities, and wider HAB dispersal
c) growing anthropogenic stresses on the marine environment, particularly nutrient enrichment on land and at sea and habitat disruption along the coast
d) improved HAB monitoring systems and greater awareness.

22.7 SOURCES AND DETERMINANTS OF MICROBIAL TOXINS IN AQUATIC ECOSYSTEMS

MTs, produced by microorganisms present in aquatic habitats, pose significant ecological and human health risks. These toxins can have both immediate and long-term effects, causing diseases, and reduce the overall productivity and biodiversity of aquatic ecosystems. MTs can contaminate aquatic environments like freshwater and marine ecosystems, posing a significant threat to public health, aquatic life, and water resources. The sources of MTs in aquatic environments are diverse and can be natural or anthropogenic (Häder et al., 2020).

HABs are one of the major sources of MTs in aquatic environments. Algal blooms can be caused by the overgrowth of certain species of microalgae in aquatic environments, leading to the production of toxins known as phycotoxins (Lopes et al., 2019). These toxins can cause a range of health problems in aquatic organisms and humans who consume contaminated seafood or come into contact with contaminated water. HABs can be triggered by environmental factors such as temperature, sunlight, and nutrient availability, which can promote the growth of other harmful algal species (Fu et al., 2008). MTs in aquatic environments are also produced by different bacterial and fungal communities. Some bacteria, such as *Vibrio cholerae*, produce cholera toxin, which can cause severe diarrhoeal disease in humans. Similarly, certain mushrooms and fungi produce mycotoxins that can contaminate water bodies (Mohammad et al., 2018). Similarly, fungal genera such as *Aspergillus* and *Fusarium* produce mycotoxins and contaminate aquatic environments.

Anthropogenic activities such as agricultural runoff, industrial pollution, and wastewater discharges can also be contributing factors in the proliferation of microbial toxins in aquatic environments. These activities can introduce pathogenic bacteria and pollutants that lead to the colonization of microbial communities, some of which can produce toxins that contaminate nearby water bodies (Bukola et al., 2015). Currently, many factors play a significant role in the introduction, translocation, and proliferation of MTs in aquatic ecosystems. Climate change is expected to impact the distribution and abundance of toxin-producing microorganisms in aquatic habitats. Changing temperatures, precipitation patterns, and ocean acidification can promote the growth of specific microorganisms and increase toxin production (Cavicchioli et al., 2019). Shifts in the abundance and distribution of toxin-producing microorganisms are likely to impact the overall ecosystem function and lead to an increased incidence of microbial toxin exposure in aquatic habitats.

Similarly, the emergence of novel toxin-producing microorganisms and toxins is a growing concern in aquatic habitats. Recent outbreaks of novel toxins, such as the MCs from cyanobacteria, have highlighted the potential for new toxins in aquatic habitats (Cheung et al., 2013). Additionally, the increasing use of biocides and other antimicrobial agents to control microbial growth in aquaculture and agriculture may lead to the emergence of antibiotic-resistant microorganisms that produce toxins (Ahmad et al., 2021). Furthermore, fish culture or aquaculture, which provides a critical source of protein for a growing human population, can, however, also contribute to the proliferation of toxin-producing microorganisms. The use of antibiotics and other agricultural practices in aquaculture can lead to the emergence of antibiotic-resistant microorganisms, promoting the production of toxins (Bentzon-Tilia et al., 2016). Additionally, aquaculture facilities can contaminate surrounding aquatic habitats with wastes and other pollutants, leading to an increase in toxin-producing microorganisms.

22.8 IMPACT OF MICROBIAL TOXINS ON AQUATIC ORGANISMS

As already mentioned, microbial toxins in aquatic environments have significant impacts on aquatic organisms and their environment. The impacts of these toxins on aquatic organisms like fish, plankton, and aquatic invertebrates can range from acute to chronic effects, leading to a variety of short-term and long-term damages. Plankton, being the base of the aquatic food chain, play a crucial role in aquatic ecosystems. However, their exposure to microbial toxins can impact their survival and reproduction, leading to a decline in their population (Brown et al., 2020). This, in turn, can affect the survival of higher trophic levels and can lead to imbalances in the ecosystem. For fish, exposure to microbial toxins have shown to cause a wide range of damages, including gill damage, reproductive impairment, and abnormal behaviour. Additionally, high doses of toxins can lead to fish mortality, which can impact ecosystem dynamics and food web interactions (Lopes et al., 2019).

Aquatic invertebrates, such as bivalves, shrimps, and lobsters are also vulnerable to microbial toxins' impacts, which can cause reduced feeding rates, growth impairment, and reduced reproduction rates. These organisms feed on detritus and algae in the environment, and as they accumulate toxins in their tissues, they become prey to predators in the food chain (Turner et al., 2021). The accumulation of toxins in the tissues of aquatic invertebrates can have severe impacts on higher trophic levels, leading to a decline in certain species' population.

Furthermore, some microbes can produce toxic compounds that can accumulate in the shellfish. These harmful compounds can cause significant health problems in humans who consume contaminated shellfish. They can cause neurological effects, stomach ache, and even death in some cases. Moreover, the accumulation of microbial toxins in fish tissue can make them inedible, causing financial loss to both the fishing industry and local communities (Farabegoli et al., 2018).

Overall, microbial toxins in aquatic organisms can cause significant harm, including altered behaviour, abnormal growth, reproductive impairment, and death. These toxins can accumulate in the food chain, leading to severe consequences in the aquatic ecosystem. Therefore, proper monitoring and management of aquatic environments are crucial to detect and prevent the impacts of microbial toxins on aquatic organisms.

22.9 EFFECT OF MICROBIAL TOXINS ON THE PHYSIOLOGICAL AND MOLECULAR MECHANISMS IN AQUATIC ORGANISMS

Microbial toxins in aquatic organisms are found to interfere with the homoeostasis of aquatic organisms and affect their molecular functions, especially by changing the expressions of various anti-oxidative, immunological, DNA damage, apoptotic, and functional transcriptional factors or genes. These changes ultimately affect the physiological functions of the organisms, particularly suppression of immune system and enhanced susceptibility to infections and diseases.

For instance, phycotoxins produced by HABs can have a significant molecular impact on aquatic organisms. STX, a potent neurotoxin produced by dinoflagellates, can cause the downregulation of genes related to nerve conduction channels. Another example is domoic acid, a neurotoxin produced by diatoms, which can cause the upregulation of genes related to oxidative stress, leading to neuronal damage in aquatic invertebrates like mussels (Guillotin et al., 2021; Detree et al., 2016). Bacterial toxins, lipopolysaccharides (LPS), can cause dysregulation of inflammatory genes (tumour necrosis factor [TNF]-α and interleukin [IL]-6), antioxidant genes, and stress genes in fish, leading to inflammation and abnormal systemic immune responses (Li et al., 2020).

Fungal toxins, such as aflatoxins, produced by *Aspergillus* fungi, can have a significant molecular impact on aquatic organisms. Aflatoxins can cause the downregulation of genes related to liver function, leading to liver damage and impaired metabolism in fish (Li et al., 2022). They can also cause oxidative stress in the kidney of aquatic organisms, leading to kidney damage and impaired renal function.

Moreover, microcystins, a class of cyanobacterial toxins, can cause the upregulation of genes related to early lymphoid development, proinflammation, apoptosis, and oxidative stress in other aquatic organisms (Wei et al., 2018; Lin et al., 2021). The toxins cause oxidative stress in the kidney of fish and other aquatic organisms, leading to tubular damage and impaired renal function (Jos et al., 2005). Plasma oxidative stress of fish has also been reported to be caused by the toxins, leading to membrane damage and impaired plasma function. For example, LPS produced by *Vibrio* bacteria can cause oxidative stress in the plasma of fish, leading to membrane damage and impaired plasma function (Sandamalika et al., 2021). Moreover, biotoxins also cause oxidative stress in the muscle of aquatic organisms in terms of structural damage to muscle and impaired muscle function. For instance, mycotoxins produced by fungi can cause oxidative stress in the muscle of fish, leading to impaired muscle antioxidants and lipid peroxidation (Matejova et al., 2017).

22.10 MCS INDUCE IMMUNOTOXICITY IN FISHES

In fishes, the suppression of the protein phosphatases 1 and 2A (PP1 and PP2A) is the conventional harmful mechanism of the MC microbial toxin class. Immunity is regarded as one of the most important physiological processes in fish for preventing infections and maintaining internal homoeostasis in the neuroendocrine-immune network. The concentrations, dosages, times, and methods of administration all affect the immunomodulatory effects of MCs in fish. Previous studies provided substantial proof of the associations between MC-induced immunotoxicity and fish mortality. Inhibition of PP1 and PP2A, oxidative stress, immune cell damage, inflammation, and apoptosis are the main characteristics of immunotoxicity due to MCs. Enhancements in fish immunoreaction towards MCs will aid in improving fish health monitoring and prediction, supporting ecotoxicological goals and ensuring species survival.

According to various research findings, MCs mostly accumulate in the liver and are potently hepatotoxic, although immunological organs like the spleen and head kidney are also prone to MC accumulation. Studies have shown that MCs can have harmful effects on both phagocytes and lymphocytes, leading to immunological disparities in fish. The proliferation of isolated splenic lymphocytes of rainbow trout (*Oncorhynchus mykiss*) showed a substantial decrease in the 40 mg mL^{-1} MC-LR group, while a substantial rise was seen in the 1 mg mL^{-1} MC-LR group (Rymuszka et al., 2007). Similarly, on injecting grass carp (*Ctenopharyngodon idella*) with 50 g kg^{-1} MC-LR, a considerable downregulation of the immune-related genes in the spleen and head kidney was recorded (Wei et al., 2018). Consequently, it has been demonstrated that MCs impair fish immunological function, which might make them more vulnerable to viruses and ultimately cause fish mortality. Theoretically, MCs have the ability to penetrate immunological organs and produce immunotoxicity. The specifics of the MC-induced immunotoxicity in fish are still not well understood, though (Lin et al., 2021).

22.11 BIOACCUMULATION OF MTS IN AQUATIC ORGANISMS

Bioaccumulation and transgenerational impact of microbial toxins in aquatic organisms are concerning issues that have severe consequences for both aquatic organisms and human health. Various aquatic organisms are found to possess the ability to bioaccumulate MTs, as described in the following subsections.

22.11.1 Fish and Shellfish

Fish forms a major source of protein for humans and predatory marine animals. However, some of the toxins produced by HABs can accumulate in fish tissues, leading to bioaccumulation. This can cause different health problems, including gastrointestinal disorders, memory loss, and respiratory distress, in the consumers of such fish. The toxins can also affect the reproductive system of fish and can have transgenerational impacts on the offspring (Barbosa et al., 2019). Similarly, shellfish like oysters, mussels, and clams are filter feeders that can take in microbe-generated toxins along the water currents, which can eventually bioaccumulate in them. The major impact of this phenomenon is shellfish poisoning in humans (Visciano et al., 2016).

22.11.2 Plankton

Plankton, being at the base of the food web, are highly vulnerable to microbial toxins. The toxins can accumulate in them and cause bioaccumulation, leading to deleterious consequences such as reduced feeding rates, growth impairment, and mortality. Furthermore, the transgenerational effects of these toxins can impact plankton populations, which can lead to an imbalance in the ecosystem (Sha et al., 2021). Therefore, bioaccumulation and the resulting transgenerational transfer and impacts of microbial toxins in aquatic organisms are significant concerns. They can cause a range of deleterious impacts, including altered behaviour, abnormal growth, reproductive impairment, and even death, ultimately leading to an imbalance in the ecosystem. Proper monitoring, control, and management of aquatic environments and seafood consumption are necessary to mitigate the impacts of microbial toxins on aquatic organisms.

22.12 DISEASES CAUSED BY MICROBIAL TOXINS IN AQUATIC ORGANISMS

MTs in aquatic organisms can lead to the development of several pathologies and diseases, affecting both the aquatic organisms and human health. HABs can generate toxins that cause numerous diseases in fish, such as fin rot, haemorrhagic septicaemia, and skin lesions (Bernardet, 1998). The accumulation of these toxins in tissues can lead to reduced growth rates, reproductive impairment, and abnormal behaviour. Moreover, some toxins can act as nerve poisons and cause acute or chronic neurological symptoms, including paralysis, muscle tremors, and brain damage. MTs can accumulate in plankton and lead to the development of a range of diseases such as reduced growth rate, altered feeding patterns, and mortality. In extreme cases, toxic algae can cause mass mortality of plankton species, leading to an imbalance in the food web dynamics (Chai et al., 2020).

Shellfish such as shrimps, crabs, mussels, oysters, and other filter feeders are the most critical players in accumulating and translocating the MTs from contaminated water to humans. Shellfish, being filter and detrital feeding in habit, can accumulate toxins produced by phytoplankton, leading to a variety of human illnesses. For instance, some toxins can cause PSP, neurotoxic shellfish poisoning (NSP), and amnesic shellfish poisoning (ASP). PSP can cause tingling, numbness, and respiratory paralysis, while NSP can cause nausea, vomiting, and neurological symptoms. ASP can cause gastrointestinal and neurological problems, leading to memory loss and brain damage (Farabegoli et al., 2018).

22.13 HEALTH-RELATED HAZARDS OF MICROBIAL TOXINS IN HUMANS

Consumption of MT-contaminated aquatic food such as fish, shellfish, and by-products can pose several health hazards to humans. The severity of these hazards is dependent on the type and level of contamination in the food and the frequency of consumption. The following list summarizes the published reports on the health risks of consuming MT-contaminated aquatic food in humans.

a) Neurological Effects: Consumption of fish and shellfish contaminated with certain toxins, such as ciguatoxin and STX, can cause a range of neurological effects in humans, including dizziness, numbness, tingling, and loss of coordination. High doses of these toxins can cause respiratory failure and lead to death. Several cases of ciguatera and STX poisoning have been reported worldwide (Grattan et al., 2016).
b) Hepatotoxicity: Many types of toxins produced by microorganisms can cause liver damage when consumed in sufficient quantities. For example, consumption of shellfish contaminated with hepatotoxic microcystins can cause liver damage, liver cancer, and even death. Several cases of microcystin toxicity have been reported in Asia, Africa, and South America (Greer et al., 2017).
c) Gastrointestinal Illness: Consumption of fish and shellfish contaminated with bacterial pathogens, such as *Vibrio* and *Salmonella*, can cause gastrointestinal illnesses in humans. Symptoms can include diarrhoea, abdominal pain, and vomiting. In severe cases, the toxins can cause septicaemia, which can be fatal. Reports of *Vibrio* and *Salmonella* poisoning through seafood consumption have been documented worldwide (Elbashir et al., 2018).

In conclusion, MT-contaminated aquatic foods pose significant health hazards to humans. It is essential to monitor the levels of microbial toxins in fish, shellfish, and their products to prevent contamination and reduce the risk of negative health outcomes. Additionally, proper storage, handling, and cooking methods can also help minimize the risk of contamination.

22.14 GLOBAL MICROBIAL TOXIN POISONING EVENTS

There have been recent accounts of MT-related diseases in humans caused by consuming contaminated seafood. Here are a few examples:

a. In August 2021, health officials in Florida, United States, issued a warning about the consumption of raw oysters in the state. Several people were infected with vibriosis, a disease caused by the *Vibrio* bacteria found in contaminated oysters. Symptoms of vibriosis include vomiting, diarrhoea, and skin infections (Brumfield et al., 2021).

b. In 2020, an outbreak of ciguatera poisoning occurred in the Florida Keys, caused by consuming contaminated barracuda fish. Ciguatera is caused by a toxin produced by certain types of algae and can lead to neurological symptoms such as tingling, numbness, and muscle weakness (Chinain et al., 2021).
c. In May 2019, the Centers for Disease Control and Prevention (CDC) reported an increase in cases of *Salmonella* infections linked to contaminated crab meat from Venezuela. The outbreak affected people from several states in the United States and caused symptoms such as fever, diarrhoea, and abdominal pain (Seelman et al., 2023).
d. In 2018, an outbreak of Hepatitis A occurred in Hawaii, caused by consuming contaminated scallops. The Hawaii Department of Health issued a warning to consumers and urged restaurants to remove the affected scallops from their menus (Foster, 2021).

These incidents highlight the importance of proper monitoring and control of seafood to prevent the spread of MTs and protect human health. It's essential to follow safe food handling procedures to avoid contamination and cook seafood thoroughly to prevent the spread of illness-causing microbes and toxins.

22.15 FUTURE SCENARIO AND PROSPECTS

Microbial toxins are naturally produced by a variety of microorganisms in aquatic habitats, including freshwater and marine environments, and can have significant ecological and economic impacts in the future. They can alter the composition and diversity of microbial communities and disrupt the food web structure by affecting the metabolism and growth of aquatic organisms. For example, blooms of toxic cyanobacteria can cause extensive fish kills, which can lead to the depletion of fish stocks and loss of biodiversity in the affected habitats (Adams et al., 2020). This effect will likely have severe repercussions on the future availability and sustainability of natural fish catches.

Microbial toxins in aquatic habitats can also have adverse economic impacts on industries that depend on them. Fishery closures due to harmful algal blooms can lead to significant economic losses for the fishing industry, as well as for the tourism and recreational industries that rely on healthy and productive aquatic environments (Sanseverino et al., 2016). Additionally, the cost of monitoring, managing, and mitigating microbial toxin contamination in aquatic habitats is substantial, both locally and globally. Furthermore, microbial toxins in aquatic ecosystems can endanger human health, most notably through the intake of contaminated fish, shellfish, and drinking water. Several types of microbial toxins have been associated with serious human health problems, including hepatotoxicity, neurotoxicity, and allergies, as previously discussed. In future, it may become mandatory to check food and water for microbial toxin contamination before its use in human consumption, fish culture, or other industrial practices. This may lead to a substantial increase in expenditure on public health management and monitoring, industrial applications, and environmental management.

The presence of microbial toxins in aquatic habitats creates opportunities for research and innovation in several areas. These include the development of new tools for detecting and monitoring microbial toxins in aquatic ecosystems, as well as more research into the ecological and genetic factors that influence the creation and distribution of these toxins (Gavrilescu et al., 2015). Also, researchers can try to identify the potential medicinal benefits of these toxins for humans. While microbial toxins in aquatic habitats can create several challenges, including economic and public health impacts, they also present opportunities for research and innovation. It is vital to continue monitoring and managing microbial toxin contamination in aquatic habitats to mitigate adverse ecological and public health effects and to ensure the sustainable use of these essential resources.

22.16 CONCLUSION

Microbial toxins are naturally occurring toxic compounds produced by microorganisms in aquatic environments. They can have serious consequences for both the environment and human health. Toxins are created by microorganisms of numerous sorts, including bacteria, cyanobacteria, algae, and dinoflagellates. These microorganisms can be found in both freshwater and saltwater. The abundance and quantities of microbial toxins in aquatic habitats are affected by a variety of parameters, including nutrient availability, temperature, salinity, and pH. Microbial toxins can have significant ecological and economic impacts on aquatic environments. They can alter the composition and diversity of microbial communities and disrupt the food web structure. HABs can lead to the depletion of fish stocks, loss of biodiversity, and fishery closures. HABs can also cause mass mortality events in aquatic organisms, leading to disruptions in the ecosystem.

Toxins produced by microorganisms can reach aquatic habitats via a variety of routes, including agricultural runoff, wastewater discharge, and atmospheric deposition. They can collect in sediments and bioaccumulate in aquatic organisms such as plankton, fish, and shellfish once in the water. Consuming contaminated seafood or drinking water can also expose humans to microbial toxins. Exposure to microbial toxins in aquatic environments can cause severe health problems in humans, including neurological, gastrointestinal, and hepatic effects. Some toxins are potent neurotoxins and can cause numbness, tingling, and loss of coordination. Certain toxins can cause liver damage, liver cancer, and respiratory failure, including cases of death. Consumption of contaminated seafood or water can potentially result in severe allergic reactions, including anaphylaxis, which can be fatal. To summarize, microbial toxins in aquatic habitats can have far-reaching consequences for the ecosystem and human health. To protect human health and ensure the sustainable use of the essential aquatic resources, it is critical to monitor and regulate the levels of microbial toxins in aquatic ecosystems.

REFERENCES

Adams, J. B., Susan Taljaard, Lara Van Niekerk, and D. A. Lemley. "Nutrient enrichment as a threat to the ecological resilience and health of South African microtidal estuaries." *African Journal of Aquatic Science* 45, no. 1–2 (2020): 23–40.

Ahmad, Iqbal, Hesham A. Malak, and Hussein H. Abulreesh. "Environmental antimicrobial resistance and its drivers: A potential threat to public health." *Journal of Global Antimicrobial Resistance* 27 (2021): 101–111.

An, JiSi, and Wayne W. Carmichael. "Use of a colorimetric protein phosphatase inhibition assay and enzyme linked immunosorbent assay for the study of microcystins and nodularins." *Toxicon* 32, no. 12 (1994): 1495–1507.

Aráoz, Rómulo, Hoàng-Oanh Nghiêm, Rosmarie Rippka, Nicolae Palibroda, Nicole Tandeau de Marsac, and Michael Herdman. "Neurotoxins in axenic oscillatorian cyanobacteria: Coexistence of anatoxin-a and homoanatoxin-a determined by ligand-binding assay and GC/MS." *Microbiology* 151, no. 4 (2005): 1263–1273.

Barbosa, Vera, Marta Santos, Patrícia Anacleto, Ana Luísa Maulvault, Pedro Pousão-Ferreira, Pedro Reis Costa, and António Marques. "Paralytic shellfish toxins and ocean warming: Bioaccumulation and ecotoxicological responses in Juvenile Gilthead Seabream (Sparus aurata)." *Toxins* 11, no. 7 (2019): 408.

Bentzon-Tilia, Mikkel, Eva C. Sonnenschein, and Lone Gram. "Monitoring and managing microbes in aquaculture–Towards a sustainable industry." *Microbial Biotechnology* 9, no. 5 (2016): 576–584.

Bernardet, J.-F. "Cytophaga, Flavobacterium, Flexibacter and Chryseobacterium infections in cultured marine fish." *Fish Pathology* 33, no. 4 (1998): 229–238.

Bergey, David Hendricks. *Bergey's Manual of Determinative Bacteriology*. Lippincott Williams & Wilkins, 1994.

Biscardi, D., A. Castaldo, O. Gualillo, and R. De Fusco. "The occurrence of cytotoxic Aeromonas hydrophila strains in Italian mineral and thermal waters." *Science of the Total Environment* 292, no. 3 (2002): 255–263.

Brooks, W. P., and G. A. Codd. "Immunological and toxicological studies on Microcystis aeruginosa peptide toxin." *British Phycological Journal* 22 (1987): 301.

Brown, Andrew Ross, Martin Lilley, Jamie Shutler, Chris Lowe, Yuri Artioli, Ricardo Torres, Elisa Berdalet, and Charles R. Tyler. "Assessing risks and mitigating impacts of harmful algal blooms on mariculture and marine fisheries." *Reviews in Aquaculture* 12, no. 3 (2020): 1663–1688.

Brumfield, Kyle D., Moiz Usmani, Kristine M. Chen, Mayank Gangwar, Antarpreet S. Jutla, Anwar Huq, and Rita R. Colwell. "Environmental parameters associated with incidence and transmission of pathogenic Vibrio spp." *Environmental Microbiology* 23, no. 12 (2021): 7314–7340.

Bukola, Dawodu, A. Zaid, E. I. Olalekan, and A. Falilu. "Consequences of anthropogenic activities on fish and the aquatic environment." *Poultry, Fisheries & Wildlife Sciences* 3, no. 2 (2015): 1–12.

Campàs, Mònica, Dorota Szydłowska, Marek Trojanowicz, and Jean-Louis Marty. "Enzyme inhibition-based biosensor for the electrochemical detection of microcystins in natural blooms of cyanobacteria." *Talanta* 72, no. 1 (2007): 179–186.

Campos, Alexandre, and Vitor Vasconcelos. "Molecular mechanisms of microcystin toxicity in animal cells." *International Journal of Molecular Sciences* 11, no. 1 (2010): 268–287.

Castenholz, Richard W., Annick Wilmotte, Michael Herdman, Rosmarie Rippka, John B. Waterbury, Isabelle Iteman, and Lucien Hoffmann. "Phylum BX. cyanobacteria." In *Bergey's Manual® of Systematic Bacteriology*, pp. 473–599. New York, NY: Springer, 2001.

Cavicchioli, Ricardo, William J. Ripple, Kenneth N. Timmis, Farooq Azam, Lars R. Bakken, Matthew Baylis, Michael J. Behrenfeld et al. "Scientists' warning to humanity: Microorganisms and climate change." *Nature Reviews Microbiology* 17, no. 9 (2019): 569–586.

Chai, Zhao Yang, Huan Wang, Yunyan Deng, Zhangxi Hu, and Ying Zhong Tang. "Harmful algal blooms significantly reduce the resource use efficiency in a coastal plankton community." *Science of the Total Environment* 704 (2020): 135381.

Cheli, Federica, Luciano Pinotti, Luciana Rossi, and Vittorio Dell'Orto. "Effect of milling procedures on mycotoxin distribution in wheat fractions: A review." *LWT-Food Science and Technology* 54, no. 2 (2013): 307–314.

Cheung, Melissa Y., Song Liang, and Jiyoung Lee. "Toxin-producing cyanobacteria in freshwater: A review of the problems, impact on drinking water safety, and efforts for protecting public health." *Journal of Microbiology* 51 (2013): 1–10.

Chinain, M., C. M. I. Gatti, H. T. Darius, J.-P. Quod, and P. A. Tester. "Ciguatera poisonings: A global review of occurrences and trends." *Harmful Algae* 102 (2021): 101873.

Clement, Alejandro, Lenadro Lincoqueo, Marcela Saldivia, C. G. Brito, Francisca Muñoz, C. Fernández, F. Pérez et al. "Exceptional summer conditions and HABs of Pseudochattonella in Southern Chile create record impacts on salmon farms." *Harmful Algae News* 53 (2016): 1–3.

Cochlan, William P., Vera L. Trainer, Charles G. Trick, Mark L. Wells, Brian D. Bill, and Bich-Thuy L. Eberhart. "Heterosigma akashiwo in the Salish Sea: defining growth and toxicity leading to fish kills." In *Proceedings of the 15th International Conference on Harmful Algae* (2013).

Detree, Camille, Gustavo Núñez-Acuña, Steven Roberts, and Cristian Gallardo-Escarate. "Uncovering the complex transcriptome response of Mytilus chilensis against saxitoxin: Implications of harmful algal blooms on mussel populations." *PloS One* 11, no. 10 (2016): e0165231.

Elbashir, S., S. Parveen, J. Schwarz, T. Rippen, M. Jahncke, and A. DePaola. "Seafood pathogens and information on antimicrobial resistance: A review." *Food Microbiology* 70 (2018): 85–93.

Falconer, Ian Robert. *Cyanobacterial Toxins of Drinking Water Supplies.* CRC Press, 2004.

Farabegoli, Federica, Lucía Blanco, Laura P. Rodríguez, Juan Manuel Vieites, and Ana García Cabado. "Phycotoxins in marine shellfish: Origin, occurrence and effects on humans." *Marine Drugs* 16, no. 6 (2018): 188.

Foster, Monique A. "Hepatitis A." In *Foodborne Infections and Intoxications*, pp. 307–315. Academic Press, 2021.

Fu, Fei-Xue, Yaohong Zhang, Mark E. Warner, Yuanyuan Feng, Jun Sun, and David A. Hutchins. "A comparison of future increased CO2 and temperature effects on sympatric Heterosigma akashiwo and Prorocentrum minimum." *Harmful Algae* 7, no. 1 (2008): 76–90.

Gaget, Virginie, Simonetta Gribaldo, and Nicole Tandeau de Marsac. "An rpoB signature sequence provides unique resolution for the molecular typing of cyanobacteria." *International Journal of Systematic and Evolutionary Microbiology* 61, no. 1 (2011): 170–183.

Gavrilescu, Maria, Kateřina Demnerová, Jens Aamand, Spiros Agathos, and Fabio Fava. "Emerging pollutants in the environment: Present and future challenges in biomonitoring, ecological risks and bioremediation." *New Biotechnology* 32, no. 1 (2015): 147–156.

Grattan, Lynn M., Sailor Holobaugh, and J. Glenn Morris Jr. "Harmful algal blooms and public health." *Harmful Algae* 57 (2016): 2–8.

Greer, Brett, Ronald Maul, Katrina Campbell, and Christopher T. Elliott. "Detection of freshwater cyanotoxins and measurement of masked microcystins in tilapia from Southeast Asian aquaculture farms." *Analytical and Bioanalytical Chemistry* 409 (2017): 4057–4069.

Guillotin, Sophie, and Nicolas Delcourt. "Marine neurotoxins' effects on environmental and human health: An OMICS overview." *Marine Drugs* 20, no. 1 (2021): 18.

Gunthardt, Barbara F., Juliane Hollender, Konrad Hungerbuhler, Martin Scheringer, and Thomas D. Bucheli. "Comprehensive toxic plants–phytotoxins database and its application in assessing aquatic micropollution potential." *Journal of Agricultural and Food Chemistry* 66, no. 29 (2018): 7577–7588.

Häder, Donat-P., Anastazia T. Banaszak, Virginia E. Villafañe, Maite A. Narvarte, Raúl A. González, and E. Walter Helbling. "Anthropogenic pollution of aquatic ecosystems: Emerging problems with global implications." *Science of the Total Environment* 713 (2020): 136586.

Hernández-Cortez, Cecilia, Ingrid Palma-Martínez, Luis Uriel Gonzalez-Avila, Andrea Guerrero-Mandujano, Raúl Colmenero Solís, and Graciela Castro-Escarpulli. "Food poisoning caused by bacteria (food toxins)." *Poisoning: From Specific Toxic Agents to Novel Rapid and Simplified Techniques for Analysis* vol. 33. IntechOpen, London, UK, (2017): IntechOpen, London, UK.

Herrero Galán, Elías, Elisa Álvarez García, Nelson Carreras Sangrà, Javier Lacadena, Jorge Alegre Cebollada, Álvaro Martínez del Pozo, Mercedes Oñaderra, and José G. Gavilanes. *Fungal Ribotoxins: Structure, Function and Evolution.* Proft, T., Ed.; Caister Academic Press: Wymondham, UK, (2009).

Jos, Angeles, Silvia Pichardo, Ana I. Prieto, Guillermo Repetto, Carmen M. Vázquez, Isabel Moreno, and Ana M. Cameán. "Toxic cyanobacterial cells containing microcystins induce oxidative stress in exposed tilapia fish (Oreochromis sp.) under laboratory conditions." *Aquatic Toxicology* 72, no. 3 (2005): 261–271.

Landsberg, Jan H. "The effects of harmful algal blooms on aquatic organisms." *Reviews in fisheries science* 10, no. 2 (2002): 113–390.

Li, Mu-Yang, Wan-Qing Guo, Gui-Liang Guo, Xin-Ming Zhu, Xiao-Tian Niu, Xiao-Feng Shan, Jia-Xin Tian, Gui-Qin Wang, and Dong-Ming Zhang. "Effects of dietary astaxanthin on lipopolysaccharide-induced oxidative stress, immune responses and glucocorticoid receptor (GR)-related gene expression in Channa argus." *Aquaculture* 517 (2020): 734816.

Li, Min, Yidi Kong, Wanqing Guo, Xueqin Wu, Jiawen Zhang, Yingqian Lai, Yuxin Kong, Xiaotian Niu, and Guiqin Wang. "Dietary aflatoxin B1 caused the growth inhibition, and activated oxidative stress and endoplasmic reticulum stress pathway, inducing apoptosis and inflammation in the liver of northern snakehead (Channa argus)." *Science of the Total Environment* 850 (2022): 157997.

Lin, Wang, Tien-Chieh Hung, Tomofumi Kurobe, Yi Wang, and Pinhong Yang. "Microcystin-induced immunotoxicity in fishes: A scoping review." *Toxins* 13, no. 11 (2021): 765.

Lopes, V. M., P. R. Costa, and R. Rosa. "Effects of harmful algal bloom toxins on marine organisms." *Ecotoxicology of Marine Organisms.* Boca Raton, FL: CRC Press, 2019, pp. 42–88.

Marijani, Esther, Emmanuel Kigadye, and Sheila Okoth. "Occurrence of fungi and mycotoxins in fish feeds and their impact on fish health." *International Journal of Microbiology* (2019): 6743065.

Matejova, Iveta, Zdenka Svobodova, Josef Vakula, Jan Mares, and Helena Modra. "Impact of mycotoxins on aquaculture fish species: A review." *Journal of the World Aquaculture Society* 48, no. 2 (2017): 186–200.

Mihali, Troco K., Ralf Kellmann, and Brett A. Neilan. "Characterisation of the paralytic shellfish toxin biosynthesis gene clusters in Anabaena circinalis AWQC131C and Aphanizomenon sp. NH-5." *BMC Biochemistry* 10 (2009): 1–13.

Mohammad, Al-Mamun, Tuhina Chowdhury, Baishakhi Biswas, and Nurul Absar. "Food poisoning and intoxication: A global leading concern for human health." In *Food Safety and Preservation*, pp. 307–352. Academic Press, 2018.

Moreira, Cristiana, António Martins, Joana Azevedo, Marisa Freitas, Ana Regueiras, Micaela Vale, Agostinho Antunes, and Vitor Vasconcelos. "Application of real-time PCR in the assessment of the toxic cyanobacterium Cylindrospermopsis raciborskii abundance and toxicological potential." *Applied Microbiology and Biotechnology* 92 (2011): 189–197.

Namikoshi, Michio, Tomokazu Murakami, Mariyo F. Watanabe, Taiko Oda, Junko Yamada, Shigeo Tsujimura, Hiroshi Nagai, and Shinshi Oishi. "Simultaneous production of homoanatoxin-a, anatoxin-a, and a new non-toxic 4-hydroxyhomoanatoxin-a by the cyanobacterium Raphidiopsis mediterranea Skuja." *Toxicon* 42, no. 5 (2003): 533–538.

Picardo, Massimo, Daria Filatova, Oscar Nunez, and Marinella Farré. "Recent advances in the detection of natural toxins in freshwater environments." *TrAC Trends in Analytical Chemistry* 112 (2019): 75–86.

Quattrocchi, Umberto. *CRC World Dictionary of Medicinal and Poisonous Plants: Common Names, Scientific Names, Eponyms, Synonyms, and Etymology (5 Volume Set).* CRC Press, 2012.

Rocha, Maria João, Catarina Cruzeiro, Mário Reis, Miguel Ângelo Pardal, and Eduardo Rocha. "Pollution by oestrogenic endocrine disruptors and β-sitosterol in a south-western European river (Mira, Portugal)." *Environmental Monitoring and Assessment* 188 (2016): 1–15.

Rymuszka, Anna, Anna Sierosławska, Adam Bownik, and Tadeusz Skowroński. "In vitro effects of pure microcystin-LR on the lymphocyte proliferation in rainbow trout (Oncorhynchus mykiss)." *Fish & Shellfish Immunology* 22, no. 3 (2007): 289–292.

Sandamalika, W. M. Gayashani, Hyukjae Kwon, Chaehyeon Lim, Hyerim Yang, and Jehee Lee. "The possible role of catalase in innate immunity and diminution of cellular oxidative stress: Insights into its molecular characteristics, antioxidant activity, DNA protection, and transcriptional regulation in response to immune stimuli in yellowtail clownfish (Amphiprion clarkii)." *Fish & Shellfish Immunology* 113 (2021): 106–117.

Sano, T., K. Nohara, F. Shiraishi, and K. Kaya. "A method for micro-determination of total microcystin content in waterblooms of cyanobacteria (blue-green algae)." *International Journal of Environmental Analytical Chemistry* 49, no. 3 (1992): 163–170.

Sanseverino, Isabella, Diana Conduto, Luca Pozzoli, Srdan Dobricic, and Teresa Lettieri. "Algal bloom and its economic impact." *European Commission*, Joint Research Centre Institute for Environment and Sustainability (2016).

Seelman, Sharon L., Brooke M. Whitney, Erin K. Stokes, Elisa L. Elliot, Taylor Griswold, Kane Patel, Steven Bloodgood et al. "An outbreak investigation of Vibrio parahaemolyticus infections in the United States linked to crabmeat imported from Venezuela: 2018." *Foodborne Pathogens and Disease* 20, no. 4 (2023): 123–131.

Sha, Jun, Haiyan Xiong, Chengjun Li, Zhiying Lu, Jichao Zhang, Huan Zhong, Wei Zhang, and Bing Yan. "Harmful algal blooms and their eco-environmental indication." *Chemosphere* 274 (2021): 129912.

Sheng, Jian-Wu, Miao He, Han-Chang Shi, and Yi Qian. "A comprehensive immunoassay for the detection of microcystins in waters based on polyclonal antibodies." *Analytica Chimica Acta* 572, no. 2 (2006): 309–315.

Shumway, Sandra E., JoAnn M. Burkholder, and Steven L. Morton, eds. *Harmful algal blooms: a compendium desk reference*. John Wiley & Sons, 2018.

Smayda, Theodore J., David G. Borkman, Gregory Beaugrand, and Andrea Belgrano. "Responses of marine phytoplankton populations to fluctuations in marine climate." In *Marine Ecosystems and Climate Variation: The North Atlantic: A Comparative Perspective*. Oxford University Press, Oxford (2004): 49–58.

Smith, John E., Gerald Solomons, Chris Lewis, and John G. Anderson. "Role of mycotoxins in human and animal nutrition and health." *Natural Toxins* 3, no. 4 (1995): 187–192.

Svendsen, Morten Bo Søndergaard, Nikolaj Reducha Andersen, Per Juel Hansen, and John Fleng Steffensen. "Effects of harmful algal blooms on fish: insights from Prymnesium parvum." *Fishes* 3, no. 1 (2018): 11.

Terao, K., S. Ohmori, K. Igarashi, I. Ohtani, M. F. Watanabe, K. I. Harada, E. Ito, and M. Watanabe. "Electron microscopic studies on experimental poisoning in mice induced by cylindrospermopsin isolated from blue-green alga Umezakia natans." *Toxicon* 32, no. 7 (1994): 833–843.

Turner, Andrew D., Adam M. Lewis, Kirsty Bradley, and Benjamin H. Maskrey. "Marine invertebrate interactions with harmful algal blooms–implications for one health." *Journal of Invertebrate Pathology* 186 (2021): 107555.

Valério, Elisabete, Sandra Chaves, and Rogério Tenreiro. "Diversity and impact of prokaryotic toxins on aquatic environments: A review." *Toxins* 2, no. 10 (2010): 2359–2410.

Vasconcelos, Vítor, Joana Azevedo, Marisa Silva, and Vítor Ramos. "Effects of marine toxins on the reproduction and early stages development of aquatic organisms." *Marine Drugs* 8, no. 1 (2010): 59–79.

Visciano, Pierina, Maria Schirone, Miriam Berti, Anna Milandri, Rosanna Tofalo, and Giovanna Suzzi. "Marine biotoxins: Occurrence, toxicity, regulatory limits and reference methods." *Frontiers in Microbiology* 7 (2016): 1051.

Wei, LiLi, Yi Liu, Shengwei Zhong, Huadong Wu, Jiming Ruan, Mingyue Liu, Qiubai Zhou, and Qiwang Zhong. "Transcriptome analysis of grass carp provides insights into the immune-related genes and pathways in response to MC-LR induction." *Aquaculture* 488 (2018): 207–216.

World Health Organization. *Guidelines for Safe Recreational Water Environments: Coastal and Fresh Waters*. Vol. 1. World Health Organization, 2003.

Yang, Huijuan, Rui Dai, Huiyan Zhang, Chenglong Li, Xiya Zhang, Jianzhong Shen, Kai Wen, and Zhanhui Wang. "Production of monoclonal antibodies with broad specificity and development of an immunoassay for microcystins and nodularin in water." *Analytical and Bioanalytical Chemistry* 408 (2016): 6037–6044.

Zaki, Manal M., S. A. El-Midany, H. M. Shaheen, and Laura Rizzi. "Mycotoxins in animals: Occurrence, effects, prevention and management." *Journal of Toxicology and Environmental Health Sciences* 4, no. 1 (2012): 13–28.

23 Applications of Geospatial Technology in the Mapping of Aquatic Microbes and Risk Assessment

Ganesan Kantharajan, R. Bharathi Rathinam, Arur Anand, Ayyathurai Kathirvelpandian, Ajey Kumar Pathak, and Uttam Kumar Sarkar

23.1 INTRODUCTION

Geospatial technology is an emerging tool that broadly covers remote sensing (RS), geographic information system (GIS), and global positioning system (GPS) which are widely employed for spatiotemporal mapping and monitoring of the environment and public health planning (Anas et al., 2021; Singh et al., 2021). RS is defined as the 'sensing of any object without having any physical contact through sensors or any means.' This process collects the attributes of any objects and is spatially tagged to a particular geographical coordinate using various sensors. GIS, on the other hand, is a computer system that arranges, analyses and visualises the geotagged information on spatial scales to derive meaningful information. The applications of geospatial technology are promising; it is widely applied in various fields of science, including biology, geography and engineering. With the advent of modern tools and algorithms along with open-source RS data, it plays a vital role in fisheries ecology, biodiversity and conservation by facilitating evidence-based scientific management at various spatiotemporal scales (Sarkar et al., 2004; Chakraborty et al., 2022; Krishnan et al., 2022).

Microbes are ubiquitous, reported from diverse habitats across the globe, and their environmental effects are reported to be both beneficial and harmful (Gupta et al., 2017). Understanding their biogeography and dynamics in the environment is essential to harness the maximum benefits from microbial resources and to avoid any negative impact. Aquatic environments harbour a diverse group of microbial populations, which contribute immensely to sustaining the ecosystem structure and functioning (Zinger et al., 2012). Further, the transmission of microbial pathogens through water reportedly causes various waterborne diseases, including diarrhoea and gastrointestinal illness (Pandey et al., 2014). The distribution, diversity and abundance of microbial populations are influenced by many factors, including physico-chemical and nutrient parameters, weather conditions and geography. The spatial integration of such environmental factors, which determine the biogeography of microbial populations, is essential to understand the patterns of diversity for efficient monitoring and management. The geotagging of microbial diversity and their unique physiological characteristics is essential to exploit these resources for bioprospecting and bioremediation applications (Sivasankar et al., 2018).

Applications of GIS and RS in aquatic microbiology have been explored as a tool for microbial risk assessment of drinking water in reservoirs (Kistemann et al., 2001), mapping (Miller et al., 2005), HAB detection (Coffer et al., 2020) and disease forecasting (Anas et al., 2021). The RS and GIS tools

DOI: 10.1201/9781003408543-23

offer huge scope for the assessment and characterisation of the structure and dynamics in both natural and laboratory conditions by overcoming the methodological challenges posed in conventional studies. Further, these tools are capable of studying the microbial ecology of remote locations in polar regions and vast oceans at a broader scale (Capone and Subramaniam, 2005). The growth and dynamics of microbes depend on various physico-chemical and nutrient parameters, which vary spatially from one location to another. Spatial integration of all these parameters is possible with GIS, which enables the visualisation and interpretation of the complex dataset for better decision-making by deriving meaningful conclusions. Further, the microbial population in the natural aquatic environment changes very rapidly in association with these parameters. The overall applications of geospatial technology, i.e., RS, GPS and GIS, are depicted in Figure 23.1. The temporal coverage of the RS dataset ensures the periodical monitoring and assessment of microbial dynamics in aquatic ecosystems. The spatial analysis, data integration and network empower the rapid detection of blooms in inland and coastal waterbodies. The spatial network analysis of GIS advances our understanding of the environmental microbiology of fish pathogens and seafood-borne microbes to predict disease outbreaks and contamination. The spatial output generated from the geospatial tools disseminates information on microbial biogeography and helps in ecosystem health management. This chapter deals with the applications of geospatial technology in the mapping of microbes and risk assessment in various aquatic ecosystems.

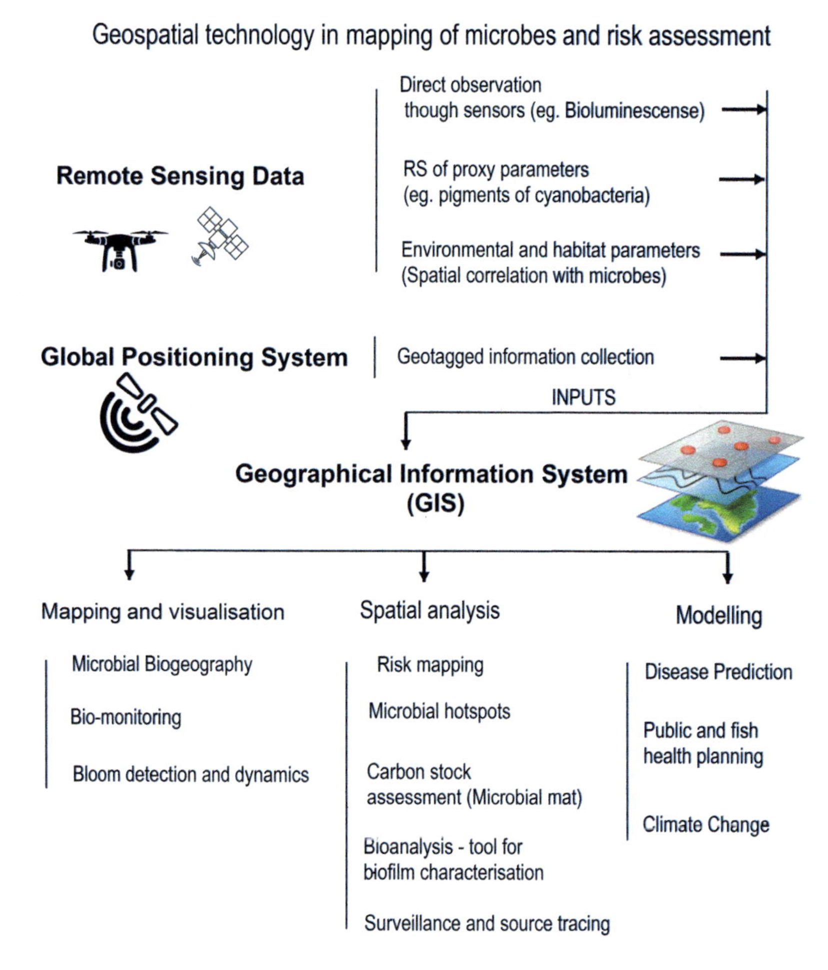

FIGURE 23.1 Applications of geospatial technology in mapping of microbes and risk assessment in an aquatic ecosystem.

23.2 MICROBIAL BIOGEOGRAPHY

The microbial composition of an ecosystem regulates autotrophic and heterotrophic production and nutrient cycling. Understanding the distribution and abundance of the microbes is very important to sustain the ecosystem structure and functions and for the identification of novel pharmaceuticals (Martiny et al., 2006). Biogeography is the study of the distribution of organisms over space and time and assesses the influence of ecological and evolutionary processes on their distribution (Hanson et al., 2012). GIS tools are widely applied for studying the distribution pattern of biological organisms, including microbes, across regions, which sheds light on their diversity and ecosystem processes. Geotagging of microbiological data is essential for understanding and visualising the spatiotemporal patterns of the microbial richness of an ecosystem, and it allows us to visualise the existence of biogeographic patterns of microbes at a large spatial scale. Sivasankar et al. (2018) used the GIS tool to map and catalogue the data of the cultivable actinobacterial diversity and their distribution pattern in the Polar Frontal Region of the Southern Ocean in Antarctica in various gradients of the environment, which was expected to aid researchers working in the field of the microbial biogeography of Antarctica in the future.

The basin-wise GIS mapping of microbial diversity of the Ganga River in India for Ecosystem Services is being undertaken by the Council of Scientific and Industrial Research (CSIR) under the National Mission for Clean Ganga. The project aims to reveal the native microbial inhabitants and their role in maintaining the health of the riverine ecosystem and the influence of anthropogenic activities on microbial diversity (www.csir.res.in/slider/neeri-completes-first-ever-gis-based-mapping-ganga). This study covers 21 locations along the Ganga River and is intended to generate baseline data on the influence of microbial diversity on ecosystem services, river health, antibiotic resistance threats, source identification of *Escherichia coli*, and identification of sewage and industrial waste contamination indicators (Kapley, 2022).

23.3 MAPPING AND BIOMONITORING OF MICROBES

The understanding of the spatial distribution of microbes is vital to decipher their role in the environment. Remote sensing (RS) sensors-based satellite images are currently used for mapping the microbial bloom in a waterbody (Bhattacharjee et al., 2021). The occurrence mapping of microbes in the ecosystem can be done using RS-based sensors through three different approaches (Grimes et al., 2014): (i) direct observation, in which the identification is based on the unique optical properties of the microbes and bioluminescence as observed from the imagery; (ii) indirect method, where the quantification is based on physical and physiological characteristics, e.g., chlorophyll or phycoerythrin in cyanobacteria; and (iii) inference or model-based method, exploiting the correlation between the microbial density and particle concentration, which is in turn measured through RS data in GIS platforms. The changes occurring in the aquatic ecosystem due to natural environmental and anthropogenic factors can be measured through the responses of the biological organisms, which is called biomonitoring (Guerrero-Aguilar et al., 2022). GIS is a promising tool for biodiversity monitoring in the ecosystem, which accommodates both spatial and non-spatial data, used for monitoring studies. The threshold value, fixed in the reflectance value of remotely sensed imagery and the spectral indices generated, allows monitoring of adverse events over space and time.

23.3.1 MARINE ECOSYSTEMS

Microbes in marine and coastal environments play a vital role in structuring the marine ecosystem. Mapping and detection of *Vibrio cholerae*, *V. parahaemolyticus* and *V. vulnificus* in the coastal water is based on their correlation with environmental parameters like sea surface temperature (SST), chlorophyll, rainfall, suspended particulate matter, dissolved organic carbon, and salinity, which are obtained through RS methods (Grimes et al., 2014). Miller et al. (2005) reported the first satellite observation of bioluminescence produced by bacterial colonies along with microalgal bloom in the surface waters in an area of ≈15,400 km^2 of the north-western Indian Ocean using unprocessed Operational Linescan System (OLS) nighttime visible imagery from the U.S. Defense Meteorological Satellite Program constellation of satellites.

Trichodesmium is a bloom-forming cyanobacterium distributed in the ocean, and the formation of bloom indicates nitrogen limitation in the surface water. The ocean colour sensor-based images were used for the detection of *Trichodesmium* on the synoptic scale, and the factors linked with the occurrences/formation in the ocean were evaluated, including SST, photosynthetically active radiation, surface wind, sea surface currents, mean sea level anomaly, wave height, mixed layer depth, sea surface salinity and chlorophyll-a content using RS datasets available online (Jyothibabu et al., 2017).

23.3.2 Polar Ecosystems

Cyanobacterial mat communities are considered sentinel organisms, which respond rapidly to climate changes. The study by Levy et al. (2020) demonstrated the application of hyperspectral imageries, captured using an Unmanned Aerial Vehicle (UAV), for assessing the microbial mat presence, type and activity in McMurdo Dry Valleys in Antarctica and monitoring the extent of microbial mat communities for early warning of ecosystem change. In another study, the Normalized Difference Vegetation Index (NDVI), developed for McMurdo Dry Valleys, Antarctica from WorldView-2 multispectral satellite images, revealed a strong correlation with microbial mat cover, biomass on ground and chlorophyll-a content. This shows the potential of geospatial tools in occurrence mapping and biological profiling. The estimated carbon stock of the microbial mats in the Canada Glacier Antarctic Specially Protected Area was 21,715 kg (Power et al., 2020), which implies their role in climate change mitigation studies.

23.3.3 Freshwater Ecosystems

HABs due to cyanobacteria (CyanoHABs) are the most prominent HABs that occur in freshwater ecosystems across the globe, which impact the aquatic biota by producing toxins and the formation of dead zones. Further, they also cause deterioration in the quality of drinking water. RS is proven to be an efficient technology to detect and monitor cyanobacterial blooms in freshwater ecosystems, including lakes and reservoirs. The vast spatial coverage and temporal frequency allow the resource managers to monitor the broad spatial extent through repeated monitoring. A study carried out by Coffer quantified the occurrence of cyanobacterial bloom in lakes of the United States weekly from 2008 to 2011, 2017 and 2018, using the Medium Resolution Imaging Spectrometer (MERIS) of Envisat and Ocean and Land Color Instrument (OLCI) of Sentinel-3, which had a temporal resolution of 2–3 days and spatial resolution of 300 m at nadir (Coffer et al., 2020). In another study, Mishra et al. (2021) developed the Cyanobacteria Index (CIcyano) using the MERIS and OLCI sensor RS data supplemented with Microcystins data measured in the field to validate the effectiveness of bloom detection in lakes of the United States over 11 bloom seasons. The overall precision of 84% obtained using this method proves its utility for synoptic and repetitive monitoring of toxic cyanobacterial blooms.

23.4 MICROBIAL RISK ASSESSMENT AND PUBLIC HEALTH PLANNING

The integration of data on various aspects is essential to understand the factors influencing microbial communities and their activities. For instance, the microbial diversity of a waterbody is closely associated with the ongoing activities in its catchment. The microbes sourced from human settlements and the animal and agriculture sector entering any waterbody impact the native microbial community in such an ecosystem. Further, the influx of harmful microbes into a waterbody poses a risk to biodiversity and resource users. The microbial community structure and mechanisms of a freshwater lake are mainly attributed to the catchment-based anthropogenic activities, especially nutrient influx and inflow of other substances (Obieze et al., 2022). GIS and RS tools offer wide scope for modelling and mapping the water flow and nutrient and sediment export from the catchment. Integration of microbial occurrence with the above-mentioned dataset will help in understanding their source distribution, which is essential to implement management measures. Microbial risk mapping of bacterial and parasitic contamination in reservoirs in Northrhine-Westfalia (Germany) was done through comprehensive geo-ecological characterisation of their catchment areas for tracing the origin of microbial loads to reduce diffuse and point pollution originating from human and animal wastes (Kistemann et al., 2001).

The spatial correlation of microbes' distribution with water quality and nutrient parameters enables us to understand the factors influencing their occurrence and proliferation in a waterbody. For instance, the indirect approach, which exploits the correlation between the annual cycle of SST and cholera cases and between sea surface height (SSH) and the spread of plankton in the Bay of Bengal predicts the incidence of waterborne diseases like cholera and other *Vibrio*-caused diseases in border countries and helps in disease preparedness and management (Capone and Subramaniam, 2005; Grimes et al., 2014). The relationship between the satellite image-derived phytoplankton biomass (chlorophyll as a proxy) and the occurrence of *V. cholerae* was explored in Vembanad Lake, a tropical lake-estuarine system on the west coast of India, for developing a risk map of cholera outbreaks (Anas et al., 2021). Likewise, the spatial integration of occurrence and distribution data of *V. parahaemolyticus,* which was linked with large outbreaks in Rias of Galicia, Spain, was evaluated with environmental and oceanographic parameters through the GIS tool. The study results revealed that the spatiotemporal distribution is closely linked with salinity, followed by temperature, which moderates its abundance (Martinez-Urtaza et al., 2008).

Geospatial tools are employed in identifying the targets and predicting the spatiotemporal patterns of vector-borne diseases in the environment, enabling policymakers to implement prevention and management measures. Applications of GIS in food safety and public health sector have proven to be an effective tool, in particular for tracing the sources of the pathogens associated with seafood. The prediction of infection sources along with the seasonal pattern of seafood-related health risks in a given area is possible through GIS. For instance, the norovirus outbreak reported in 2016 in the Gulf of Mexico Coast regions of the United States was found to be linked with the consumption of raw oysters by employing prediction models through spatial integration of epidemiological data, oyster collection sites and their transportation route (Wang and Deng, 2016). Likewise, GIS tools have been employed to track outbreaks and identify the potential sources of contamination for public health incidences associated with *V. parahaemolyticus* infections linked with raw seafood consumption in Asia.

23.5 MAPPING OF ANTI-MICROBIAL RESISTANCE AND SURVEILLANCE

The GIS approach is used for integrating biotic and abiotic variables at the landscape level to derive meaningful inferences at the spatial scales. Resistance Map is an interactive spatial data visualisation tool that comprises the data collection of anti-microbial resistance (AMR) and antibiotic use for various countries across the globe. At present, this dataset visualises the distribution of resistance and antibiotic consumption data from 46 and 76 countries, respectively (CDDEP, 2018). Incidence or occurrence data of AMR only reveal the distribution pattern over the spatial scale and prepare the cluster-based hotspot maps. However, linking these datasets with other variables, including anthropogenic and climate variables, sheds more light on tracking its sources and transmission paths. The field data and their attributes are geotagged, collated and arranged in a spatial database to explore the spatial and environmental dimensions. Akinduti et al. (2022) used the geotagged data of *E. coli* O157 strains and interpolated them for analysis in the GIS platform concerning administrative boundaries, which revealed the spatial spread of the high-level resistance encoded strains of resistant *E. coli* O157 within and around the urban area in southwest Nigeria. Further, the GIS tool can be used as an initial screening tool to design the spatial sampling framework, which is economical and an alternative to regular sampling campaigns. Additionally, the extrapolation of AMR observed at the local scale to a larger area is possible through RS and GIS. The GIS tool allows the user to overlay various complex data layers, including agriculture (livestock density), demographics, groundwater wells and monitoring stations, healthcare (hospitals and anti-microbial consumption), hydrology of inland waters, coastal aquaculture activities (farms), water and wastewater facilities, and solid waste management facilities to assess the potential sources, transmission pathways and impacts on local populations (Chique et al., 2019).

23.6 GIS IN AQUATIC ANIMAL DISEASE MANAGEMENT

Aquatic animal diseases pose a significant threat to the aquaculture industry and can cause substantial economic losses to farmers. Rapid and efficient disease surveillance strategies are essential for the early detection of disease outbreaks that would help us to control and prevent the spread of several pathogens. In recent times, GIS has gained significance in the field of animal disease surveillance.

23.6.1 Mapping and Outbreak Investigation

A primary application of GIS is fish disease mapping by using the available disease occurrence data on different spatiotemporal scales. The information on disease spread patterns, risk zones and vulnerable areas at a spatial scale is easy to visualise and can be communicated to various stakeholders to alert them of any disease outbreak and help the policy managers in making swift decisions. Many of the environmental parameters, such as salinity, temperature, nutrient level and dissolved oxygen, influence the dynamics of disease outbreaks in the culture system. Further, the correlation between the disease occurrence and these parameters can be made possible through a variety of spatial analyses in the GIS platform. The existing information on disease occurrence and integration of closely associated causes, i.e., water source, migratory routes of wild fish, etc., aid in understanding the actual spreading pattern of pathogens to different farms in the region. The geospatial tools are used to map the disease transmission routes, which are essential to understand the pattern of disease spread.

GIS aids in the integration of geotagged information of the culture systems, i.e., water and soil quality parameters, feeding rate, and growth parameters, at various stages of disease occurrence/spread. This data integration gives an inclusive view of the culture environment, assists us to identify the risk factors associated with disease outbreaks for the chosen location, and also aids in the early detection of a disease outbreak through regular monitoring of real-time data associated with the culture system. This will facilitate the timely containment of the disease outbreak.

23.6.2 Prediction and Surveillance

The early detection of emerging aquatic animal diseases and the knowledge base available on existing diseases is vital for developing control measures and strategies for an efficient aquatic disease management process (Sood et al., 2021). The accumulation of geotagged information on disease occurrence and culture conditions through a citizen science approach by employing the stakeholders (passive disease surveillance through fish farmers [Sood et al., 2021]) is a viable option to generate voluminous data. Further, this can be used for mapping the potential fish disease hotspots and also to predict the disease occurrence and spread in any geographical area by employing artificial intelligence (AI) tools. The integration of data from human infections associated with aquatic animals can allow us to identify and predict the potential hotspots of fish-borne zoonotic diseases. The GIS-based spatial risk assessment of aquaculture farms for various aquatic animal diseases helps in mitigating the risk of fish diseases to avert economic losses.

Apart from the above-mentioned applications, GIS can also play a major role in data acquisition, sharing and collaboration with farmers, researchers, industries and government agencies. The application of GIS in aquatic animal disease surveillance will transform the existing disease management approach, which would help finfish and shrimp farmers to reduce the economic losses associated with diseases. The establishment of a spatial decision support system (SDSS) by integrating geotagged field data and output of models through network communication, query, statistics and spatial analysis helps the policymakers in the formulation of suitable guidelines and management measures. Further, dissemination of information through web-based GIS portals on a real-time basis allows stakeholders and resource managers to visualise disease occurrence and spread, thus facilitating surveillance.

23.7 MICROBIAL DYNAMICS ASSESSMENT

Geospatial technology enables us to characterise aquatic waterbodies at various spatial scales on a regular seasonal interval basis; this led to deciphering the pattern and dynamics of primary producers by observing the ocean colour, which helps us to understand the microbes in the global carbon cycle. Further, RS datasets of various sensors, available in the public domain, allow us to assess the dynamics of primary production of microalgae, including cyanobacteria, in the vast open ocean at regular intervals for estimating their ecological and biogeochemical significance. Primary productivity in the ocean is also closely associated with other parameters, like particulate inorganic carbon, total suspended solids, particulate organic carbon and coloured dissolved organic matter, which correlates with bacterial activity. Hence, algorithms based on these parameters for estimating the concentrations describe the microbial dynamics in the ecosystem (Capone and Subramaniam, 2005).

Applications of geospatial tools for characterising the microbial dynamics in seagrass infected with wasting disease outbreaks can be tested through variation in spectral indices, including NDVI, obtained from UAV-based high-resolution images and geotagging of in-situ parameters like microbiomes and root metabolomes, and their correlation with RS-based environmental variables help in characterising the interactive effects of multiple stressors on pathogenic microbial dynamics in seagrass (Beatty et al., 2021).

Bhattacharjee et al. (2021), using satellite images, observed the sudden occurrence of carotenoid pigmentation in the crater-based saline lake called 'Lonar Lake' in Maharashtra due to *Haloarchaea* microbes. The dynamics of the bloom associated with these bacteria and possible causes of this colour change were studied by employing various spectral indices, i.e., Salinity Index, Surface Algal Bloom Index, and Normalized Difference *Haloarchaea* Index, during January–June 2020. The precise mapping of occurrence and dynamics was possible due to the applications of RS datasets. The view of carotenoid pigmentation in Lonar Lake, as observed in two different images acquired on 25 May 2020 and 20 June 2020 with an Operational Land Imager (OLI) sensor by Landsat 8, is depicted in Figure 23.2.

Climate change affects temperature, rainfall patterns, formation of dead zones, changes in ocean water current and acidification, which impacts all life on earth. Microbes play a predominant role in global biogeochemical cycles and depend on temperature, which adversely impacts the microbial community. Global warming is expected to cause the spread of a potent pathogen, *V. harveyi*, in the marine microbiota (Abirami et al., 2021). Hence, understanding the diverse weather factors, i.e., temperature, humidity, precipitation, wind, etc., is essential to predict the future impact of climate change on spatiotemporal scales. GIS is a vital tool for the documentation and visualisation of dynamic weather parameters and for assessing their correlation with microbiota in the aquatic ecosystem.

Evaluation and quantification of biofilm image structure, obtained through scanning confocal laser microscopy with fluorescent lectin probes, has been conducted by exploring geospatial tools such as ERDAS Imagine and GIS software. The complete characterisation of the biofilm section

FIGURE 23.2 View of carotenoid pigmentation due to *Haloarchaea* microbes in Lonar Lake, Maharashtra, India, as observed by Landsat 8. (From https://earthobservatory.nasa.gov/images (Hansen, 2020).)

image was performed after the image enhancement and pre-processing. The quantification of bacterial cells in the biofilm was assessed through supervised classification of the image. This shows the potential of GIS in quantification and variability assessment within biofilms related to their formation, structure and growth dynamics (Petrisor et al., 2003). Further, a comparison between the GIS-based approach and the existing computer program-based biofilm structure assessment tool, COMSTAT, revealed that the GIS-based approach generates consistent results close to theoretical values in estimating the biovolume of *Pseudomonas aeruginosa* biofilm, though some limitations were reported in this process (Petrisor et al., 2004).

23.8 CONCLUSION

Microbes are a vital component of any ecosystem, involved in nutrient cycling and energy transfer. Understanding the dynamics and biogeography of microbial populations is essential to harness the potential of such organisms and avoid any harmful effects on human communities. Geospatial technology covers GIS and RS, which are being widely used in various arenas of environmental science for monitoring and assessment. It has immense potential in public health management by detecting HABs and microbial pathogens in inland and coastal waterbodies. The role of GIS in data collection through mobile applications and geotagging of other biological and microbial information from aquaculture and wild environment has huge scope in disease surveillance, environmental monitoring, and assessment and prediction modelling at a spatial scale. The spatial integration of field-collected data and laboratory analysis results, derived from advanced microbial tools through geospatial technology, enables us to understand and visualise the complex datasets in spatial perspectives and derive meaningful conclusions for aquatic environment management.

REFERENCES

Abirami, Baskaran, Manikkam Radhakrishnan, Subramanian Kumaran, and Aruni Wilson. "Impacts of global warming on marine microbial communities." *Science of the Total Environment* 791 (2021): 147905.

Akinduti, Akinniyi Paul, Oluwafunmilayo Ayodele, Babatunde Olanrewaju Motayo, Yemisi Dorcas Obafemi, Patrick Omoregie Isibor, and Olubukola Wuraola Aboderin. "Cluster analysis and geospatial mapping of antibiotic resistant Escherichia coli O157 in southwest Nigerian communities." *One Health* 15 (2022): 100447.

Anas, Abdulaziz, Kiran Krishna, Syamkumar Vijayakumar, Grinson George, Nandini Menon, Gemma Kulk, Jasmin Chekidhenkuzhiyil et al. "Dynamics of Vibrio cholerae in a typical tropical lake and estuarine system: Potential of remote sensing for risk mapping." *Remote Sensing* 13, no. 5 (2021): 1034.

Beatty, Deanna S., Lillian R. Aoki, Olivia J. Graham, and Bo Yang. "The future is big—And small: Remote sensing enables cross-scale comparisons of microbiome dynamics and ecological consequences." *Msystems* 6, no. 6 (2021): e01106–21.

Bhattacharjee, Rajarshi, Abhinandan Choubey, Nilendu Das, Anurag Ohri, and Shishir Gaur. "Detecting the carotenoid pigmentation due to haloarchaea microbes in the Lonar lake, Maharashtra, India using Sentinel-2 images." *Journal of the Indian Society of Remote Sensing* 49 (2021): 305–316.

Capone, Douglas G., and Ajit Subramaniam. "Seeing microbes from space." *Asm News* 71, no. 4 (2005): 179–186.

Center for Disease Dynamics, Economics & Policy. "ResistanceMap: Antibiotic resistance." *The Center for Disease Dynamics Economics & Policy* (2018). https://resistancemap.OneHealthTrust.org (Accessed 2 August 2023).

Chakraborty, H., T. Kayal, L. Lianthuamluaia, U. K. Sarkar, A. K. Das, S. Chakraborty, B. K. Sahoo, K. Mondal, S. Mandal, and B. K. Das. "Use of geographical information systems (GIS) in assessing ecological profile, fish community structure and production of a large reservoir of Himachal Pradesh." *Environmental Monitoring and Assessment* 194, no. 9 (2022): 643.

Chique, Carlos, John Cullinan, Brigid Hooban, and Dearbhaile Morris. "Mapping and analysing potential sources and transmission routes of antimicrobial resistant organisms in the environment using geographic information systems—An exploratory study." *Antibiotics* 8, no. 1 (2019): 16.

Coffer, Megan M., Blake A. Schaeffer, John A. Darling, Erin A. Urquhart, and Wilson B. Salls. "Quantifying national and regional cyanobacterial occurrence in US lakes using satellite remote sensing." *Ecological Indicators* 111 (2020): 105976.

Grimes, D. Jay, Tim E. Ford, Rita R. Colwell, Craig Baker-Austin, Jaime Martinez-Urtaza, Ajit Subramaniam, and Douglas G. Capone. "Viewing marine bacteria, their activity and response to environmental drivers from orbit: Satellite remote sensing of bacteria." *Microbial Ecology* 67 (2014): 489–500.

Guerrero-Aguilar, Armando, Ulises Emiliano Rodriguez-Castrejón, Alma Hortensia Serafín-Muñoz, Christoph Schüth, and Berenice Noriega-Luna. "Bioindicators and biomonitoring: Review of methodologies applied in water bodies and use during the Covid-19 pandemic." *Acta Universitaria* 32 (2022): e3388.

Gupta, Ankit, Rasna Gupta, and Ram Lakhan Singh. "Microbes and environment." In *Principles and Applications of Environmental Biotechnology for a Sustainable Future*, edited by Ram, Lakhan Singh, Springer Singapore, pp. 43–84, 2017.

Hansen, Kathryn. "Lonar lake tries on a Rosy Color." NASA Earth Observatory (2020). https://earthobservatory.nasa.gov/images/146859/lonar-lake-tries-on-a-rosy-color (Accessed 9 August 2023).

Hanson, China A., Jed A. Fuhrman, M. Claire Horner-Devine, and Jennifer B. H. Martiny. "Beyond biogeographic patterns: Processes shaping the microbial landscape." *Nature Reviews Microbiology* 10, no. 7 (2012): 497–506.

https://www.csir.res.in/slider/neeri-completes-first-ever-gis-based-mapping-ganga (Accessed 2 August 2023).

Jyothibabu, R., C. Karnan, L. Jagadeesan, N. Arunpandi, R. S. Pandiarajan, K. R. Muraleedharan, and K. K. Balachandran. "Trichodesmium blooms and warm-core ocean surface features in the Arabian Sea and the Bay of Bengal." *Marine Pollution Bulletin* 121, no. 1–2 (2017): 201–215.

Kapley, Atya. "Mapping microbial diversity of Holy Ganga." India Science Wire (2022). https://registration.geospatialworld.net/index/media/2022/geosmartindia/presentation/Mapping-Microbial-Diversity-of-Holy-Ganga-Dr_Atya_Kapley.pdf.

Kistemann, Thomas, Friederike Dangendorf, and Martin Exner. "A Geographical Information System (GIS) as a tool for microbial risk assessment in catchment areas of drinking water reservoirs." *International Journal of Hygiene and Environmental Health* 203, no. 3 (2001): 225–233.

Krishnan, P., G. Kantharajan, Rejani Chandran, A. Anand, and Vindhya Mohindra. "Geo-spatial tools for science-based management of inland aquatic habitats and conservation of fish genetic resources." *Indian Journal of Plant Genetic Resources* 35, no. 3 (2022): 312–316.

Levy, Joseph, S. Craig Cary, Kurt Joy, and Charles K. Lee. "Detection and community-level identification of microbial mats in the McMurdo Dry Valleys using drone-based hyperspectral reflectance imaging." *Antarctic Science* 32, no. 5 (2020): 367–381.

Martinez-Urtaza, Jaime, Antonio Lozano-Leon, Jose Varela-Pet, Joaquin Trinanes, Yolanda Pazos, and Oscar Garcia-Martin. "Environmental determinants of the occurrence and distribution of Vibrio parahaemolyticus in the rias of Galicia, Spain." *Applied and Environmental Microbiology* 74, no. 1 (2008): 265–274.

Martiny, Jennifer B. Hughes, Brendan J. M. Bohannan, James H. Brown, Robert K. Colwell, Jed A. Fuhrman, Jessica L. Green, M. Claire Horner-Devine et al. "Microbial biogeography: Putting microorganisms on the map." *Nature Reviews Microbiology* 4, no. 2 (2006): 102–112.

Miller, Steven D., Steven H. D. Haddock, Christopher D. Elvidge, and Thomas F. Lee. "Detection of a bioluminescent milky sea from space." *Proceedings of the National Academy of Sciences* 102, no. 40 (2005): 14181–14184.

Mishra, Sachidananda, Richard P. Stumpf, Blake Schaeffer, P. Jeremy Werdell, Keith A. Loftin, and Andrew Meredith. "Evaluation of a satellite-based cyanobacteria bloom detection algorithm using field-measured microcystin data." *Science of the Total Environment* 774 (2021): 145462.

Obieze, Chinedu C., Gowher A. Wani, Manzoor A. Shah, Zafar A. Reshi, André M. Comeau, and Damase P. Khasa. "Anthropogenic activities and geographic locations regulate microbial diversity, community assembly and species sorting in Canadian and Indian freshwater lakes." *Science of the Total Environment* 826 (2022): 154292.

Pandey, Pramod K., Philip H. Kass, Michelle L. Soupir, Sagor Biswas, and Vijay P. Singh. "Contamination of water resources by pathogenic bacteria." *Amb Express* 4 (2014): 1–16.

Petrisor, Alexandru I., Adrian Cuc, and Alan W. Decho. "Reconstruction and computation of microscale biovolumes using geographical information systems: Potential difficulties." *Research in Microbiology* 155, no. 6 (2004): 447–454.

Petrisor, Alexandru-Ionut, Tomohiro Kawaguchi, and A. W. Decho. "An introduction to bacterial geography." In *Proceedings of the 27th Annual Congress of the American-Romanian Academy of Arts and Sciences*, vol. 1, pp. 158–162. Polytechnic International Press, 2003.

Power, Sarah N., Mark R. Salvatore, Eric R. Sokol, Lee F. Stanish, and J. E. Barrett. "Estimating microbial mat biomass in the McMurdo Dry Valleys, Antarctica using satellite imagery and ground surveys." *Polar Biology* 43, no. 11 (2020): 1753–1767.

Sarkar, Uttam K., Ajey K. Pathak, Dhurendra Kapoor, Samir K. Paul, and Lakhan L. Mahato. "Use of geographical information systems in developing freshwater aquatic sanctuary management strategies." *GIS/spatial Analysis in Fishery and Aquatic Sciences* 2 (2004).

Singh, Ram Kumar, Pavan Kumar, Semonti Mukherjee, Swati Suman, Varsha Pandey, and Prashant K. Srivastava. "Application of geospatial technology in agricultural water management." In *Agricultural Water Management*, pp. 31–45. Academic Press, 2021.

Sivasankar, P., K. Priyanka, Bhagwan Rekadwad, K. Sivakumar, T. Thangaradjou, S. Poongodi, R. Manimurali, P. V. Bhaskar, and N. Anilkumar. "Actinobacterial community structure in the Polar Frontal waters of the Southern Ocean of the Antarctica using Geographic Information System (GIS): A novel approach to study Ocean Microbiome." *Data in Brief* 17 (2018): 1307–1313.

Sood, Neeraj, Pravata K. Pradhan, T. Raja Swaminathan, Gaurav Rathore, J. K. Jena, and Kuldeep K. Lal. "National surveillance programme for aquatic animal diseases–A stepping stone for establishing disease governance system in India." *Current Science* 120, no. 2 (2021): 273.

Wang, Jiao, and Zhiqiang Deng. "Modeling and prediction of oyster norovirus outbreaks along Gulf of Mexico coast." *Environmental Health Perspectives* 124, no. 5 (2016): 627–633.

Zinger, Lucie, Angelique Gobet, and Thomas Pommier. "Two decades of describing the unseen majority of aquatic microbial diversity." *Molecular Ecology* 21, no. 8 (2012): 1878–1896.

24 Statistical Aspects of Aquatic Microbiology

V. Ramasubramanian and H. Sanath Kumar

24.1 INTRODUCTION

Statistics is a branch of science (read mathematics) that deals with the collection, analysis and interpretation of data. In this chapter, the applications of statistical tools and techniques specific to the field of microbiology in general and aquatic microbiology in particular will be discussed. Having said that, the methods can generally be applied to other related fields as well in a straightforward manner. As microbiological experiments typically yield data on microorganisms, the magnitude of datapoints will be, more often than not, large when the counts of such organisms are noted down (even though, as such, the organisms will be in micro-sizes). Such sizes and counts parlance reminds us of some of the quizzical quotes that have by now become clichés – "Everything that can be counted does not necessarily count while everything that counts cannot necessarily be counted/ Not everything that counts can be counted, and not everything that can be counted counts" (William Bruce Cameron); "Size doesn't matter, fast data is better than big data" (Hilary Mason) – emphasizing the importance of available measurable count data, even though the number of such observations (sample size) may sometimes be small. As the data values are large, they are often transformed using logarithms (typically to the base 10 or to the base as exponential constant e in the case of natural logarithms) before proceeding to analysis, as the statistical analysis or model fitting that ensues requires the data to have constant variance, a condition that can be restored by this logarithmic (variance stabilizing) transformation.

24.2 STATISTICAL METHODS APPLIED IN MICROBIOLOGY

Despite the importance of statistical applications in microbiology, very few review articles exist on this subject. Ilstrup (1990) has discussed statistical methods in microbiology with an introduction to the types of variables, both in terms of the scales of measurement and also from the viewpoint of their dependent or independent nature, hypothesis testing and certain diagnostic tests. Statistical tools are indispensable in microbial ecology studies. Paliy and Shankar (2016) elaborately discussed the application of multivariate statistical techniques in microbial ecology. They described and compared the most widely used multivariate statistical techniques, including exploratory, interpretive and discriminatory procedures, also considering several important limitations and assumptions of these methods while considering their applications to the ecology of the microbial world. Several statistical tools have been developed that aid in ecological research. For example, GUide to STatistical Analysis (GUSTA ME) is a dynamic, web-based resource providing accessible descriptions of numerous multivariate techniques relevant to microbial ecologists (Buttigieg and Ramette, 2014). Schreiber et al. (2022) employed generalized linear mixed-effect models (GLMMs) and performed Bayesian data analyses for water quality assessment and monitoring in river ecosystems, thus emphasizing the use of multilevel models for analysing ecological datasets.

Fukushima et al. (2022) employed statistical methods, namely, Least Absolute Shrinkage and Selection Operator (LASSO) in combination with Bootstrap and also LASSO in combination with

DOI: 10.1201/9781003408543-24

Sparse Principal Component Regression (SPCR), for identification of microorganisms responsible for wastewater treatment. Microbial community analysis was also performed with DNA extracted from the biomass from the samples and amplified by Polymerase Chain Reaction (PCR). Subsequently, sequencing of the PCR was performed, and sequences with a distance-based similarity of 97% or greater were grouped into operational taxonomic units (OTUs), which were then regarded as the microbial species. The microbial communities were determined from the OTUs. The relative abundances of the microbial species (OTUs) were plotted, enabling several dozen species of microorganism to be visually recognized from among the several thousand microbial species. However, in this manner, the identification of important microorganisms can only be done subjectively. Hence, they proceeded to LASSO in combination with the Bootstrap method for fitting the regression model between wastewater treatment performance (degradation rate of pollutants) and microorganisms (species) by identifying important microorganisms responsible for wastewater treatment that degrade major constituents of pollutants. In this way, LASSO contributed to both regression coefficient estimation simultaneously performing variable selection. This was also coupled with stability analysis via Bootstrap. Thus, three to six types of significant microorganisms could be identified, which led to the removal of six types of pollutants. The model was validated, and the predicted values were found to be close to the observed values. To understand the relationships between the microorganisms and environmental factors and between the microorganisms themselves and to create regression modelling using these relationships, a SPCR model was applied to the selected microbial species, and the fitting improved.

24.3 STATISTICAL MODELS FOR ASSESSING FACTOR EFFECTS ON MICROBIOLOGICAL KINETICS

Predictive microbiology usually follows a two-step fitting process. As a first step, a primary model describing bacterial growth/inactivation/survival over time under constant environmental conditions (say, temperature, water activity, humidity, etc.) is fitted. This is followed by fitting an appropriate secondary model to assess the effect of temperature and other environmental conditions on the kinetic parameters of the primary models (say, lag time, growth rate, D value, etc.). Thereafter, the fitted primary and secondary models are combined into a tertiary model to predict the final number of microorganisms over time under the different environmental conditions considered. Alternatively, a one-step approach has also been widely used in kinetic analysis, wherein the primary and secondary models are fitted simultaneously during the estimation of kinetic parameters.

Knipper et al. (2023), while modelling the survival of *Campylobacter jejuni* (which causes campylobacteriosis, a gastrointestinal infection in humans) in raw milk, considering the viable but non-culturable cells (VBNC), investigated them, taking into consideration colony-forming units (CFUs) and VBNC cells. Data on CFU from two different strains of *C. jejuni* were collected at three temperatures (5, 8 and 12 °C). Simultaneously, a viability real-time PCR test was also conducted to quantify intact and putatively infectious units, i.e., IPIUs, that comprise CFU and VBNC bacteria. The generated data were used to model the viability of *C. jejuni* during raw milk storage. They performed a one-step fitting approach using parameter estimates from an intermediate two-step fit to get starting values for fitting tertiary models. They tried various primary models from among the models considered by, but not limited to, Buchanan and Golden (1995); Geeraerd et al. (2005) and van Boekel (2002). Buchanan and Golden (1995) studied the effect and interactions of temperature, lactic acid concentration, sodium chloride content and sodium nitrite concentration on the survival of a three-strain mixture of *Listeria monocytogenes* in acidified foods using inactivation curves; among other models, a trilinear model. Geeraerd et al. (2005) compared nine different types of microbial survival models on user-specific experimental data relating to the evolution of the microbial population with time to cover all known survivor curve shapes for vegetative bacterial cells, including classical log-linear curves, curves displaying a so-called shoulder before a log-linear decrease, a so-called tail after a log-linear decrease, both shoulder and tailing behaviour, convex/

concave curves followed by tailing, etc. These models can be selected by seeing which of the fitted models' sum of squared errors (among other measures considered) is the minimum, for which they have developed a freeware computer tool named Geeraerd and Van Impe Inactivation Model Fitting Tool, in short, GInaFiT. For the initial two-step fitting approach, Knipper et al. (2023), while fitting different primary model equations to all individual CFU ($\log_{10}$ CFU mL^{-1} versus time) and IPIU data ($\log_{10}$ IPIU mL^{-1} versus time) for the three temperatures considered, finally chose the trilinear model of Buchanan and Golden (1995), which best describes the experimental data. The trilinear model employed on CFU data is given by

$$\log_{10}(N_t) = \log_{10}(N_0), t < S_l;$$

$$\log_{10}(N_t) = \log_{10}(N_0) - \frac{k_{\max}}{\ln(10)}(t - S_l), S_l \leq t < S_t;$$

$$\log_{10}(N_t) = \log_{10}(N_{res}), t \geq S_t$$

where N_0 is the initial concentration (CFU mL^{-1}), N_t is the bacterial concentration (CFU/mL^{-1}) at time t, N_{res} is the residual population density, $k_{\max}$ is the maximum specific inactivation rate, and S_l and S_t are the durations of shoulder and tail effects, respectively. Note that S_t was computed as $\left[S_l + \{\log_{10}(N_0) - \log_{10}(N_{res})\}\frac{\ln(10)}{k_{\max}}\right]$.

While the primary model was trilinear to fit the CFU, the Weibull model (van Boekel, 2022) was selected as the best for fitting the IPIU data. This is given as $\log_{10}(N_t) = \log_{10}(N_0) - \left(\frac{t}{\delta}\right)^{\rho}$, where δ is the time for the first decimal reduction and ρ is the shape parameter. Linear secondary models for the effect of temperature T on the above-mentioned $k_{\max}$ and δ were considered as $\log_{10}k_{\max}$ = Intercept$_1$ + Slope$_1$ × T and $\log_{10}\delta$ = Intercept$_2$ + Slope$_2$ × T. In this way, the parameters estimated in the two-step fitting procedure were used as starting values for the one-step fitting approach. For instance, the tertiary model for the IPIU data becomes

$$\log_{10}(N_t) = \log_{10}(N_0) - \left(\frac{t}{10^{(\text{Intercept}_2 + \text{Slope}_2 + \text{T})}}\right)^{\rho}.$$

Thus, strain-specific linear secondary models were fitted to assess the effect of storage temperature on the maximum specific inactivation rate of CFU. The time of the first decimal reduction parameter of IPIU was modelled by a strain-independent linear secondary model. The developed tertiary models for CFU and IPIU differed significantly in their predictions on the survival kinetics of *C. jejuni* in raw milk for the time required to effect a one $\log_{10}$ reduction. Their results underlined the importance of considering IPIU, and not only CFU, to avoid underestimation of the survival of *C. jejuni* in raw milk by taking into account the fact that VBNC could revert to a culturable state during the raw milk storage.

24.4 USE OF LOGISTIC REGRESSION IN FOOD SAFETY-RELATED MICROBIOLOGICAL STUDIES

Logistic regression (binomial and multinomial) has been used by microbiologists for studying food safety with respect to the presence or inactivation of microorganisms. To cite one, Obeso et al. (2010) utilized logistic regression for prediction of the fate of *Staphylococcus aureus* in pasteurized milk in the presence of two lytic bacteriophages. Probabilistic models for predicting *S. aureus* inactivation by the phages in pasteurized milk were developed. A linear logistic regression procedure was used to describe the survival/death interface of *S. aureus* as a function of the initial phage titre, initial bacterial contamination and temperature. Besides, the model also provided the minimum phage concentration required to inactivate *S. aureus* in milk at different temperatures, irrespective of the bacterial contamination level.

Regression analysis *per se* is a method for investigating functional relationships among variables. The relationship is expressed in the form of an equation or a model connecting the response or dependent variable and one or more explanatory (i.e., predictor) variables. Generally, conventional theory of multiple linear regression (MLR) analysis has been applied for a quantitative response variable. If our response variable in the model is qualitative in nature, the usual MLR model does not hold good due to violation of its standard assumptions. However, the probabilities of this response variable falling into various categories can be modelled in place of the response variable itself, using a logistic regression model that has a non-linear form, developed primarily by a researcher named Cox during the late 1950s. A good account of logistic regression can be found in Fox (1984) and Kleinbaum et al. (2002).

A multivariate technique called discriminant analysis could be used for addressing each of the above situations. However, because the independent variables are a mixture of categorical and continuous variables, the multivariate normality assumption may not hold good. In such cases, the preferable technique is either probit or logistic regression analysis, as it does not make any assumptions about the distribution of the independent variables. The dependent factor is known as the response factor, as it is qualitative or categorical. In this model-building process, various log odds related to response factors are modelled. As a simple case, if the response factor has only two categories with probabilities p_1 and p_2, respectively, then the odds of getting category one are (p_1/p_2). In a real sense, logit and logistic are names of transformations. In the case of logit transformation, a number p between values 0 and 1 is transformed with log $\{p/(1-p)\}$, whereas in the case of logistic transformation, a number x between $-\infty$ and $+\infty$ is transformed with the $\{e^x/(1+e^x)\}$ function. It can be seen that these two transformations are the reverse of each other, i.e., if logit transformation is applied to the logistic transformation function, it provides value x, and similarly, if logistic transformation is applied to the logit transformation function, it provides value p.

Thus, logistic regression models are more appropriate when the response variable is qualitative and a non-linear relationship can be established between the response variable and the qualitative and quantitative factors affecting it. Processes producing sigmoidal or elongated S-shaped curves are quite common in microbiology data. This approach addresses the same questions that discriminant function analysis and multiple regression do, but with no distributional assumptions about the predictors. In a logistic regression model, the predictors do not have to be normally distributed, the relationship between response and predictors need not be linear, the observations need not have equal variance in each category of the response variable, etc.

The logistic response function resembles an S-shaped curve (Figure 24.1). Here, the probability π initially increases slowly with increase in X, and then the increase accelerates, finally stabilizes, but does not increase beyond 1.

The shape of the S-curve can be reproduced if the probabilities can be modelled with, say, only one predictor variable as follows:

$$\pi = P(Y = 1 \mid X = x) = 1/(1+e^{-z})$$

where $z = \beta_0 + \beta_1 x$, and e is the base of the natural logarithm. Thus, for more than one (say r) explanatory variables, the probability π is modelled as

$$\pi = P(Y = 1 \mid X_1 = x_1, X_2 = x_2, \ldots X_r = x_r,) = 1/(1+e^{-z})$$

where $z = \beta_0 + \beta_1 x_1 + \beta_2 x_2 + \ldots + \beta_r x_r$

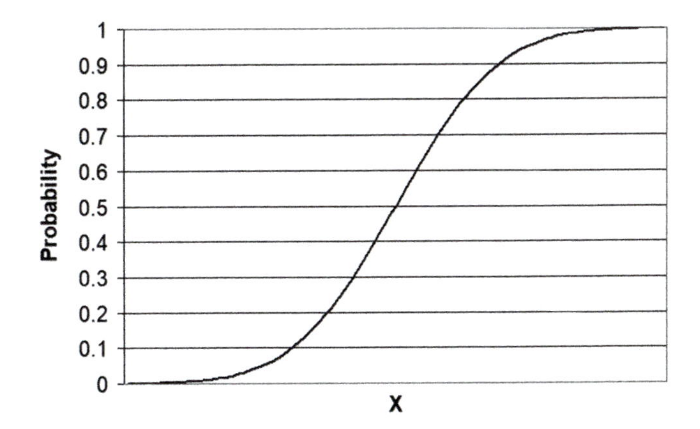

FIGURE 24.1 Logistic response curve.

This equation is called the logistic regression equation. It is non-linear in the parameters β_0, β_1, …, β_r. Modelling the response probabilities by the logistic distribution and estimating the parameters of the model constitutes fitting a logistic regression. The method of estimation generally used is the maximum likelihood estimation method.

To explain the popularity of logistic regression, let us consider the mathematical form on which the logistic model is based. This function, called $f(z)$, is given by

$$f(z) = 1/(1 + e^{-z}), -\infty < z < \infty$$

Now, when $z = -\infty$, $f(z) = 0$, and when $z = \infty$, $f(z) = 1$. Thus, the range of $f(z)$ is 0 to 1. So, the logistic model is popular because the logistic function, on which the model is based, provides estimates that lie in the range between zero and one, and an appealing S-shaped description of the combined effect of several explanatory variables on the probability of an event.

Wald, likelihood ratio and score tests are three commonly used tests for testing the overall significance of the logistic regression model. The Hosmer–Lemeshow goodness-of-fit test is one of the most common tools conveniently used in logistic regression analysis for testing whether or not the model is a good fit.

Comparison between various logistic regression models fitted and with other classification methods, such as discriminant function and decision tree methods, can be made with respect to their classifying ability with the help of (2 × 2) classification tables in the case of a binary response group variable. The columns are the two observed values of the dependent variable, while the rows are the two predicted values of the dependent. In a perfect model, all cases will be on the diagonal, and the overall percent correct will be 100%.

Table 24.1 provides the confusion matrix of any fitted logistic regression model. In this table, the abbreviations TN, FN, FP and TP represent True Negative (actual is "0" and model also predicts it

TABLE 24.1
Confusion Matrix for the Fitted Model

		Actual Status	
		Does Not Have Disease	**Has Disease**
Predicted Status	Identified as Not Having disease	True Negative (TN)	False Negative (FN)
	Identified as Having Disease	False Positive (FP)	True Positive (TP)

as "0"), False Negative (actual is "1" but model predicts it as "0"), False Positive (actual is "0" but model predicts it as "1") and True Positive (actual is "1" and model also predicts it as "1"), respectively, computed from the outcomes of the fitted logistic regression model. It should be noted here that "0" represents "absence", "no", "Not having disease", etc., and "1" represents "presence", "yes", "having disease", etc.

The accuracy of prediction is discussed subsequently. For this, several measures have been used. Correct Classification Rate (CCR), also called the hit rate, is the number of correct predictions divided by sample size. Thus, from the Table, CCR = (TN + TP)/(TN + FN + FP + TP). On the other hand, the misclassification rate is (1 – CCR). The false positive rate (FPR) is the number of cases that do not have disease but were predicted as having the disease divided by the total number of cases that do not have disease (includes all TNs and FPs). Thus, FPR = FP/(TN + FP). The false negative rate (FNR), also called the miss rate, is the number of cases that have disease but were predicted as not having disease (all FNs) divided by the total number of cases that have disease (includes all FNs and TPs). Thus, FNR = FN/(FN + TP). True Positive Rate (TPR), also known as sensitivity (the ability of the model to predict "1" correctly), is the proportion of cases that belong to "having disease" and were also predicted as "having disease". Thus, TPR = TP/(FN + TP). True Negative Rate (TNR), also known as specificity (the ability of the model to predict "0" correctly), is the proportion of cases that belong to "not having disease" and were also predicted as "not having disease". Thus, TNR = TN/(TN +FP). Note that (FPR + TNR) and (TPR + FNR) each total to 100%, and thus, the pairs within the respective brackets complement each other. For a good model, it is expected that both sensitivity and specificity are large enough, and the FP and false negative rates are small. It is noted here that the left-hand side of the logistic regression model considered was P(Y = 1), which was nothing but the probability that the "Having disease" has occurred. Thus, the classification (confusion) matrix gives an idea about the predictive power of the fitted model.

While interpreting the fitted logistic regression, the concept of "odds ratios" needs to be understood. An association exists between two variables if the distribution of one variable changes when the level (or value) of the other variable changes. If there is no association, the distribution of the first variable is the same regardless of the level of the other variable. The odds ratio is usually used for measuring such associations. For example, consider the following situation:

The odds of an event is the ratio of the probability of an event occurring to the probability of it not occurring. That is, Odds = P(event)/{1 – P(event)} = P(event = 1)/P(event = 0) = $\pi/(1-\pi)$. In the above situation, there is 82% probability that the food safety will be "Good" in the case of "Female". The odds of "Good" in "Female" category = 0.82/0.18 = 4.5. The odds of "Good" in "Male" category = 0.60/0.40 = 1.5. The odds ratio of "Female" to "Male" equals (4.5/1.5) = 3, indicating that the odds of getting "Food safety in good category" from "Female" is three times greater than from "Male". Also, there is 18% probability that Food safety will be "Bad" in the case of "Female"; the

TABLE 24.2
Two Attributes, "Gender" and "Food Safety", Each at Two Levels

	Food Safety	
	Good	Bad
Gender		
Female	82	18
Male	60	40

odds of "Bad food safety" in "Female" = 0.18/0.82 = 0.22. Thus, as the probability is very small (0.18 in this case), there is no appreciable difference between describing it as probability or odds (Table 24.2).

The importance of odds ratio in the case of logistic regression modelling can be further explained by taking a simple case of influence of the aforementioned attribute "Gender" X with two levels (Male or Female) along with the attribute "Food safety" Y with two levels (Good = 1, Bad = 0). Logistic regression, when written in its linearized form, takes the following "logit" form: logit{$Y = 1 \mid X = x$} = $\log(\pi/(1 - \pi))$ = log(odds) = $\beta_0 + \beta_1 \times x$. Hence, log(odds) = $\exp(\beta_0 + \beta_1 \times x)$. Now, substituting $x = 1$ for Females, Odds (Females) = $\exp(\beta_0 + \beta_1)$, and $x = 0$ for Males, Odds (Males) = exp(β0). Hence, odds ratio = $\exp(\beta_0 + \beta_1)/\exp(\beta_0) = \exp(\beta_1)$. Thus, here, the regression coefficient of Y on X, i.e., β_1, is not interpreted directly but after taking its exponent.

24.5 APPLICATIONS OF DIVERSITY INDICES IN THE CONTEXT OF AQUATIC ECOLOGY AND MICROBIOLOGY

Diversity index is a statistic that combines species richness (number of different kinds of species present/list of species) and evenness/equitability (homogeneity of species abundances) as a measure of biodiversity of an ecosystem and provides more information about community composition. Species richness, even though it very much depends on sample size, as a measure on its own, takes no account of the number of individuals of each species present. It gives as much weight to those species that have very few individuals as to those that have many individuals. Evenness is a measure of the relative abundance of different species. If Sample-1, say, is quite evenly distributed between various species in comparison to, say, Sample-2, then Sample-1 can be said to be more even and Sample-2 can be said to be less diverse as compared with the former (see Table 24.3). A community dominated by one or two species is considered to be less diverse than one in which several different species have a similar abundance. Thus, diversity indices provide answers in relation to different species in a group as to whether they have richness or evenness, and which species are relatively abundant and hence, more (or less) diverse and thus, can be considered as rare or more common.

Various studies have been conducted to understand the floral and faunal diversity that exist in many ecosystems (agricultural, avian, ecology, etc.) at both national and international levels (Rathod et al., 2011). Bibi and Ali (2013) measured diversity indices of avian communities at Taunsa Barrage wildlife sanctuary in Pakistan. Kumar and Khan (2013) studied the distribution and diversity of benthic macroinvertebrate fauna in Pondicherry mangroves in India. Bhattacharyya et al. (2015) studied the diversity and distribution of *Archaea* in the mangrove sediment of Sundarbans in India. Dangan-Galon et al. (2016) studied the diversity and structural complexity of mangrove forest along Puerto Princesa Bay, Palawan Island, Philippines.

In general, it has been observed that an increase in environmental stress decreases diversity, leading to a decrease in both richness and evenness – in other words, increased dominance of one or a few species. However, it has also been found that if disturbance situations are minimal, then also there will be decreased diversity due to competitive exclusion between species. Hence, a moderate (i.e., intermediate) level or frequency of disturbance (for example, pollution stress or other anthropogenic

TABLE 24.3
Application of Diversity Indices

Species Id.	1	2	3	4	5	6	7	8	9	10
Sample 1	10%	10%	10%	10%	10%	10%	10%	10%	10%	10%
Sample 2	91%	1%	1%	1%	1%	1%	1%	1%	1%	1%

effects) is also effective for increased diversity to be present in the environment. For performing analysis of diversity studies, first, one has to define the boundaries of the ecosystem under consideration and take samples to find and list the species. The number of species found in a given sample is strongly dependent upon the size of that sample. Comparable samples are needed to ensure that most of the species have been found, and for this, all regions should be intensively sampled.

Now, the construction of various species diversity indices is discussed briefly. Let S denote the number of species in an ecosystem, n_i denote the number of individuals of the i-th species ($i = 1, 2, \ldots, S$) that are counted, $N = \sum_{i=1}^{S} n_i$ be the total number of all individuals counted, and $p_i = (n_i/N)$ be the fraction of all organisms that belong to the i-th species ($i = 1, 2, \ldots, S$).

A more popular and commonly used diversity index is Shannon's index, given by Claude Shannon. It assumes that all species are represented in the sample and that the sample was obtained randomly. It is also called the Shannon–Weaver index (the attribution to Robert Weaver is a misnomer, since he only performed the philanthropic activity of granting money for the information theory research of which diversity studies are a part). It is also sometimes called Shannon–Wiener (again, this is a misnomer, as Norbert Wiener worked on information theory *per se* and not directly on diversity studies). Shannon's index is given by $H = -\sum_{i=1}^{S} p_i \log(p_i)$. Usually, logarithms to the base 2 are used. It needs to be multiplied or divided by a constant factor while comparing indices, as one needs to check and correct for the same base. The identity $\log_2 x = \log_e x \log_e 2$ comes in handy while doing such conversions. Sometimes, $H' = \exp(H)$ is used. The most important source of error in this index is failing to include all species from the community in the sample. This makes a species-area curve assessment (discussed subsequently) very important at the beginning of a study. A greater number of species and a more even distribution increase this diversity index. Values of the Shannon diversity index for real communities typically fall between 1.5 and 3.5, even though theoretically, the value can be between 0 and ($\log S$). Hence, evenness can be measured by the ratio ($H/\log S$), which is constrained between 0 and 1.

The next famous index is the Simpson's diversity index, given by George Gaylord Simpson. It is given under two cases, With Replacement (WR) and Without Replacement (WOR) as $D = \sum_{i=1}^{S} p_i^2 = \sum_{i=1}^{S} \left(\frac{n_i}{N}\right)\left(\frac{n_i}{N}\right)$ if WR and $D = \sum_{i=1}^{S} \left(\frac{n_i}{N}\right)\left(\frac{n_i - 1}{N-1}\right)$ if WOR. The WOR formula of Simpson corrects biasness while taking samples and also is reliable when N is small. The limits of D are 0 and 1. If D is near to zero, then the species composition is highly diverse or heterogeneous, and if D is near to one, then it refers to more homogeneous ecosystems. It is a dominance index because it is leveraged towards the abundance of the most common species. It gives the probability of any two individuals belonging to the same species if drawn at random from an infinitely large community having many species. Rarely, the formula expression is also called the Herfindhal index by some authors (Rathod et al., 2011). Sometimes, $D' = (1/D)$ is used for Simpson's. To make it comparable with Shannon's diversity index, $\tilde{D} = (1 - D)$ is used; $\tilde{D}$ represents the probability that two individuals randomly selected from a sample will belong to different species; $\tilde{D}$ is more of an evenness index. Both Shannon H and Simpson $\tilde{D}$ diversities increase as richness increases, for a given pattern of evenness, and increase as evenness increases, for a given richness, but they do not always rank communities in the same order. Simpson is less sensitive to richness and more sensitive to evenness than Shannon; for the latter, it is vice versa.

There are numerous other diversity indices available in the literature. These include Berger Parker or index of maximal proportion $\max_{1 \le i \le S}(p_i)$; Ogive $\sum_{i=1}^{S} \frac{\left(p_i - \frac{1}{N}\right)^2}{\frac{1}{N}}$; Margelef (for measuring richness) $\frac{S-1}{\log N}$; Pielou (for measuring evenness) $\frac{H}{\log S}$; Brillouin $\frac{1}{N}\log\left(\frac{N!}{n_1 n_s \ldots n_S}\right)$;

Hill numbers (unification of several indices) / Renyi entropy $\frac{1}{1-\alpha}\log\left(\sum_{i=1}^{S} p_i^{\alpha}\right)$. For $\alpha = 0, 1, 2$ and ∞, these Hill numbers reduce to $N_0 = S$; $N_1 = \exp(H)$; $N_2 = \frac{1}{\sum_{i=1}^{S} p_i^2}$; …; $N_\infty = \frac{1}{\max_{1 \le i \le S(p_i)}}$.

The choice of a particular diversity index depends on the goals of the study (for example, the emphasis may be either on abundance or alternatively, on rare species), and to what extent sampling can be assured to be random. It is noted here that any index is sensitive to degree of sampling effort. There is generally no relationship between one index and another. For comparison, indices should be compared across equivalent sampling designs. An ideal index would discriminate clearly and accurately between samples, not be greatly affected by differences in sample size, and be relatively simple to calculate. Biologists often use a combination of several indices to take advantages of the strengths of each and develop a more complete understanding of community structure.

Diversity indices rely on species richness but can be controlled for the effects of sample size by what is called rarefaction. Rarefaction avoids incompatibility of measurements resulting from samples of different sizes. It is the statistical expectation of the number of species in a survey or collection (on the y-axis or vertical axis) as a function of the accumulated number of individuals or samples (on the x-axis or horizontal axis), based on resampling from an observed sample set. It can be produced by drawing 1, 2, 3, …, N samples (or individuals) at a time (WOR) from the full set of samples, then plotting the averages of many such draws. Alternatively, mathematics of combinations allows it to be computed directly. Rarefaction curves will be steeper and more elevated if the community is more diverse. Samples sizes for different sites or times may differ, but relevant sections of rarefaction can still be compared.

$E(S) = \sum_{i=1}^{S}\left[1 - \left\{\binom{M = m_i}{m}\binom{M}{m}\right\}\right]$ where $E(S)$ is expected number of species in the rarefied sample; m is standardized sample; M and m_i are, respectively, total number of individuals and number of individuals in the i-th species in the sample to be rarefied.

Now, the graphical and distributional plots used in diversity studies are discussed. The species accumulation curve is the observed number of species in a survey or collection as a function of the accumulated number of individuals or samples. It is sometimes called the collector's curve. The common kinds of species are usually first to be found, and the rate of discovery of kinds that are new to the collection declines steadily, while the order in which samples are added is arbitrary. The curve first rises quickly as the common species are recorded, levelling off as the rarest species are finally included. The number of species accumulated at that point is called "alpha diversity".

Species abundance (dominance) curves are of many kinds. According to Whittaker (1965), the rank abundance curve is obtained when each species is represented by a vertical bar proportional to its abundance (Colwell, 2009). According to Preston (1948), the log abundance plot is obtained by counting up the number of species in each abundance category, starting with the rarest species, and plotting these frequencies against abundance categories in powers of two. Thus, the number of species is plotted in geometric abundance classes as a means of detecting stress (environmental/anthropogenic). For this, plot the number of species with 1 individual in the sample (2^0 in class 1), 2–3 individuals (2^1 in class 2), 4–7 individuals (2^2 in class 3), etc.

In unpolluted environments, there will be many rare species, and the curve will be smooth, with the mode well to the left. In polluted environments, there will be fewer rare species and therefore more of the abundant species. Hence, higher classes will be strongly represented, with the curve more irregular. It has been observed that intermediate classes 3 to 5 are more sensitive to pollution-induced changes, suggesting indicator species to compare between sites. When relative abundance distributions approximate a bell-shaped curve in a log abundance plot, it follows a lognormal

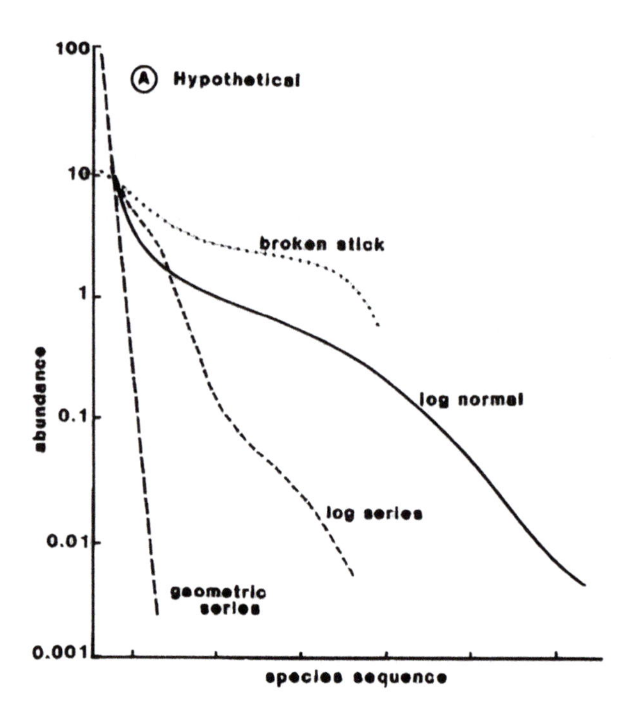

FIGURE 24.2 Species abundance curves.

distribution. The abundance of each species can be plotted against its rank order abundance from 1 (i.e., the highest) to S (i.e., the lowest). The curves thus obtained will have one of the lognormal, geometric, logarithmic or "McArthur's broken stick" distributional forms (see Figure 24.2). Thus, these are ranked abundances expressed as percentage of total abundances of all species plotted against relevant species rank. Sometimes, log transformations of one or more axes are used to emphasize or down-weight different sections of the curve. Logging the *x*- (rank) axis enables the distribution of commoner species to be better visualized. Species are not equally common in a given community, with rare species occurring if they are localized geographically or have sparse density or are habitat specific.

k-dominance curves are cumulative ranked abundances plotted against species rank or log species rank, which has a smoothing effect on curves. The ordering of curves on a plot will obviously be the reverse of rarefaction curves, with the most elevated curve having the lowest diversity. To compare dominance separately from the number of species, the *x*-axis (species rank) may be rescaled from 0 to 100 (relative species rank) to produce Lorenz curves. Mostly, *k*-dominance curves are transformed, as very often, *k*-dominance curves approach a cumulative frequency of 100% for a large part of their length, and it may be difficult to distinguish between the forms of these curves. In that case, transformation of the *y*-axis is done so that the cumulative values are closer to linearity by using modified logistic transformation:

$$y_i' = \log\left[\frac{1 + y_i}{101 - y_i}\right]$$

In certain situations, partial dominance curves are preferred. A problem with the cumulative nature of k-dominant curves is that the visual information is over-dependent on the single most dominant species. In that case, the dominance of every consecutive ordered ranked species over the remainder is computed as follows:

$$p_1 = \frac{100a_1}{\sum_{j=1}^{S} a_j};\, p_2 = \frac{100a_2}{\sum_{j=2}^{S} a_j};\, p_{S-z} = \frac{100a_{S-1}}{a_{S-1} + a_S}$$

where a_i is the absolute (or percent) abundance of the i-th species. When ranked in decreasing abundance order, the partial dominance curve is a plot of p_i against log i ($i = 1, 2, \ldots, S - 1$).

The Abundance/Biomass Comparison (ABC) plot involves plotting of separate k-dominant curves for species abundances and species biomasses on the same graph and making a comparison of their forms. The species are ranked in the order of importance in terms of abundance or biomass on the x-axis (log scale) with percent dominance on the y-axis (cumulative scale). In undisturbed communities, the biomass is dominated by one of a few large species, leading to an elevated biomass curve; each of these species, however, is represented by rather few individuals, so they do not dominate the abundance curve, which shows a typical diverse, equitable distribution. Thus, the k-dominance curve for biomass lies above the curve of abundance for its entire length. Under moderate pollution (disturbance), the large competitive dominants are eliminated, and the inequality is reduced, so that the biomass and abundance curves are closely coincident and may cross each other one or more times. As pollution becomes more severe, communities become increasingly dominated by one or a few opportunistic species, which while they dominate the numbers, do not dominate the biomass, because they are very small-bodied. Hence, the abundance curve lies above the biomass curve throughout its length (Figure 24.3).

There is a striking regularity in the pattern of increase in species count as larger geographic areas are considered. When the number of species or its logarithm (depending on the case) is plotted against the logarithm of area, an approximately linear relationship is revealed. With a log-log power curve or a semi-log exponential curve, the pattern on arithmetic axes is a decelerating but ever-increasing number of species as the area increases. A graph of the total number of species found along a transect or as the number of "quadrats" increases explains the relationship. One is assured that sufficient samples have been taken when the curve begins to "plateau" (Figures 24.4 and 24.5).

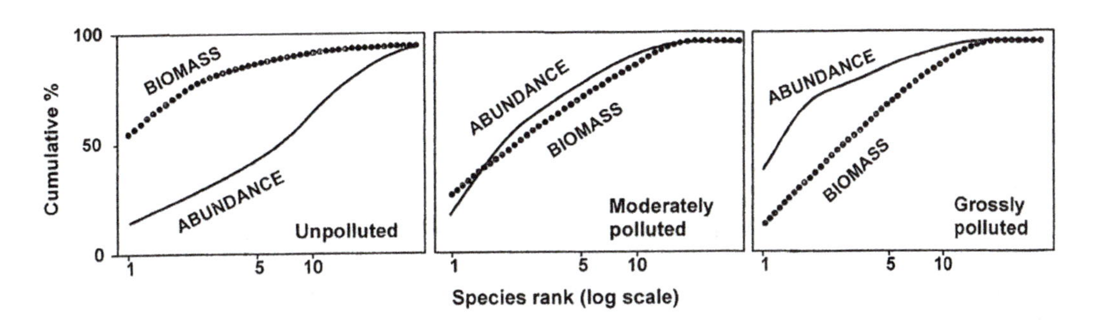

FIGURE 24.3 Hypothetical k-dominance curves for species biomass and abundance showing "unpolluted", "moderately polluted" and "grossly polluted" conditions. (From Clarke, K.R. and Warwick, R.M., *Change in Marine Communities. An Approach to Statistical Analysis and Interpretation*, 2, 1–168, 2001, PRIMER-E Ltd, Plymouth UK.)

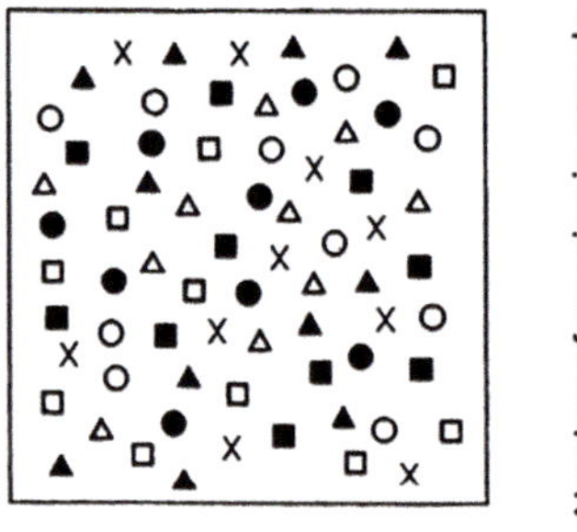

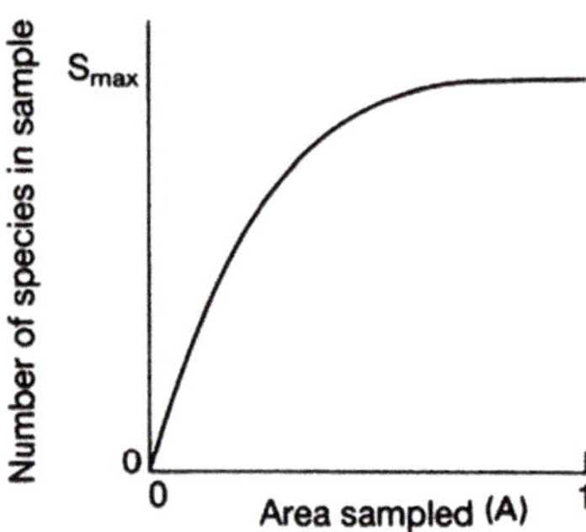

FIGURE 24.4 Species-area curve: high equitability.

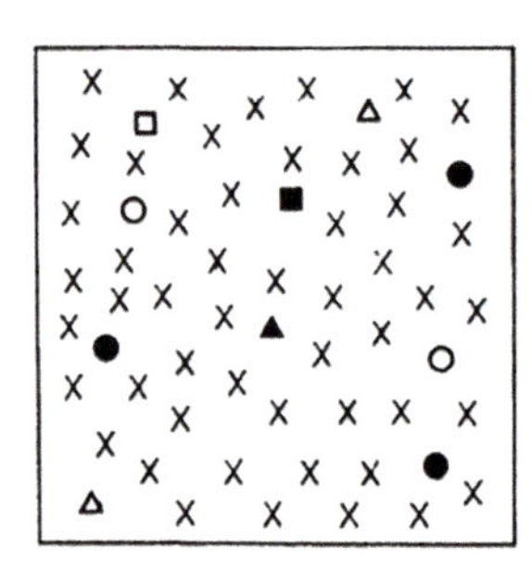

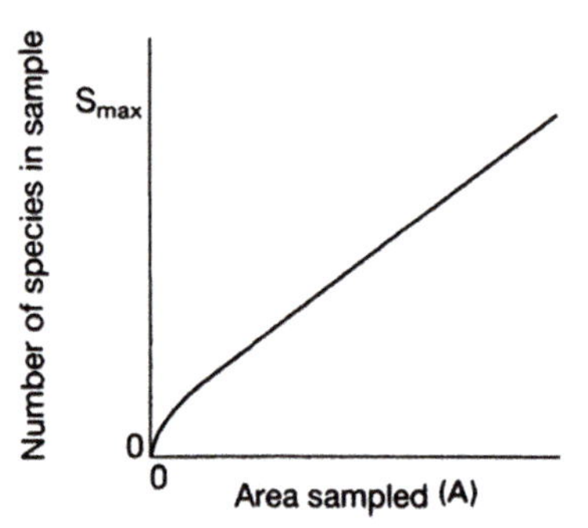

FIGURE 24.5 Species-area curve: low equitability.

Statistical analyses and modelling can be done on diversity-related data in many ways. Significant differences between indices over sampling sites or times can be done using one-way ANOVA, t test or multiple comparison procedures for individual pairs. The literature abounds with diversity studies using various statistical analyses and modelling techniques, such as non-metric multi-dimensional scaling (NMDS), correspondence analysis, principal coordinate analysis, cluster analysis, principal component analysis (PCA), redundancy analysis, etc. (Clarke and Warwick, 2001; Kindt and Coe, 2005). There is umpteen software available (both free and proprietary) for statistical analyses and modelling in diversity-related studies, such as PAST (PAleontological STatistics), Biodiversity. R, vegan for R, estimates, Biodiversity Pro, EcoSim, MVSP, Ecology software, Krebswin/ Pices Species Diversity and Richness II, etc.

24.6 MICROBIAL GROWTH CURVES

Bacteria multiply exponentially by binary fission, and therefore, their growth occurs on a geometric scale. For microbiologists, the rate of growth of bacteria is of particular interest due to its importance in fermentation, pathogenesis, food spoilage and bacterial communication mechanisms

(Tonner et al., 2017). Various methods are available for estimation of the growth rates of microbial groups. Methods such as the frequency of dividing cells and computation of rates of cell or biomass production exist. The generation time or the doubling time is a measure of bacterial growth rate which has numerous applications in the bacterial population dynamics and is defined as the time taken for a bacterium to double its number. Since bacteria grow geometrically, the generation time is the average time required for a bacterium to double its number. For example, *Escherichia coli* doubles its number every 20 minutes, meaning that a single cell of *E. coli* becomes 2, 4, 8, 16, 32, and so on after every 20 minutes. Generation time can vary from a few minutes to a few hours depending on the growth conditions, such as the nutrient medium used, temperature, pH, water activity, etc. The marine bacterium *Pseudomonas natriegens* has a generation time of 9.8 min (Eagon, 1962). The spore-forming *Clostridium perfringens* is a fast-growing bacterium with a generation time of 10 minutes, while the slow-growing *Mycobacterium tuberculosis* has a generation time of 12–16 hours. The generation time (G) can be calculated using the formula

$$G = \frac{t}{n} = \frac{t}{3.3\log\left(\frac{b}{B}\right)}$$

where B is the number of bacteria present at the initial period, b is the number of bacteria present after a specific time period t, and n is the number of generations.

In the traditional spectrophotometric method of growth determination, the absorbance of the growth medium at specific time intervals is plotted against the absorbance to obtain a sigmoidal curve with lag, exponential and stationary phases of growth. Alternatively, the natural logarithm [$y(t) = \ln(N)$] of cell counts obtained by plating at each time interval can be plotted against time (t) to obtain the growth curve, the slope (μ_{max}) of which gives the maximum specific growth rate of the bacterium in defined growth conditions (Baranyi and Pin, 1999; Longhi et al., 2017; Zwietering et al., 1990). Growth rate is fundamental to estimating the microbial productivity of an ecosystem, based on the rate at which microbial populations increase exponentially over time under prevailing environmental conditions.

Baranyi and Pin (1999) proposed that the detection time based on the initial physiological state of the inoculum is critical while estimating growth parameters of a homogeneous bacterial population, and developed a "physiological state theorem" for estimating the specific growth rate of a bacterial population, which is unaffected by differences in the initial lag phases of individual bacteria. This method of estimation has potential application in food microbiology to understand the dynamics of low populations of spoilage or pathogenic bacteria.

Different methods are used to estimate bacterial growth rates, such as the increase in the cellular biomass or the RNA content (Hagström et al., 1979; Kemp et al., 1993). Although estimating the growth rates of individual bacteria in laboratory media might appear straightforward, it is challenging to estimate the changes in mixed populations of bacteria in an ecosystem. According to (Kirchman, 2002), earlier, the estimates of changes in microbial biomass (dB_t/dt) were based on $dB_t/dt = \mu B_t$, where B_t is the microbial biomass (in units of cells or mass per unit volume) at time t and μ the specific growth rate. This differential equation, when integrated on both sides, yields $B_t = B_0 e^{\mu t}$, where B_0 is the initial microbial biomass when $t = 0$. This estimate could be erroneous, as it assumes a homogeneous microbial population that divides exponentially with no mortality. Thus, he argued that production P_t is the increase in microbial biomass over time in the absence of any mortality and can be expressed mathematically as the slope of the curve on a graph of biomass versus time, which is nothing but the first derivative of the biomass versus time curve that appears in the left-hand side of the aforementioned differential equation. Thus, growth rate becomes, simply, $[(dB_t/dt)/B_t] = (P_t/B_t)$. Therefore, (P/B) is the simplest way to compute growth rates, which can be achieved by dividing the rate of cell production P by an estimate of cellular abundance (standing stock) B, i.e., (P/B). This is especially applicable for mixed assemblages with lowest uncertainty, as it is based on the direct relationship between incorporation rates per cell and growth rates.

24.7 MOST PROBABLE NUMBER (MPN)

The MPN technique (also known as the multiple tube or dilution tube method) is used to estimate the bacterial numbers in food, water, soil, etc., particularly for samples with low bacterial counts of <10–100 per gram or ml, without an actual count of single cells or colonies. Introduced by McCrady (1915), the technique makes an estimate of bacterial numbers based on specific growth characteristics rather than the actual growth or colony counts, and is based on the principle of "dilution to extinction"; hence, it is sometimes called the extinction dilution method (McCrady, 1915). In other words, the actual counts can be obtained only when at least one of the dilution tubes is negative for bacterial growth. The method is performed in three, five or ten tubes (replicates), inoculated with different volumes of samples such as 10, 1 and 0.1 ml (i.e., in tenfold increasing dilution series, 10, 10/10 and 10/100, or in other words, decreasing sample volumes). In a typical three-replicate design (refer to Figure 24.1), if two of the three 10-ml tubes are positive; two of the three 1-ml tubes are positive; and none of the three 0.1-ml tubes is positive, then the result of "220" would suggest a MPN of 21 with a 95% confidence level of 4.5–42 (values taken from a table, which will be discussed subsequently). If all the tubes exhibit growth, then the results will be noted as 333 (the corresponding table value of MPN is ">1100" with 420–4200 as 95% confidence level); if only one tube of each replicate shows growth, then the results will be 111 (MPN 11 with 3.6–38 as 95% confidence level).

The inoculated tubes are incubated at a suitable temperature for 24 to 48 hours (Silliker et al., 1979). Microbial growth characteristics such as turbidity and gas formation (or those bringing about some characteristic and readily recognizable transformation in the medium) are used to score a tube as positive, and based on the number of tubes positive from different series of dilutions, the number of bacteria is estimated using a MPN table (available for use with particular combinations of the number of tubes inoculated and the relative quantities of the original sample in the inoculum applied). The MPN method is useful for estimating quantitative bioburden if surface plating for CFU is not advised, as such samples contain particulate material that interferes with plate count enumeration methods (Sutton, 2010). Bacterial estimates obtained using the MPN method are usually higher than the counts obtained by the surface plating method. The MPN method is widely followed to determine fecal contamination of water and food, which still remains a reliable and cost-effective method of determining the potability of water, although the entire process takes 3–4 days. MPN methods are usually applied in the case of *Salmonella*, *Enterococcus*, *Vibrio parahaemolyticus*, *Staphylococcus*, *Pseudomonas aeruginosa*, sulfur reducing and nitrifying bacteria (Figure 24.6).

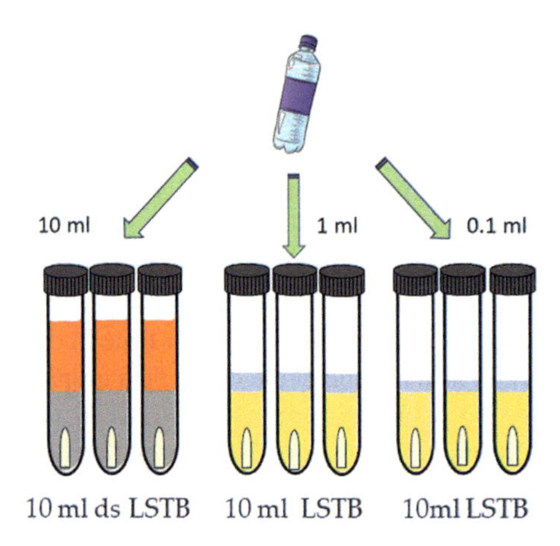

FIGURE 24.6 A three-tube MPN test scheme (ds = double strength).

The basic concept of MPN works on the principle of a "growth versus no-growth" pattern of low numbers of organisms in tubes by diluting the sample to such a degree that inoculum in the tubes will sometimes (not always) contain viable organisms. This information is due to sampling error; for instance, if one replicate tube of medium receives a dilution of the sample that contains a bacterial cell, the tube will turn turbid, while its replicate (which has same conditions as its neighbour) may not receive any bacteria in its sample due to pipetting or sampling and will not turn turbid. The accuracy of MPN of cells in the sample is ensured by having replicates and dilution series, and accordingly, the choice of the number of tubes and of dilution ratios is determined by the permissible sampling error (Alexander, 1983). This basic premise was further elaborated as early as in Woodward (1957). As the quantity of sample in each tube decreases by a factor of ten from one dilution to the next, the number of positive tubes per dilution normally decreases from dilution to dilution, but not always. That is to say, it is extremely unlikely that all tubes containing 0.1 ml of sample will show positive results when none of the tubes containing 10 ml or 1 ml of sample was positive, unless some other influencing factor was involved.

24.8 STATISTICS IN FOOD SAFETY MANAGEMENT

24.8.1 Sampling Plans

Food safety is compromised by the presence of microbial contaminants, which are either naturally present in the food itself or introduced from outside as secondary contaminants. Numerous bacteria are known to cause food poisoning, and the national and international agencies have prescribed tolerance levels for these bacteria based on the level of risk to consumer health. In fish and fishery products, several bacteria, such as *E. coli*, *Salmonella*, *L. monocytogenes*, *Vibrio cholerae*, *V. parahaemolyticus*, and *S. aureus*, represent hazards to consumers' health and therefore, need to be strictly monitored throughout the food production chain till the products reach the consumer. Importing countries have limits of tolerance for these bacterial pathogens, with certain pathogens, such as *Salmonella* and *L. monocytogenes*, being zero-tolerance pathogens in some importing countries. There have been numerous instances of seafood export consignments from India being rejected on account of the presence of zero-tolerance pathogens. Considering the volume of fishery products exported by individual suppliers, robust sampling methods have been developed that unambiguously represent the whole consignment with regard to the distribution of target pathogen(s). It is essential that the samples/subsamples collected represent the whole consignment uniformly and allow the detection of sparsely dispersed pathogens or toxins in the sample. A statistically significant number of samples should be collected so that target pathogens that are not evenly distributed in the food or are present in low numbers are not missed.

For fish and fishery products, depending on the regulatory limits of the bacteria in question, a two- or three-class sampling plan is followed. The two-class sampling plan is designed to accept or reject a lot based on a microbiological criterion (ICMSF, 1986). The parameters of a two-class sampling plan include n, the number of sample units to be collected randomly from a product lot; m, the microbiological limit (CFU g^{-1}), and the sample is rejected if bacterial counts exceed this limit; and c, the maximum allowable number of sample units positive for a particular microbiological parameter or exceeding a set microbiological limit (CFU ml^{-1}) (Figure 24.7). For zero-tolerance pathogens such as *Salmonella* and *L. monocytogenes*, c is zero.

Therefore, in a two-class sampling plan, the results of sample testing will lead to two possibilities, $\leq m$ and $>m$. Figure 24.7 illustrates the outcomes of a two-class sampling plan (CFS, 2014). For pathogens whose presence is not acceptable in food at any level, a two-class sampling plan is advocated.

In a three-class sampling plan, n is the number of sample units to be collected randomly from a product lot, m is the microbiological limit (CFU g^{-1}) that separates good quality samples from marginally acceptable quality ones, M is the microbiological limit above which the samples are

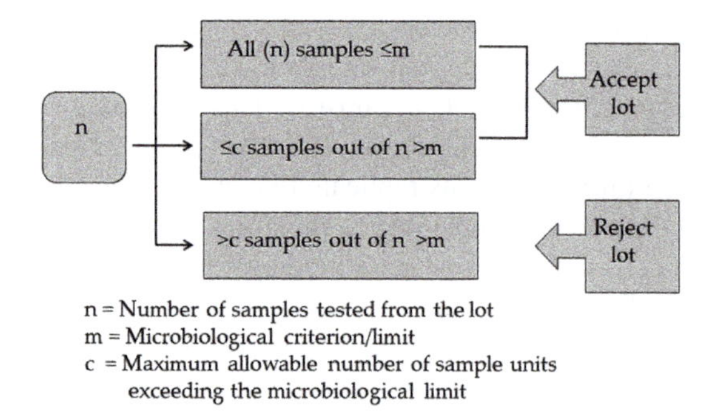

FIGURE 24.7 Two-class sampling plan and the presence/absence criteria that determine the acceptability of a lot.

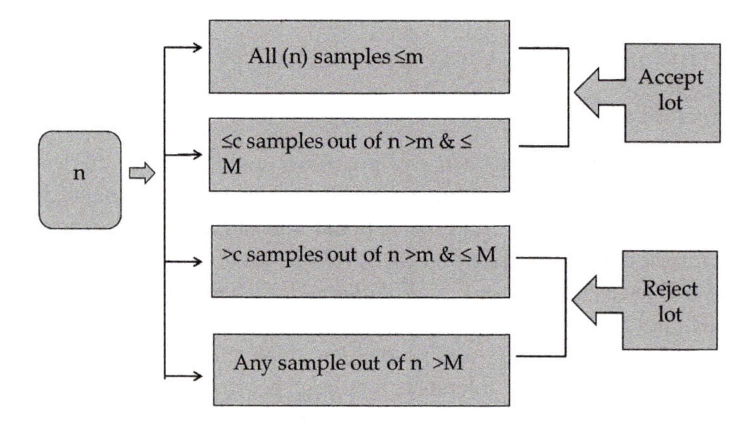

FIGURE 24.8 Three-class sampling plan and the criteria used for the interpretation of results of microbiological analysis.

considered unacceptable or defective, and *c* is the maximum allowable number of samples falling between *m* and *M*. When it exceeds *M*, the lot is rejected (Figure 24.8).

The sample is rejected when more than *c* number of samples fall between *m* and *M*, and/or when any one sample exceeds *M*. The two-class sampling plan is presence/absence based, while the three-class sampling plan is concentration based (CFS, 2014).

REFERENCES

Alexander, M. "Most probable number method for microbial populations." *Methods of Soil Analysis: Part 2 Chemical and Microbiological Properties* 9 (1983): 815–820.

Baranyi, József, and Carmen Pin. "Estimating bacterial growth parameters by means of detection times." *Applied and Environmental Microbiology* 65, no. 2 (1999): 732–736.

Bhattacharyya, Anish, Niladri Shekhar Majumder, Pijush Basak, Shayantan Mukherji, Debojyoti Roy, Sudip Nag, Anwesha Haldar et al. "Diversity and distribution of Archaea in the mangrove sediment of Sundarbans." *Archaea* 2015 (2015): 1–14.

Bibi, F., and Z. Ali. "Measurement of diversity indices of avian communities at Taunsa Barrage Wildlife Sanctuary, Pakistan." *The Journal of Animal & Plant Sciences* 23, no. 2 (2013): 469–474.

Buchanan, R. L., and M. H. Golden. "Model for the non-thermal inactivation of *Listeria monocytogenes* in a reduced oxygen environment." *Food Microbiology* 12 (1995): 203–212.

Buttigieg, Pier Luigi, and Alban Ramette. "A guide to statistical analysis in microbial ecology: A community-focused, living review of multivariate data analyses." *FEMS Microbiology Ecology* 90, no. 3 (2014): 543–550.

Centre for Food Safety (CFS) Food and Environmental Hygiene Department of Hong Kong (HKSAR). "Microbiological guidelines for ready-to-eat food." *Food and Environmental Hygiene Department of Hong Kong* HKSAR) (2014): 1–38

Clarke, K. Robert, and R. M. Warwick. *Change in Marine Communities. An Approach to Statistical Analysis and Interpretation* 2. Primer-E Ltd, Plymouth Marine Laboratory, Plymouth, UK, (2001): 1–168.

Colwell, Robert K. "Biodiversity: Concepts, patterns, and measurement." *The Princeton Guide to Ecology* 663 (2009): 257–263.

Dangan-Galon, Floredel, Roger G. Dolorosa, Jeter S. Sespene, and Nelly I. Mendoza. "Diversity and structural complexity of mangrove forest along Puerto Princesa Bay, Palawan Island, Philippines." *Journal of Marine and Island Cultures* 5, no. 2 (2016): 118–125.

Eagon, Robert Garfield. "*Pseudomonas natriegens*, a marine bacterium with a generation time of less than 10 minutes." *Journal of Bacteriology* 83, no. 4 (1962): 736–737.

Fox, John. *Linear Statistical Models and Related Methods: With Applications to Social Research.* Wiley, New York, (1984).

Fukushima, Toshikazu, Junichi Nakagawa, Shuichi Kawano, and Mamoru Oshiki. *Development of Statistical Method for Identification of Microorganisms Responsible for Wastewater Treatment.* Technical Report 127, Nippon Steel Technical Report, 2022.

Geeraerd, A. H., V. P. Valdramidis, and J. F. Van Impe. "GInaFiT, a freeware tool to assess non-log-linear microbial survivor curves." *International Journal of Food Microbiology* 102, no. 1 (2005): 95–105.

Hagström, U. Larsson, P. Hörstedt, and S. Normark. "Frequency of dividing cells, a new approach to the determination of bacterial growth rates in aquatic environments." *Applied and Environmental Microbiology* 37, no. 5 (1979): 805–812.

ICMSF (International Commission on Microbiological Specifications for Foods). *Sampling for Microbiological Analysis: Principles and Specific Applications.* Springer, NY, (1986).

Ilstrup, Duane M. "Statistical methods in microbiology." *Clinical Microbiology Reviews* 3, no. 3 (1990): 219–226.

Kemp, P. F., S. Lee, and Julie LaRoche. "Estimating the growth rate of slowly growing marine bacteria from RNA content." *Applied and Environmental Microbiology* 59, no. 8 (1993): 2594–2601.

Kindt, Roeland, and Richard Coe. *Tree Diversity Analysis: A Manual and Software for Common Statistical Methods for Ecological and Biodiversity Studies.* World Agroforestry Centre, 2005.

Kirchman, David L. "Calculating microbial growth rates from data on production and standing stocks." *Marine Ecology Progress Series* 233 (2002): 303–306.

Kleinbaum, David G., Mitchel Klein, and Erica Rihl Pryor. *Logistic Regression: A Self-Learning Text.* Vol. 94. New York: Springer, 2002.

Knipper, Anna-Delia, Carolina Plaza-Rodríguez, Matthias Filter, Imke F. Wulsten, Kerstin Stingl, and Tasja Crease. "Modeling the survival of *Campylobacter jejuni* in raw milk considering the viable but non-culturable cells (VBNC)." *Journal of Food Safety* (2023): e13077.

Kumar, Palanisamy Satheesh, and Anisa Basheer Khan. "The distribution and diversity of benthic macroinvertebrate fauna in Pondicherry mangroves, India." *Aquatic Biosystems* 9, no. 1 (2013): 1–18.

Longhi, Daniel Angelo, Francieli Dalcanton, Gláucia Maria Falcão de Aragão, Bruno Augusto Mattar Carciofi, and João Borges Laurindo. "Microbial growth models: A general mathematical approach to obtain μ max and λ parameters from sigmoidal empirical primary models." *Brazilian Journal of Chemical Engineering* 34 (2017): 369–375.

McCrady, Mac H. "The numerical interpretation of fermentation-tube results." *The Journal of Infectious Diseases* (1915): 183–212.

Obeso, José M., Pilar García, Beatriz Martínez, Francisco Noé Arroyo-López, Antonio Garrido-Fernández, and Ana Rodriguez. "Use of logistic regression for prediction of the fate of *Staphylococcus aureus* in pasteurized milk in the presence of two lytic phages." *Applied and Environmental Microbiology* 76, no. 18 (2010): 6038–6046.

Paliy, Oleg, and Vijay Shankar. "Application of multivariate statistical techniques in microbial ecology." *Molecular Ecology* 25, no. 5 (2016): 1032–1057.

Preston, Frank W. "The commonness, and rarity, of species." *Ecology* 29, no. 3 (1948): 254–283.

Rathod, Santosha, H. S. Surendra, R. Munirajappa andD.M. Gowda. "Statistical appraisal on the extent of agriculture diversification in different district of Karnataka." *Mysore Journal of Agricultural Sciences* 45, no. 4 (2011): 788–794.

Schreiber, Stefan G., Sanja Schreiber, Rajiv N. Tanna, David R. Roberts, and Tim J. Arciszewski. "Statistical tools for water quality assessment and monitoring in river ecosystems–a scoping review and recommendations for data analysis." *Water Quality Research Journal* 57, no. 1 (2022): 40–57.

Silliker, J. H., D. A. Gabis, and A. May. "ICMSF methods studies. XI. Collaborative/comparative studies on determination of coliforms using the most probable number procedure." *Journal of Food Protection* 42, no. 8 (1979): 638–644.

Sutton, Scott. "The most probable number method and its uses in enumeration, qualification, and validation." *Journal of Validation Technology* 16, no. 3 (2010): 35–38.

Tonner, Peter D., Cynthia L. Darnell, Barbara E. Engelhardt, and Amy K. Schmid. "Detecting differential growth of microbial populations with Gaussian process regression." *Genome Research* 27, no. 2 (2017): 320–333.

van Boekel, Martinus A. J. S. "On the use of the Weibull model to describe thermal inactivation of microbial vegetative cells." *International Journal of Food Microbiology* 74, no. 1–2 (2002): 139–159.

Whittaker, Robert H. "Dominance and diversity in land plant communities: Numerical relations of species express the importance of competition in community function and evolution." *Science* 147, no. 3655 (1965): 250–260.

Woodward, Richard L. "How probable is the most probable number?." *Journal (American Water Works Association)* 49, no. 8 (1957): 1060–1068.

Zwietering, Marcel H., Ida Jongenburger, Frank M. Rombouts, and K. Van't Riet. "Modeling of the bacterial growth curve." *Applied and Environmental Microbiology* 56, no. 6 (1990): 1875–1881.

Index

C

F

G

H

Z

9781032526805